Multicomponent Polymer Systems

Multicomponent Polymer Systems

A symposium co-sponsored by
the Division of Industrial
and Engineering Chemistry,
the Division of Polymer
Chemistry, and the Division
of Cellulose, Wood, and Fiber
Chemistry at the 159th Meeting
of the American Chemical
Society, Houston, Tex.,
Feb. 23-26, 1970

Norbert A. J. Platzer

Symposium Chairman

ADVANCES IN CHEMISTRY SERIES **99**

AMERICAN CHEMICAL SOCIETY

WASHINGTON, D. C.   1971

Coden: ADCSHA

Copyright © 1971

American Chemical Society

All Rights Reserved

Library of Congress Catalog Card 70-159768

ISBN 8412-0113-7

PRINTED IN THE UNITED STATES OF AMERICA

Advances in Chemistry Series
Robert F. Gould, *Editor*

Advisory Board

Paul N. Craig

Thomas H. Donnelly

Gunther Eichhorn

Frederick M. Fowkes

Fred W. McLafferty

William E. Parham

Aaron A. Rosen

Charles N. Satterfield

Jack Weiner

AMERICAN CHEMICAL SOCIETY PUBLICATIONS

FOREWORD

ADVANCES IN CHEMISTRY SERIES was founded in 1949 by the American Chemical Society as an outlet for symposia and collections of data in special areas of topical interest that could not be accommodated in the Society's journals. It provides a medium for symposia that would otherwise be fragmented, their papers distributed among several journals or not published at all. Papers are refereed critically according to ACS editorial standards and receive the careful attention and processing characteristic of ACS publications. Papers published in ADVANCES IN CHEMISTRY SERIES are original contributions not published elsewhere in whole or major part and include reports of research as well as reviews since symposia may embrace both types of presentation.

CONTENTS

Preface xi

POLYBLENDS

1. The Concept of Compatibility in Polyblends 2
   Arthur J. Yu
2. Compatible High Polymers: Poly(vinylidene fluoride) Blends with Homopolymers of Methyl and Ethyl Methacrylate 15
   J. S. Noland, N. N.-C. Hsu, R. Saxon, and J. M. Schmitt
3. Physical Properties of the System: Poly(2,6-dimethylphenylene oxide)-Polystyrene 29
   W. J. MacKnight, J. Stoelting, and F. E. Karasz
4. Critical Phenomena in Multicomponent Polymer Solutions 42
   W. Borchard and G. Rehage
5. Viscosity of a Solution of Two Mutually Incompatible Polymers .. 53
   H. L. Doppert and W. S. Overdiep
6. Interfacial Bonding between Different Elastomer Phases 68
   R. L. Zapp
7. The Theory of Rubber Toughening of Brittle Polymers 86
   C. G. Bragaw
8. Permeability and Phase Structure of PVC/EVA Systems 107
   Helge Storström and Bengt Rånby
9. Blends of Chlorinated and Normal PVC 119
   G. Ajroldi, G. Gatta, P. D. Gugelmetto, R. Rettore, and G. P. Talamini

COPOLYMERS

10. Copolymerization in the Presence of Depolymerization Reactions .. 140
    Paul Wittmer
11. Demixing Phenomena in Copolymers 175
    Fritz Kollinsky and Gerhard Markert
12. Dynamic and Stress-Optical Properties of Polyblends of Butadiene-Styrene Copolymers Differing in Composition 189
    Gerard Kraus and K. W. Rollmann

13. Terpolymerization of Cyclopentene, Sulfur Dioxide, and Acrylonitrile 211
    Yuyu Yamashita, Shouji Iwatsuki, and Kozo Sakai

14. High Impact Polystyrene by Prepolymerization in a Water-in-Oil Emulsion Followed by Suspension Polymerization 221
    Rudolf B. DeJong

15. Comparison of Methyl Methacrylate–Butadiene–Styrene with Acrylonitrile–Butadiene–Styrene Graft Copolymers 229
    Rudolph D. Deanin, Ismail S. Rabinovic, and Antonio Llompart

16. New Transparent Impact-Resistant Polymer Blends from Resins and Rubbers Having Minimized Compositional Heterogeneities. I. Blends of Resins with Diene Copolymer Elastomers 237
    R. G. Bauer, R. M. Pierson, W. C. Mast, N. C. Bletso, and L. Shepherd

17. New Transparent Impact-Resistant Polymer Blends from Resins and Rubbers Having Minimized Compositional Heterogeneities. II. Blends of Resins with Acrylic Copolymer Elastomers 251
    R. G. Bauer, R. M. Pierson, W. C. Mast, N. C. Bletso, and L. Shepherd

18. Grafting in an Aqueous Suspension of Vinyl Chloride on Ethylene–Propylene Copolymers 260
    F. Severini, E. Mariani, A. Pagliari, and E. Cerri

19. Polymeric Systems as Poly(vinyl chloride) Modifiers 278
    J. M. Michel

20. Thermal Stabilization of Poly(vinyl chloride) through Graft Copolymerization of cis-1,4-Polybutadiene 302
    Norman G. Gaylord and Akio Takahashi

21. Properties of Graft and Block Copolymers of Fibrous Cellulose .. 321
    Jett C. Arthur, Jr.

22. Substrate Particle Size in ABS Graft Polymers 340
    Charles F. Parsons and Edmond L. Suck, Jr.

23. Kinetics of Aggregation and Dimensions of Supramolecular Structure in Noncrystalline Block Copolymers 351
    M. Hoffmann, G. Kampf, H. Kromer, and G. Pampus

24. Dispersion of Solid Particles in Organic Media 379
    G. E. Molau and E. H. Richardson

25. Time–Temperature Superposition in Block Copolymers 397
    C. K. Lim, R. E. Cohen, and N. W. Tschoegl

26. Multicomponent Systems from Copolymers of Maleic Anhydride and Vinyl Monomers 418
    Raymond B. Seymour, Hing Shya Tsang, Elvis E. Jones, Patrick D. Kincaid, and Ashwin K. Patel

27. Block and Random Copolymers by Oxidative Coupling of Phenols 431
    Glenn D. Cooper, James G. Bennett, Jr., and Arthur Katchman

28. Polymer–Fibrous Glass Composites: Advances and Potential Properties 452
    Fred G. Krautz

29. Halogenated Epoxy Matrix Plastics in Filament-Wound Composites 471
    James R. Griffith, Arthur G. Sands, and Jack E. Cowling

30. Inorganic Reinforcements for High-Performance Structural Composites 482
    Walter H. Gloor

31. Mechanisms of Reinforcement of Elastomers by Polymeric Fillers 490
    Maurice Morton

32. Short Fiber–Elastomer Composites 510
    George C. Derringer

33. Reinforcement of Thermosetting Cycloaliphatic Epoxy Systems with Elastomers 531
    A. C. Soldatos and A. S. Burhans

34. The Preparation and Characteristics of Concrete–Polymer Composites 547
    M. Steinberg, L. E. Kukacka, P. Colombo, and B. Manowitz

35. Carboxylated Polyester Additives for Improving the Adhesion of Coatings 562
    W. J. Jackson, Jr., and J. R. Caldwell

36. Segregated Metallic Particles in Polymers 572
    J. E. Scheer and D. T. Turner

37. Luminescence of Thin Plastic Scintillators. A Report on Energy Transfer in Plastic Scintillators 581
    R. Kosfeld and K. Masch

Index 589

PREFACE

Materials composed of several components are replacing natural materials, metals, and homopolymers in increasing amounts. They are used in the building, automotive, marine, aircraft, appliance, and electrical industries. Fabricators are willing to change from a single material to a multicomponent system if the latter is either better, more attractive, less expensive, or easier to fabricate into a product. They look at a material as a bundle of desirable properties and care little about its chemical composition. Metallurgists define alloys as multicomponent systems. Polymer chemists include polyblends, copolymers, and reinforced composites under this term, as illustrated in Figure 1.

Manufacturers of multicomponent polymer systems are modifying the properties of homopolymers to obtain superior materials. These materials are either stronger, tougher, more flexible, will withstand environmental influences, or have other desirable characteristics.

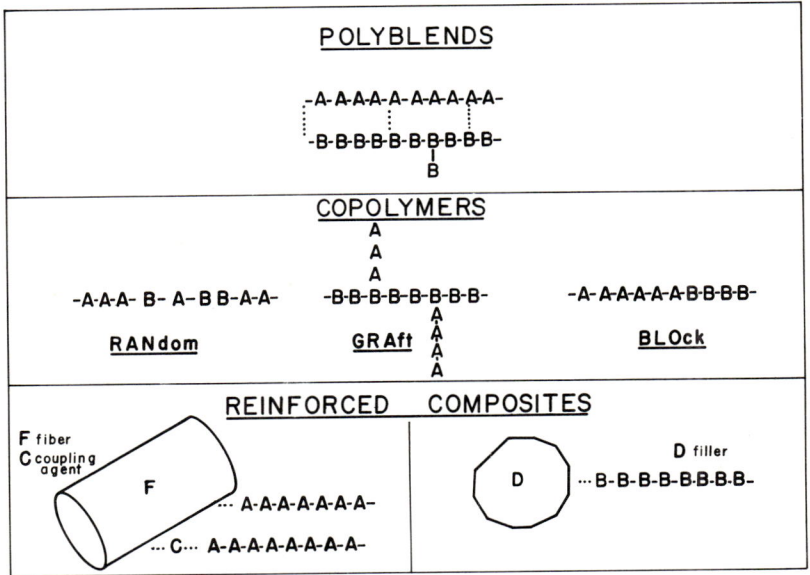

Figure 1.   Multicomponent polymer systems

Polyblends

Polyblends may be made from polymeric components by mixing and compounding them together on mill rolls, in extruders, Banbury mixers, or other thermoprocessing equipment.

Twenty-five years ago, T. A. Grotenhuis observed that a rubber of superior flexibility and abrasion resistance could be obtained from a heterogeneous polyblend of a hard unmasticated rubber as disperse phase and an elastic rubber as continuous phase. It was recognized that a heterogeneous blend was also necessary for making tough and flexible plastics. To obtain an impact-resistant polystyrene, butadiene and styrene were polymerized separately in aqueous emulsion, and the two latices were blended together, coagulated, and processed. In an analogous process, an ABS polyblend was made. A nitrile rubber latex and a SAN latex were made separately by emulsion polymerization, blended, coagulated, and compounded together. Today, most rigid PVC is still produced by polyblending. Vinyl chloride homopolymer is compounded with an elastomer, such as chlorinated polyethylene, ABS, MBS, EVA, or others. In 1970 the domestic production of rigid PVC was expected to amount to 560 million pounds.

Blending high polymers consisting of long chain molecules is different from mixing liquids. In the first chapter of this volume, A. J. Yu defines the term compatibility between high polymers. True compatibility between different polymers is rare. It was observed in amorphous blends of poly(vinylidene fluoride) with poly(methyl methacrylate) and poly(ethyl methacrylate) by J. S. Noland and co-workers. It was also found in blends of PPO with polystyrene, as reported by F. E. Karasz, W. J. MacKnight, and J. Stoelting. Incompatibility or partial compatibility is more frequent.

To understand the mechanism of polyblending, experiments have been carried out with polymeric solution. W. Borchard and G. Rehage mixed two partially miscible polymer solutions, measured the temperature dependence of the viscosity, and determined the critical point of precipitation. When two incompatible polymers, dissolved in a common solvent, are intimately mixed, a polymeric oil-in-oil emulsion is formed. Droplet size of the dispersed phase and its surface chemistry, along with viscosity of the continuous phase, determine the stability of the emulsion. Droplet deformation arising from agitation has been measured on a dispersion of a polyurethane solution with a polyacrylonitrile solution by H. L. Doppert and W. S. Overdiep, who calculated the relationship between viscosity and composition.

In a heterogeneous polyblend of two dissimilar elastomers, such as chlorinated butyl rubber and polybutadiene, a certain interfacial bonding may exist. By using two different solvents—one which is a good solvent for the first elastomer but a poor solvent for the second elastomer, and one which is a good solvent for the second elastomer but a poor solvent for the first one—R. L. Zapp was able to prove the evidence of interfacial bonding between heterogeneous blends.

The stress needed to break a covalent C–C bond is about 2 million psi. However, the tensile strength of commercial thermoplastics is only in the order of 3000–17,000 psi owing to the presence of flaws and surface cracks. Incorporating a dispersed rubber phase can significantly raise impact toughness of brittle polymers. C. G. Bragaw explains this phenomenon through broadening of the energy absorbing area and branching of cracks and crazes dynamically along imbedded rubber particles.

Certain rigid polyblends of PVC with EVA elastomer were investigated by B. Rånby and co-workers who observed an increase of heterogeneity with higher EVA contents. PVC may be added as a processing aid to chlorinated PVC. G. P. Talamini and co-workers found that the added PVC acts like a reinforcing agent in the continuous phase of chlorinated PVC and that heterogeneity increases with higher chlorine content.

Copolymers

In polyblends the components adhere together through secondary bond forces, either van der Waals forces, dipole interaction, or hydrogen bonding. In copolymers the components are linked by strong covalent bonds. We distinguish between random, graft, and block copolymers. In random copolymerization, two or more monomers are present simultaneously in the reactor. In graft copolymerization, one polymer is prepared, and one or two monomers are grafted onto this polymer. The final product consists of a polymeric backbone with one or more side branches. In block copolymerization, one monomer is polymerized, and another monomer is polymerized on to the living ends of the polymeric chains. The final block copolymer is a linear chain with a sequence of different polymer segments.

Random Copolymers. Fifty years ago the first copolymer was prepared from isoprene and dimethylbutadiene and evaluated as a possible synthetic rubber. During the 1930's systematic work on copolymerization was started at Ludwigshafen, Germany. Styrene, vinyl chloride, vinyl acetate, and acrylics were available for copolymerization.

From Ludwigshafen comes the first chapter of the second section of this volume. It summarizes some of the theories and vast experiences gathered there. It has been assumed that polymerization occurs only in one direction. Compositions and molecular weights of random copolymers had been calculated from the addition rates of the individual monomers during the propagation steps. Depolymerization, however, can accompany polymerization, and a reaction equilibrium is reached. P. Wittmer discusses copolymerization in the presence of depolymerization reactions. In his experiments, he used α-methylstyrene as a monomer which forms an equilibrium during polymerization, and methyl methacrylate and acrylonitrile as comonomers.

Heterogeneity, as in polyblends, has also been observed in random copolymers. F. Kollinsky and G. Markert found phase separation in binary mixtures of copolymers of methyl methacrylate and butyl acrylate. C. Kraus and K. W. Rollmann discovered heterogeneity in blends of random copolymers of butadiene and styrene if they differ by more than 20% in composition.

A terpolymer has been prepared from cyclopentene, sulfur dioxide, and acrylonitrile by Y. Yamashita and co-workers. The mechanism was recognized as a binary copolymerization between a cyclopentene/SO_2 complex and free acrylonitrile.

Graft Copolymers. Although discovered in 1933, the term "graft copolymer" was not defined until 1952, when the principle of grafting was disclosed. It consists of placing one polymer into a reactor, adding one or two monomers, and completing the copolymerization by free-radical initiation. One portion of the added monomers polymerizes as a side branch upon the polymeric backbone. The other portion undergoes either homopolymerization or random copolymerization. It is difficult to determine how much graft copolymer results and how much of the second portion is present because some of the latter is occluded in the graft copolymer particles. As in the polyblends, heterogeneity prevails.

Today, a large part of the more than one billion lbs/year of impact polystyrene and 500 million lbs/year of ABS produced domestically is made by graft copolymerization. Impact polystyrene may be synthesized by dissolving a diene rubber in styrene monomer, in the presence or absence of another solvent, prepolymerizing the solution, and completing the polymerization in bulk, solution, or suspension. R. B. deJong describes a process wherein he prepolymerizes in emulsion with styrene as the continuous phase and the water as the dispersed phase and completes polymerization in aqueous suspension.

Most ABS is made by emulsion polymerization. A polybutadiene or nitrile rubber latex is prepared, and styrene plus acrylonitrile are grafted upon the elastomer in emulsion. The effect of rubber particle size in ABS graft copolymer on physical properties is the subject Chapter 22 by C. F. Parsons and E. L. Suck. Methyl methacrylate was substituted for acrylonitrile in ABS by R. D. Deanin and co-workers. They found a better thermoprocessability, lighter color, and better ultraviolet light stability.

Elastomeric graft copolymers of methyl methacrylate upon diene and acrylic rubbers were prepared by R. G. Bauer and co-workers. These elastomers are compatible with rigid methyl methacrylate–styrene copolymers of identical refractive index, yielding transparent polyblends.

Vinyl chloride graft copolymers are less available commercially than the styrene graft copolymers. F. Severini discusses grafting vinyl chloride upon an ethylene–propylene backbone in aqueous suspension. Using PVC as the backbone and grafting various monomers upon it is the subject of the next two chapters. J. M. Michel reports on grafting of diene and acrylate monomers upon PVC by irradiation and chemical initiation. N. Gaylord and A. Takahashi grafted butadiene upon PVC with ionic catalysts and discovered the graft copolymer as a thermal stabilizer for vinyl chloride homopolymer. Monomers can be grafted upon natural materials as well as synthetic backbones. J. C. Arthur discusses the improvement of cotton fibers through grafting with vinyl monomers.

Block Copolymers. The first block copolymers were prepared in 1960. While free radical initiation is generally used in graft copolymerization, ionic catalysts are used in block copolymerization. They allow the polymer formed to retain its ability to grow. In a second consecutive step, this living polymer reacts with another monomer resulting in a block copolymer. Ionic block copolymerization is used to manufacture special diene and olefin rubbers. This process may be combined with graft copolymerization also. The elastomeric block copolymer can be used as backbone upon which acrylonitrile and/or styrene are grafted when producing ABS or impact polystyrene.

An amorphous block copolymer of butadiene and styrene, made by anionic polymerization at Leverkusen, was used as a model to study the mechanism of aggregation and reaggregation. M. Hoffmann and co-workers measured the critical concentration of phase separation and compared it with that of polyblends. They also determined the shape and the size of the aggregates. A block copolymer of styrene with carboxylated butadiene, made also by anionic polymerization, was evaluated by G. E. Molau and E. H. Richardson as a dispersant for pigments. The carboxylic groups in the butadiene sequence provided sites for selective adsorption of pigments, such as TiO_2. A terblock copolymer of styrene/butadiene/styrene was analyzed by N. W. Tschoegl, and its thermal and mechanical properties were compared with those of diblock copolymers.

In block copolymerization, the second step is also ionically catalyzed. However, it is possible for the living homopolymer to react with a peroxide initiator first and to complete the process by free-radical polymerization. R. B. Seymour and co-workers carried out both steps by free-radical polymerization. They obtained live polymers in free-radical solution polymerization of styrene and maleic anhydride and used these to initate further polymerization with selected monomers.

Block copolymers may also be made by condensation polymerization. Elastomer fibers are produced in a three-step operation. A primary block of a polyether or polyester of a molecular weight of 1000–3000 is prepared, capped with an aromatic diisocyanate, and then expanded with a diamine or dihydroxy compound to a multiblock copolymer of a molecular weight of 20,000. The oxidative coupling of 2,6-disubstituted phenols to PPO is also a condensation polymerization. G. D. Cooper and co-workers report the manufacture of a block copolymer of 2,6-dimethylphenol with 2,6-diphenylphenol. In the first step, a homopolymer of diphenylphenol is preformed by copper–amine catalyst oxidation. In the second step, oxidation of dimethylphenol in the presence of the first polymer yields the block copolymer.

Composites

Reinforced Polymers. Strength and rigidity can be raised 10 to 100 times by reinforcement with glass fibers, glass mats, and glass cloth, as shown by the comparative data of Table I. In 1970 the U.S. production of glass-fiber reinforced polymers reached 1.2 billion lbs and is expected to pass the 3 billion lb mark in 1975. Reinforced polyesters amount to 85% of the market, reinforced epoxies to 5%, and the remaining 10% is shared by the reinforced thermoplastics. In reinforced polyesters and epoxies, the average polymer content is 45% with 25% glass fibers and 30% fillers, such as kaolin, calcium carbonate, asbestos or talc. A review of the potential properties of glass-fiber reinforcement thermosets and thermoplastics is given by F. G. Krautz.

The market distribution of reinforced polymers is as follows:

boats and marine uses	26%
automobiles, trailers & other vehicles	22%
building industry	13%
corrosion resistant equipment	11%
electrical components	8%
aircraft and aerospace	4%
appliances	4%
consumer products & miscellaneous	12%

Today, 29% of the total number of boats built are constructed of reinforced polymers. Water absorption represents an important property. J. R. Griffith, A. Sands, and J. E. Cowling describe reduction of water absorption through the use of halogenated epoxies in glass fiber reinforced composites.

Table I. Increase of Strength and Rigidity through Reinforcement

	Tensile Strength, 10^3 psi	Modulus of Elasticity, 10^5 psi
Unreinforced thermoplastics	3–12	1–6
Glass-fiber reinforced thermoplastics	7–30	8–10
Glass-fiber reinforced thermosets	4–30	8–55
Glass-cloth reinforced thermosets	20–70	13–250
Novel composites	100–200	170–500

Reinforced Elastomers. The U.S. rubber production amounts to 6.1 billion lbs, of which 3.7 billion lbs are fabricated into tires. Active fillers are used to reinforce tires against abrasion, tearing, cut growth, flex cracking, and tensile failure. To understand the mechanism of reinforcing rubber with carbon black or pigments, M. Morton carried out experiments with well-defined particles of styrene–diene block copolymers as model fillers. W. H. Gloor discusses inorganic fibrous reinforcements used to obtain high strength composites. Modifying rubber with appropriate resin formers to achieve better coupling between the tire yarn and rubber is the subject of G. C. Derringer's chapter. A. C. Soldatos and A. S. Burnhans found that the toughness of epoxy composites can be significantly improved by modifying them with elastomers.

Interfacial Bonding between Polymers and Other Materials. As polymers can be reinforced by fillers, inorganic materials can be reinforced by polymers. M. Steinberg and co-workers found that monomers can be soaked into concrete and polymerized *in situ* by either radiation or thermal catalytic techniques. Increase in strength by as much as four times of untreated concrete and reduction of water absorption and of freeze cracking have been observed. Concrete–polymer significantly increases the bond strength and flexural strength of steel-reinforced and fiber-reinforced concrete.

In surface coating, it has been customary to coat metal with a primer to effect better adhesion of a polymeric coating to the substrate. W. J. Jackson and J. R. Caldwell report that a single coat is sufficient if carboxylated polyesters are added to the polymeric coating. These carboxylated polyesters are soluble in volatile lacquer type solvents and improve the adhesion of the common polymeric lacquers and varnishes.

It is also possible to reinforce polymers with metallic particles. D. T. Turner and one of his students observed that good electrical conductivity can be measured even at very low fillings, such as only 6% by volume. Microscopic examinations showed that the metallic particles formed continuous chains segregated around zones of unpenetrated polymer.

Finally, it is possible to imbed luminescent substances in transparent polymers. The interesting effect of scintillation of these composites has been studied by R. Kosfeld and K. Masch.

The field of multicomponent systems has grown quickly, and these materials will replace existing ones and be used in new engineering applications. For the scientist and technologist, this volume should represent a source of helpful information and inspire him to build even stronger and tougher polyblends, copolymers, and composites of higher heat resistance, greater clarity, better endurance, or other desirable properties.

Monsanto Co.
Indian Orchard, Mass.
August 1970

NORBERT A. J. PLATZER

Polyblends

1

The Concept of Compatibility in Polyblends

ARTHUR J. YU

Stauffer Chemical Co., Dobbs Ferry, N. Y. 10522

> *When liquids are mixed to form a homogeneous and single-phase mixture, they are said to be compatible. This mixing is straightforward, and the result is easily observed; mixing polymers is quite different. From thermodynamic, kinetic, and mechanical considerations, a homogeneous single-phase polyblend is highly improbable. No polyblend is compatible if compatibility means homogeneity and single-phase morphology. Conceptually, compatibility is a representation of how close a polyblend can approach this ultimate state of molecular mixing. Phenomenologically, it is a relative measure of the degree of heterogeneity of the polyblend— i.e., how fine one polymer is dispersed in another. Compatibility of a polyblend is subject to differences in experimentation and interpretation of result. One should describe it by giving the sample history and method and instrument used to determine its degree of heterogeneity.*

In studying polyblends, one is frequently involved with the "compatibility" of the polymeric components comprising the polyblends. A survey of the polymer literature revealed that the word "compatible" is used widely to describe certain morphology, solution property, or mechanical behavior of the system. However, there seems to be a lack of consistency in what "compatibility" really means. As used in polyblends, compatibility has its origin in the mixing of two liquids. When two liquids on mixing give a single homogeneous phase, they are described as compatible. Thus, compatibility signifies homogeneity and no phase separation. Whereas the result of mixing two liquids can be determined by visual observation and phase separation usually occurs quite noticeably, the mixing of two polymers is more complex. As shown later, compatibility as applied to polyblends cannot possibly mean the same as when it is applied to the mixing of simple liquids. Mixing polymers to give a polyblend involves not only the thermodynamic consideration of the components but also the thermal, mechanical, and kinetic aspects associated with the attainment of equilibrium. Even when true equilibrium conditions have been reached, the determination of the degree of heterogeneity of the mixture often is subject to variation in test methods and interpreting the results. Thus, to state that a polyblend is compatible without qualification conveys little useful information.

Perhaps another reason contributing to this imprecise use of the term "compatibility" can be traced to the history of the development of polyblends. Polyblend is still a new material. Although the first recorded preparation occurred in 1912 (1), most exploratory work was done in the late 1940's by industrial laboratories. As in the case of most industrial developments in the early stage, there was more art than science being applied, and the proper choice of words was not diligently considered and followed. The first commercial polyblend was introduced by Dow Chemical in 1948. It was an impact-resistant polystyrene containing 5 wt % of a poly(butadiene–co-styrene) rubber. As more polyblend patents were issued and more polyblends became commercially successful, more and more research into the nature of polyblends was started. Today, a large body of information is available on compatibility, properties, and performance of polyblend systems.

For this discussion, a polyblend is defined as a single entity of material containing within its physical boundary at least two thoroughly mixed polymers which are not linked covalently. Although in some literature sources (2) a polyblend is described simply as a physical mixture of two or more polymers, such a broad definition is not practical. Consider mixing polystyrene powder and polyethylene powder in a Waring blender. Should the powdery product be called a polyblend? It will not be useful to call such a mixture a polyblend because there is little one can do with it without first forming it into a single piece. Now consider the definition of alloy. A metallic alloy is not a mixture of two metals or a metal and a non-metal but rather the fusion product thereof. Thus, from these two considerations polyblend should be limited to a single entity of material in its physical state. The size and shape of the single entity is not important as long as within the boundary of that single piece there are at least two thoroughly mixed polymers. Mixing will exclude such unrelated systems as a laminate of two different materials and painted plastic objects. Thorough mixing will ensure reproducible results of experimentation. The condition that the components of a polyblend should not be linked covalently will exclude such block copolymers of the ABA type, some of which have two-phase morphology like most polyblends. This definition will restrict the consideration of compatibility to a more limited but more relevant set of multicomponent polymer systems. Take the previous example of polystyrene powder mixed with polyethylene powder; the mixture is not a polyblend, according to this definition, because the whole mixture is not a single entity but a collection of many powdery particles. Within each single particle of this powdery material, the polymer is either polystyrene or polyethylene, so each particle is not a polyblend. If one dissolves the mixture in a solvent and precipitates it into a nonsolvent in a blender, the resulting powdery product is not a polyblend because again the product is not a single entity. Each single particle is a polyblend because it contains within its physical boundary two well-mixed polymers. If these two powdery mixtures are fused separately into a single piece, each of these two single pieces of material is a polyblend. In fact, if the components are mixed thoroughly, they will be indistinguishable.

Polyblends can be prepared by four methods: (1) mechanical mixing on rubber mills or extruders, (2) polymerization of one monomer in the presence of another polymer, (3) evaporation or precipitation from mixture of polymer solutions, (4) coagulation of mixture of polymer latices. Notice that the products from Methods 2, 3, and 4 are not polyblends according to the definition above. They may be in a thoroughly mixed state and not covalently linked, but they have not been formed into a single piece. The forming process, either through solution or fusion, could result in demixing of the components. Frequently, the components of a polyblend have such low affinity for each other that their fusion product will readily break on gentle handling, such as removing it from a press or mold. In the extreme cases a fusion product cannot even be obtained by mixing the components on a hot roll-mill; the two molten components will stay on separate rolls.

Mixing of Nonpolymeric Liquids

No attempt is made to review solution theory. For a thorough treatment on this subject, numerous authoritative monographs (3, 4, 5, 6, 7) are available. We discuss only those thermodynamic considerations which have some bearing on the discussion of compatibility of polymer blends.

Homogeneous Liquid Mixtures. Although most compatible liquid pairs give nearly random and complete mixing at the molecular level, this condition is not a necessary requirement for compatibility. If liquids with different mutual interaction are mixed—*e.g.*, liquids capable of hydrogen bonding—various degrees of molecular association will occur. As long as the association does not result in phase separation, the liquid mixture is compatible. Homogeneity is the only requirement for compatibility.

To understand what constitutes the condition leading to homogeneous liquid mixtures, it may be desirable to start with the fundamental question: what determines whether a reaction, a process, or a change will take place? If it does take place, to what extent will it proceed? In our specific case, we want to know if N_A moles of liquid A and N_B moles of liquid B are mixed at room temperature and atmospheric pressure, will a one-phase homogeneous mixture or a two-phase heterogeneous mixture result? In the latter case what will be the composition and relative amount of the two phases? Thermodynamics has taught that if the change in free energy ΔF of the reaction or process is negative, the process will proceed spontaneously. It will proceed to the extent that maximum negative ΔF is obtained. At that point, any further change will involve expenditure of free energy—i.e., positive ΔF, so the system will remain at this equilibrium position.

In Figure 1, the free energy of N_A moles of A and N_B moles of B before mixing is at point 1. If the free energy for the homogeneous mixture is at point 2, the reaction will proceed to give a compatible mixture since $\Delta F = F_2 - F_1$ is negative. If the free energy of the homogeneous mixture is at point 3, the mixing of N_A moles of A and N_B moles of B will proceed to point 2 giving a two-phase mixture consisting of two liquid mixtures of A and B of different compositions in dynamic equilibrium. For the system to go from point 2 to 3 ($\Delta F = F_3 - F_2$ is positive), the reaction will not proceed on its own without adding energy, equal to $F_3 - F_2$, into the system. At point 2, the position of equilibrium, the chemical potentials of either component in both phases are equal.

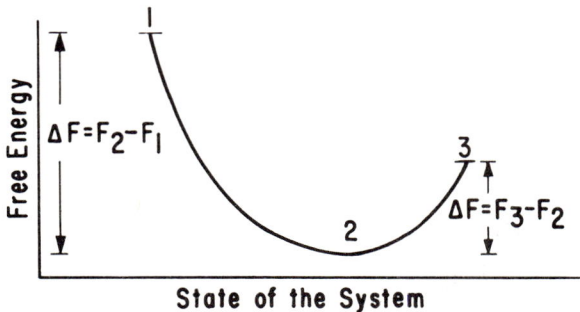

Figure 1. Thermodynamic model for determining compatibility

There will be no net change in the relative amount and in the composition of the two phases in time. As to the exact composition of the two phases and their relative amounts, one would need a phase diagram relating temperature and mole fraction of the two liquids. Thus, ΔF determines whether or not a reaction will proceed. Thermodynamics has further taught that $\Delta F = \Delta H - T\Delta S$. ΔH is the change in heat content, and ΔS is the change of entropy of the process. For our purpose, if we know the magnitude of ΔH and ΔS of mixing and the temperature at which mixing takes place, we can calculate ΔF and predict if a homogeneous mixture can be obtained. ΔH for mixing is a measure of the attraction between the molecules to be mixed. A negative ΔH means that heat is evolved during mixing and lost to the surroundings. Heat is evolved on mixing if the molecules to be mixed attract each other more than they attract their own kind. This naturally favors mixing, as shown by the contribution of a negative ΔH to the negative ΔF. As shown later, ΔS of mixing is always positive and, therefore, $-T\Delta S$ will always be negative. Thus, the sign of ΔF will be determined by the sign of ΔH. If it is negative or zero, ΔF will be negative. If it is positive, it must not be larger than $T\Delta S$ for ΔF to be negative.

For a quantitative calculation of heat of mixing, the Scatchard-Hildebrand equation (8) gives:

$$\Delta H_M = V_M (\delta_A - \delta_B)^2 \phi_A \phi_B \tag{1}$$

where V_M is the total volume of the mixture, ϕ is the volume fraction of the liquids, and δ is the so-called solubility parameter of the liquids. The solubility parameter of a liquid is a measure of the cohesive energy of the liquid—i.e., how strong the molecules attract themselves. It is related to the heat of vaporization of the liquid. ΔH_M depends on concentration. $\phi_A \phi_B$ has a maximum value when $\phi_A = \phi_B = 0.5$ and approaches zero when the liquid mixture is infinitely dilute. Notice that $(\delta_A - \delta_B)^2$ is always positive when $\delta_A \neq \delta_B$, and ΔH_M is zero when $\delta_A = \delta_B$. A positive ΔH_M will contribute to phase separation unless overcome by a large $T\Delta S$. If the force field of A is very different from B, one may obtain a two-phase system. On the other hand, if A and B have equal force fields, ΔH_M will be zero to promote compatibility. Thus, the rule of "like dissolves like" has its theoretical origin in $\delta_A = \delta_B$ giving $\Delta H_M = 0$. The equation is unable to explain a negative ΔH_M—i.e., heat is evolved on mixing. This is caused by the fact that some assumption made in deriving Equation 1 was erroneous. A negative ΔH_M will always result in a homogeneous mixture. This explains why exothermic reactions will proceed spontaneously.

The entropy of a system can be interpreted in various ways. For mixing purposes, statistical mechanics provides the best conceptual understanding. In statistical mechanics, entropy is a measure of the randomness or disorder of a system. High entropy is associated with disorder. Since a spontaneous process goes from order to a random state, all spontaneous processes are accompanied by a gain in entropy. When N_A moles of A and N_B moles of B are mixed, the change in the entropy of mixing is represented by Equation 2.

$$\Delta S_M = -R(N_A \ln X_A + N_B \ln X_B) \qquad (2)$$

where X_A and X_B are the mole fractions of A and B in the mixture. Equation 2 represents the entropy change in a completely random mixing of A and B. For the equation to hold, the molecules of A and B must have similar size and shape. ΔS_M is also known as the configurational entropy change. We can now write, since $\Delta F_M = \Delta H_M - T \Delta S_M$,

$$\Delta F_M = V_M(\delta_A - \delta_B)^2 \phi_A \phi_B + RT(N_A \ln X_A + N_B \ln X_B)$$

All terms on the right of the equal sign are either characteristic of the liquids before mixing or related to the composition of the mixture. Therefore, we can estimate, at least in principle, the sign and magnitude of ΔF_M when N_A moles of A and N_B moles of B are mixed.

There are four kinds of solution behavior as shown in Figure 2.

		ΔH_M	
		zero	non-zero
ΔS_M	ideal	ideal	regular
	non-ideal	athermal	irregular

Figure 2. Diagram of the four kinds of solution behavior

An ideal solution is characterized by zero heat of mixing and ideal entropy of mixing—*i.e.*, random and complete mixing of the molecules at the molecular level. Ideal solution requires the molecules of the liquids to have the same size, the same shape, and the same force field. They are compatible in all proportions at all temperatures and pressures. Glasstone (9) has discussed how large deviations from ideal solution behavior will lead to incomplete solubility of one liquid in another as the temperature is changed. Systems showing deviation from ideal behavior are common. Glasstone stated that if the two liquid constituents in force field of interaction, polarity, length of hydrocarbon chain of molecule, and degree of association in liquid state, the mixture will show nonideal solution behavior. Thus, even among simple liquids, compatible mixtures in all composition and temperature ranges are exceptions rather than the rule. Therefore in describing a liquid mixture as compatible, it is necessary to specify the composition and temperature of the mixture.

Metastable Homogeneous Liquid Mixtures. We need to appreciate the importance of the history of a sample of liquid mixture and how homogeneity is measured. We refer to Figure 1. Suppose point 1 represents the free energy state of a sample of oil and water before mixing, point 2 represents the free energy state of oil and water at equilibrium as a heterogeneous mixture, and point 3 represents the free energy state of an emulsion of oil and water after the components are mixed in an intensive mixer, homogenizer, or colloid mill. We choose to reach equilibrium from points 1 through 3 to 2. That is, the mixture of oil and water is mixed to a fine dispersion, with particle sizes smaller than the wavelength of visible light, such that it gives a transparent appearance. We know from thermodynamics that change in free energy depends only on the initial and final states and is independent of the path from the initial state to the final state. There is no question in this case that oil and water form a two-phase incompatible system at equilibrium and the oil-in-water emulsion is at a metastable state. If water and oil are simply mixed—going from point 1 to 2 directly—one can ascertain that an incompatible two-phase mixture results. However, if the components of the same mixture are homogenized first—going from point 1 to 3—one may have to wait a long time before the system will separate into two phases. If the sample history is known, and there is reason to suspect that the apparent homogeneity may not be real, one may want to wait a little longer before deciding if the mixture is indeed homogeneous, but, how long should one wait? Alternatively, one may choose to use an instrument to detect the presence of a second phase. If an optical microscope is used to examine a transparent oil-in-water emulsion, the small oil droplets may not be visible. On the other hand, electron microscopy or light scattering may show the heterogeneous nature of the emulsion. Finally, mild agitation, shaking, or freezing can result in phase separation of the metastable emulsion. These examples, which are not unlikely, show that even for mixing simple nonpolymeric liquids like oil and water the conclusion cannot always be clear cut. Whether a seemingly homogeneous liquid mixture from mixing of liquids can be considered and concluded as truly homogeneous depends on:

(1) The time scale in determining homogeneity
(2) The size scale in determining homogeneity
(3) The amount of disturbance allowed to upset the homogeneity

To summarize, even mixing simple liquid molecules, one-phase, homogeneous, and compatible mixture is by no means prevalent, and the demarkation of homogeneity and heterogeneity is not always unequivocal.

Polyblends

To study the problem of compatibility of a polyblend, one can extend the thermodynamic treatment of the mixing of liquids to systems involving two or more polymers. Since we should deal with systems at equilibrium, we must consider first the problems involved in attaining the equilibrium state of mixing. First consider the kinetic aspect of mixing. How fast can a polymer diffuse through a highly viscous medium, even with the application of heat and mechanical work? Short distance interpenetration involving segments of polymer chains is conceivable, but long distance migration of a whole polymer molecule through the viscous matrix within the duration of the mixing process intuitively is slow and difficult. Consider next the mechanical problem. The mixing equipment for polymer blends is large compared with the size of the molecules. It is difficult to expect that individual polymer molecules are actually being disentangled and separated during mixing. More likely, the volume elements in the blend which are being mixed involve clusters of polymer containing many chains. This conclusion was drawn by Leigh-Dugmore (10), based on his interpretation of the literature results. Eventually, mixing will reach an equilibrium state at which any further mixing will not reduce the degree of heterogeneity. At this point, we need to find out if the polyblend has a two-phase heterogeneous structure or a one-phase homogeneous structure. We must bear in mind that excessive mechanical mixing may cause degradation, chain scission, block or graft copolymer formation. All these side reactions will complicate the study of polymer compatibility since new species are introduced into the system. Thus, mixing of two polymers to reach an equilibrium state of distribution of one polymer in another is more complex than mixing of liquids. If one can separate the chains of the polymers before mixing, one may alleviate the kinetic and mechanical problems as discussed above. This can be accomplished by dissolving the polymers in a single solvent or solvent mixture and then mix the solutions. In the classic experiment by Dobry and Boyer-Kawenoki (11) involving two polymers and a solvent, it was found that even in dilute solutions of a few percent concentration, most mixtures showed phase separation with each polymer dissolved in the same common solvent occupying a separate phase. When chains of two polymers are free to migrate, they do not want to intermix.

What does thermodynamics say on the mixing of polymers? The equation $\Delta F_M = \Delta H_M - T \Delta S_M$ still applies. Since mixing polymers, as compared with mixing simple liquids, involves a smaller number of molecules, ΔS_M will be very small. Thus, the sign of ΔF_M is affected strongly by ΔH_M. Flory (12, 13) has proposed methods to calculate these terms. In the case where calculations can be made, ΔF_M was always positive

(*14*). Even long before a method was developed to make possible such calculations Flory (*15*) stated that:

Two high polymers are mutually compatible with one another only if their free energy of interaction is favorable, *i.e.*, negative. Since the mixing of a pair of polymers, like the mixing of simple liquids, in the great majority of cases, is endothermic, incompatibility of chemically dissimilar polymers is observed to be the rule and compatibility is the exception.

Of course, Flory has not defined the criteria for compatibility. If compatibility is to mean single homogeneous phase and is associated with nearly complete mixing of the molecules at the molecular level, as it does for simple liquids, no polyblends are known to be truly compatible. So what does compatibility mean when applied to polyblends? A search of the literature revealed that various methods have been used to determine compatibility of polyblends and each method has its own standard and sensitivity.

Solution in Common Solvent. The method of Dobry and Boyer-Kawenoki (*11*) showed that compatibility of a polymer pair can be determined by its solution in a common solvent. If phase separation occurs, the pair is incompatible. Since phase separation is affected by polymer concentration and by temperature, this test is quite arbitrary and gives only relative results.

Film Casting. Frequently films are cast from homogeneous dilute solution of two polymers. An opaque and crumbly film indicates incompatibility, and a clear and self-supporting film suggests better compatibility. Since there is a continuous change in clarity and opacity and transition from crumbly state to self-supporting is also gradual, it is difficult to judge where compatibility leaves off and incompatibility starts.

Appearance of Fused Product. From the fabrication point of view, if two polymers give a "smooth band" on a two-roll mill, the polyblend is said to be "compatible." If the fused product is "cheezy," it is said to be "incompatible." Frequently, the fused product is pressed into a flat sheet. Transparency of the sheet signifies compatibility, whereas an opaque appearance means incompatibility. Obviously, these criteria are arbitrary and crude. They are subject to great variation owing to difference in individual judgement. In addition, they give no information on the morphological feature of the system.

Glass Transition Temperature. If the glass transition temperatures of the polymeric components are known and the glass transition temperature of the polyblend is determined, one of two things can happen. If the polyblend shows two distinct transitions corresponding to the parent polymers, it is incompatible. If the polyblend shows one transition only, the system is compatible. Since the glass transition temperature is a measure of the segmental mobility of a polymer, it must be sensitive to the environment of the segments. If a polyblend shows glass transition temperatures similar to the parent components, the chains of the parent polymers must be very much within its own kind. Normally, one does not always get either a sharp single transition or two transitions. These represent the extreme situations with the real results falling between these two. Thus, various degrees of "partial compatibility" were proposed as an interpretation. Glass temperatures can be determined in many ways, and many different methods were actually used to determine polyblend compatibility. Jenkel (*16*) described the use of refractive index–temperature relationship to determine polyblend compatibility. Bartenev (*17*) determined glass temperature by plotting specific volume as a function of temperature. Again, when the results of the parent components and the polyblend are compared, compatibility or the lack of it is readily determined. Differential thermal analysis, differential scanning calorimetry, and thermal mechanical analysis can also be used to detect glass transition temperature; no new principle is involved.

Dynamic-Mechanical Measurement. This is a very sensitive tool and has been used intensively by Nielsen (*17*) and by Takayanagi (*18*). When the damping curves from a torsion pendulum test are obtained for the parent components and for the polyblend and the results are compared, a compatible polyblend will show a damping maximum between those of the parent polymers whereas the incompatible polyblend gives two damping maxima at temperatures corresponding to those of the parent components. Dynamic mechanical measurement can also give information on the moduli of the parent polymer and the polyblend. It can be shear modulus or tensile modulus. If the modulus–temperature curve of a polyblend locates between those of the two parent polymers, the polyblend is compatible. If the modulus–temperature curve shows multiple transitions, the polyblend is incompatible.

Since dynamic mechanical tests measure the response of a material to an applied stress at different temperature and frequency, they measure the transition of the material from glassy to leathery to rubbery state. If the frequency is kept constant and low (about one cycle/sec), the results are related to measurements of transition by other techniques. Thus, some cross-checking is possible.

Microscopy. This is a powerful tool for studying visually the distribution of the two phases in the polyblend. One can tell not only the domain size of the dispersed phase but also which polymer forms the dispersed phase from refractive index. A phase contrast light microscope can detect heterogeneity at the 0.2–10 μ level. If the sample can be stained preferentially and sectioned with microtome, then under favorable conditions electron microscopy can show heterogeneity to a very fine scale. In a study of PVC–poly(butadiene–co-acrylonitrile) blend,

Matsuo, Nozaki, and Jyo (20) showed that heterogeneity at 100 A scale and under can be detected readily. Thus, microscopy can offer a measure of heterogeneity down to 0.01 μ scale which is much smaller than the domain size of most polyblends. Results of microscopy have established convincingly that nearly all polyblends are heterogeneous two-phase systems. How does one describe the results? Obviously, heterogeneity as revealed by microscopy is a relative property. If compatibility is used in a qualitative sense, a polyblend with a finer domain size will be more compatible than one with a larger size, provided equilibrium size distribution has been attained in both cases.

Of the methods listed, glass transition temperature, dynamic mechanical measurement, and microscopy undoubtedly give the most useful information on the compatibility of polyblends. However, what is described as compatible by one method may be incompatible by another. Thus Stoelting, Karasz, and Macknight (21) found that blends of poly(2,6-dimethyl-1,4-phenylene ether) with atactic polystyrene gave two relaxation peaks corresponding to the parent polymers in dynamic mechanical measurement, indicating incompatibility by this method, and yet the same polyblend gave only a single glass transition temperature by differential scanning calorimetry, suggesting compatibility. Similarly, Matsuo (20) found in the blend of poly(vinyl chloride) and poly(butadiene–co–acrylonitrile) (60/40 by mole) that whereas dynamic mechanical measurement gave only one peak indicating a compatible system, a thin section of the blend under an electron microscope showed heterogeneity clearly at the 100 A scale and under.

We may now conclude from both thermodynamic considerations and experimental observations that when polymers of different chemical compositions are mixed, they will not intermix down to the molecular level and will not give a homogeneous single-phase structure. This ultimate state of molecular mixing, attainable by many liquid mixtures, can only be approached by polyblends as a limit. Therefore, the criterion of polyblend compatibility should not depend on the attainment of a single phase homogeneous structure. Conceptually, compatibility can be a representation of how close a polyblend approaches the ultimate state of molecular mixing as a limit. Phenomenologically, compatibility can best be described by the degree of homogeneity of the polyblend and measured and compared by the domain size of the dispersed phase. The finer the size of the dispersed phase in the continuous phase, the better the compatibility. Since the fineness of dispersion is relative, compatibility will have to be a relative and not an absolute attribute of the polyblend systems. Just like small molecules, if two polymers have a strong affinity for each other owing to strong intermolecular interaction, they will mix more intimately. In other words, compatibility is a relative measure of the affinity of two polymers for each other. Shultz (22) has reviewed rigorously the thermodynamic nature of this interaction. We can say that when two polymers show strong affinity for each other, they can be dispersed to small domain size. The smaller the domain size, the more compatible the system. Since compatibility as used in polyblends is a relative property and no scale has been established for comparison, it cannot be measured quantitatively as the degree of crystallinity of a crystalline polymer or the purity of a chemical compound. To avoid confusion, it is desirable to describe compatibility of a polyblend by giving: (1) the composition of the polyblend and the characterization of its components, (2) the sample history as to method of preparation, (3) the method and instrument used to determine compatibility, and (4) the experimental results and conclusions. Merely to state that two polymers are compatible is not sufficient to convey useful information.

Literature Cited

(1) Matthews, E. F., British Patent **16,278** (June 6, 1912).
(2) Gesner, B. D., "Encyclopedia of Polymer Science and Technology," Vol. 10, p. 694, Interscience, New York.
(3) Hildebrand, J. H., Scott, R. L., "The Solubility of Nonelectrolytes," Dover, 1964.
(4) Guggenheim, E. A., "Mixtures," Oxford, 1952.
(5) Prigogine, I., "The Molecular Theory of Solutions," North-Holland, Amsterdam, 1957.
(6) Rowlinson, J. S., "Liquids and Liquid Mixtures," Academic, New York, 1959.
(7) Hildebrand, J. H., Scott, R. L., "Regular Solutions," Prentice-Hall, Englewood Cliffs, 1962.
(8) Scatchard, G., *Chem. Rev.* **8**, 321 (1931).
(9) Glasstone, S., "Textbook of Physical Chemistry," 2nd ed., Chap. 10, Van Nostrand, New York, 1951.
(10) Leigh-Dugmore, C. H., "Microscopy of Rubber," p. 41, W. Heffer, Cambridge, 1961.
(11) Dobry, A., Boyer-Kawenoki, F., *J. Polymer Sci.* **2**, 90 (1947).
(12) Flory, P. J., Orwoll, R. A., Vrij, A., *J. Am. Chem. Soc.* **86**, 3507, 3515 (1964).
(13) Flory, P. J., *J. Am. Chem. Soc.* **87**, 1833 (1965).
(14) Flory, P. J., Eichinger, E. E., Orwoll, R. A., *Macromolecules* **1**, 287 (1968).
(15) Flory, P. J., "Principles of Polymer Chemistry," p. 555, Cornell University Press, Ithaca, 1953.
(16) Jenckel, E., Herwing, H. U., *Kolloid-Z.* **148**, 57 (1956).
(17) Bartenev, G. M., Kongarov, G. S., *Vysokom. Soed.* **2**, 1692 (1960).
(18) Nielsen, L. E., *J. Am. Chem. Soc.* **75**, 1453 (1953).
(19) Takayanagi, M., *Mem. Faculty Eng. Kyosho Univ.* **23** (1), 11 (1963).
(20) Matsuo, M., Nozaki, C., Jyo, Y., *Polymer Eng. Sci.* **9** (3), 197 (1969).
(21) Stoelting, J., Karasz, F. E., Macknight, W. J., *Polymer Eng. Sci.* **10** (3), 133 (1970).
(22) Shultz, A. R., *Am. Chem. Soc. Div. Polymer Chem., Preprints* 1455 (Sept. 1967).

Bibliography

Keskkula, H., Ed., "Polymer Modification of Rubbers and Plastics," *J. Appl. Polymer Sci. Polymer Symp.* **7** (1968).

Mellan, I., "Compatibility and Solubility," Noyes Development Corp., 1968.

Nielsen, L. E., "Mechanical Properties of Polymers," Chap. 6 and 7, Reinhold, New York, 1962.

SPE RETEC Technical Papers, "Polymer Modified Polymers," New York, N. Y., March 1968.

Buckley, D. J., *Trans. New York Acad. Sci., Ser. II*, **29** (6), 735–747 (1967).

Bohn, L., *Kolloid Z., Z. Polymere* **213**, 55 (1966).

Rosen, S. L., *Polymer Eng. Sci.* **7**, 115 (1967).

Pazonyi, T., Dimitrov, M., *Rubber Chem. Technol.* **40**, 1119 (1967).

Slonimskii, G. L., *J. Polymer Sci.* **30**, 625 (1958).

Petersen, R. J., Corneliussen, R. D., Rozelle, L. T., *Am. Chem. Soc. Div. Polymer Chem., Preprints* **10** (1), 385 (1969).

RECEIVED August 27, 1970.

2

Compatible High Polymers: Poly(vinylidene fluoride) Blends with Homopolymers of Methyl and Ethyl Methacrylate

J. S. NOLAND, N. N.-C. HSU, R. SAXON, and J. M. SCHMITT

Plastics Division, American Cyanamid Co., Stamford, Conn. 06904

The homopolymers poly(methyl methacrylate) and poly-(ethyl methacrylate) are compatible with poly(vinylidene fluoride) when blended in the melt. True molecular compatibility is indicated by their transparency and a single, intermediate glass transition temperature for the blends. The T_g results indicate plasticization of the glassy methacrylate polymers by amorphous poly(vinylidene fluoride). The T_g of PVdF is consistent with the variation of T_g with composition in both the PMMA–PVdF and PEMA–PVdF blends when T_g is plotted vs. volume fraction of each component. PEMA/PVdF blends are stable, amorphous systems up to at least 1 PVdF/1 PEMA on a weight basis. PMMA/ blends are subject to crystallization of the PVdF component with more than 0.5 PVdF/1 PMMA by weight. This is an unexpected result.

In the early 1960's it was discovered in our laboratory (20) that poly(methyl methacrylate) (PMMA) and poly(vinylidene fluoride) (PVdF) were compatible when blended in the melt. Similarly, compatibility was found for poly(ethyl methacrylate) (PEMA) with PVdF. Blends of the fluorinated polymer with higher alkyl methacrylate polymers, however, were nonhomogeneous.

At about the same time, Koblitz and co-workers (14) observed the formation of a "homogeneous physical mixture of PVdF and a solid PMMA resin," although they placed primary emphasis upon minor amounts of PMMA as a processing aid for PVdF. Our work, which is described below, was chiefly concerned with PMMA-rich systems, in recognition of the possible technological advantages inherent in having a light-stable, durable, polymeric plasticizer for PMMA (17).

Experimental

Materials. The PMMA used was Acrylite H-12 molding compound (American Cyanamid Co.). This material was available as pellets. PVdF was Kynar 401 (Pennwalt Chemicals Co.), a powder of high but unspecified molecular weight. PEMA was polymerized in bulk in the laboratory from ethyl methacrylate (Rohm & Haas) by a laboratory scale adaptation of the PMMA process (22).

Preparation of Blends. Blends were most conveniently made by fluxing a weighed amount of the acrylic component on a two-roll mill at 165°–170°C, followed by the addition of the PVdF and working for a few minutes until the blend was homogeneous. It was then removed from the mill, cooled, granulated in a Cumberland chopper, and either used in the form of irregular granules or passed through a small single- or twin-screw extruder to provide pellets or thin film.

Differential Thermal Analysis (DTA) and Thermochemical Analysis (TMA). Differential Thermal Analysis (DTA) measurements were carried out on a du Pont model 900 differential thermal analyzer. Measurements on PMMA–PVdF polyblend samples annealed at room temperature were carried out by heating at 10°/minute over the range 25° to ca. 170°C. Samples of melt extruded PVdF, annealed several weeks at ambient temperature, exhibited no definitive changes characteristic of the glass transition when heated at 10°/minute from −100° to 100°C, apparently owing to the large portion of crystalline material. The samples of PVdF were heated to 190°C (which is above the melting region observed at 156°C) and quenched rapidly to −195°C. These quenched samples were heated through the recorded range of −110° to 20°C. In four heating cycles, transitions in the range −43° to −47°C were observed. The average of −46°C was chosen as T_g.

A du Pont model 940 thermomechanical analyzer was employed for the T_g measurements on the PEMA/PVdF polyblends by the TMA technique. The sample specimens were cut from 30-mil, melt pressed sheets of the respective blends and examined by heating from −60° to 100°C at a rate of 5°C/minute. Sharp, unambiguous transitions were observed.

Dilatometry. Conventional dilatometric equipment was used (10), J shaped with a 60 cm long, 1 mm diameter, graduated capillary, with the specimen section in an inverted position to reduce hydrostatic pressure on the sample. After introduction of the sample, the sample section was sealed, and the dilatometer was evacuated to 0.01 mm Hg and heated to 190°C at 1°–2°C/minute. It was removed from the bath and cooled rapidly to the specified quench temperature, at which point the mercury confining liquid was allowed to flow into the evacuated dilatometer. In cases where the quench temperature was lower than the freezing point of mercury, ethanol was used as the confining liquid, and straight dilatometers of the same size were used. The dilatometric data for T_g—i.e., the point of inflection of the resulting volume vs. temperature plots for the PVdF–PMMA blends—are presented in Table I.

Table I. Dilatometric Data for PVdF/PMMA Blends

PVdF, %	Run No.	Quench Temp., °C	T_g, °C	Average T_g, °C
15	a	−30	73.5	
	b	−30	80.0	
	c	−30	80.0	78.1
	d	−30	75.0	
	e	26	82.0	
35	a	26	61.0	
	b	46	66.8	
	c	26	53.0	62.0
	d	50	67.0	
	e	26	62.0	
50	a	−30	39.5	
	b	−30	38.0	40.2
	c	26	43.0	
70	a	−30	0.0	
	b	−30	5.0	
	c	−70	1.5	2.4
	d	−70	3.0	

Ultraviolet Exposure. A combination fluorescent ultraviolet and fluorescent blacklight lamp assembly (2) was used which yields an ultraviolet spectrum similar to that of the ultraviolet portion of sunlight (9).

Light Transmission. This was determined on a GE recording spectrophotometer using thin films, followed by integration of the transmission over the visible range.

Mechanical Tests. These were determined on an Instron model TM using thin films.

Discussion

Physical Properties of PMMA–PVdF and PEMA–PVdF Polyblends. The compatibility of these pairs of homopolymers is illustrated both by their clarity and by the fact that single glass transition temperatures, intermediate between the two homopolymer values, are observed.

T_g data for the blends are shown in Figure 1. Results obtained by DTA on PMMA–PVdF blends which had been melt extruded and "annealed" at room temperature appeared to be anomalous in terms of the theory for glass transition of copolymers or compatible polymer blends (6). These data indicated a limiting value for T_g of *ca.* 40°–45°C. However, x-ray examination showed that samples with more than *ca.* 35% PVdF exhibited a crystalline phase, indicating that some of the PVdF had precipitated. When these systems were re-examined by dilatometry

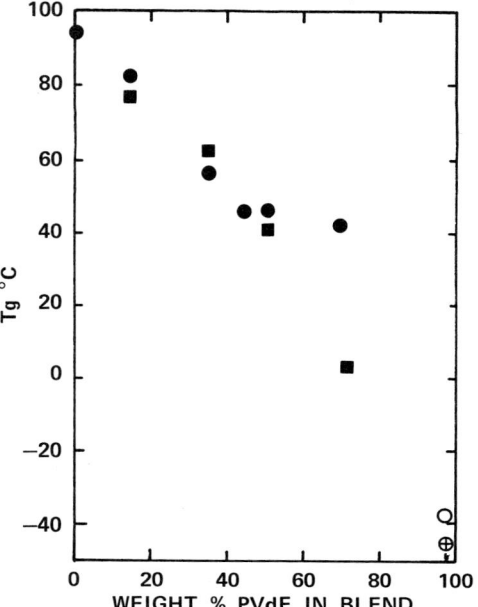

- ■ DETERMINED DILATOMETRICALLY
- ● DETERMINED BY DTA ON ANNEALED SAMPLES
- ○ MANDELKERN et al
- ⊕ DTA QUENCHED SAMPLE

Figure 1. Glass transition temperature of PMMA–PVdF blends

with rapid quenching from the melt to $-30°$ or $-70°C$, a more continuous variation with composition was observed.

At least two different glass transition temperatures have been reported for PVdF homopolymer. Owing to the large proportion of crystalline structure in this polymer and the rapid crystallization which occurs while heating quenched amorphous samples, it is difficult experimentally to obtain an unambiguous, well-defined second-order transition. Mandelkern, Martin, and Quinn (16) reported a value "below $-40°C$" based upon an extrapolation of the T_g data for vinylidene fluoride–chlorotrifluoroethylene copolymers in accordance with the Fox equation (6),

$$\frac{1}{T_g} = \frac{W_1}{T_{g_1}} + \frac{W_2}{T_{g_2}}$$

where W_1 and W_2 are the weight fractions of the respective comonomers in the copolymer. They arrived at a somewhat higher T_g, $-35°$ to

−40°C, by dilatometric experiments on annealed PVdF. Peterlin and Holbrook (*18*) have reported a value of 13°C. Our own data, established by DTA on samples quenched rapidly from the melt, gave an average value of −46°C, in agreement with the extrapolation of Mandelkern and co-workers.

Using this experimental value for PVdF and the value of 95°C for PMMA, and plotting the experimental T_g data for the blends *vs.* wt % composition, the observed values show large deviation from the behavior which would be predicted by the Fox equation (Figure 2). It appeared

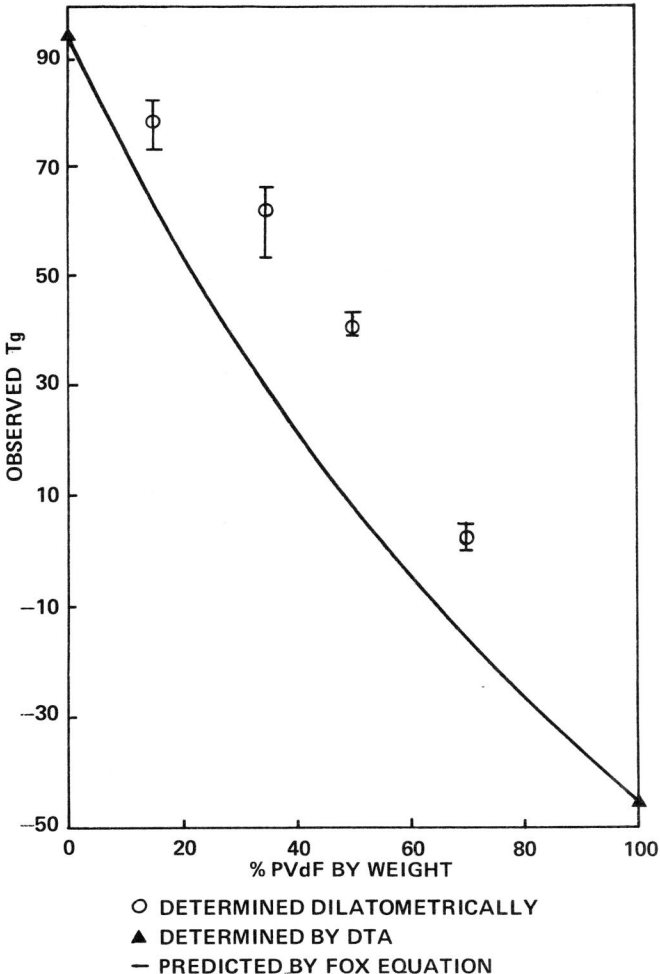

Figure 2. Comparison of experimental T_g *values of PMMA–PVdF blends with theoretical curve based on Fox equation*

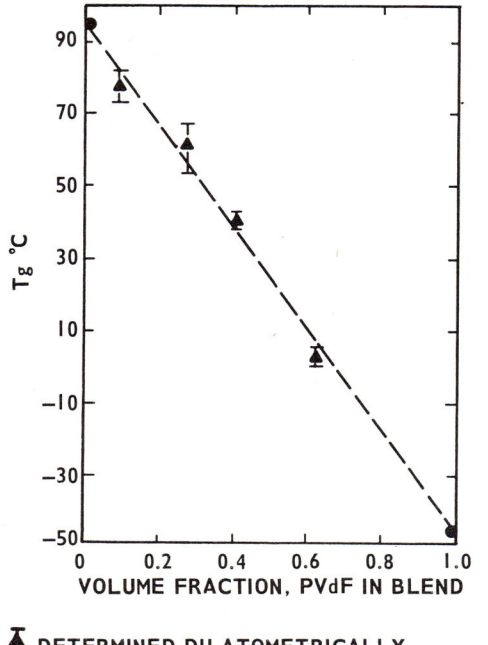

Figure 3. Observed vs. *calculated* T_g *of PMMA–PVdF blends*

to us that the present situation involving a polymer blend might be described more accurately by equations developed for polymer–diluent systems and represented a special case of the Kelley-Bueche (*13*) equation:

$$T_g = \frac{\nu_p T_{g_p}(\alpha_l - \alpha_g) + \nu_d T_{g_d}\alpha_d}{\nu_p(\alpha_l - \alpha_g) + \nu_d\alpha_d}$$

where ν is the volume, α is the thermal expansion coefficient, the subscript p refers to the polymer, d to the diluent, l to liquid, and g to glass. This equation is derived on the assumption that free volumes of the polymer and diluent are additive and that the free volume has the same critical value for polymer, diluent, and their mixtures at their respective glass temperatures. [T. G. Fox has called the authors' attention to a publication of L. A. Wood (*24*), in which a generalized expression for T_g of polymer mixtures is presented, from which the Fox equation and the modified Kelley-Bueche equations, below, can also be derived.]

Since the diluent is itself a polymer, the same "standard" free volume at T_g would be assigned to both components, and α_d should be replaced with $(\alpha_{dl} - \alpha_{dg})$. Following Bueche's precedent that $(\alpha_l - \alpha_g)$ is a constant for all polymers, we may write $(\alpha_l - \alpha_g) = (\alpha_{dl} - \alpha_{dg})$. The Kelley-Bueche equation then reduces to

$$T_g = \frac{v_p T_{g_p} + v_d T_{g_d}}{v_p + v_d}$$

or, since $v_p + v_d = 1$, and p and d are simply two different polymers,

$$T_g = v_1 T_{g_1} + v_2 T_{g_2}$$

Thus, T_g should be a linear function of the volume fraction of each component rather than the weight fraction as implied by the Fox equation. When the T_g data are plotted vs. volume fraction, a reasonable fit is obtained for the PMMA–PVdF polyblends (Figure 3) and excellent agreement for PEMA–PVdF blends (Figure 4). [For these calculations,

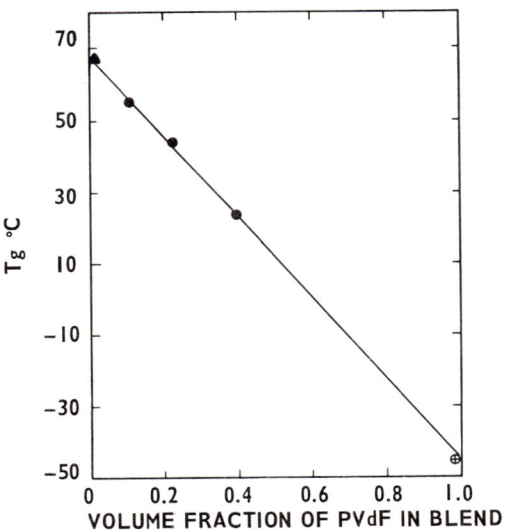

▲ PEMA Tg FROM LITERATURE
⊕ PVdF Tg BY DTA
● DETERMINED BY THERMOMECHANICAL ANALYSIS
— CALCULATED BY MODIFIED KELLEY–BUECHE EQUATION

Figure 4. Observed vs. calculated T_g of PEMA–PVdF blends

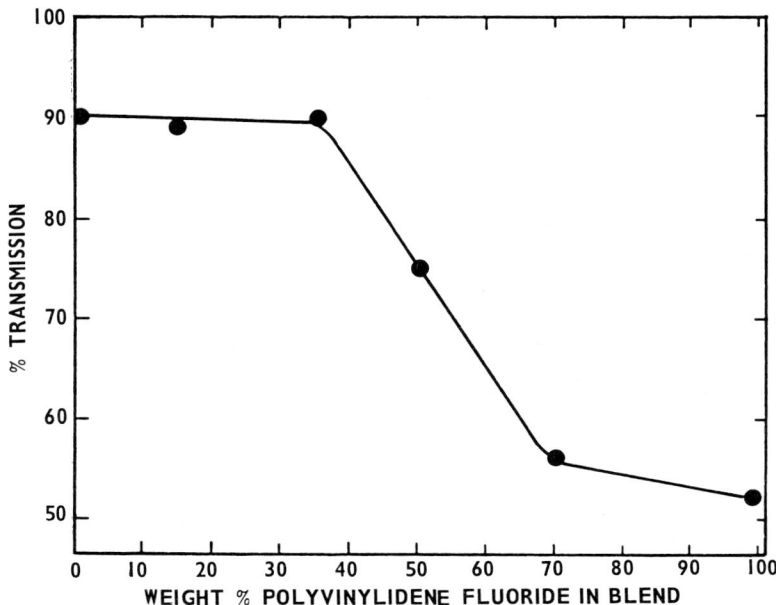

Figure 5. Light transmission of PMMA–PVdF blends

the amorphous density of PVdF = 1.74, the density of PMMA = 1.19, the density of PEMA = 1.12, and the T_g of PEMA = 66°C, were all taken from Lewis (15).]

Thus, the PVdF behaves as a homogeneous plasticizer for the methacrylate polymers. For the PMMA–PVdF system, the compositions with more than 40 wt % PVdF, whose T_g's are below 50°C, permit crystallization of PVdF as a separate phase. Consequently, the T_g is elevated since the concentration of the diluent is effectively reduced in the amorphous phase. At ambient temperature, this produces the phenomenon of limiting T_g as observed in the original room temperature annealed samples (Figure 1).

It is interesting that the PEMA–PVdF blends are amorphous up to at least 50 wt % PVdF even though the T_g of the latter is 24°C. The crystallization of PVdF observed in the analogous PMMA blend does not occur under the same conditions with PEMA–PVdF. This suggests that there is a specific interaction between the fluoropolymer and the methacrylate polymer which is sufficient to "dissolve" PVdF in the PMMA and PEMA, and that this specific interaction is superimposed on the conventional diluent–crystalline polymer interactions. The complexity of the rate processes involved with high molecular weight systems arising from molecular mobility makes it impossible to elucidate the nature of such specific interaction without a considerably more detailed investigation.

Other properties of the PMMA–PVdF system are consistent with the amorphous–crystalline blend model. Thus, the light transmission of films of these blends at room temperature (Figure 5) decreases fairly abruptly as the the PVdF content increases, coincident with the appearance of a crystalline phase.

The tensile properties at ambient temperatures are also consistent with this view (Figure 6).

Ultraviolet Stability of PMMA–PVdF Polyblends. The potential ultraviolet resistance of this polyblend was of considerable interest. Let us first consider what is known of the photochemical stability of the components.

The degradation of PMMA in the short wavelength ultraviolet region, principally with the 2537-A mercury line, has been studied by several investigators (4, 7, 8, 12, 21). In general, exposure to this high energy radiation has caused molecular weight degradation by random chain scissions, the quantum yield depending on the purity of the polymer and the nature of the exposure (air or vacuum). This chain scission may be accompanied by fragmentation of the ester group and chain unzipping; the latter also depends upon the temperature of the exposure.

For practical purposes the resistance of PMMA to near ultraviolet is more important since it is the region which is responsible for the

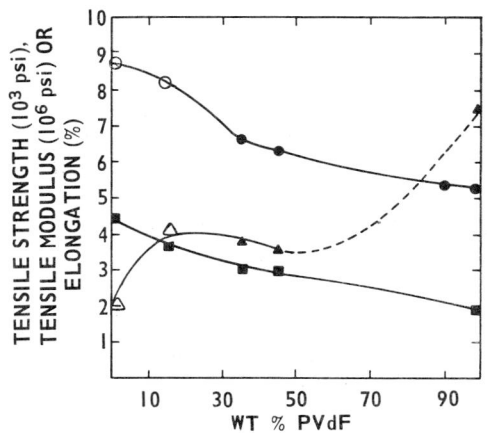

○ TENSILE STRENGTH AT RUPTURE, ● AT YIELD
△ ELONGATION (%) AT RUPTURE, ▲ AT YIELD
■ TENSILE MODULUS

Figure 6. Mechanical properties of films of PMMA–PVdF blends

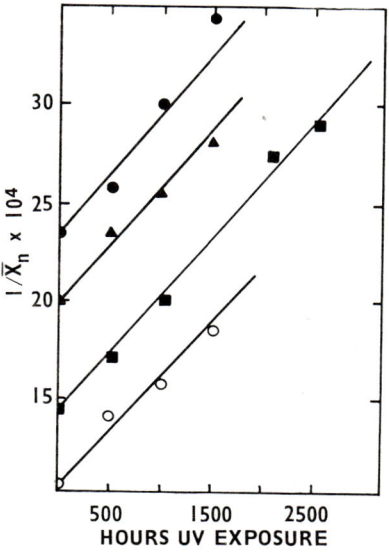

INITIAL $\overline{M}_w$:
188,000 (○), 135,000 (■), 100,000 (▲), 85,000 (●)

Figure 7. $\overline{M}_w$ degradation of PMMA by ultraviolet exposure

photochemical deterioration of plastics in exposure to weather. In terrestrial sunlight more than 99% of the ultraviolet energy occurs at wavelengths longer than 3000 A. Therefore, we have carried out exposure tests in a device (2) which combines fluorescent ultraviolet and fluorescent blacklight lamps to yield a spectrum similar to the ultraviolet spectrum of sunlight (9). These experiments have established that the same kind of molecular weight degradation occurs in PMMA with ultraviolet energy above 3000 A as in the 2537-A region but at a slower rate and with no observable by-products. The color retention and general appearance of the acrylic polymer samples depended upon the purity—*i.e.*, freedom from ultraviolet-absorbent species such as aromatic residues, peroxides, etc.—but the samples also undergo molecular weight degradation at a slow rate which is nearly constant with exposure time.

The rate of polymer degradation was determined by measuring the intrinsic viscosity of the initial and exposed samples in chloroform. The molecular weight ($\overline{M}_w$) was calculated from the equation (3):

$$[\eta] = 4.3 \times 10^{-5} M^{0.80}$$

for which the constants had been obtained from log-log plots of intrinsic viscosity vs. M_w, determined by light scattering.

It was established, in unpublished results, that the PMMA polymerization process yielded a random, most probable distribution of molecular weights, and therefore it was assumed that the relationship

$$\overline{M}_w = 2\overline{M}_n$$

was valid.

The degree of polymerization $\overline{X}_n$ was calculated from the relationship

$$\overline{X}_n \equiv \frac{\overline{M}_n}{\text{monomer molecular weight}} = \frac{\overline{M}_n}{100}$$

It is well known (*11*) that for degradation by random chain scission, the following relationship is valid:

$$\frac{1}{\overline{X}_{n,\,t}} - \frac{1}{\overline{X}_{n,\,o}} = kt$$

Hence a graph of $\frac{1}{\overline{X}_{n,\,t}}$ vs. t should produce a linear plot.

A series of four samples of different initial $\overline{M}_w$, but prepared by analogous procedures, were subjected to near-ultraviolet degradation (Figure 7). The expected family of straight, parallel lines is obtained. Therefore, we could predict an ultimate tensile failure for PMMA when the molecular weight $\overline{M}_w$ had degraded beyond a certain threshold—*ca.* 70,000. This behavior was indeed observed (Figure 8).

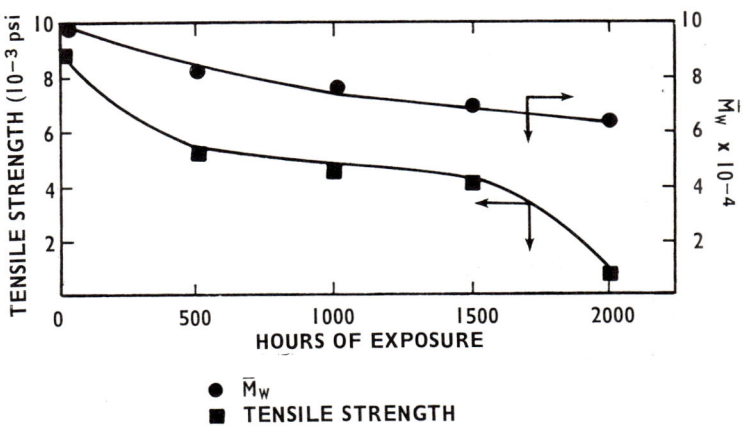

Figure 8. Tensile strength and $\overline{M}_w$ of PMMA as functions of duration of fluorescent ultraviolet irradiation

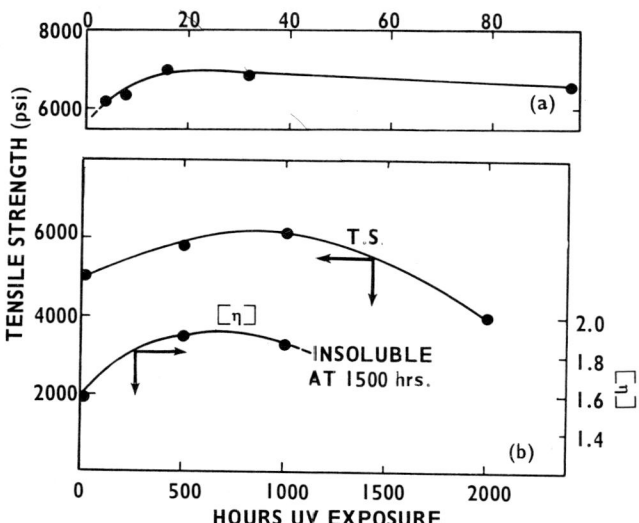

Figure 9. (a) Effect of electron beam irradiation on tensile strength of PVdF
(b) Effect of ultraviolet irradiation on tensile strength and intrinsic viscosity of PVdF

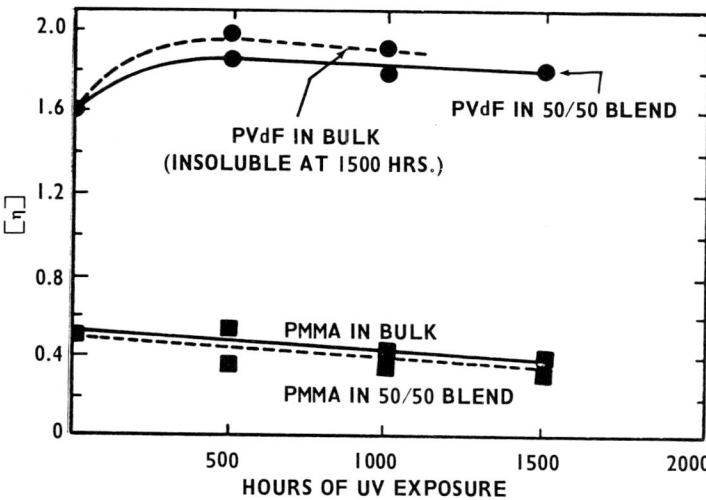

Figure 10. Intrinsic viscosity of PMMA and PVdF, exposed to ultraviolet radiation separately and in blend

The ultraviolet degradation of PVdF has been studied much less extensively. We observed, in studies with the same energy sources as mentioned above, that PVdF homopolymer exhibits an initial increase in $\overline{M}_w$ but eventually becomes insoluble. This appears to be consistent with a published report (23) that PVdF increases in tensile strength and decreases in elongation when exposed to irradiation with electrons. These results suggest that PVdF undergoes both a crosslinking and a chain scission mechanism. Our data are summarized in Figure 9, along with an analogous plot of strength vs. electron irradiation dose.

The performance of polyblends is quite consistent with that of the two components alone. This might be expected owing to the high degree of ultraviolet transparency of each of the component's polymers (Figure 10). Their retention of tensile strength and elongation is limited by the $\overline{M}_w$ of the PMMA portion, which is the more ultraviolet-sensitive of the two polymers. The PVdF portion also crosslinks, though at a slightly reduced rate compared with PVdF homopolymer alone. (The crosslinking is deduced from the fact that the PVdF fraction in the blend retains its solubility after longer exposure than does PVdF exposed alone.)

In practical terms, very good durability can be achieved by increasing the $\overline{M}_w$ of the PMMA component. Since PVdF plasticizes the PMMA, the melt viscosity of the blend was not appreciably changed by increasing the $\overline{M}_w$ of the PMMA component from 100,000 to 200,000. This change did, however, effectively double the exposure lifetime through which the maximum tensile properties of the polyblend were maintained.

Conclusions

Mixtures of poly(vinylidene fluoride) with poly(methyl methacrylate) and with poly(ethyl methacrylate) form compatible blends. As evidence of compatibility, single glass transition temperatures are observed for the mixtures, and transparency is observed over a broad range of composition. These criteria, in combination, are acceptable evidence for true molecular intermixing (1, 19). These systems are particularly interesting in view of Bohn's (1) review, in which he concludes that a "compatible mixture of one crystalline polymer with any other polymer is unlikely except in the remotely possible case of mixed crystal formation." In the present case, the crystalline PVdF is effectively dissolved into the amorphous methacrylate polymer melt, and the dissolved, now amorphous, PVdF behaves as a plasticizer for the glassy methacrylate polymers.

This unusual compatibility between homopolymers which can be readily characterized opens the way for studying the criteria of polymer compatibility in more detail. Other aspects, such as the observed melting point depression for crystalline PVdF upon dilution with PMMA and crystallization isotherms for PVdF-rich systems will be reported later.

Acknowledgment

DTA experiments were carried out by Elspeth C. Eberlin and Joan Z. Whiting. TMA experiments were carried out by Joan K. Lucas and Marie T. Borghetti.

Literature Cited

(1) Bohn, L., *Rubber Chem. Technol.* **41**, 495 (1968).
(2) Cipriani, L. P., Giesecke, P., Kinmonth, R., *Plastics Technol.* **11** (5), 34 (1965).
(3) Cohn-Ginsberg, E., Fox, T. G., Mason, H. F., *Polymer* **3**, 97 (1962).
(4) Crowley, P. R. E. J., Melville, H. W., *Proc. Roy. Soc. (London)* **A210**, 461 (1957).
(5) *Ibid.*, **A211**, 320 (1958).
(6) Fox, T. G., *Bull. Am. Phys. Soc.* **1**, 123 (1956).
(7) Fox, R. B., Isaacs, L. G., Stokes, S., *U. S. Naval Res. Lab. Rept.* **5720** (1961); *J. Polymer Sci.* **A, 1**, 1079 (1963).
(8) Frolova, M. I., Nevskii, L. V., Ryabov, A. V., *Vysokomol. Soedin.* **3**, 877 (1961).
(9) Hirt, R. C., Schmitt, R. G., Searle, N. D., Sullivan, A. P., *J. Opt. Soc. Am.* **50**, 706 (1960).
(10) Hsu, N. N.-C., Ph.D. Thesis, University of Akron (1966).
(11) Jellinek, H. H. G., "Degradation of Vinyl Polymers," p. 21, Academic, New York, 1955.
(12) Jellinek, H. H. G., Wang, I. C., *Kolloid-Z.* **202**, 1 (1965).
(13) Kelley, F. N., Bueche, F., *J. Polymer Sci.* **50**, 549 (1961).
(14) Koblitz, F. F., Petrella, R. G., Dukert, A. A., Christofas, A., U. S. Patent **3,253,060** (1966).
(15) Lewis, O. G., "Physical Constants of Linear Homopolymers," Springer-Verlag, New York, 1968.
(16) Mandelkern, L., Martin, G. M., Quinn, Jr., F. A., *J. Res. Natl. Bur. Std.* **58**, 137 (1957).
(17) Miller, Jr., C. H., U. S. Patent **3,458,391** (1969).
(18) Peterlin, A., Holbrook, J. D., *Kolloid-Z.* **203**, 68 (1965).
(19) Rosen, S. L., *Polymer Eng. Sci.* **7**, 155 (1967).
(20) Schmitt, J. M., U. S. Patent **3,459,834** (1969).
(21) Schultz, A. R., *J. Phys. Chem.* **65**, 967 (1961).
(22) Terenzi, J. F., Schmitt, J. M., U. S. Patent **3,252,950** (1966).
(23) Timmerman, R., Greyson, W., *J. Appl. Polymer Sci.* **6**, 456 (1962).
(24) Wood, L. A., *J. Polymer Sci.* **28**, 319 (1958).

RECEIVED January 23, 1970.

3

Physical Properties of the System: Poly(2,6-dimethylphenylene oxide)–Polystyrene

W. J. MacKNIGHT, J. STOELTING,[1] and F. E. KARASZ

Polymer Science and Engineering Program and Department of Chemistry, University of Massachusetts, Amherst, Mass. 01002

> *The dielectric relaxation of bulk mixtures of poly(2,6-dimethylphenylene oxide) and atactic polystyrene has been measured as a function of sample composition, frequency, and temperature. The results are compared with earlier dynamic mechanical and (differential scanning) calorimetric studies of the same samples. It is concluded that the polymers are miscible but probably not at a segmental level. A detailed analysis suggests that the particular samples investigated may be considered in terms of a continuous phase-dispersed phase concept, in which the former is a PS-rich and the latter a PPO-rich material, except for the sample containing 75% PPO–25% PS in which the converse is postulated.*

Several criteria have been used to examine the miscibility or compatibility of two or more polymers in the bulk phase (2). These include the presence in the mixture of mechanical integrity, optical transparency, a single glass transition temperature (1, 12), and homogeneity on a submicroscopic level as revealed by electron microscopy. In each case the presence of the property listed has been taken as evidence for compatibility. However, it is by no means clear that if a system satisfied any given criterion, it would satisfy one or more of the other criteria, nor, in most cases, is it obvious to what extent the mixing of two polymers would have to be carried to satisfy any of these criteria. One may visualize a spectrum of miscibility ranging from the most intimate, at the polymer

---

[1] Present address: BASF, Ludwigshafen-am-Rhein, Germany.

segment (monomer unit) level, to the comparatively coarse in which the placement of the individual polymer molecules themselves is randomized. At a still coarser level, of course, one obtains segregation into clusters of polymer molecules, and microphase separation occurs at various dimensional levels. In the latter event, however, the rather diffuse boundary between compatibility and incompatibility probably has been crossed, and we shall not consider this case further. For a pair of polymers to exhibit thermodynamic compatibility as the term is presently understood, therefore, $\Delta G_m < 0$ for some value of s. ΔG_m is the free energy of mixing per unit mass; s is a parameter representing the level of mixing in terms of the average normalized size of the segmental clusters in the system. If the latter corresponds to the polymer molecule itself, $s = 1$; if alternatively mixing should occur at a segmental level, then $s = 1/N \simeq 0$, where N is the degree of polymerization. The condition $s >> 1$ corresponds to phase separation. ΔG_m will clearly be negative for all values of s if the heat of mixing $\Delta H_m < 0$, but the size of the average cluster, sN, will depend also on the relative contribution of ΔS_m, the entropy of mixing per segment. The latter factor will also determine whether miscibility occurs at all in the more prevalent situation, $\Delta H_m > 0$. In the latter case, ΔG_m can still become negative, with a minimum for $0 < s < 1$, if ΔH_m is not too large. This occurs because as s decreases, ΔS_m at first increases as a result of the greater number of segment configurations available. However, ΔS_m eventually declines because of the increasing restrictions in the number of possible configurations imposed by the condition of contiguity of polymer segments. If we restrict ourselves, therefore, to the discussion of compatibility at levels between the mixing of individual polymer molecules (crudely approximated, perhaps, by considering these as soft spheres) and the most intimate segmental mixing, several questions arise. The most obvious is, can we determine what miscibility level we have in a particular system? It seems probable that in favorable circumstances this might be accomplished by direct electron microscopic observation, but for an indirect miscibility criterion such as the presence of a single T_g, we cannot specify the size of the average compositional fluctuation because it is not known at present how large a particular homogeneous region may have to be to display a glass transition. In the case of the single T_g criterion, also, it may be possible that this in itself is not enough to specify the level of miscibility because the glass transition can be observed by various experimental techniques which may differ in their sensitivities to heterogeneity. In the present paper we present evidence for this latter effect.

The question of compatibility in polymers, therefore, is much more complicated than the corresponding problem in low molecular weight systems. In addition, the kinetics of mixing must inevitably be considered.

It cannot be assumed *a priori* that an apparently compatible mixture is in thermodynamic equilibrium (*13*).

In the present paper we examine the behavior of mixtures of poly(2,6-dimethyl-1,4-phenylene oxide) (PPO) and atactic polystyrene (PS) with some of the above points in mind; in particular, we are concerned with a comparison of the dielectric and dynamic mechanical relaxation spectra. This system satisfies many of the criteria discussed above—the mixture forms a strong and optically transparent polyblend which in calorimetric studies displays a single glass transition at a temperature corresponding approximately to the weighted mean of the T_g's of the constituents, 208°C (PPO) and 90°C (PS) (*3*). Earlier work indicates, however, that in dynamical mechanical measurements this loss peak, corresponding to the glass transition, is partially resolvable, at least for certain compositions and thermal histories (*11*). This is taken as evidence for the presence in these samples of two distinct phases, one rich in PPO and the other in PS. However, it was also shown that the same samples which displayed the resolvable loss peak in dynamical mechanical studies showed only a single T_g in DSC measurements; this latter agreed generally with the lower T_g, corresponding to the PS-rich phase, of the mechanical data. In this paper, therefore, we discuss dielectric relaxation studies of these samples as a function of composition, measurement frequency, and temperature and compare the results with the earlier DSC and dynamic mechanical measurements.

Experimental

Materials. High molecular weight, additive-free PPO was obtained from the General Electric Co. (courtesy of A. Katchman). The atactic polystyrene was also an additive-free resin manufactured by the Monsanto Co. These polymers had number average molecular weight of 5.8×10^4 and 17.3×10^4, respectively.

Measurements. Dielectric relaxation measurements were made using a General Radio 1620-A bridge at 100, 500, 1000, 5000, and 10,000 Hz over the temperature range $-180°$ to 200°C. A Balsbaugh LD-3 dielectric cell and a specially designed stainless steel high temperature cell were used in these experiments. The data were reduced in the customary fashion to yield ϵ' and ϵ'', the real and imaginary components of the complex dielectric constant ϵ^*, and the ratio $\epsilon''/\epsilon' = \tan \delta$. The data at the higher temperatures were corrected for contributions from dc conductivity.

The dynamical mechanical measurements were made using a Vibron dynamic viscoelastomer, model DDV II (Toyo Instrument Co.) at two frequencies, 3.5 and 110 Hz, over the temperature range $-180°$ to 240°C.

Calorimetric measurements were made using a Perkin-Elmer DSC-1B differential scanning calorimeter. A uniform heating rate of 10°C per minute was employed for all measurements.

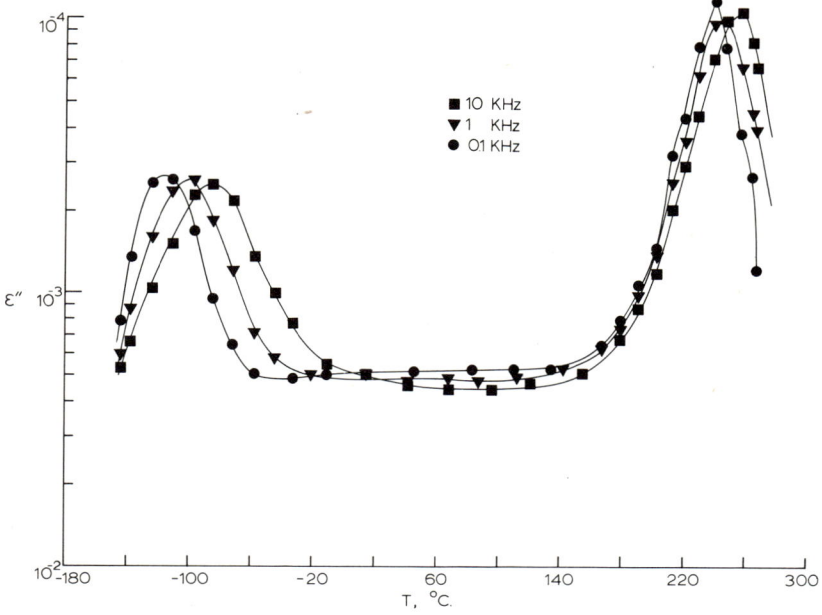

Figure 1.   Dielectric loss of PPO as a function of temperature and frequency

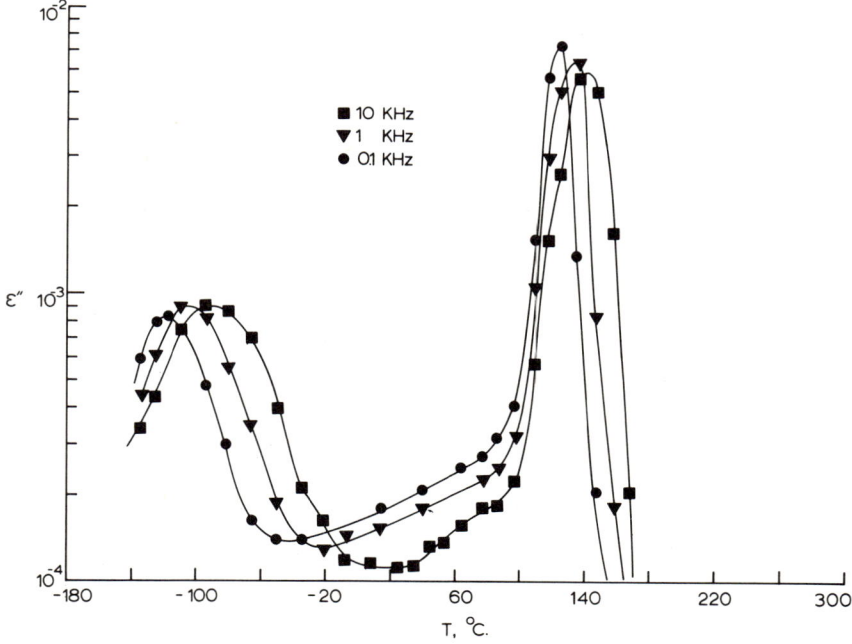

Figure 2.   Dielectric loss of pure PS as a function of temperature and frequency

Sample Preparations. Three different blends were prepared by melting appropriate quantities of the solid polymers together and then compression molding into film suitable for dynamic mechanical and dielectric testing. A variety of mixing times and temperatures and annealing conditions were investigated. The thermal treatments finally adopted were such that further annealing did not produce any changes in the relaxational or calorimetric results. The details of these treatments are as follows:

(1) 75 PS–25 PPO: the mixture was heated to 300°C, maintained at this temperature for 6 minutes, compression molded into a film, and cooled slowly to room temperature in the press.

(2) 50 PS–50 PPO: the mixture was heated to 300°C, maintained at this temperature for 10 minutes, compression molded into a film, and cooled slowly to room temperature in the press. The film was then annealed at 180°C for 12 hours.

(3) 25 PS–75 PPO: the mixture was treated in an identical fashion to the 75 PS–25 PPO blend.

Results and Discussion

Results of the dielectric relaxation measurements of the pure constituents and of the polyblends in terms of ϵ'' as a function of temperature

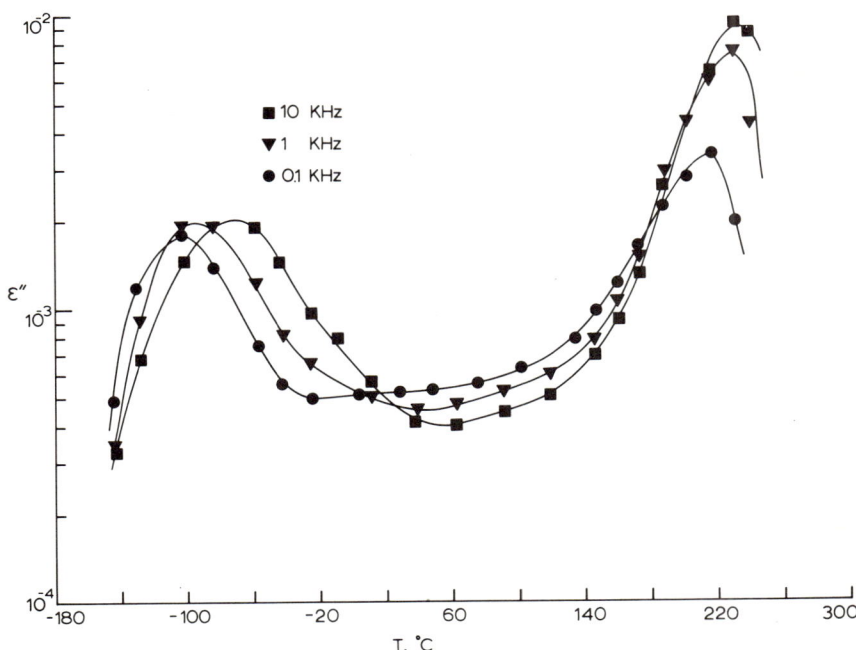

Figure 3. Dielectric loss of 75 PPO–25 PS mixture as a function of temperature and frequency

at 0.1, 1, and 10 kHz are shown in Figures 1–5. The result for PS is in good agreement with those of other workers and is included here merely for reference, while a detailed analysis of dielectric relaxation in pure PPO is presented elsewhere (6). Hence in this paper we are concerned mainly with the PPO–PS mixtures. Figure 6 shows dynamical mechanical relaxation spectra at 110 Hz for the three mixtures studied. The dielectric and mechanical data both show relatively strong α relaxations corresponding to the glass transition and low temperature γ loss peaks, but whereas the latter are very weak in the mechanical spectra, they are roughly comparable in intensity to the α relaxation in the dielectric measurements. A more detailed analysis elicits additional points, and for convenience we discuss the α and γ loss peaks separately.

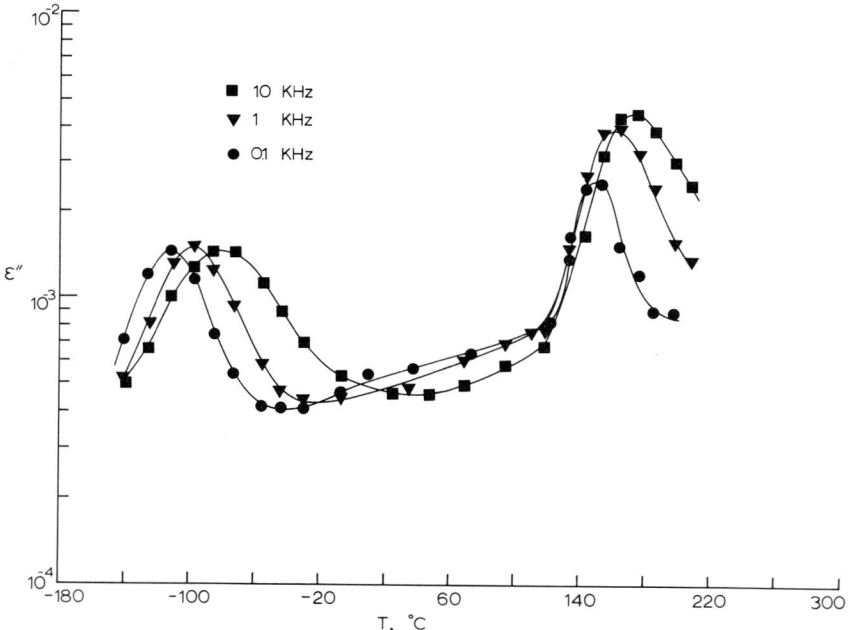

Figure 4. Dielectric loss of 50 PPO–50 PS mixture as functions of temperature and frequency

The α Relaxation. One of the principal findings in the earlier study of the PPO–PS mixtures (11) was that samples which appeared to be homogeneous when studied calorimetrically, inasmuch as they indicated only a single T_g, showed a dynamical mechanical α relaxation which could be resolved into two components. These were interpreted in terms of PS-rich (lower T_g) and PPO-rich (upper T_g) phases. It is clear from

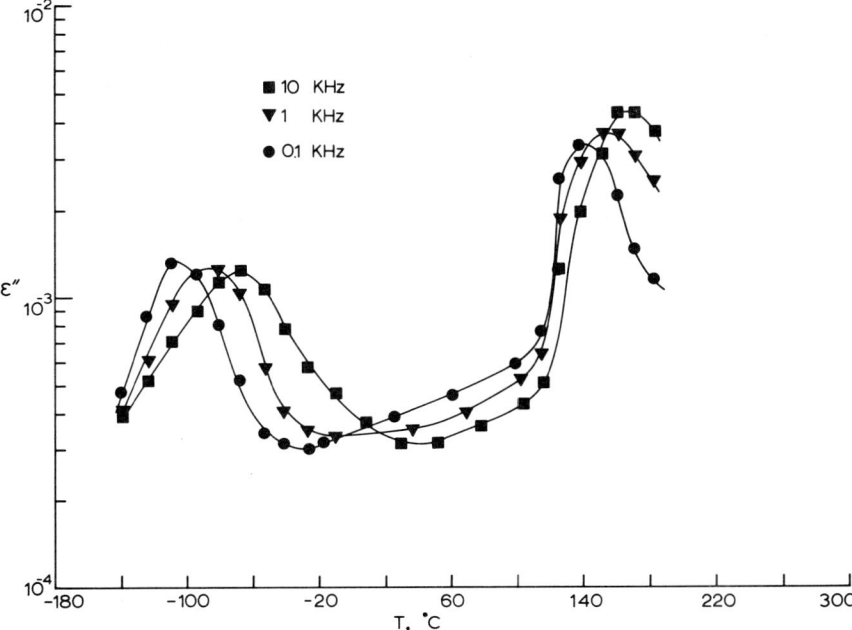

Figure 5. Dielectric loss of 25 PPO–75 PPO mixture as functions of temperature and frequency

Figures 3–5 that the dielectric relaxation again reveals only a single α relaxation for the mixtures. These are, however, noticeably broader than the α relaxation of the pure polymers. The temperatures of the loss maxima, when plotted (Figure 7) as a function of w_1, the weight fraction of PPO in the mixtures, do not display the smooth monotonic increase in T_g vs. w_1 that was shown by both the Vibron and the DSC results. Instead, there is a pronounced increase in T_g above $w_1 = 0.5$ to give a sigmoid curve for this relation. Some reservations should be attached to this observation inasmuch as data for only three polyblend compositions are available; nevertheless a qualitatively similar phenomenon is observed in the analysis of the intensity of the γ peak (below). Further, if only the stronger maxima in the dynamical mechanical data are considered—*i.e.*, if the secondary peaks and shoulders which led to the identification of two phases are omitted—then a similar sigmoid curve is found. The significance of this observation is discussed later.

The dielectric and mechanical relaxation data together span approximately 3.5 decades of frequency. Arrhenius plots (Figure 8) are linear and show that the two sets of data are mutually consistent; apparent α relaxation energies for the mixtures as well as the pure polymers lie in

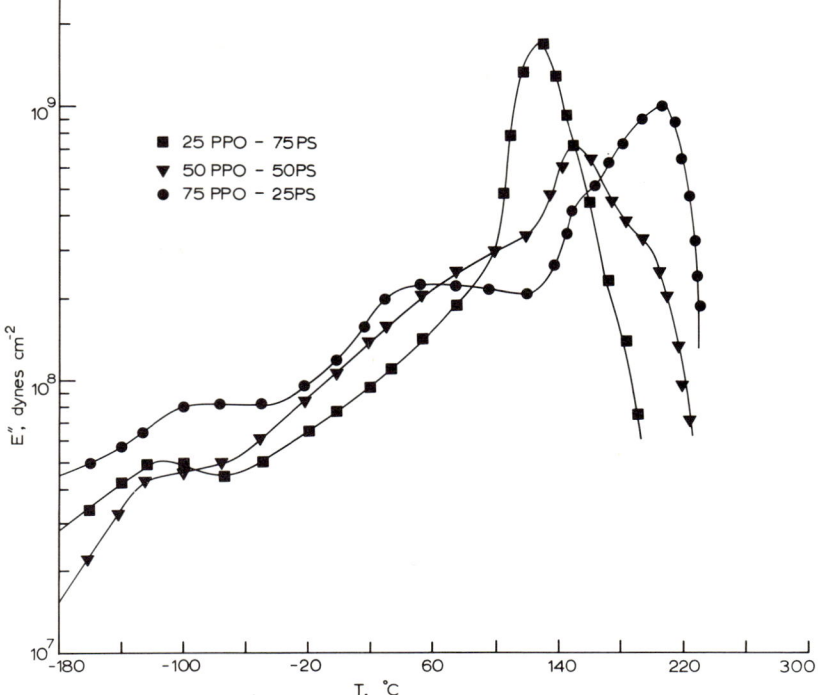

Figure 6. Dynamical mechanical loss at 110 Hz for three PPO–PS mixtures as a function of temperature

the range 80–150 kcal/mole. These values are typical for α loss mechanisms in this frequency range (8) we would expect; however, significant departures from linearity will be observed when data for a wider frequency range become available.

The γ Relaxation. In common with many other polymers (8) both PPO and PS display significant loss maxima below room temperature at the frequencies under consideration. Whereas the process responsible for the α loss is at least qualitatively understood in terms of a main chain relaxation associated with the glass transition, γ losses can often only tentatively be attributed to specific mechanisms. In PPO, for example, it does not seem unreasonable to propose that the γ loss is associated with librations in the two pendant methyl groups; this view is somewhat reinforced by the observation that in the dielectric measurements the relaxational strengths of the γ and α loss processes are comparable. As the latter can be well interpreted (6) in terms of a dipolar relaxation of the main chain in which the entire dipolar contributions arise from the methyl groupings, it seems plausible to assume that the same dipoles are responsible for the γ loss mechanism. In polystyrene there is a similar

3. MAC KNIGHT ET AL. *Physical Properties* 37

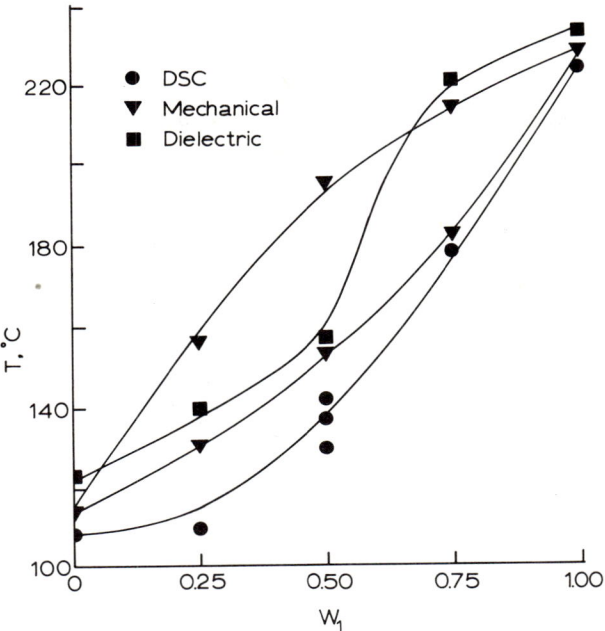

Figure 7. Glass transition of PPO–PS mixtures measured by different techniques.

The upper and lower curves for the dynamical mechanical data (110 Hz) correspond to the maxima in the resolvable loss curves. Dielectric data at 100 Hz.

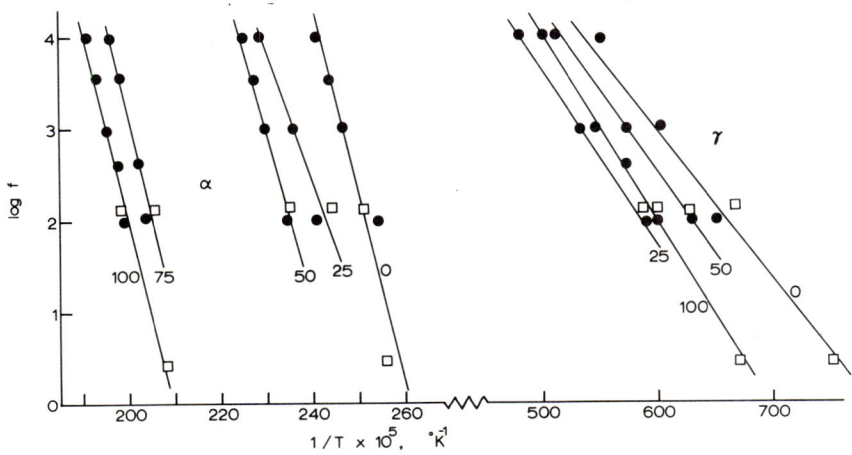

Figure 8. Loss maxima as a function of reciprocal temperature for dynamical mechanical [open squares] and dielectric loss [filled circles] measurements of α and γ peaks. The numbers correspond to the composition, wt % of PPO, in the PPO–PS mixtures.

lack of a definite assignment of the process responsible for the γ loss, though several mechanisms have been put forward (7).

The present dielectric results show that for corresponding frequencies the temperatures of the γ loss maxima for pure PPO and PS are extremely close. At 100 Hz, for example, these occur at $-116°$ and $-119°C$, respectively. Further, the temperature but not intensity of the PPO γ peak is somewhat sensitive to sample preparation and could be shifted upwards by $5°$–$10°$ by increasing the annealing temperature from $180°$ to $210°C$. Even though annealing was conducted *in vacuo*, this indicates the possibility of the γ peak's arising at least in part from polar species introduced as a result of oxidation. As has already been observed, the dynamic mechanical γ loss peaks are uniformly weak, but as far as can be observed, the peak temperatures again are consistent with the dielectric data.

Because the γ loss maxima of the constituent polymers are so close in temperature, we would not expect to find any change in this parameter in the mixtures, and indeed none was observed. The maxima in pure PPO are somewhat stronger than in PS; at 1 kHz, the ϵ''_{max} are 2.5×10^{-3} and 9×10^{-4}, respectively. Consequently, it is not unexpected also to find a uniform change in the ϵ''_{max} values for the mixtures roughly proportional to the respective compositions. More information, however, can be obtained by considering the relaxational strengths themselves—*i.e.*, of the areas under the loss peaks. The values of $\epsilon_R - \epsilon_U$ were obtained from the following equation (*10*):

$$\epsilon_R - \epsilon_U = \frac{2\Delta H}{\pi R} \int \epsilon'' d(1/T) \qquad (1)$$

In Equation 1, ϵ_R and ϵ_U refer to the relaxed (low frequency) and unrelaxed (high frequency) dielectric constants, and ΔH is the measured activation energy for the γ process. The latter was nearly independent of blend composition; an average value of 8.7 kcal/mole was used. The integral in Equation 1 was found to be approximately independent of frequency in the range studied. The loss peak in absolute terms is rather weak, and values of $\epsilon_R - \epsilon_U$ were of the order of 10^{-2} and less. From these values, it was also possible to calculate the apparent dipolar density, $N\mu^2$, using the Onsager relation (*9*):

$$N\mu^2 = \frac{3kT}{4\pi} \left(\frac{2\epsilon_R + \epsilon_U}{3\epsilon_R} \right) \left(\frac{3}{2 + \epsilon_U} \right)^2 (\epsilon_R - \epsilon_U) \qquad (2)$$

N and μ are the number of dipolar groups per unit volume and the dipole moment of each, respectively. Equation 2 can be used only when orientational correlation between the dipoles is not present or has been corrected for. Although, in the present case, neither the nature of the dipoles in-

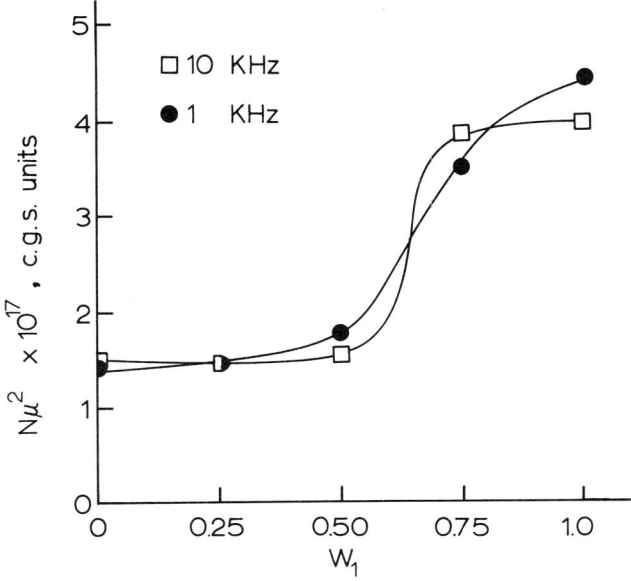

Figure 9. Apparent dipolar density calculated from the γ loss peak as a function of composition

volved in the relaxation nor the appropriate correlation factor is known, it is still valid to calculate an "effective" $N\mu^2$. Figure 9 shows a plot of this (for the 10-kHz measurements) as a function of polyblend composition. Within the limits of the available data, a sigmoid relation is again observed. Since it is generally believed that the motions responsible for low temperature losses are comparatively localized and hence relatively unaffected by longer range environmental fluctuations, this result is quite unexpected. In other words, if mixing in our samples of PPO–PS occurs at a level consistent with the observations of two phases in the dynamical mechanical measurements and with the appearance of a sigmoid ϵ'' vs. composition curve for the α process, it would nevertheless be expected that for the supposedly localized γ relaxation an approximately linear relation between $N\mu^2$ and composition would be found. Such a linear relation for the γ relaxation has actually been observed in dynamical mechanical studies of compatible blends of poly(vinyl chloride) and ethylene vinyl acetate copolymers (4). Our finding suggests that longer range correlations may be involved and that the length of polymer chains involved in the α and γ relaxational losses may even be roughly comparable.

Such an analysis, of course, still does not explain why such a relation should be obtained at all. One hypothesis, based on a consideration of the sample preparation, is the following. Since the T_g's of the parent polymers are more than 100°C apart, under typical sample preparation temperatures used (280° to 330°C) PS is a much more mobile fluid than the PPO. Hence, it may be possible that in most of the mixtures studied, the PS forms first what is essentially a continuous phase while the PPO is in a dispersed phase. Mixing then occurs by interdiffusion to some particular level, but upon cooling there remains a residual PPO-rich phase dispersed in a PS-rich matrix. For the 75% PPO–25% PS mixture, however, the PPO is present to such an excess that it in turn forms the basis for the "continuous phase." It must then be further supposed that the correlations responsible for the various loss mechanisms can be obtained only in the continuous phase—the polymer clusters constituting the dispersed phase being too small to contribute effectively to the losses, at least for those kinds of measurements in which the sigmoid curve was obtained. This description envisages that the compositional fluctuations in the mixtures correspond to a much finer "particle size" than normally implied when the continuous-dispersed phase descriptive terminology is used. If this were not so, no mixing by any criterion would be observed, and each mixture, regardless of composition, would show the two T_g's characteristic of the constituent polymers.

Conclusions

The study of what may be assumed to be a prototype of an apparently compatible polymer pair, PPO and PS, has revealed considerable complexity. It seems fairly certain that mixing does not occur at the segmental level, at least for the particular samples studied, and therefore that PPO- and PS-rich phases are present. Nevertheless, the mixing is "fine" enough so that the T_g's characteristic of the constituents are not seen. By utilizing the observation that different types of measurements appear to vary in their sensitivity for detecting compositional heterogeneity, it may be possible to clarify basic questions pertinent not only to the study of mixtures but to single component systems also. These would include, for example, the problem of what is the minimum size of a polymer aggregation necessary to display a glass transition.

Acknowledgments

We are grateful to the General Electric Co. for gifts of samples. This work was supported by AFOSR Grant 68-1434 (FEK) and NSF Grant GP 8581 (WJMacK). Acknowledgment is made to the donors of the Petroleum Research Fund, administered by the American Chemical Society for partial support (4362AC56–FEK; 2914-A5 WJMacK) of this research.

Literature Cited

(1) Bartenov, G. M., Kongarov, C. S., *Vyskomolekul. Soedin.* **2,** 1692 (1969).
(2) Bohn, L., *Kolloid Z.* **213,** 55 (1966).
(3) Cizek, E. P., U. S. Patent **3,383,435** (May 14, 1968).
(4) Hammer, C. F., *Bull. Am. Phys. Soc.* **15,** 351 (1970).
(5) Ishida, Y., *J. Polymer Sci.* **A2, 11,** 1835 (1969).
(6) Karasz, F. E., MacKnight, W. J., Stoelting, J., *J. Appl. Phys.* (Oct. 1970).
(7) McCammon, R. D., Saba, R. G., Work, R. N., *J. Polymer Sci.* **A2, 11,** 1721 (1969).
(8) McCrum, N. G., Read, B. E., Williams, G., "Anelastic and Dielectric Effects in Polymeric Solids," Wiley, New York, 1967.
(9) Onsager, L., *J. Am. Chem. Soc.* **58,** 1486 (1936).
(10) Read, B. E., Williams, G., *Trans. Faraday Soc.* **57,** 1979 (1961).
(11) Stoelting, J., Karasz, F. E., MacKnight, W. J., *Polymer Eng. Sci.* **10,** 133 (1970).
(12) Takayanagi, M., *Mem. Faculty Eng., Kyushu Univ.* **23,** 11 (1963).
(13) Voyutskii, S. S., Kamenskii, A. N., Fodiman, N. M., *Kolloid Z.* **215,** 36 (1966).

RECEIVED June 1, 1970.

4

Critical Phenomena in Multicomponent Polymer Solutions

W. BORCHARD and G. REHAGE

Physikalisch-Chemisches Institut der Technischen Universität
Clausthal, Germany

The maximum (or minimum) precipitation temperature of a partially miscible polymer solution cannot be identified with the critical point. These macromolecular solutions must be treated as multicomponent systems. Thus, the thermodynamic properties can be described in the critical region. The critical point can be determined by measuring phase–volume ratios as a function of concentration at temperatures near the cloud point. Light scattering measurements on the system polystyrene–cyclohexane show that the maximum dissymmetry of the scattering envelope arising from the critical opalescence lies at polymer concentrations lower than the critical point. The scattered intensities exhibit a typical function by which the cloud points can be determined. Concentration fluctuations persisting over large distances near the critical point lead to anomalous behavior of the transport coefficients.

In the critical region of mixtures of two or more components some physical properties such as light scattering, ultrasonic absorption, heat capacity, and viscosity show anomalous behavior. At the critical concentration of a binary system the sound absorption (13, 26), dissymmetry ratio of scattered light (2, 4–7, 11, 12, 23), temperature coefficient of the viscosity (8, 14, 15, 18), and the heat capacity (15) show a maximum at the critical temperature, whereas the diffusion coefficient (27, 28) tends to a minimum. Starting from the fluctuation theory and the basic considerations of Ornstein and Zernike (25), Debye (3) made the assumption that near the critical point, the work which is necessary to establish a composition fluctuation depends not only on the average square of the amplitude but also on the average square of the local gradient of the fluctuation. Fixman considers the existence of an intense spectrum of long wavelength composition variations as the basis of a theory of critical viscosities that relies on macroscopic transport equations to describe the interaction of a velocity gradient with composition fluctuations. As far as the above-mentioned theories were applied to binary low molecular compounds, comparisons with experiments were completely conclusive. However, applying the theoretical predictions to solutions of a solvent and a polymer the maximum in the dissymmetry ratio was at a different polymer concentration than the maximum of the relative temperature coefficient of the viscosity (8). This paper reports measurements of light scattering and viscosity of the system polystyrene–cyclohexane, which has a miscibility gap with an upper critical point.

Critical Opalescence of Polystyrene in Cyclohexane

According to the fluctuation theory of Einstein (10), the additional amount of light scattered by a solution compared with that scattered by a pure solvent is given by the mean square of the concentration fluctuations in small volume elements. Derivations for the scattered light of a polydisperse polymer in a solution were given by Brinkman and Hermans (1) and soon afterwards by Kirkwood and Goldberg (19) and by Stockmayer (33). Zimm and Doty (35) found that the weight average molecular weight can be obtained from light scattering measurements.

Near a critical point strong scattering, called the critical opalescence, is observed. This scattering phenomenon can be described only if the influence of the concentration gradients is taken into consideration. The intensity of light scattered at an angle θ between the secondary and primary beam at large distances R from the scattering volume V has been formulated by Debye and others (6, 9, 24):

$$\frac{I_\theta}{I_0} = \frac{4\pi^2 V \alpha}{R^2 \lambda^4} \cdot \frac{K\, P(\theta)}{\phi_2 \dfrac{\partial}{\partial \phi_2}\left(\dfrac{\Pi}{kT}\right) + \dfrac{16\pi^2 \phi_2^2 H}{kT\lambda^2} \cdot \sin^2 \dfrac{\theta}{2}} \qquad (1)$$

I_θ is the intensity of the scattered light at angle θ, I_o is the intensity of the primary light beam, λ is the wavelength of light in the medium, ϕ_2 is the volume fraction of the polymer, Π is the osmotic pressure. For unpolarized light α is given by $\dfrac{1 + \cos^2\theta}{2}$.

The optical constant K is a function of the refractive index and the refractive index increment, and $P(\theta)$ is the particle scattering function which depends on internal interference. This function is influenced by the particle shape and is less than 1 for molecules large compared with the wavelength of light. The quantity H depends on the interaction energies between the different molecules and the range of molecular forces. Comparing the derivation given by Debye with the well known Flory-Huggins (16, 17) treatment one can conclude that H depends on concentration if the Flory-Huggins interaction parameter is a function of the polymer concentration.

If the incident beam is polarized at right angles to the plane of observation, according to the formula of Debye, the reciprocal scattered intensity is proportional to $\lambda^2 \sin^2 (\theta/2)$ at the spinodal composition. When scattering intensities are extrapolated to zero angle at low concentrations of not too large molecules, Einstein's formula is obtained. An equation for the light scattered by a polydisperse system in the critical region has not yet been derived. From the above mentioned considerations it follows that in the case of Debye scattering, however, the scattering intensities extrapolated to zero angle depend in the same way on the concentration fluctuations as in Rayleigh scattering, so that the theories for polydisperse polymer solutions can be used (32).

Experimental

The light scattering measurements were performed with a Sofica photogoniometer. The temperature was controlled by a special bath, which was mounted around the index vat, in which benzene was stirred to achieve a good temperature uniformity in the measuring cell. By means of a heated head-piece, which was kept at bath temperature, the temperature remained constant to 0.01°C.

The solutions from a sample of anionic polystyrene (courtesy of BASF, Ludwigshafen) and the solvent were prepared by weighing. Polystyrene is characterized by the data in Table I. [The average molecular weights were determined in the Central Laboratory of the N.V. Staatsmijnen/DSM (Geleen, The Netherlands).]

Solutions of polystyrene and purified cyclohexane after filtration still exhibited appreciable dissymmetry of scattered intensities at high temperatures where the influence of the critical opalescence was precluded. Therefore all solutions were freed from dust by centrifugation. After a

Table I. Properties of Polystyrene Sample Investigated

$M_n = (4.4 \pm 0.2) \times 10^5$ grams mole^{-1}

$M_w = (4.7 \pm 0.2) \times 10^5$ grams mole^{-1}

$M_z = \phantom{(4.7 \pm 0.2) \times {}} 6.5 \times 10^5$ grams mole^{-1}

$\dfrac{M_w}{M_n} = 1.07 \pm 0.07 \; ; \dfrac{M_z}{M_w} = 1.4 \; ; \dfrac{M_{z+1}}{M_z} = 1.6$

$T_c = 27.63°C \; \lceil w_{2c} = 8.5$ wt %

$T_{thr} = 28.43°C \; \lceil w_{2thr} = 2.0$ wt %

run of 1 hour in a centrifugal field of 17000 g no dissymmetry could be observed. When the light scattering measurements were finished, the concentration of each solution was checked by weighing a small portion before and after evaporation of the solvent or by measuring the refractive index at several temperatures.

The phase separation temperatures T_p were measured in two ways. Solutions of various concentrations in sealed glass tubes were placed in a water bath. In the previously roughly determined temperature range the temperature was lowered in steps about 0.1°C. Phases which separated out became clear within several hours. The highest phase-separation temperature was considered to be the cloud point. The second method is described below.

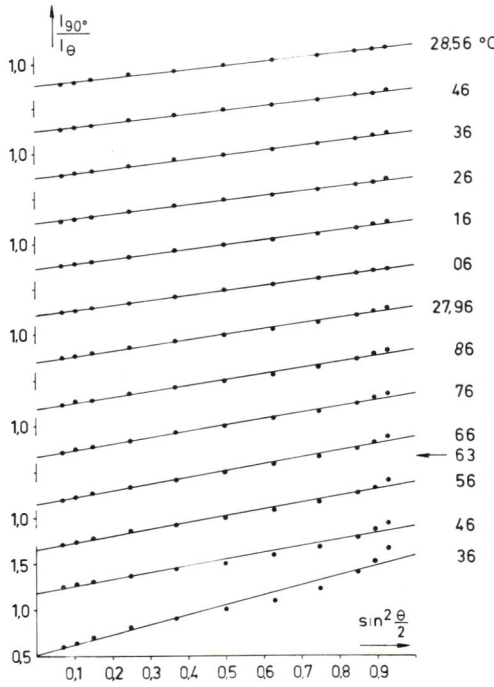

Figure 1. I_{90}/I_θ vs. $sin^2(\theta/2)$ for a solution of 8.5 wt % polystyrene in cyclohexane at different temperatures. Cloud-point temperature is indicated by an arrow.

The critical concentration was obtained from measurements of the phase-volume ratio of coexisting phases near the critical temperature. At an over-all concentration of 8.5 wt % this ratio was unity. According to the lever rule this concentration is the critical one.

Results and Discussion

Figures 1–3 show the observed reciprocal excess intensities of scattered light multiplied by sin θ (to correct for the irradiated volume observed at each angle) plotted against $\sin^2(\theta/2)$ at constant polymer concentration for several temperatures above the phase-separation temperature. To give a clearer presentation the intensities are expressed in relation to intensities at the scattering angle $\theta = 90°$, as was also done by Eskin and Nesterow (11). In accord with the Debye theory, the plots give straight lines and can be represented by

$$I_\theta^{-1} = A + B \sin^2 \frac{\theta}{2} \qquad (2)$$

The temperature at which phase separation occurs is marked by an arrow. All the plots show pronounced deviations from a straight line when phase separation has begun. All straight lines were calculated by the method of least squares. Figure 4 shows a plot of A/B vs. temperature. The plots are presented only for a small temperature range above

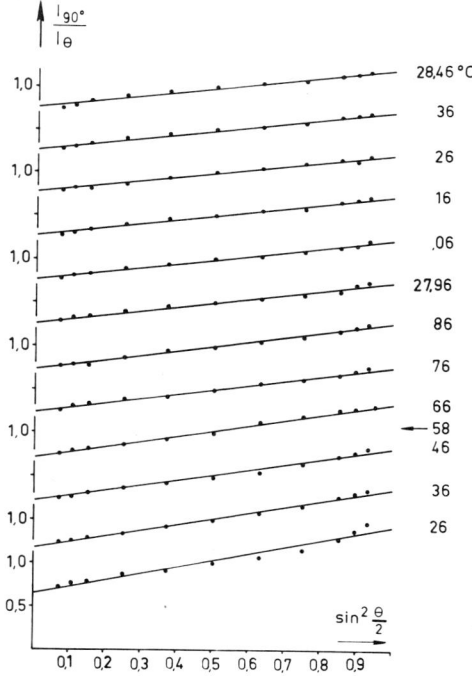

Figure 2. Same as Figure 1 for 10 wt % polystyrene

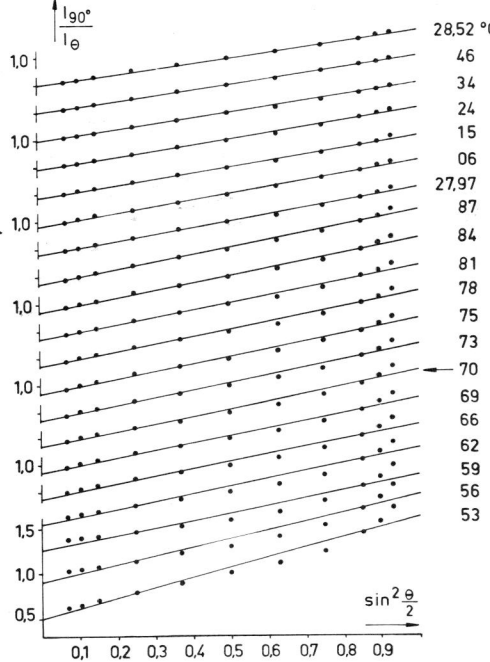

Figure 3. Same as Figure 1 for 7 wt % polystyrene

the cloud-point curve. In every curve there is a typical break at the cloud-point temperature. Arrows mark the phase-separation temperatures determined visually as described above. Since the scattered light immediately registers the beginning of the formation of the new phase, the determination of the break is a rapid method for detecting phase separation even in highly viscous solutions. We mention that this is the only method applicable for determining the cloud points at very high and very small concentrations of polystyrene in cyclohexane.

According to the Debye equation extrapolating A/B to zero for the critical concentration (8.5 wt %) should give the critical temperature T_c. Figure 4 shows that for this polymer the function should at least be of the form $A/B = f_1$ (conc.) $+ f_2 (T - T_c)$. Even at greater distances from the critical temperature there did not exist a straight line portion for the function A/B vs. T which could give the critical temperature. Therefore, no attempts were made to calculate molecular parameters from the data.

Figure 5 shows the dissymmetries z plotted against the concentration of polystyrene for several temperatures. At a temperature 8.6°C above

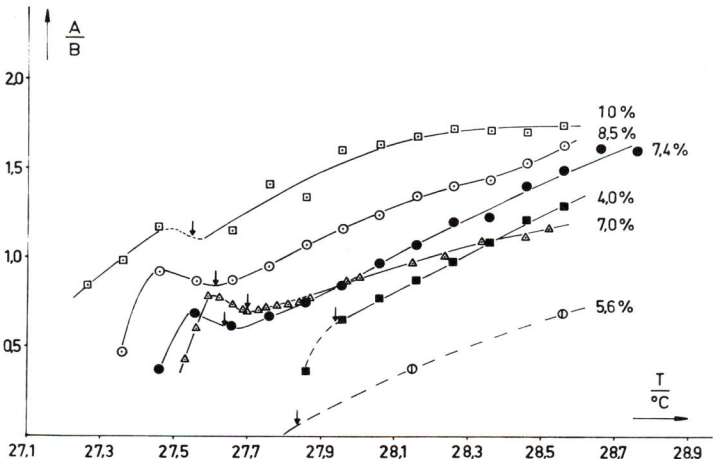

Figure 4. Plot of A/B *vs. temperature. Cloud points are indicated by arrows.*

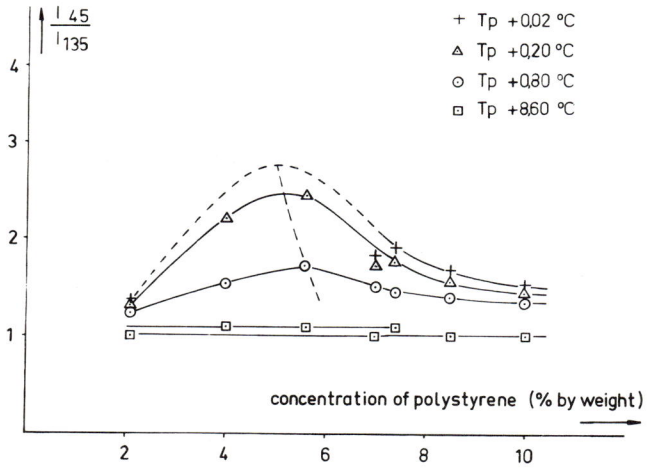

Figure 5. Dissymmetry ratio I_{45}/I_{135} *vs. concentration of polystyrene in cyclohexane at different temperatures above the phase-separation temperature*

the phase-separation temperature no dissymmetry from the solutions in the polished measuring cells could be observed ($z \approx 1.01$). For three unpolished cells a slight dissymmetry was measured. At all concentrations z increases if the temperature is lowered. The maximum of the curves lies at a lower concentration than the critical and shifts to lower concentrations for smaller values of $T - T_p$.

The cloud-point curve is shown in Figure 6. It is typical for the demixing of polydisperse polymer in a low molecular solvent. The values in Table I show that the critical point lies at a higher concentration of the polymer (w_{2c}) and lower temperature (T_c) than the maximum of the cloud-point curve (w_{2thr}, T_{thr}), called by Tompa the precipitation threshold. These phenomena have been treated for the system polystyrene–cyclohexane (28–31) and for the system polyethylene–diphenyl ether (20, 21). Although the ratio of the weight average to the number average molecular weight is 1.07, the temperature difference between the temperatures of the precipitation threshold and the critical point is $T_{thr} - T_c = 0.8°C$, and the concentration difference $w_{2c} - w_{2thr} = 6.5\%$ by weight. Recently the different values of the above mentioned coordinates of the precipitation threshold and the critical point were shown to be a sensitive criterion for polydispersity (31).

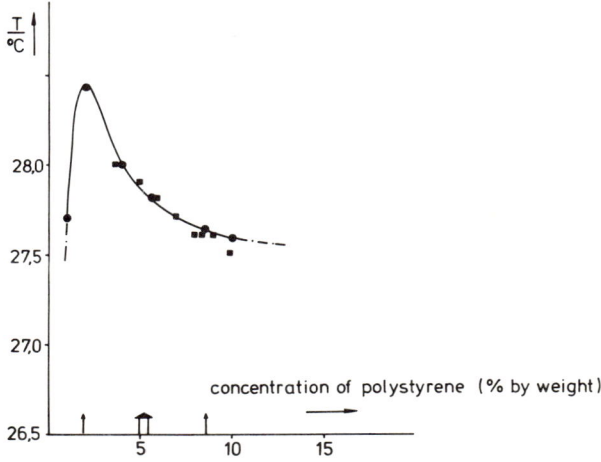

Figure 6. *Cloud-point curve for polystyrene in cyclohexane (●) determined from light scattering and (■) determined visually. The arrows indicate (from left to right): maximum of the cloud-point curve, maximum of the dissymmetry ratio, critical point.*

The maximum of the dissymmetry lies at 5–6 wt % of the polymer near the quasi-binary spinodial. All maxima are indicated by arrows in Figure 6. In our opinion polydispersity is the main reason that the maximum of critical opalescence is not found at the critical point. In a system consisting of a polydisperse polymer and a solvent the shape of the spinodial surface may be such that highly unsymmetrical fluctuations may occur in the critical region and give rise to the above mentioned anomalous behavior. These fluctuations are not expected to be described by the Debye theory. McIntyre and co-workers (23) conclude that strong deviations from straight lines in the plot of I_θ^{-1} vs. $\sin^2 (\theta/2)$ are caused by polydispersity or by higher order correlations in monodisperse polymer solution. Further work will be required to decide whether polydispersity is the only reason.

Viscosity Measurements of the Critical Mixtures

A solution at a high concentration of the polystyrene sample was diluted in steps with cyclohexane after viscosity determinations at sev-

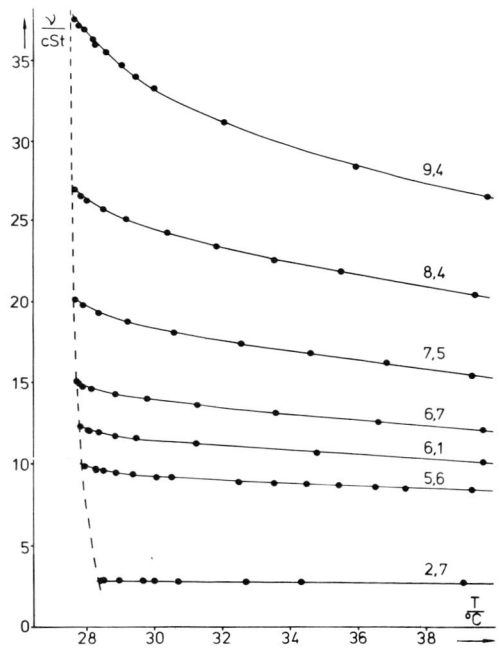

Figure 7. Kinematic viscosity vs. temperature of polystyrene in cyclohexane. Numbers indicate wt % polystyrene.

eral temperatures. In an Ostwald viscosimeter, calibrated with pure solvents, the time of the flow could be measured within ±0.01 sec automatically. Temperatures were controlled to ±0.005°C. We assumed that in the concentration range up to 9 wt % of polymer the solutions were Newtonian. Very close to the phase separation temperature the viscosity cannot be expected to be independent of shear gradient. Plots of kinematic viscosity *vs.* temperature at several concentrations are shown in Figure 7.

All curves end at the dashed curve which corresponds to the cloud-point curve. Phase separation in the diluted solutions could be detected by a rapid decrease in the viscosity. Under certain conditions the cloud points could be obtained by extrapolating the viscosities in the homogeneous and heterogeneous phase regions. At low concentrations the temperature coefficient of the viscosity is low and increases with polymer concentration. The increase near the cloud point is strongest for a polymer concentration of 6.7 wt %. The relative temperature coefficient of the kinematic viscosity, which is only slightly different from the temperature coefficient of the dynamic viscosity, is plotted *vs.* the composition of the solution in Figure 8. Up to 2°C above the phase-separation temperature no anomalous phenomena could be observed. At a temperature $T = T_p + 0.2°C$ a maximum is found which shifts for $T = T_p + 0.02°C$ to lower polymer concentration. The maximum lies between 6.2 and 7%. Like the maximum of the dissymmetry, this maximum is at a smaller concentration than the critical one. We think that both maxima are caused by the above mentioned fluctuations near the maximum of the quasi-binary spinodal. With the formula given by Koningsveld (*22*) for the spinodal of a multicomponent system and the concentration and temperature function for the Flory-Huggins parameter

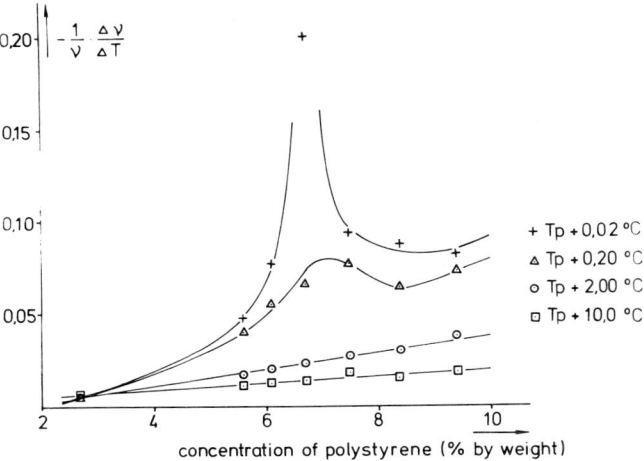

Figure 8. *Relative temperature coefficient of kinematic viscosity,* $\frac{1}{v}\frac{\Delta v}{\Delta T}$, *vs. concentration of polystyrene in cyclohexane at different temperatures above the phase-separation temperature*

in that paper we calculated a maximum spinodial composition of 6.4 wt % of the polymer. This seems to be in a qualitative agreement with the observed phenomena in the critical region.

Acknowledgment

The authors are indebted to B. Meyer for the light scattering measurements and to the Deutsche Forschungsgemeinschaft for supporting this research. Thanks are due to R. Koningsveld and Th. G. Scholte for discussions and to Th. G. Scholte and A. J. Pennings for determining the average molecular weights in Table I.

Literature Cited

(1) Brinkman, H. C., Hermans, J. J., *J. Chem. Phys.* 17, 574 (1949).
(2) Chu, B., *J. Phys. Chem.* 69, 2329 (1965).
(3) Debye, P., *J. Chem. Phys.* 31, 680 (1959).
(4) Debye, P., Coll, P., Woermann, D., *J. Chem. Phys.* 33, 1746 (1960).
(5) Debye, P., Woermann, D., Chu, B., *J. Chem. Phys.* 36, 851 (1962).
(6) Debye, P., Chu, B., Woermann, D., *J. Chem. Phys.* 36, 1803 (1962).
(7) Debye, P., Woermann, D., Chu, B., *J. Polymer Sci., Pt. A,* 1, 255 (1963).
(8) Debye, P., Chu, B., Woermann, D., *J. Polymer Sci., Pt. A,* 1, 249 (1963).
(9) De Gennes, P. G., *Phys. Letters* 26 A, 313 (1968).
(10) Einstein, A., *Ann. Physik.* 33, 1275 (1910).
(11) Eskin, V. Ye., Nesterow, A. E., *J. Polymer Sci., Pt. C,* 16, 1619 (1967).
(12) Eskin, V. Ye., Serduyk, J. N., *Vysokomol. Soyed.* A11, 372 (1969).
(13) Fixman, M., *J. Chem. Phys.* 36, 1961 (1962).
(14) *Ibid.,* p. 310.
(15) *Ibid.,* p. 1957.
(16) Flory, P. J., *J. Chem. Phys.* 18, 58 (1950).
(17) Huggins, M. L., *Ann. N.Y. Acad. Sci.* 43, 1 (1942).
(18) Kawasaki, K., Tanaka, M., *Proc. Phys. Soc.* 90, 791 (1967).
(19) Kirkwood, J. G., Goldberg, R. J., *J. Chem. Phys.* 18, 54 (1950).
(20) Koningsveld, R., Staverman, A. J., *J. Polymer Sci., Pt. C,* 16, 1775 (1967); *Pt. A, 2,* 6, 349 (1968).
(21) Koningsveld, R., Staverman, A. J., *Kolloid-Z. Z. Polymere* 218, 114 (1967).
(22) Koningsveld, R., Shultz, A. R., *J. Polymer Sci.,* in press.
(23) McIntyre, D., Wims, A., Green, M. S., *J. Chem. Phys.* 37, 3019 (1962).
(24) Muenster, A., Schneeweiss, Ch., *Z. Phys. Chem. N.F.* 37, 353 and 369 (1963).
(25) Ornstein, L. S., Zernike, F., *Physik. Z.* 19, 134 (1918); 27, 761 (1926).
(26) Pings, C. J., Anantaraman, A. V., *Phys. Rev. Letters* 14, 781 (1965).
(27) Rehage, G., Ernst, O., *Dechema-Monographie* 49, 157 (1964).
(28) Rehage, G., Moeller, D., Ernst, O., *Makromol. Chem.* 88, 232 (1965).
(29) Rehage, G., Moeller, D., *J. Polymer Sci.* 16, 1787 (1967).
(30) Rehage, G., Koningsveld, R., *Polymer Letters* 6, 421 (1968).
(31) Rehage, G., Wefers, W., *J. Polymer Sci., Pt. A,* 2, 1683 (1968).
(32) Scholte, Th. G., *European Polymer J.,* in press.
(33) Stockmayer, W. H., *J. Chem. Phys.* 18, 58 (1950).
(34) Tompa, H., *Trans. Faraday Soc.* 46, 970 (1950).
(35) Zimm, B. H., Doty, P. M., *J. Chem. Phys.* 12, 203 (1944).

RECEIVED May 4, 1970.

5

Viscosity of a Solution of Two Mutually Incompatible Polymers

H. L. DOPPERT and W. S. OVERDIEP

AKZO Research and Engineering N.V., Arnhem, Netherlands

> *The viscosity of emulsions obtained from two mutually incompatible polymers dissolved in a common solvent was studied by a falling ball viscometer, a cone-plate viscometer, and a capillary viscometer. The two polymers are polyacrylonitrile and polyurethane, and the solvent is N-methylpyrrolidone. The measurements are compared with theory, and a model is proposed for the development of a stationary pressure flow of an emulsion in a capillary.*

In the last decades many attempts have been made to obtain attractive materials by intimate mixing of two polymers with opposite or complementary properties. For example, the impact resistance of brittle polystyrene is increased by mixing with a rubber; the wettability of polyacrylonitrile fiber is increased by mixing with hydrophilic saponified cellulose acetate, and the inconvenient flat-spotting of nylon-reinforced tires is suppressed by mixing stiffer polyester fibrils into the nylon fibers. In practically all cases these products acquire their final shape *via* the liquid state. Thus, the viscous properties of these liquid mixtures are important.

At present, with a few exceptions concentrated solutions of two polymers contain two liquid phases as a result of the positive free energy of mixing (4, 6). These two phases can be mixed intimately by stirring to form a so-called polymeric oil-in-oil emulsion, which is stable compared with the classical oil-in-water emulsions. A characteristic difference between an oil-in-oil emulsion and an oil-in-water (detergent-free) emulsion is the relatively low interfacial tension (*ca.* 0.1 dyne/cm) of the former compared with that of the latter (*ca.* 40 dynes/cm). Owing to the low interfacial tension in the emulsions considered here, in many cases gentle stirring is enough to obtain a stable emulsion. After prolonged stirring, an equilibrium between breaking and coalescence of the dispersed drops is reached.

Since both components are generally non-Newtonian liquids and the droplet shape of the dispersed phase depends on shear stress, the rheological properties of these emulsions must be rather complicated. Consequently, the shear–stress dependence of the viscosity of the emulsions is caused not only by the characteristic properties of the macromolecular substances but also by the behavior of the dispersed phase in a shear field.

In two regions the influence of droplet shape is not important:

(1) The region of very low shear stresses, where the shape of the droplets remains spherical.

(2) The region of high shear stresses, where the droplets are deformed into very long streaks.

In this paper we present experimental data demonstrating the existence of these two regions and their viscous properties. A model is derived for pressure flow through a cylindrical tube, and the calculated results are compared with experimental results.

Experimental

The emulsions were prepared from two different polymers dissolved in a common solvent. Because of its good solvent properties and high boiling point (204°C: no evaporation during experiments) N-methylpyrrolidone (from BASF) was chosen as solvent. The two polymers were polyurethane (PU) from Goodrich Chemical Co. (Estane 5707 F1) and polyacrylonitrile (PAN) supplied by Cyanenka. The viscosity average molecular weights were 48,400 and 60,300, respectively. Two solutions were prepared, each containing 15 wt % of either polymer.

On top of the PU solution the same quantity of PAN solution was poured with great care to prevent emulsification at the liquid interface. Thermodynamic equilibrium between the two phases was unlikely. After three days the volume of the PU phase had decreased by about 5%. After the solution had stood for eight days, the two layers were separated, and the different emulsions were prepared by vigorous stirring of different proportions of the two liquids. The emulsions thus obtained were stable for at least one day.

To obtain information on droplet deformation in shear flow, the following experiment was carried out. One of the polymer solutions was poured into the annular space between two concentric glass beakers of different diameters. The large beaker could be rotated; the smaller beaker within the larger one was stationary. A droplet of the other polymer solution (blackened by a small amount of carbon black to make it visible) was injected with a syringe into the first solution near the core of the annulus. When the beaker started rotating, a shear flow was created in the liquid, filling the annulus. The droplets were photographed at different shear rates in the liquid (Figure 1).

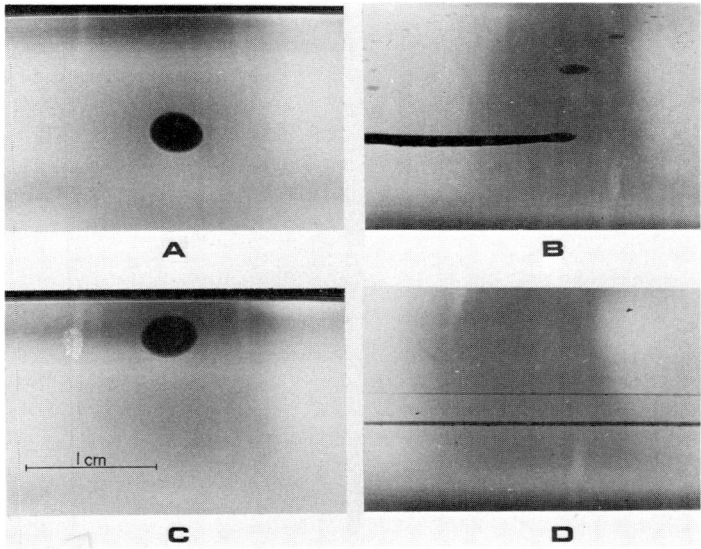

Figure 1. Difference in deformability between a PAN droplet in a PU solution (A, C) and a PU droplet in a PAN solution (B, D) at corresponding shear rates

Shear Stress, dynes/cm²		Shear Rate, sec⁻¹
A:	2.7	6.8×10^{-2}
B:	10	6.8×10^{-2}
C:	86	2.15
D:	350	2.15

Viscosity measurements on emulsions were carried out with three types of viscometers. Figure 2 shows the flow curves of emulsions with different volume ratios of the two solutions, as measured with a Ferranti-Shirley cone-plate viscometer. The ratio between the viscosities of the two pure polymer solutions is about 3 at low shear rates but only 2 at the highest shear rates.

To obtain a larger range in viscosities determined with a capillary viscometer, polymers from different batches were used to prepare the emulsions. The results obtained with the capillary viscometer are given in Figure 3. The ratio between the viscosities of the two components of the emulsions is about 10.

Figures 4 and 5 represent the viscosities of the emulsions measured with a falling-ball viscometer. When one ball is used (a steel ball 0.8 mm in diameter), only one value of the viscosity is measured for each emulsion and not a flow curve. The emulsions in Figure 4 are rich in PU solution, and those in Figure 5 are rich in PAN solution.

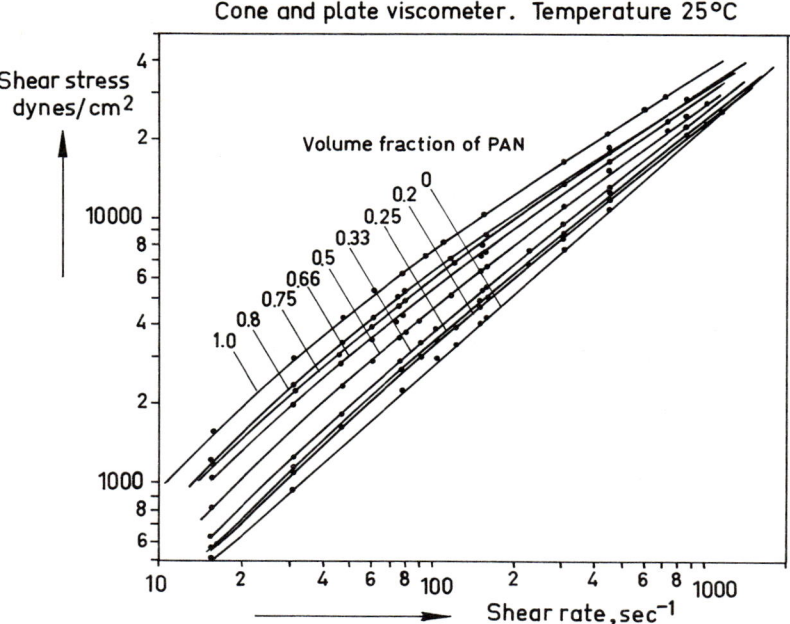

Figure 2. Flow curves of emulsions of different compositions and the pure components as measured with a cone-plate viscometer

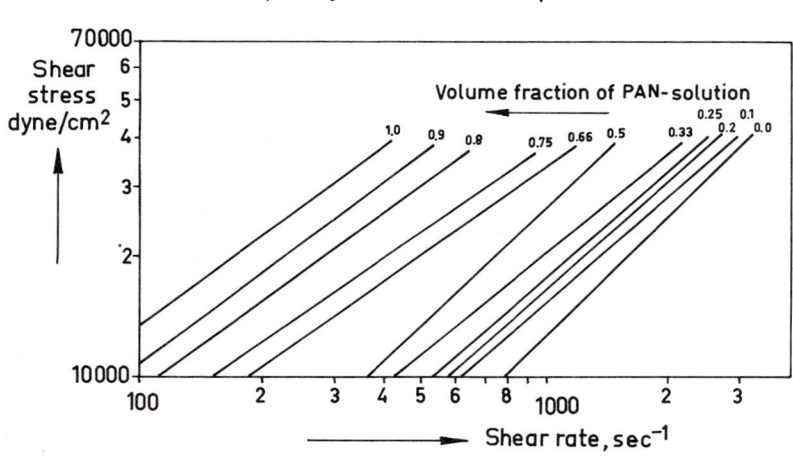

Figure 3. Flow curves of emulsions of different compositions and the pure components as measured with a capillary viscometer

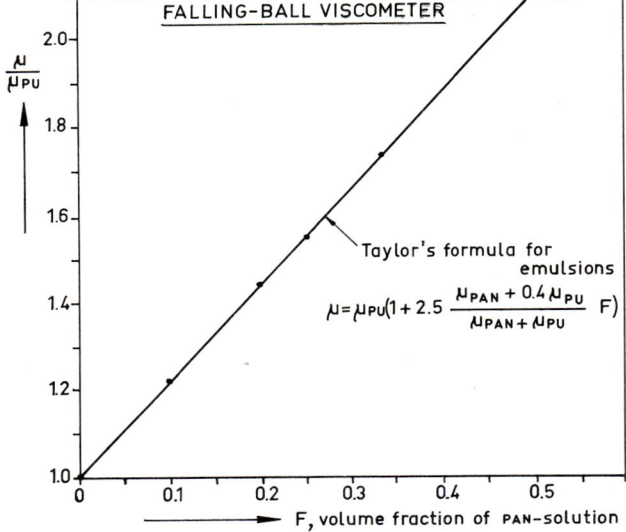

Figure 4. Measurements with a falling-ball viscometer. Ratio between viscosities of emulsions and matrix vs. volume fraction of dispersed PAN solutions.

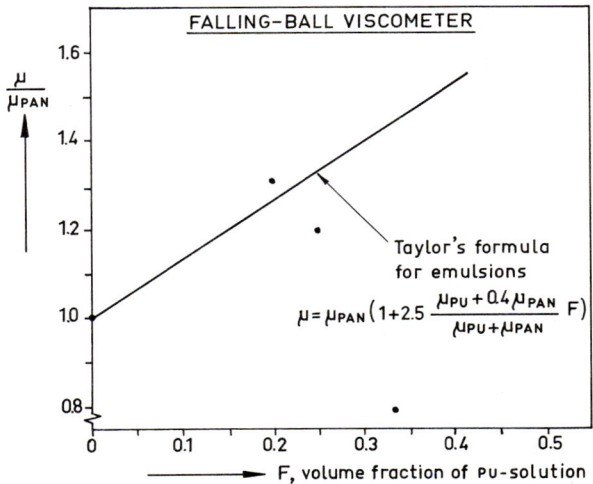

Figure 5. Measurements with a falling-ball viscometer. Ratio between viscosities of emulsions and matrix vs. volume fraction of dispersed PU solution.

Discussion

Falling-Ball Viscometer. Taylor (5) has developed a formula for the viscosity of emulsions:

$$\mu = \mu_o \left(1 + 2.5 \frac{\mu_i + 0.4\mu_o}{\mu_i + \mu_o} F\right) \quad (1)$$

where μ_o = viscosity of the continuous phase
μ_i = viscosity of the dispersed phase
F = volume fraction of the dispersed phase

For this formula to be valid, the dispersed droplets must remain spherical during the measurement. The measurements on the dispersed PAN solution (Figure 4) obey Taylor's law, whereas those on the emulsions of dispersed PU solution (Figure 5) do not. The gravity force and the viscous force acting on the sinking steel ball cancel when the sinking velocity becomes constant. The viscous force on the ball is the shear stress on the ball surface integrated over the ball surface. Hence,

$$1/6\pi d^3(\rho_b - \rho_m)g = \int_0^s \tau ds \quad (2)$$

where d = diameter of the ball
ρ_b = density of the ball
ρ_m = density of the medium
g = acceleration of gravity
τ = shear stress on the ball surface
s = surface area of the ball

Taking for the average shear stress the total force divided by the surface area, we get

$$\tau_{ave} = 1/6d(\rho_b - \rho_m)g \quad (3)$$

For a steel ball 0.8 mm in diameter this average shear stress is about 100 dynes/cm^2. Referring to Figure 1 we see that at this shear stress PAN droplets in a PU solution are spherical and that PU droplets in a PAN solution are highly deformed into oblong bodies. Thus, it may be expected that only the dispersed PAN emulsions will conform to Taylor's law (Figure 4) and not the dispersed PU emulsions (Figure 5). Our observations confirmed this.

Cone-Plate Viscometer. Measurements with the cone-plate viscometer were made between 500 and 30,000 dynes/cm^2. Evidently, the PU droplets in the emulsions are deformed into long streaks during the experiments at these high stresses. To understand the measured results, the Taylor model for the emulsions must be replaced by another model. For

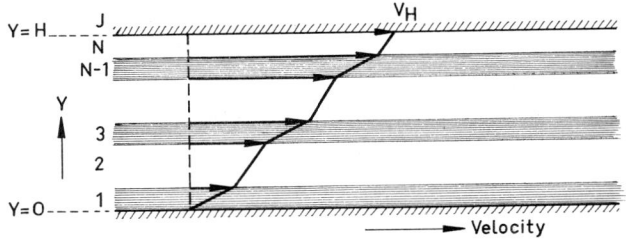

Figure 6. The alternating layers model

a bicomponent system, as considered here, Heitmiller, Naar, and Zabusky (1) proposed a model consisting of alternating layers of the two liquids (Figure 6). For couette flow this model leads to a simple expression for the viscosity of the mixture as a function of the viscosity of the components and their volume ratio. Stress equilibrium leads to

$$\frac{\partial \tau}{\partial y} = 0 \qquad (4)$$

in stationary flow. In the jth layer the flow velocity is

$$V = V_{j-1} + \int_{y_{j-1}}^{y} \frac{\partial v}{\partial y} dy =$$

$$= V_{j-1} + \int_{y_{j-1}}^{y} \frac{\tau}{\mu_j} dy =$$

$$= V_{j-1} + \frac{\tau}{\mu_j} (y - y_{j-1})$$

When $j = 0$ $V_j = 0$. Thus, the velocity at $y = H$ is

$$V_H = \sum_{j=1}^{N} \frac{\tau}{\mu_j} (y_j - y_{j-1}) = \tau \sum_{j=1}^{N} \frac{y_j - y_{j-1}}{\mu_j}$$

Since the effective shear rate in the system is $\frac{V_H}{H}$, the reciprocal effective viscosity Φ is

$$\Phi = \sum_{j=1}^{N} \frac{1}{\mu_j} \frac{y_j - y_{j-1}}{H} \qquad (5)$$

The ratio between the volume of the jth layer and the total volume is $y_j - y_{j-1}/H$.

If this layer is occupied by component A, the sum of $y_j - y_{j-1}/H$ over all the layers occupied by component A equals F_A, the volume fraction of component A. Consequently, the same summation for component B leads to $1 - F_A$. Equation 5 then becomes:

$$\Phi = \Phi_A F_A + \Phi_B(1 - F_A) \qquad (6)$$

According to this model

$$F_A = \frac{\Phi - \Phi_B}{\Phi_A - \Phi_B} \qquad (7)$$

Equation 7 gives a simple linear relation between the fluidity Φ and F. It is not valid if Φ_A and Φ_B depend on F, and this is the case if A and B are non-Newtonian. Hence, we apply Equation 7 only to values for Φ, Φ_A, and Φ_B measured at the same shear stress.

In Figure 7

$$\frac{\Phi - \Phi_{pu}}{\Phi_{PAN} - \Phi_{pu}}$$

is plotted against the volume fraction of PAN solution at different shear stresses. At these high shear stresses the measurements deviate strongly from Taylor's law but roughly follow Equation 7.

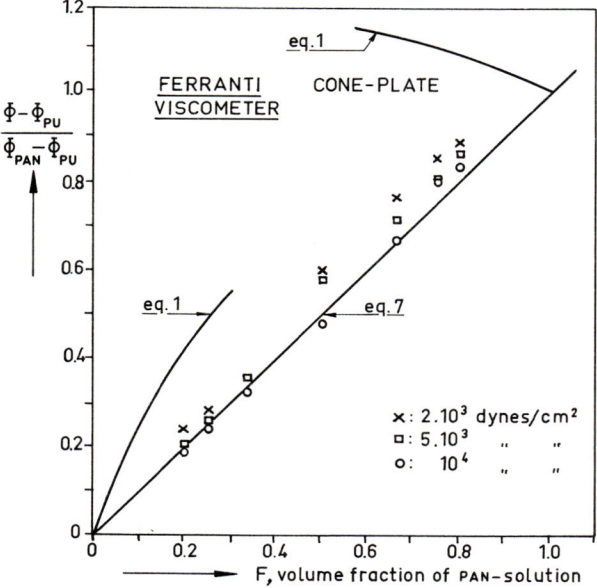

Figure 7. *Measurements with a cone-plate viscometer: reduced fluidity of the emulsions vs. their composition*

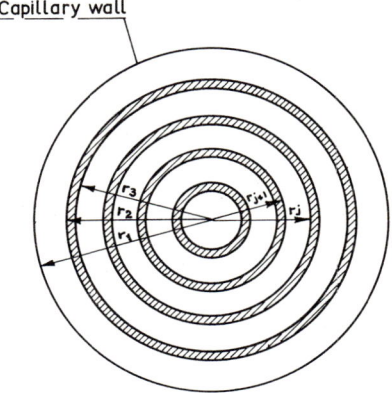

ANNULI OCCUPIED BY THE FLUID COMPONENTS
IN AN ALTERNATING MANNER

Figure 8. The alternating annuli model

Capillary Viscometer. The end products from the liquid mixtures are usually obtained by extruding the liquid mass through narrow tubes or slits (*e.g.*, spinning of fibers, injection molding, or film extrusion). Therefore, the pressure flow through a capillary is of technological interest. Hence, we analyzed the flow of a liquid mixture through a capillary with circular cross-section and compared the results of theory and measurement.

The measurements were obtained in a stress range between about 10^4 and 4×10^4 dynes/cm^2 (Figure 3)—*i.e.*, the stress at the capillary wall was of this order of magnitude. In stationary flow the stress decreases linearly with the radius, reaching zero at the capillary center. In general, then, we must use Taylor's model for emulsions near the center of the capillary and the layer model in the outer part. However, because of the extremely high stresses at the wall compared with the stresses at which Taylor's formula is valid, we neglect the fact that we should use this formula for the capillary center.

The alternate layers in the capillary are considered to be concentric tubes (*see* Figure 8). This picture of flow of a two-liquid system through a capillary was also the basis for the analysis of Heitmiller *et al.* (*1*). However, our analysis differs fundamentally from theirs in two respects:

(1) Heitmiller *et al.* assume that all the layers of a given fluid component have the same cross-sectional area. We hold that all the layers of a given fluid component have the same thickness. Moreover we assume that the tubular layers of the component corresponding to the

dispersed phase in the real mixture have a thickness of the same order of magnitude as the diameter of the dispersed droplets in the real mixture.

(2) Heitmiller *et al.* (*1*) assume that the basis for their analysis as given in point 1 is valid in the capillary where the flow is fully developed into a stationary flow. In the following analysis, however, we assume that the supposition given in point 1 holds only at the capillary entrance, where the velocity profile is flat (Figure 10). [A more detailed elucidation of this figure is given later.] The idea underlying this assumption is that the mixture entering the capillary is homogeneous but that the distribution of the two liquids in the developed flow is a result of the rheological properties of the emulsion. Figure 10 shows that the layer thicknesses in the capillary differ from those at the entrance. This redistribution of the layer thicknesses over the cross-section depends on the developed velocity profile.

The number N of concentric layers having thickness d at the entrance and corresponding to the fluid with volume fraction F is

$$N = \frac{FR}{d}$$

$$= \frac{F}{\delta} \tag{8}$$

where R is the capillary radius, and δ is the reduced layer thickness. It follows from Equation 8 that the reduced thickness ϵ of the layers corresponding to the other fluid component is

$$\varepsilon = \frac{1 - F}{N} \tag{9}$$

provided that N is large compared with 1. Knowing the layer thicknesses, we can calculate any radius r_j^o of the interfaces between the layers at the capillary entrance. The volume rate of flow of a tubular layer is:

$$Q_j = \pi (r_j^{o2} - r_{j+1}^{o2}) V_o$$

The total volume rate of flow is

$$Q = \pi R^2 V_o$$

So the reduced volume rate of flow

$$q_j = Q_j/Q = \rho_j^{o2} - \rho_{j+1}^{o2} \tag{10}$$

of any tubular layer can be calculated.

In the capillary, where the flow is stationary, the shear stress at the capillary wall is

$$\tau_w = \mu_{app} \dot{\gamma}_{nw}$$

where

$$\dot{\gamma}_{nw} = \frac{4V_o}{R}$$

is the apparent shear rate at the wall and equal to the true shear rate if the liquid were Newtonian, and μ_{app} is the apparent viscosity. Hence, the apparent viscosity of the liquid mixture, which will be calculated, is

$$\mu_{app} = \frac{R\,\tau_w}{4\,V_o} \qquad (11)$$

At distance r from the capillary center the shear stress is

$$\tau = -\mu \frac{\partial v}{\partial r}$$

where μ is the viscosity of the liquid at distance r. It follows that

$$\frac{\tau R}{V_o} = -\mu \frac{\partial \nu}{\partial \rho} \qquad (12)$$

where $\nu = \dfrac{V}{V_o}$

From Equation 12 it follows:

$$\nu_{j+1} - \nu_j = \frac{R}{V_o} \int_{\rho_{j+1}}^{\rho_j} \frac{\tau}{\mu}\, d\rho$$

$$= \frac{\tau_w R}{V_o} \int_{\rho_{j+1}}^{\rho_j} \frac{\rho}{\mu}\, d\rho$$

(By stress equilibrium $\dfrac{\tau}{\tau_w} = \rho$).

$$= 4\mu_{app} \int_{\rho_{j+1}}^{\rho_j} \frac{\rho}{\mu}\, d\rho \qquad (13)$$

The volume rate of flow of an annulus is

$$Q_j = \int_{r_{j+1}}^{r_j} v 2\pi r\, dr$$

Consequently the reduced rate of flow is

$$q_j = 2 \int_{\rho_{j+1}}^{\rho_j} \nu \rho\, d\rho = \nu_j \rho_j^2 - \nu_{j+1}\rho_{j+1}^2 - \int_{\rho_{j+1}}^{\rho_j} \rho^2 \frac{\partial \nu}{\partial \rho}\, d\rho$$

$$= \nu_j \rho_j^2 - \nu_{j+1}\rho_{j+1}^2 + 4\mu_{app} \int_{\rho_{j+1}}^{\rho_j} \frac{\rho^3}{\mu}\, d\rho \qquad (14)$$

Combining Equations 13 and 14 gives

$$v_j = v_j (\rho_j^2 - \rho_{j+1}^2) + 4\mu_{app} \left(\rho_{j+1}^2 \int_{\rho_{j+1}}^{\rho_j} \frac{\rho}{\mu} d\rho + \int_{\rho_{j+1}}^{\rho_j} \frac{\rho^3}{\mu} d\rho \right) \quad (15)$$

By using Eyring's equation, for example,

$$\frac{1}{\mu} = \frac{1}{\mu_o} (1 + k\tau^2) \quad (16)$$

and the condition for stress equilibrium

$$\frac{\tau}{\tau_w} = \rho$$

Equations 13 and 15 can be rewritten as:

$$v_{j+1} = v_j + \frac{2\mu_{app}}{\mu_o} (P_j - P_{j+1} + k\tau_w^2 (P_j^2 - P_{j+1}^2)) \quad (17)$$

and

$$A P_{j+1}^3 + P_{j+1}^2 - B P_{j+1} + C = 0 \quad (18)$$

where

$$P_j = \rho_j^2$$

$$A = \frac{1}{3} k\tau_w^2$$

$$B = v_j \frac{\mu_o}{\mu_{app}} + 2P_j$$

$$C = \frac{2}{3} k\tau_w^2 P_j^3 + P_j^2 + (v_j P_j - q_j) \frac{\mu_o}{\mu_{app}}$$

At the capillary wall ($j = 1$) holds:

$$v_j = 0$$
$$P_j = 1$$

At the capillary center ($j = 2N$) holds:

$$P_{2N} = 0$$

Starting with an arbitrary value for μ_{app}, every P_j and v_j can now be calculated by Equations 10, 17, and 18 at a given shear stress at the wall. By an iterative procedure the correct value for μ_{app} can now be approached by approximating P_{2N} to zero. To calculate μ_{app} the following constants describing the rheological behavior of the components were used:

	PU solution	PAN solution
μ_o	12.6	222
k	0	2.5×10^{-9}

Microscopic examination revealed that phase inversion of the emulsions occurred at a volume fraction of ca. 0.5 and that the average size of the dispersed droplets was about 5 microns.

There are two ways to calculate μ_{app}. First, decide which fluid component occupies the layer adjoining the capillary wall. Carrying out the calculations for the two cases we obtain two values of μ_{app}. The most probable will be the average of the two.

The calculations were carried out for a shear stress at the wall of 2×10^4 dynes/cm². The two values for μ_{app} and their average were plotted together with the results from the measurements (Figure 3) against the volume fraction of the PAN solution in Figure 9. The calculated mean viscosity line coincides with the measurements on the emulsions that are rich in PU solution. On the other hand, for PAN rich

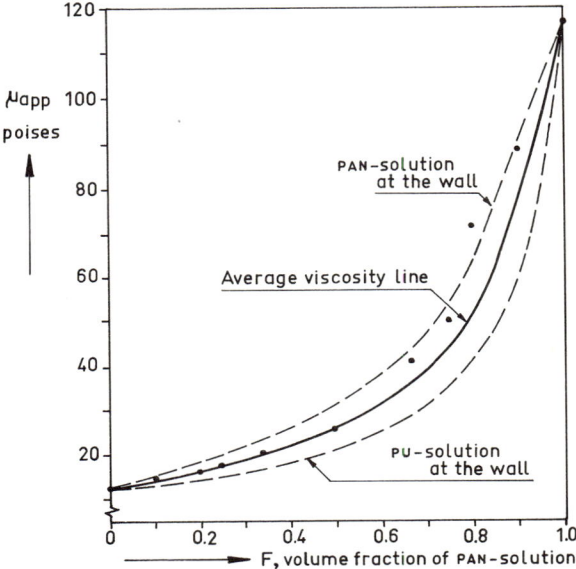

Figure 9. Measurements with a capillary viscometer: viscosity of the emulsions vs. their composition

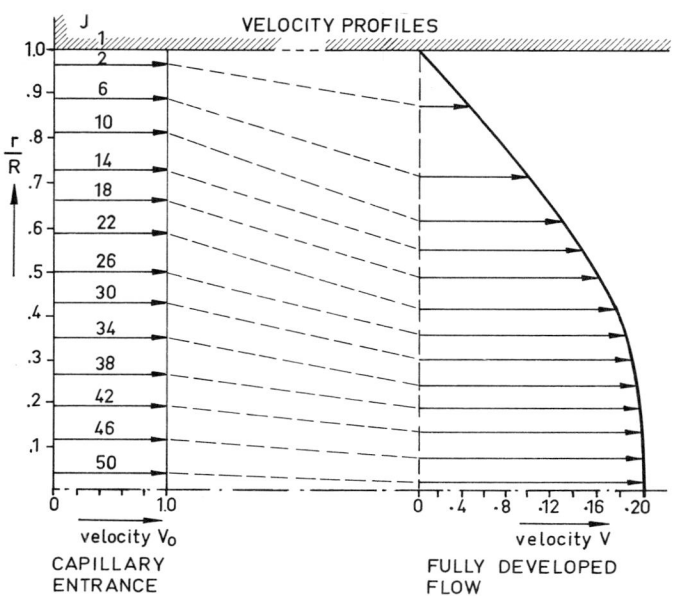

Figure 10. Calculated velocity profiles for 90 vol % PAN solutions

solutions the measurements show a better fit with the line calculated for PAN solution at the wall.

Figure 10 gives the calculated velocity profile for a volume fraction of PAN solution of 0.9 and the same solution adjoining the capillary wall. For clarity not all the velocities at the interfaces are represented. The cross-sectional area of a tubular layer increases as the average velocity of this layer in the capillary becomes smaller and vice versa.

Conclusions

The measurements presented here on emulsions of polyurethane and polyacrylonitrile dissolved in N-methylpyrrolidone can be explained well by theory. However, in the literature certain examples of comparable systems are given (*e.g.*, polyacrylonitrile and cellulose–acetate dissolved in dimethylformamide (2)), which show a much more complicated behavior. These emulsions have even lower viscosities than that of the least viscous component. Thus, the viscosity–composition curves have minima. Such behavior cannot be explained by any of the models discussed above. It seems that the basic assumptions used in our analysis are not valid for such systems.

These assumptions are: continuity of stress and velocity across the interfaces; since stress discontinuities owing to high interfacial tensions increase the viscosity (3), to explain the unexpected low viscosities we must doubt the validity of the velocity condition at the interfaces. This is supported by the fact that the two polymers are incompatible, which implies a relatively low concentration or even complete absence of entanglements at the interface. A very thin layer of solute-free solvent around the droplets would result. Such a low viscosity layer might be the cause of the anomalous behavior.

Acknowledgment

The authors thank J. E. van Laar for the elaborate preparatory experimental work and the measurements with the cone-plate viscometer.

Literature Cited

(1) Heitmiller, R. F., Naar, R. Z., Zabusky, H. H., *J. Appl. Polymer Sci.* **8**, 873 (1964).
(2) Meissner, W., Berger, W., Hoffmann, H., *Faserforsch. u. Textiltechnik* **19**, 407 (1968).
(3) Nawab, M. A., Mason, S. G., *Trans. Faraday Soc.* **54**, 1712 (1958).
(4) Scott, R. L., *J. Chem. Phys.* **17**, 279 (1949).
(5) Taylor, G. I., *Proc. Roy. Soc. (London)* **A138**, 41 (1932).
(6) Tompa, H., *Trans. Faraday Soc.* **46**, 970 (1950).

RECEIVED February 25, 1970.

6

Interfacial Bonding between Different Elastomer Phases

R. L. ZAPP

Enjay Polymer Laboratories, Esso Research and Engineering Co., Linden, N. J. 07036

> *A differential solvent-swelling technique was used to study the ability of different elastomers in a binary blend to co-vulcanize. The technique required that the two elastomer phases have sufficiently different solvent parameters so that one phase would be highly swollen while the other, below theta temperature conditions, would be lightly swollen. If interfacial bonds were present, a lightly swollen dispersed phase would restrict the swelling of a highly swollen continuous phase in a manner analogous to the restriction of swelling by reinforcing pigments. If interfacial bonds were absent, no restriction in swelling was observed. The presence or absence of interfacial bonds between chlorinated butyl rubber and polydiene rubbers depended on the type of curatives and crosslinking agents used. This suggested that certain types of chemical bonds as well as relative vulcanization rates were controlling factors.*

From thermodynamic considerations (2) supported by microscopic observations (5), two different high molecular weight polymers when blended exist in a heterogeneous state. In the case of elastomers capable of crosslinking these separate phases may crosslink in the presence of one another. However, the question arises, does bonding exist across the interfaces?

Many studies of vulcanized elastomer blends have revealed discontinuities in physical property trends attributable to poor interfacial bonding. Recently Rehner and Wei (5) have observed discontinuities in the swelling of blended crosslinked networks swollen in a common solvent. This departure from an averaged swelling behavior, based upon compositional ratios and the swelling behavior of the two homophases, reflected peculiar changes in over-all crosslink concentration at specific ratios. These authors did not mention the absence of interfacial bonding as an explanation of the discontinuities although their vulcanized systems were well crosslinked on the basis of low extractable content.

The present study addressed itself to the subject of interfacial bonds between the phases of blended elastomeric networks. For the most part these investigations have emphasized systems of chlorinated butyl rubber and polydiene elastomers. Chlorinated butyl rubber (1) possesses allylic chlorine atoms that allow this elastomer to be crosslinked by the agency of zinc oxide and the elimination of chlorine, independent of conventional sulfur vulcanization. Thus, this low functionality polymer chain molecule will crosslink in the presence of highly unsaturated rubbers contrary to experience with strictly low unsaturated elastomers. Other elastomer systems have been investigated briefly to add support to the diagnostic solvent-swelling techniques, which is the subject of this discussion. These techniques are based upon solvent-swelling observations in a pair of solvents defined as differential swelling solvents.

Adhesion Analysis in Swollen Elastomer Blended Networks

In searching for solvent-swelling techniques that might provide more positive evidence of bonding between elastomer phases, attention was drawn toward elastomer networks containing reinforcing fillers such as carbon black as well as systems with non-reinforcing mineral pigments. The phenomenon of carbon black restriction of elastomer network swelling was observed in a qualitative and graphical way quite a few years ago (7). Mineral fillers displayed much less restriction to network swelling. More recently Kraus (4) has placed a mathematical interpretation upon the restriction of network swelling by reinforcing fillers based upon the presence of an elastomer–particle bond that remains intact during swelling. Agreement between experiment and theory was satisfactory.

Development of Solvent Pairs. The restriction of elastomer network swelling by a reinforcing filler imbedded in the network is pictorially represented in Figure 1. In deriving an equation relating swelling restriction to filler loading, Kraus postulated that the elastomer network attached to the surface of the particle would remain unswollen and that swelling would increase as the distance from the particle increased. The over-all effect would be to reduce the amount of solvent imbibed in a filled system as compared with an unfilled elastomer network. The restriction in swelling expressed by the ratio of volume fractions of gel in the unfilled and filled (restricted) case was related to a function of the

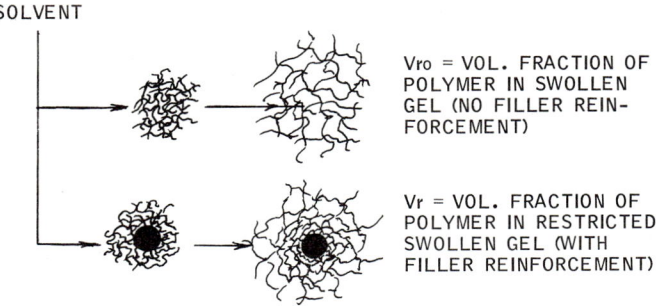

Figure 1. Restriction of network swelling (4). $V_{ro}/V_r = 1 - m(\phi/1 - \phi)$.

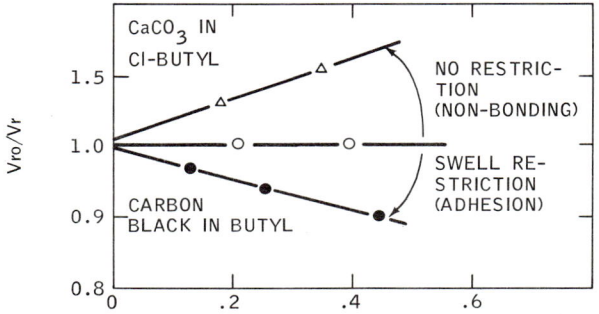

Figure 2. Filler adhesion analysis. Function $\phi/1 - \phi$ (ϕ = volume fraction of filler).

filler concentration by the equation shown in Figure 1. Volume fraction of polymer in the swollen gel, V_r is related to swelling volume increase by:

$$V_r = \frac{100}{100 + (\%\text{ vol increase})}$$

If pigment–polymer adhesion were restricting swelling, the linear plot would have a negative slope, and if no adhesion were present, a zero or positive slope would be observed. The positive slope could be associated with the presence of voids that would be filled with imbibed liquid, thus producing an abnormally high swell. A representation of the two cases, one with pigment–polymer adhesion, in this case carbon black, and the other with a mineral filler where no adhesion exists is shown in Figure 2.

In the filler–polymer adhesion analysis, the filler particles do not imbibe solvent during the swelling process, so the problem in attempting to adapt this concept to swollen elastomer blends would be to restrict the swelling of one elastomer phase to make it appear as a slightly swollen particle. If the other elastomer phase is highly swollen, considerable interfacial restriction to swelling of this phase would exist if interfacial bonding were present. Such a differential swelling condition can be found for certain pairs of elastomers, and for this study, the technique has been given the term "swelling in differential solvents."

Organic solvents exist in a wide range of solubility parameters based upon their cohesive energy densities, which are related to molar heats of vaporization (6). Elastomers and polymeric materials capable of crosslinked network formation and of being swollen will have an indirectly determined solubility parameter equivalent to the parameter of the solvent in which the polymer swells to the highest extent. If the two elastomers have sufficiently different solubility parameters ($\sqrt{CED}$ — square root of the cohesive energy density), it is possible to select solvent pairs which will swell the two phases differentially. If one superimposes reduced temperatures on the elastomer–solvent systems so that one elastomer phase is well below its "theta" temperature (3), a condition approaching the system of a filler in a swollen network can be obtained. When an elastomer–solvent system is well below its "theta" temperature, the polymer chains are coiled tightly, and the low degree of swelling has no relation to an extent of network crosslinked density. Thus, the elastomer phase that is "well below its theta temperature" can be considered to be a slightly swollen pigment or filler particle.

Table I lists the final results of solvent-swelling conditions which resulted in selecting 2,2,4-trimethylpentane and styrene at $-25°C$ for a differential solvent pair. The table also includes the published values for the solubility parameters ($\sqrt{CED}$) of the elastomers and the solvents. This table indicates that for the elastomer systems Cl–butyl–*cis*-polybutadiene or Cl–butyl–SBR excellent differentiation can be obtained.

Table I. Swelling of Various Elastomer Crosslinked Networks in Differential Solvents at $-25°C$

Elastomer Network $\sqrt{CED}^a$		2,2,4-Trimethyl-pentane $\sqrt{CED} = 6.9$	Styrene $\sqrt{CED} = 9.3$
Cl–butyl	7.8	315	73
Natural rubber	8.1	210	465
cis-Polybutadiene	8.4	65	310
SBR (1502)	8.3	60	420
L–300 Vistanex (not crosslinked)	7.8	soluble	72 (0.9% Ext.)

^{a} CED = cohesive energy density.

On the one hand, at −25°C trimethylpentane swells a chlorobutyl network to a high degree while limiting the swell of the two polydiene rubbers. Styrene at −25°C, on the other hand, highly swells the two polydiene rubber vulcanized networks but limits the swell of the chlorobutyl. To show that an elastomer network swelling in a solvent well below a "theta" temperature is essentially independent of the extent of crosslinking, a high molecular weight polyisobutylene (which cannot be vulcanized) swells to the same low level in styrene at −25°C as the chlorobutyl vulcanized system. Equilibrium swelling at these low temperatures was ensured by contacting 0.025-inch thick specimens with solvent for 3 days in a constant temperature cold box. The polyisobutylene was soluble in trimethylpentane at −25°C.

The system Cl–butyl–natural rubber (or *cis*-polyisoprene) could not be resolved by differential solvent techniques because the polymeric solubility parameters were too similar. At one end of the spectrum—*i.e.*, with styrene at −25°C—natural rubber could be highly swollen while restricting the chlorobutyl swell, but the reverse was not possible, as indicated by the swelling volumes in the trimethylpentane. As displayed in Table II, attempts to use a highly symmetrically branched hydrocarbon with a very low solubility parameter, served only to reduce both the swelling of natural rubber and chlorobutyl. (Neopentane is a gas above 10°C and a solid below −20°C). Therefore, for this report the use of differential solvents in the study of interfacial bonding in blends was limited to systems of Cl–butyl and *cis*-polybutadiene or SBR.

Table II. Swelling of Cl–Butyl and Polyisoprene Crosslinked Systems in Neopentane at −10°C

Elastomer Network $\sqrt{CED}{}^a$	% Volume Increase in Neopentane $\sqrt{CED} = 6.1$
Cl–butyl 7.8	120
Polyisoprene 8.1 (Shell)	91

^{a} CED = cohesive energy density.

Evidence of Interfacial Bonding. In these experiments compounded blends were composed of two elastomer phases that were first mixed with curative ingredients before blending in various ratios. This procedure was adapted to ensure proper initial distribution of the ingredients between the two phases. Blending was accomplished, with one exception, by mill mixing on a small laboratory two-roll mill. After blending, cured specimens in the form of small 0.025-inch thick pads were swollen in a common solvent to observe the over-all state of network formation. Only those blended systems which displayed satisfactorily low levels of extractables were subject to differential solvent swelling.

Such a series of blends swollen in a common solvent, cyclohexane, is shown in Figure 3. This system, blends of Cl–butyl and *cis*-polybutadiene, was composed of the following compounded phases: [Cl–butyl 100, zinc oxide 5, sulfur 1, TMTDS 1: and *cis*-polybutadiene 100, zinc oxide 5, sulfur 2.5, TMTDS 0.5] (TMTDS = tetramethylthiuram disulfide). All numbers represent parts by weight. In Figure 3, the separate homophases as well as all intermediate blends are well crosslinked by swelling standards. Attention is also drawn to the level of extractables; these are very similar and at an acceptably low level throughout the range of blend ratios.

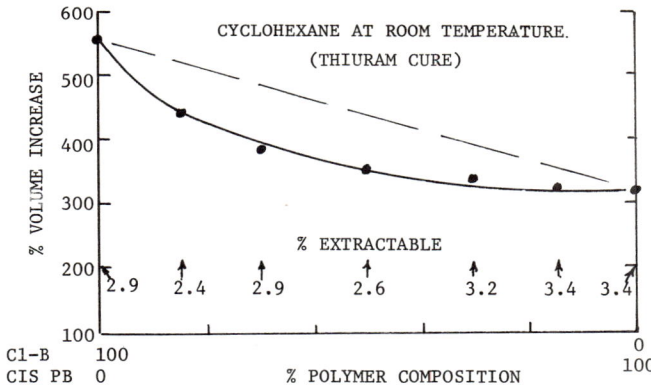

Figure 3. *Swelling in a common solvent system, Cl–butyl, cis-PB*

When this same system of blends cured in the same manner was subject to swelling in differential solvents at −25°C, plots of swelling *vs.* composition produced the two curves of Figure 4. The two curves descend sharply from high swell to nominal swell as the phase subject to theta temperature swelling conditions increases and then becomes predominant. In these specific curves the swelling of the blends is below an additive line which represents a restriction in volume swell of one phase (the highly swollen phase) by the lightly swollen phase, and is indicative of some degree of interfacial bonding. To analyze further the swelling data, on the basis that one phase is a lightly swollen pigment, it is assumed that solvent inhibition of this phase (below theta temperature conditions) does not depend on any cure state and is related only to the amount of this phase present in the blend. This is represented graphically by the two crossed straight lines near the bottom of the figure. The numbers (*i.e.*, 11.6; 23.2, etc.) near the lines are estimated volume percentages of solvent imbibed by the lightly swollen phase and must be subtracted from the total swelling directly above them to determine the swelling of the highly swollen phase. In the analysis of the data (analogous to the polymer–pigment adhesion concept of Kraus) two regions, one of *cis*-polybutadiene in chlorobutyl and the other of chlorobutyl in *cis*-polybutadiene, are considered.

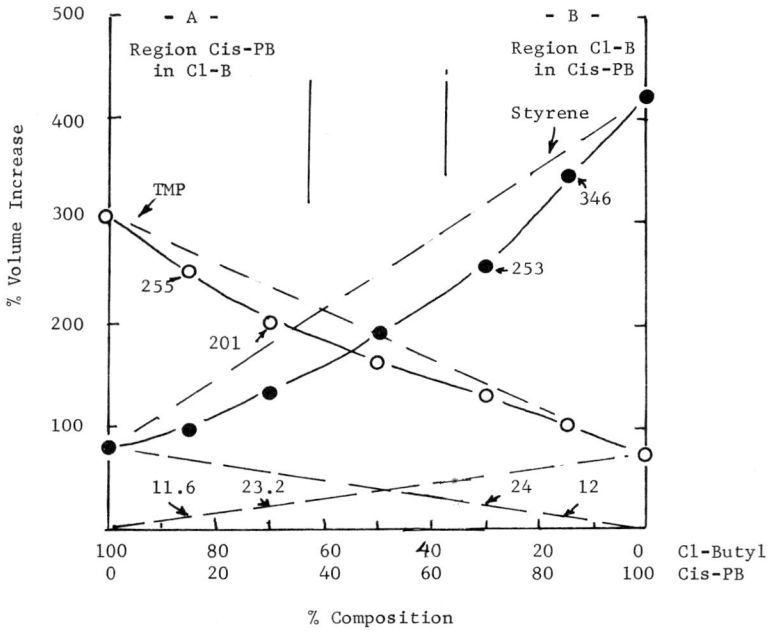

Figure 4. Blended systems of Figure 2 swollen in differential solvents at $-25°C$. Cure 30 minutes at $300°F$.

Representative calculations are shown in Table III, and % volume increase for the homophases and the blends is converted to volume fraction of the highly swollen polymer in the gel by the relationships shown for V_{ro} and V_r.

Finally, the corresponding ratios of V_{ro}/V_r are plotted as a function of the volume fraction of the dispersed lightly swollen phase $\phi/1 - \phi$ as shown in Figure 5. In this case both lines have descending slopes, reflecting the restriction of swelling of the highly swollen phase by the lightly swollen phase which can now be associated with an interfacial bonding.

Table III. Calculations for Interfacial Adhesion Analysis Shown in Figure 4

Comp.	Calc. of V_{ro} and V_r		V_{ro}/V_r
	A–Region, cis-PB in Cl–B		
100% Cl–B	$V_{ro} = \dfrac{100}{100 + 305}$	$= .247$	1
85 ClB 15 cis-PB	$V_r = \dfrac{85}{85 + (255 - 11.6)}$	$= .259$	.95
70 ClB 30 cis-PB	$V_r = \dfrac{70}{70 + (201 - 23.2)}$	$= .282$	.88
	B–Region, Cl–B in cis-PB		
100% cis-PB	$V_{ro} = \dfrac{100}{100 + 425}$	$= .190$	1
85 cis-PB 15 ClB	$V_r = \dfrac{85}{85 + (346 - 12)}$	$= .203$	.94
70 cis-PB 30 ClB	$V_r = \dfrac{70}{70 + (253 - 24)}$	$= .233$	.82

Composition Ratio

$\phi^a/(1 - \phi) =$	$\dfrac{100}{0}$	$\dfrac{85/15}{.176}$	$\dfrac{70/30}{0.429}$

[a] ϕ = Volume fraction of dispersed phase.

The question immediately raised is: "would this technique portray an opposite or negative adhesion response if applied to a polymer blended system where no interfacial bonding could be present?" Such a system would be cis-polybutadiene and high molecular weight polyisobutylene restricted to that portion of the blend system where polyisobutylene is the minor dispersed phase in cis-polybutadiene. A high molecular weight polyisobutylene [L-300 Vistanex (Enjay Chemical Co.)] was compounded with zinc oxide, sulfur, and TMTDS and then dissolved in hexane. cis-Polybutadiene (Phillips Chemical Co.) was also mixed with the same compounding ingredients and separately dissolved in hexane. The two polymer phases were then solution blended on a solids content basis of 15 and 30 wt direct polyisobutylene in 85 and 70% *cis*-polybutadiene. The blended solutions were poured onto Teflon surfaces for quick evaporation followed by high vacuum drying at 40°C. The homophases were also evaporated and dried in the same manner. The two blended

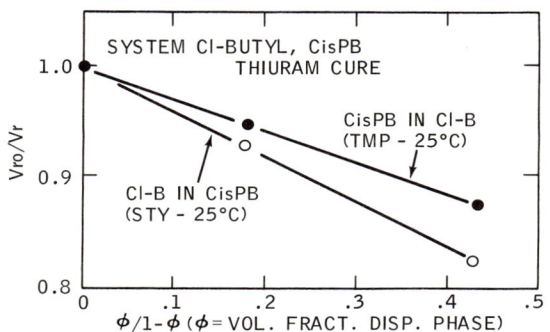

Figure 5. Adhesion analysis of blends swollen in differential solvents

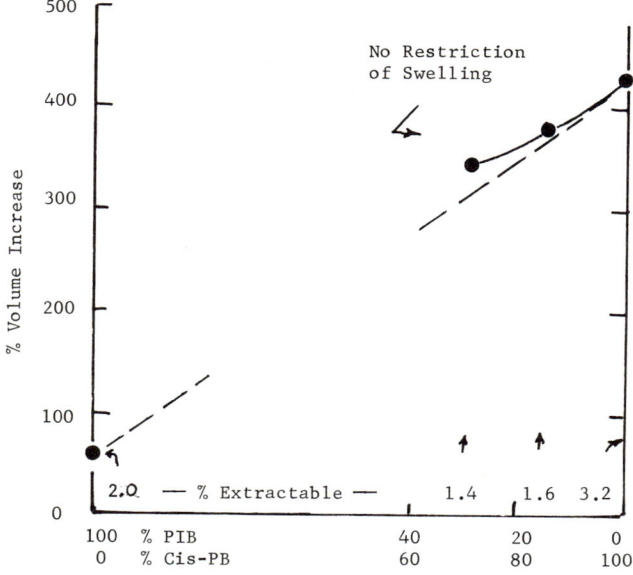

Figure 6. High molecular weight polyisobutylene in cis-polybutadiene. Differential swelling in styrene at $-25°C$. Blends cured 30 minutes at $300°F$.

ratios and the separate homophases were swollen in styrene at −25°C since only systems involving polyisobutylene in polybutadiene were considered. (Unvulcanizable polyisobutylene cannot be a continuous highly swollen phase in trimethylpentane since it is soluble.) Samples were cured 30 minutes at 300°F.

Swelling results in styrene at −25°C are shown in Figure 6, wherein there is no restriction in swelling below an average additive line even though polyisobutylene is swollen only lightly and is not soluble in styrene at this low temperature. Proceeding with an adhesion analysis as described under Figure 4, the subsequent plot in Figure 7 of the ratio of V_{ro}/V_r vs. the function of lightly swollen polyisobutylene content in cis-polybutadiene has a positive upward slope indicative of an absence of interfacial bonding.

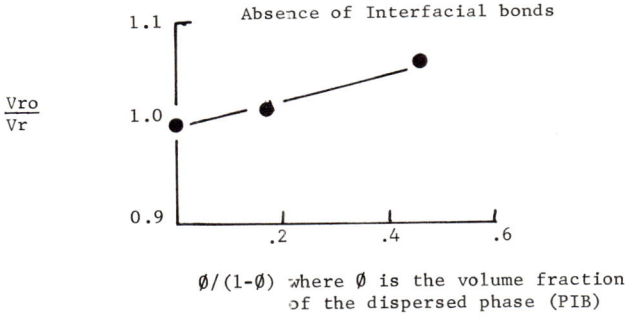

Figure 7. Adhesion analysis after swelling in styrene

The System Cl–Butyl–cis-Polybutadiene

The system Cl–butyl–cis-polybutadiene has been studied in some detail because it was suitable for the developed differential swelling technique and because this system of blends vulcanized with zinc oxide, sulfur, and thiuram disulfide first revealed the presence of interfacial bonds. This curative system has the feature of a "flat cure"—i.e., the two homophases are vulcanized rapidly, and the crosslinked density does not increase radically as vulcanization time is prolonged. This is observed in Table IV by swelling and extractable levels of a series of crosslinked networks cured at increasing times and swollen in a common solvent, cyclohexane.

All crosslinked networks in the form of 0.025-inch thick sheets had acceptably low levels of extractable contents that were relatively constant regardless of the time of vulcanization. Although states of crosslinking of the homophases changed little with prolonged vulcanization

Table IV. Swelling Volumes and Extractable Contents of Networks of Cl–Butyl and *cis*-Polybutadiene-Vulcanized at 300°F—Sulfur, Zinc Oxide, TMTDS

Network Cured	% Vol. Increase			% Extractables		
	10 min.	30 min.	60 min.	10 min.	30 min.	60 min.
100% Cl–Butyl	570	560	550	2.9	2.9	2.6
85 Cl–B 15 *cis*-PB	510	470	440	3.2	2.4	2.1
70 Cl–B 30 *cis*-PB	459	420	417	3.0	2.9	2.3
50 Cl–B 50 *cis*-PB	430	380	360	3.5	2.6	2.5
30 Cl–B 70 *cis*-PB	418	340	330	3.7	3.2	2.9
15 Cl–B 85 *cis*-PB	395	350	355	3.6	3.4	3.4
100% *cis*-PB	342	330	350	3.1	3.4	3.6

time, the possibility existed that the effect of time upon interfacial bond development would be different.

Effect of Vulcanization Time. The Cl–butyl–*cis*-polybutadiene blended systems discussed earlier with reference to Figures 4 and 5 were vulcanized for 10, 30, and 60 minutes at 300°F and subject to differential solvent swelling as shown in Figure 8. It is obvious that the 10 minute cure has a markedly different swelling response in styrene at −25°C than the other longer cures. The marked difference occurs in the region where Cl–butyl is the minor phase in *cis*-polybutadiene, and in this region swelling is not restricted at this short cure time. As cure time is prolonged, the swelling curve loses its S-shaped pattern, and swelling restrictions begin to appear. The other blended region of *cis*-polybutadiene in chlorobutyl is not as sensitive; restrictions of swell in trimethylpentane at −25°C are observed early in the curing cycle.

Adhesion analysis based on the differential solvent swelling is shown in Figure 9 where the 10-minute cures of blends of small amounts of Cl–butyl in *cis*-polybutadiene reveal the lack of interfacial bonding. As cure time is prolonged the plots of the ratios of network swelling (expressed as volume fractions of polymer) change to a descending slope pattern indicative of interfacial bonding. In the less sensitive blend area of *cis*-polybutadiene in chlorobutyl the adhesion analysis from differential swell in TMP at −25°C shows descending slope patterns for all cures with small increases in slope steepness as cure time progresses. Interfacial

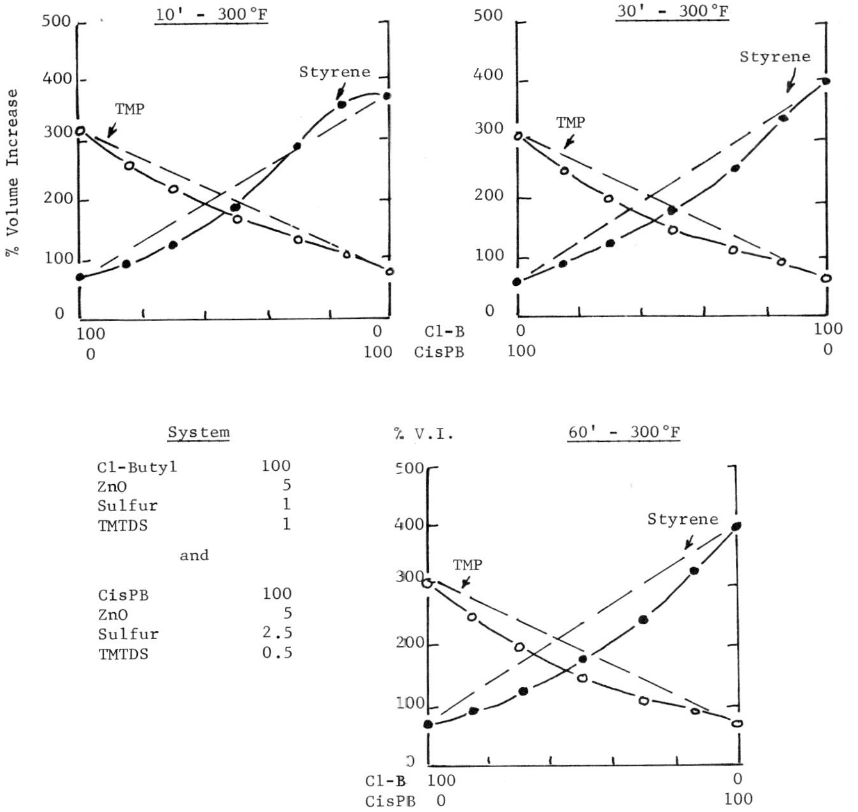

6. ZAPP *Interfacial Bonding* 79

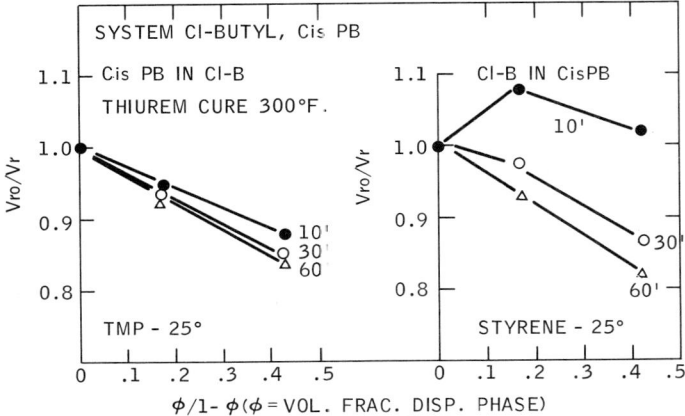

Figure 8. Changing conditions at the interface as cure time increases. Blended networks swollen in differential solvents at $-25°C$.

Figure 9. The changing condition at the interface

bonding occurs earlier in the cure cycle, in this blend area, and increases modestly with time.

Sensitivity toward Vulcanization Ingredients. The zinc oxide-induced crosslinking is fundamental for elastomers containing active combined chlorine. Other ingredients are most often added to enhance the crosslink density or accelerate the rate. One class of materials is represented by polyhydroxy aromatic compounds such as catechol or resorcinol. Polyhydroxy aromatic compounds have been used (3) to enhance the ZnO cure of Cl–butyl. The mechanism proposed is based upon an alkylation type reaction catalyzed by $ZnCl_2$ (from the reaction of ZnO with the chlorine of the polymer). This type of vulcanization system is unrelated to an accelerated sulfur system, so one would not anticipate a co-cure between elastomer phases that contained these two systems in the separate phases. To test this assumption a Cl–butyl phase and a polybutadiene phase were prepared as follows. Cl–butyl 100: zinc oxide 5, di-o-tolylguanidine salt of dicatechol borate [Permalux (Dupont)] 2; cis-polybutadiene 100, zinc oxide 5, sulfur 2.5, TMTDS 0.5. These separately compounded elastomer phases were mixed and vulcanized in a range of ratios and subjected to differential solvent swelling. Results of an adhesion analysis from differential solvent swelling are shown in Figure 10. The upward trends of the adhesion analysis plot reflect lack of

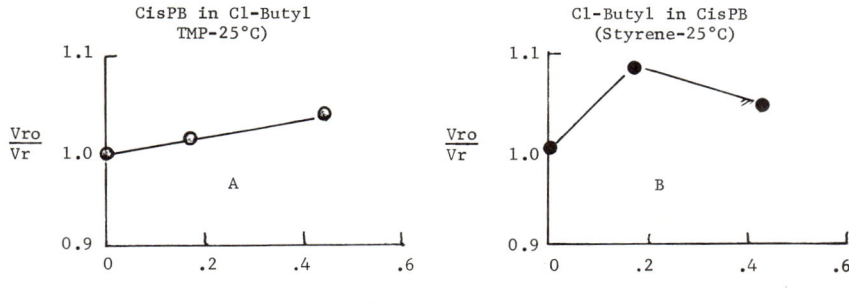

$\emptyset/(1-\emptyset)$ where $\emptyset$ = Vol. fraction dispersed phase

Figure 10. Adhesion analysis after differential solvent swelling at $-25°C$

Critical curing ingredients:			
System			
Cl-butyl 1066	100	cis-PB	100
zinc oxide	5	zinc oxide	5
Permalux	2	sulfur	2.5
		thiuram (TMTDS)	0.5

Permalux = di-o-tolylguanidine salt of dicatechol borate (duPont) Increasing the state of cure of chlorobutyl by an independent accelerator does not promote interfacial bonding. Blends mill mixed and cured at 300°F for 60 minutes

interfacial bonds whether Cl–butyl or *cis*-polybutadiene is the continuous phase.

A general pattern of vulcanization ingredients that promote interfacial bonds has emerged. This pattern appears to be based upon the use of thiuram di- or polysulfides as accelerators in both elastomer phases. For example, when a thiazole-type accelerator is substituted for the thiuram disulfide in the polybutadiene phase, there is lack of evidence for interfacial bonding with Cl–butyl. The adhesion analysis from differential solvent swelling for such a blended vulcanized system is shown in Figure 11. Compare the essentially horizontal plots of this figure with the descending slopes of the plots in Figure 5.

The System Cl–Butyl–Styrene–Butadiene Rubber

In blended systems of Cl–butyl and styrene–butadiene–rubber evidence of interfacial bonds has been more difficult to obtain when com-

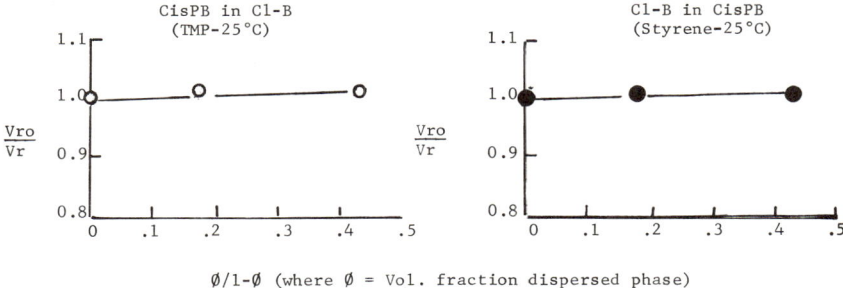

Figure 11. Compounding variations in blended elastomer phases (Cl–butyl–cis-polybutadiene). A thiazole in cis-PB phase impairs adhesion.

pared with the system Cl–butyl–*cis*-polybutadiene. Positive evidence of interfacial bonding by differential solvent swelling has been demonstrated most successfully with a thiuram sulfur donor-type of vulcanization accelerator. An example of a Cl–butyl–SBR blend system that displays an absence of interfacial bonds contains a thiuram monosulfide which is not a sulfur donor. Elemental sulfur must be added to produce crosslinks in the SBR phase. This system is shown in Figure 12. Blended vulcanized systems were swollen in the pair of differential solvents, and no restriction in swell below an additive average line is noted for the critical regions. This swelling pattern results in adhesion analysis plots that are horizontal, reflecting an absence of interfacial bonding.

In contrast, blended systems of Cl–butyl and SBR vulcanized with a thiuram sulfur donor, dipentamethylenethiuram tetrasulfide–DMTTS, in

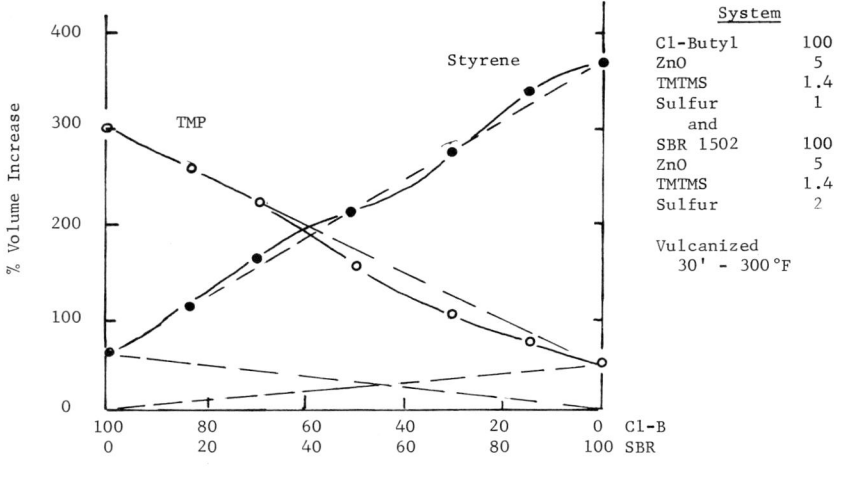

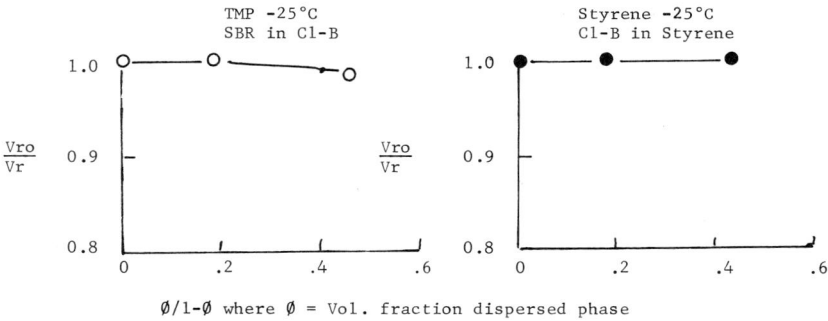

Figure 12. Swelling in differential solvents at $-25°C$ for the system SBR 1502, Cl–butyl tetramethylthiuram monosulfide formulation

both phases reflects positive evidence of interfacial bonds. In Figure 13, the compounded system given adjacent to panel A, produced volume swell restrictions in differential solvents. The 30 and 60 minute vulcanization times are almost exact duplicates in this regard. The restrictions in differential solvent swelling generate the descending slopes of the adhesion analysis plots shown in panels B and C.

Acrylonitrile Butadiene Rubber (NBR) with SBR

From practical compounding experience, we would predict that a butadiene–acrylonitrile rubber will co-vulcanize across an interface with a styrene–butadiene rubber. They both have comparable quantities and types of butadiene polymerized structures, but the acrylonitrile units impart vastly different polymeric solubility parameters to the molecular chain. Whereas an SBR hydrocarbon elastomer has a solubility parameter of 8.3 the nitrile rubber of 25% acrylonitrile has a value of 9.4 (6). Because of this difference and level of solubility parameters another differential solvent pair was selected for co-vulcanizing analysis. The solvent

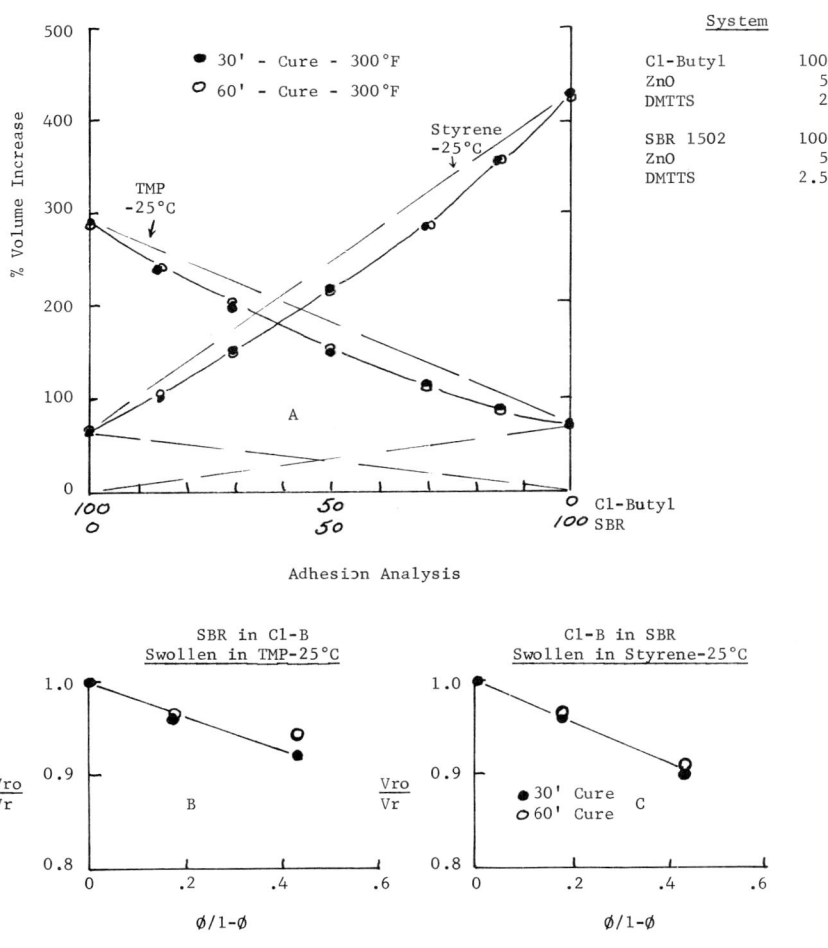

Figure 13. *Swelling in differential solvents for the system Cl–butyl, SBR 1502 dipentamethylenethiuram tetrasulfide accelerator. Sulfur donor is DMTTS.*

pair of cyclohexane and acetone at room temperature was entirely adequate as the swelling volume results shown below will testify.

% Vol Increase in:	SBR-1502	NBR (25% Acrylonitrile)
Cyclohexane	308	35
Acetone	16	257

Blends of the following formulated systems were mill mixed and cured at 300°F:

$$\begin{bmatrix} \text{SBR–1502 100, ZnO 5, S 2, TMTDS 0.5} \\ \text{NBR–100, ZnO 5, S 2, TMTDS 0.5} \end{bmatrix}$$

and then swollen in cyclohexane and in acetone at room temperature.

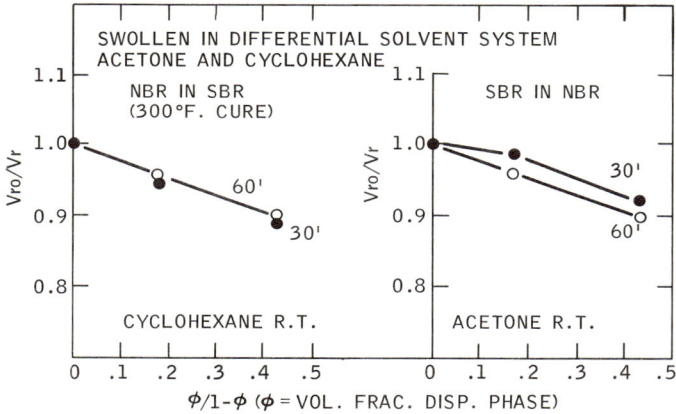

Figure 14. Blends of SBR and NBR swollen in differential solvent system acetone and cyclohexane

Adhesion analysis from this differential swelling is shown in Figure 14 wherein the descending pattern of the plots of 30 or 60-minute cure time indicates the presence of interfacial bonds.

Summary

Although two dissimilar elastomers—e.g., chlorinated butyl rubber and polybutadiene—may crosslink when in contact with one another, does bonding exist between the two interfaces? Based upon thermodynamic theory as well as microscopic observations, we know that two such elastomers are not molecularly dispersed in a blend, so the diagnostic problem is one of considering two dispersed phases.

By using swelling techniques and especially swelling in differential solvents evidence of interfacial bonding has been observed under certain conditions of vulcanization. A differential solvent is defined as one which is a good solvent (or swellent) for one phase and essentially a theta solvent (or poor swellent) for the other phase under well-defined temperature conditions. In recent studies styrene (solvent parameter 9.3) at $-25°C$ will highly swell a polydiene rubber network while having very little swelling power for a butyl rubber network. Conversely, 2,2,4-trimethylpentane (solvent parameter 6.9) at $-25°C$ will highly swell a butyl network while possessing very little swelling power for some polydiene elastomer backbones. Thus, one phase, if it is the dispersed phase, can be considered a lightly swollen pigment or filler, and a type of filler adhesion analysis based upon restrictions of solvent volume swelling can be applied. Techniques of this kind have shown that under selected conditions one elastomer phase will restrict the volume swell of another in a manner analogous to reinforcing carbon black fillers. This restriction of swelling of the continuous elastomer phase by the dispersed phase, in the presence of differential swelling solvents, is considered to be evidence of interfacial bonds. The acquisition of interfacial bonds depends upon the type of co-vulcanization agents used to promote crosslinking. This suggests that relative crosslinking rates of the two phases and the chemical nature of the crosslinks are the factors that control the formation of interfacial bonds.

Literature Cited

(1) Baldwin, F. P., Buckley, D. J., Kuntz, I., Robison, S. B., *Rubber Plastics Age* **42**, 500 (1961).
(2) Flory, P. J., "Principles of Polymer Chemistry," Chap. XIII, p. 555, Cornell University Press, Ithaca, N. Y., 1953.
(3) *Ibid.*, p. 545.
(4) Kraus, G., *J. Appl. Polymer Sci.* **7**, 861 (1963); *Rubber Chem. Technol.* **37**, 6 (1964).
(5) Rehner, J. Jr., Wei, P. E., *Rubber Chem. Technol.* **42**, 985 (1969).
(6) Sheehan, C. J., Bisio, A. L., *Rubber Chem. Technol.* **39**, 149 (1966).
(7) Zapp, R. L., Guth, E., *Ind. Eng. Chem.* **43**, 430 (1951); *Rubber Chem. Technol.* **24**, 894 (1951).

RECEIVED February 25, 1970.

7

The Theory of Rubber Toughening of Brittle Polymers

C. G. BRAGAW

Plastics Department, E. I. du Pont de Nemours & Co., Wilmington, Del.

> *The technology of ABS and impact polystyrene resins shows that incorporation of a dispersed rubber phase of appropriate properties can raise impact toughness by an order of magnitude, apparently by increasing the energy-absorbing volume in the resin. Principal existing theories of the toughening of brittle plastics by rubber dispersion are considered; all appear quite limited in ability to explain known phenomena and have substantial difficulties. It is proposed that the mechanism whereby rubber dispersions greatly increase energy-absorbing volume is by causing cracks and/or crazes to branch dynamically at rubber sites through the Yoffe mechanism (35). The theory explains anomalous morphological evidence and is supported by evidence from a variety of physical measurements.*

Polymers such as poly(methyl methacrylate) (PMMA) and polystyrene are brittle on a macro scale, but on a micro scale they are enormously tough. Crack propagation in these materials requires 3×10^5 to 2×10^6 ergs/cm^2 of new crack surface (5), far more than the theoretical value of $\sim$450 ergs/cm^2 (5) calculated assuming that fracture involves breakage of molecules oriented perpendicular to the crack surface. The formation of interference colors at the crack surface indicates the presence of a low density oriented (crazed) layer at the crack surface (6, 18), and this layer is commonly thought to have absorbed the energy measured. Assuming 2μ as the upper limit of layer thickness (18), it is calculated that the deforming layer in the plastic absorbs 5×10^8 ergs/gram. This is about half the average specific energy dissipation in a tough, ductile steel (yield stress 48,000 psi, ultimate elongation 31%).

The apparent brittleness of PMMA, polystyrene, and styrene/acrylonitrile copolymer (S/AN) arises because energy absorption is confined to thin layers of the order of a micron (*18*). In the case of rubber-toughened PMMA, polystyrene, S/AN, and PVC, deformation occurs in millimeter-thick layers, and macro energy absorption is usefully high. This layer is most readily marked by blushing. Evidence of large deformations in matrix material in ABS-type resins is (a) tensile bars elongate and neck permanently—this permanence requires permanent deformation of the S/AN matrix since it is the only continuous phase and constitutes ~75% of total sample volume, (b) electron micrographs (Figure 1) show regions where rubber spheres have been distorted into 2:1 or 3:1 ellipsoids. Independent work by Mann, Bird, and Rooney (*23*) also reaches the same conclusion.

The extremely high energy dissipation in the craze layer, 5×10^8 ergs/gram, would lead to an adiabatic temperature rise of about 24°C during crack/craze propagation, which is insufficient to cause most matrix polymers to pass through their T_g at usual environmental temperatures. Introduction of rubber particles can only lower this temperature rise since rubber secant modulus is very low and rubber deformation does not exceed the 300% or so of the deformed matrix.

The relationship between craze initiation and propagation, and crack initiation and propagation, is not well understood (*31*). Crazing clearly precedes formation of unstable cracking. For example, in tensile tests on Lustran A-21 S/AN, crazing occurred at 1.8–2.0% elongation, whereas

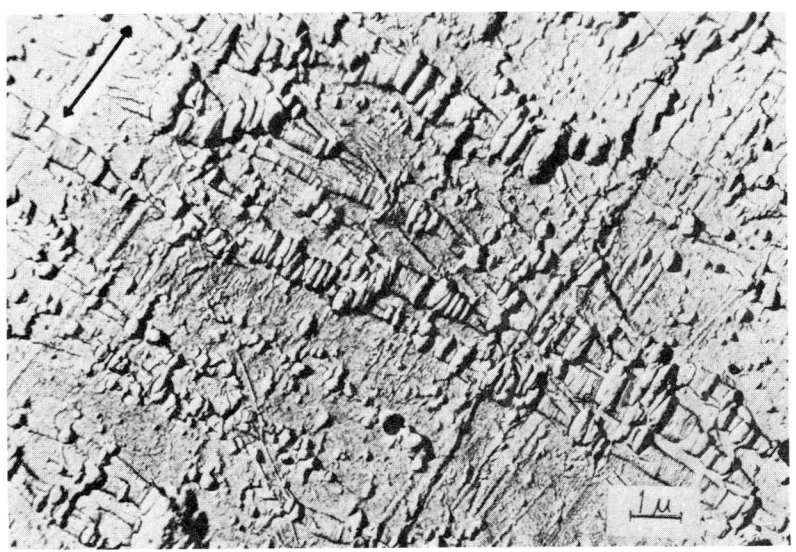

Polymer

Figure 1. Craze traces in an ABS resin (25)

catastrophic crack propagation occurred at 2.3–2.4% elongation. The role of craze in crack initiation is completely obscure at this time.

Cracks grow slowly after initiation until the crack length reaches the Griffith crack length (29). Then rapid acceleration occurs, the crack finally reaching a limiting speed of about half the transverse wave velocity (32). At this speed, the crack surface becomes rough because of branching (9). The critical crack length is given by Griffith's relation

$$T = \sqrt{\frac{2E\gamma}{\pi c}} \qquad (1)$$

where T = tensile strength, E = Young's modulus, γ = energy required to produce a unit area of new crack, and c = critical crack length. The relation is applicable to systems which are brittle or where ductile flow regions are small compared with c. For a resin like polystyrene or PMMA, $T \sim 5 \times 10^8$ dynes/cm², $E \sim 3 \times 10^{10}$ dynes/cm², $\gamma \sim 2 \times 10^5$ ergs/cm². Thus, one calculates a Griffith crack length of 50 or 100μ. As noted above, ductile effects are limited to layers of 1 or 2μ, so the Griffith relation is applicable (29). By breaking S/AN bars, it is experimentally observed that crack roughening begins when c reaches 250μ, which is entirely compatible with the calculation of 50 or 100μ Griffith crack length. The difference between the two numbers represents crack acceleration distance.

Almost all ABS resins turn white (blush) when deformed beyond the yield stress, even at the extremely high deformation rates characteristic of the notched Izod test. The relatively high energy absorption properties are thought to be caused by energy dissipation in the blush region. Electron micrographs of Haward and Mann (17) and Mann, Bird, and Rooney (23) show gross matrix deformation in the blush area. Blush density is lower than matrix density (0.91 vs. 1.01 for blushed and unblushed Cycolac H ABS resin), and blush disappears when the material is compressed; thus, voids appear to be present in blushed material. Haward and Mann (17) and Newman and Strella (28), report that much deformation beyond the yield point is recoverable upon heating the material above the glass transition temperature (T_g) of the glassy matrix; thus, molecular orientation is believed to exist in the continuous phase. Merz, Claver, and Baer (27) and Schmitt and Keskkula (33) concluded that blushed material consisted of networks of cracks, whereas Matsuo (25) and Bucknall and Smith (7) conclude that craze networks make up the blushed material. Bucknall and Smith used light microscopy and saw no voids in blush; their conclusion that no cracks exist, however, must be interpreted to mean no cracks or voids with smallest dimensions larger than the wavelength of light. Kambour indicates that craze material is 40–50% voids of 20 to 200 A in size (21). Matsuo's electron

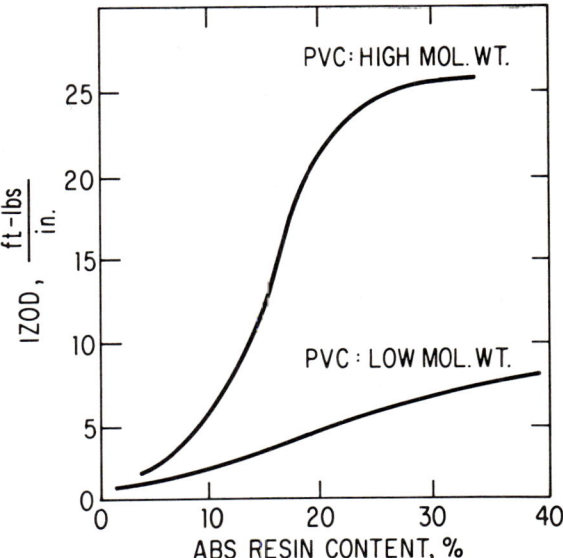

Figure 2. Effect of ABS content in toughening PVC (24)

micrographs show lines $<< 0.1\mu$ wide in craze layers, which can be interpreted to be cracks.

Existing Theories of Toughening

Rubber Energy Absorption. The earliest ideas concerning the toughening effect on a brittle resin of incorporating rubber were quite direct: the rubber absorbs energy in impact as it stretches across cracks in the brittle matrix (27). Not yet discarded by all (2, 24), the theory has a number of difficulties. Calculations by Newman and Strella (28) show that the stretching mechanism cannot account for more than a small fraction ($\sim$1/10) of the energy absorption in a toughened resin. The theory ignores the high local energy absorption and high elongation actually observed around fractures in all brittle matrix resins. Actually, the impact strength of a rubber-toughened resin can depend greatly on matrix properties; the case of rubber-toughened PVC (toughened by incorporation of a high-rubber ABS resin of the Blendex type) is shown in Figure 2.

Further, electron micrographs such as those of Matsuo (25) indicate that matrix craze bands are as wide as rubber particle diameter. Thus, energy absorption in the rubber during the important craze phase is

Table I. Hydrostatic Tension Calculated from Goodier's Equations (14)

Angle, θ	Value of Hydrostatic Tension at Interface		
	$\mu_2 = 0$ (void)	$\mu_2 = 0.1\mu_1$ ("rubber")	$\mu_2 = \infty$ (hard inclusion)
0°	$-0.45T$	$-0.32T$	$+1.2T$
90°	$+0.73T$	$+0.67T$	$-0.1T$

much less than that in the matrix since rubber secant modulus is much less than that of the matrix, and total stretch is the same for both phases.

Dilatation Theory of Yielding. Newman and Strella (28), recognizing the inadequacy of the direct energy absorption theory, proposed that the rubber particles generated a hydrostatic tensile stress state in the adjacent matrix polymer; hydrostatic tension generates dilatation— *i.e.*, an increase in free volume, and the increase in free volume facilitates yielding, which occurs instead of brittle fracture. In the original paper (28) the source of the hydrostatic pressure was the differential lateral shrinkage expected because the Poisson's ratio of the rubber (1/2) is greater than that of the matrix (1/3). In a later paper, however, Strella (34) based the theory on Goodier's analysis (14) of stresses in a system under simple tension composed of an elastic sphere embedded in an elastic matrix.

With an elementary elaboration, Goodier's solutions for matrix stresses can be expressed as:

$$HT = \frac{T}{3}\left[1 - \frac{5a^3}{2r^3} \times \frac{(1 + \sigma_1)(1 + 3\cos 2\theta)(\mu_1 - \mu_2)}{(7 - 5\sigma_1)\mu_1 + (8 - 10\sigma_1)\mu_2}\right] \quad (2)$$

where HT = hydrostatic tensile stress, T = simple tensile stress applied at $r = \infty$, r = distance from sphere center, a = sphere radius, θ = angle between radius vector r and the direction of T, μ_1 and μ_2 are shear moduli of matrix and sphere, respectively, and σ_1 is the Poisson's ratio of matrix.

We note that σ_2, the Poisson's ratio of the sphere, does not appear at all, so that differential Poisson's contraction does not exist (that is, HT is independent of σ_2). For a homogeneous material, $HT = 0.33\ T$. The magnitude of HT depends on the relationship between μ_1 and μ_2 and on angle θ. Equation 2 has been solved for a number of interesting cases, with results shown in Table I. Table I shows that the elastic inclusion does increase the hydrostatic tension maximum above the $0.33\ T$ level in a homogeneous material, that a void increases hydrostatic tension more than a rubber particle, and that a hard inclusion increases hydrostatic tension more than a void. These results assume adhesion of sphere to matrix.

Recognizing that the extensive art in rubber toughening teaches that only adherent rubbers toughen (nonadherent rubber particles act like voids), we nevertheless tested the above conclusions of the dilatation theory by attempting to toughen styrene acrylonitrile copolymer (S/AN) with dispersions of voids and with dispersions of hard particles.

In a first series of experiments, S/AN (Lustran A20) was foamed under compression using azobisformamide (Celogen AZ) as the foaming agent. Densities obtained ranged from 0.98 to 0.73, and microscopic examination showed a good dispersion of 20–33μ diameter bubbles, typically spaced about 20μ apart (note that Equation 2 implies that no particular void size is required). Izod tests on these void-filled bars showed that impact strength divided by density decreased as void content increased. Secondly, an equivalent "void" experiment using a dispersion of 20 wt % of 0.2–0.3μ fluorocarbon particles (a Teflon T6C dispersion), which are nonadherent, in poly(vinyl chloride) produced no toughening either. Electron microscopy indicated that an excellent dispersion had been obtained.

Table II. Izod Impact Strength of TiO_2-Filled S/AN[a]

TiO_2 Content, vol %	Izod, ft lb/in
0	0.24
4	0.20
10	0.23
17	0.25

[a] ASTM: D256: 23°C.

By milling, completely satisfactory dispersions of 0.3μ TiO_2 pigment particles (Ti-Pure 610) in S/AN (Lustran A21) were made. Electron micrographs show an excellent dispersion with a close resemblance to an ABS resin. This system is ideally suited to testing the dilatation theory. Particle size and distribution (1–3μ between particles) closely approximated an ABS; hence, size effects were well simulated. Since TiO_2 has a very high modulus ($\sim 10^7$ psi), a high hydrostatic tension would be produced if the particles were adherent at high levels of interfacial tension. This point is experimentally inaccessible, so the experiments described below may represent either the hard inclusion case or the void case of Table I. In either case, the hydrostatic tension produced is calculated to be higher than in the case of a rubber particle. Bars molded from the TiO_2 dispersions were brittle in hand bending and were found to be untoughened as measured by Izod impact strength, Table II. Tensile specimens were tested at various temperatures with the results

Table III. Tensile Yield and Ultimate Strength of TiO$_2$-Filled S/AN

Temperature, °C	Behavior, S/AN Alone	S/AN + 17% (vol) TiO$_2$
23	Brittle fracture @ 9300 psig	Brittle fracture @ 8600 psig
60	Yielded @ 7140 psig	Brittle fracture @ 6700 psig
70	Yielded @ 6400 psig	Brittle fracture @ 7000 psig
80	Yielded @ 4880 psig	Yielded @ 6250 psig

shown in Table III; inclusion of hard particles *raised* yield stress and decreased ductility. These effects are the opposite of those predicted by the theory.

If differential Poisson's shrinkage does not account for the enhanced hydrostatic tension, what does? In this case, the increased hydrostatic tension arises because the elastic sphere is, in fact, a kind of a notch (crack), and notches (abrupt changes in cross section) generate hydrostatic tension partly by a stress concentration effect and partly by inhibited matrix Poisson's contraction arising because of matrix cross section change (29). It is appropriate to ask if a rubber particle can generate enough extra dilatation to allow a matrix polymer with a glass transition temperature of, say, 90°C to yield instead of break in a brittle manner at temperatures of 23° to −40°C. In a homogeneous bar of S/AN, T_{max} (breaking stress) is commonly ~9500 psi, and the hydrostatic tension is ~3167 psi. In an ABS resin, T_{max} (yield stress) is ~6000 ± 1000 psi, and the corresponding hydrostatic tensile stress is ~4400 psi (assuming $\mu_2 = 0$ as a first approximation). The bulk modulus of S/AN is ca. 4 × 10^5 psi, so the two dilatations are 0.8 and 1.1%, respectively. Since the cubic coefficient of thermal expansion of S/AN is about 25 × 10^{-5}/°C (26), we calculate that the extra dilatation arising from the presence of the rubber spheres is equivalent to the dilatation which would arise by raising the temperature by 12°C. This extra dilatation appears to be too little to account for ABS toughness at room temperature, much less at −40°C. The above difficulties, together with the muteness of the theory in regard to such important features in toughening art as particle size and interphase adhesion effects, prompted search for a more satisfactory theory of toughening.

Craze Nucleation Theory. In various ways it has been suggested that the role of the rubber particles is that of stress concentrators. Thus, Schmitt and Keskkula (33) believe that the multiplicity of stress concentrators (*i.e.*, a multitude of weak points) produce a large number of small cracks rather than a few large ones; more energy is needed to propagate a large number of small cracks, and stress fields of the various cracks interfere with one another in a favorable way. Matsuo (25) visualizes the rubber particles "as voids or stress-concentrators, thus presenting a site for craze formation under stress." Orowan suggests that the particles nucleate a mass of tiny cracks ahead of the main crack and so produce a "statistical branching" effect (30).

The actual stress distribution around a rubber particle has been calculated for the case of a rubber particle embedded in a rigid matrix, following Goodier (14). Assuming the shear modulus of the matrix, μ_1, is much greater than the shear modulus of the rubber, μ_2, we find:

$$\sigma_{\theta\theta} = T\left[\frac{1}{2} + \frac{A}{r^3} + \frac{B}{r^5} - \left(\frac{1}{2} + \frac{C}{r^3} + \frac{D}{r^5}\right)\cos 2\theta\right] \quad (3)$$

where $\sigma_{\theta\theta}$ is normal stress in the θ direction, T is the simple tensile stress applied to the system, θ is the angle between radius vector r and T, and A, B, C, D are constants. The equation is valid where $r \geq a$, where a is the radius of the rubber particle.

Thermal shrinkage stresses must be superposed on the stress of Equation 3. Assuming an elastic sphere embedded in an infinite elastic matrix, the complete analytical solution for thermal stress is obtained:

$$\sigma_{rr} = -2\sigma_{\theta\theta} = -2\sigma_{\psi\psi} = \frac{-(\alpha_2 - \alpha_1)\Delta T}{\frac{1}{3B_2} + \frac{1+\nu_1}{2E_1}} \quad (4)$$

where r, θ, and ψ are the usual spherical coordinates, σ is tensile stress in the matrix at the sphere surface $r = a$, α is linear coefficient of thermal expansion, B is bulk modulus, E is Young's modulus, ν is Poisson's ratio, and ΔT is temperature change from a reference state. Subscript 1 is for matrix phase and, 2 is for rubber phase. Assuming $\Delta T = -70°C$ (e.g., matrix becomes rigid at 90°C, and stresses develop as it cools to 20°C), $\alpha_1 \approx 8.3 \times 10^{-5}$ (°C)$^{-1}$, $\alpha_2 \approx 22 \times 10^{-5}$ (°C)$^{-1}$, $B_2 \approx 485,000$ psi (26), $E_1 \approx 500,000$ psi, $\nu_1 \approx 0.3$ (28), we calculate that σ_{rr}, the tensile stress across the interface, equals +4500 psi. The rubber droplet is subjected to hydrostatic tension of +4500 psi. These stresses are higher than interphase adhesion levels measured in tension, Table IV. Adhesion measurements of Table IV were made on compression-molded sandwiches of rubber films between sheets of S/AN copolymer. Further, Gent and Lindley (13) found internal fracture in rubbers at critical values of hydrostatic tension ranging from 200 psi for a GRS rubber (Polysar S) to 720 psig for a carbon black-reinforced natural rubber. Clearly, differential thermal contraction in cooling 70°C would produce either partial interphase adhesive failure or rubber fracture or both.

Table IV. Tensile Adhesion

Rubber Interlayer[a]	Tensile Adhesion Stress, psig	
	23°C	*−40°C*
Butadiene (coagulated Pliolite latex)	90	158
Nitrile (extract of commercial ABS)	405 (cohesive failure)	1400–>1700
S/AN-graft-butadiene (extract of commercial ABS)	650	>1200

[a] Matrix layers are Lustran A20 S/AN copolymer.

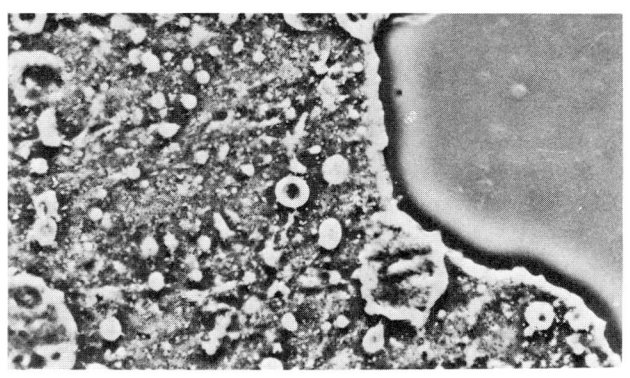

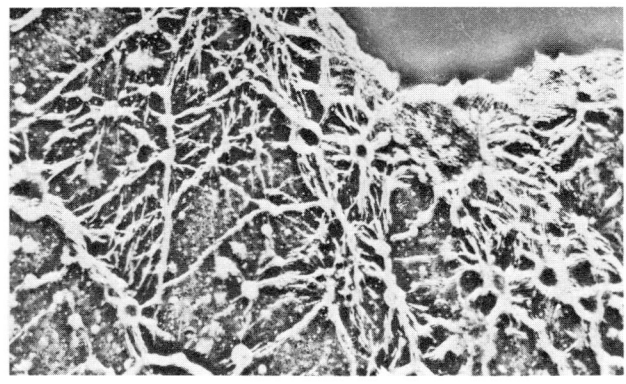

Journal of Applied Polymer Science

Figure 3. Craze traces in a toughened polystyrene resin (ca. 1100 ×) (33)

If we assume that the above processes limit the thermal shrinkage stress, σ_{rr} to 1000 psig and if we superpose thermal stresses on the mechanical stresses obtained by inserting appropriate boundary conditions in Equation 3, we obtain at $r = a$:

$$\sigma_{\theta\theta} = T\,(0.68 - 1.36 \cos 2\theta) - 500 \text{ psig} \quad (5)$$

Since T is of the order of 6000 psig during the yielding of an ABS resin, the 500-psig thermal shrinkage stress is negligible compared with the maxima in $\sigma_{\theta\theta}$.

Because of the 2θ function in Equation 5, tensile stress in the matrix resin at the phase boundary is a sharp function of angle, stress falling to zero on planes inclined at only at 30° angle to the maximum stress plane, then going into the compression region. This is an important point since no sharp dependence of cracking or crazing can be seen in micrographs of rubber-toughened resins (Figures 1 and 3); indeed, crazes are seen on the zero-stress and compression planes, which is unlikely to occur in the quasistatic propagation case.

Further, the Goodier equations predict that hard particles and voids produce higher stress concentrations (*i.e.*, stronger craze nucleation) than rubbers, and thus hard particles and voids should toughen even better than rubbers if nucleation were the operative mechanism. This is not observed experimentally. The nucleation theory is thus seen to have substantial drawbacks.

Craze/Crack Branching Theory of Toughening

As discussed above, even "brittle" glassy polymers exhibit great ductility and high energy absorption in thin layers. The central question in rubber toughening is why the energy-absorbing volume, a consequence of crack and/or craze proliferation, is so large. The theory is advanced that the mechanism for crack/craze proliferation is dynamic branching at the rubber particles by means of the Yoffe effect (35).

Crack/Craze Dynamics. The displacement relationships for particles in a crazed solid are essentially the same as in a cracked solid in the region outside the discontinuity itself except perhaps for details of tip shape (*19, 22*), and, hence, the spatial distribution of elastically stored potential energy around the two types of discontinuity is the same. Apparent surface energy for a crack is essentially that for the preceding craze deformation process (*19*) which is thought to be the origin of oriented low-density surface layers at crack surfaces. Crazes may propagate without simultaneous crack formation, of course; in tensile tests on S/AN bars (Lustran A-21), extensive crazing occurred at 1.8–2.0% elongation, whereas catastrophic cracking occurs at 2.4–2.5% elongation.

The energetic relations for growing crazes are much like those for growing cracks since the additional energy required to produce free surface is small compared with energy absorbed during the crazing phase (5).

A dynamic theory of crack propagation has been developed in detail by Griffith (15, 16), Roberts and Wells (32), Cotterell (10), and Yoffe (35). The basic works of Griffith and Yoffe refer to crack propagation but are directly applicable to craze propagation because the strain energy pattern in elastically deformed matrix material is essentially the same in both cases; differences in tip configuration such as those discussed by Kambour (18) do not change the analyses substantially. The Griffith theory of strength (15, 16) teaches that in a homogeneous solid, cracks become unstable and accelerate when crack length exceeds the critical length defined by Equation 1. The crack then accelerates to a limiting velocity which is in the neighborhood of half the shear wave velocity (8, 10, 32). As cracks approach their limiting speed, crack surface becomes rough, and crack branching occurs. This was explained in 1951 by Yoffe (35), who calculated stresses ahead of a moving crack and found that maximum stress swings out of the plane of the crack at high speeds. Her calculations are summarized in Figure 4. Tensile stress maxima exist in two planes ahead of a rapidly moving crack, so the crack deviates onto one or both (branching) of the newly preferred planes.

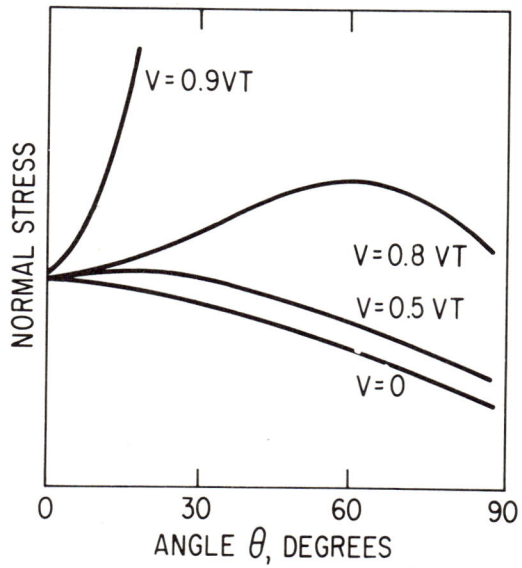

The Philosophical Magazine

Figure 4. Effect of crack/craze speed on stress distribution (35)

Cotterell (10) showed that the planes of maximum tensile stress were also principal stress planes and that crack propagation along these planes releases maximum strain energy (*i.e.*, is probabilistically favored).

In an isotropic medium, cracks do not move faster than half the shear wave velocity V_t, so the implications of the $0.8V_t$ curve in Figure 4 were not explored. In the two-phase ABS system, however, one can imagine cracks or crazes propagating rapidly in the matrix ($V_t/2 \sim 620$ meters/sec), and thence into the rubber particle [at 23°C, polybutadiene ($V_t/2 \sim 29$ meters/sec)] where violent branching would occur.

It is appropriate to ask how far a crack in the matrix resin must propagate before it accelerates to a speed permitting branching in the rubber phase. Dulaney and Brace (11) and Berry (4) have developed the following expression for crack velocity V:

$$v^2 = \frac{2\pi E}{kd}\left(1 - \frac{Co}{C}\right)\left(1 - [n-1]\frac{Co}{C}\right) \tag{6}$$

where E = Young's modulus, d = density, k = constant, Co = Griffith crack length, C = crack length at speed V, $n = 2\frac{T_g^2}{T_f^2}$ where T_g is critical Griffith stress and T_f = observed fracture stress. We note that $V = 0$ when $C = Co$, and $T_g = T_f$ ($T_g = T_f$ for unnotched specimens—*i.e.*, those containing only the intrinsic Griffith cracks). However, $V = V_t/2$ when $C = \infty$ and $T_g = T_f$. Thus, Equation 6 becomes $V = \frac{V_t}{2}$ $\left(1 - \frac{Co}{C}\right)$, as the crack starts to accelerate. If we rewrite $C = Co + \Delta$, where Δ is the acceleration distance, we obtain:

$$\frac{\Delta}{Co} = \frac{2}{\left(\frac{V_t}{V}\right) - 2} \tag{7}$$

If $V_t \approx 1240$ meters/sec in the matrix and branching will occur in the rubber at 29 meters/sec, we calculate $\Delta/Co = 0.047$. Thus, branching can occur after a matrix crack acceleration distance of only 2 to 5μ (assuming a Griffith crack length of 50–100μ); hence, ample room for the development of "fast" cracks or "fast" crazes exists in the ABS structure. Note that the expressions for craze instability, acceleration, and speed (Equations 1, 6, 7) show that the macro strain rate of the specimen is irrelevant—"fast" cracks and crazes propagate in specimens strained even at slow creep rates.

In addition to the rapid propagation of mature crazes discussed above, there is a period of extremely rapid propagation at craze initiation, as discussed by Knight (22).

Experimental Evidence. MORPHOLOGY. Figure 3 (33) shows in phase contrast microscopy the development of crack or craze patterns around rubber particles in a toughened polystyrene. The lack of dependence of crack inclination on direction of stress is especially marked in this micrograph, and can be explained only by reference to dynamic branching rather than to crack or craze nucleation by stress raisers. Schmitt and Keskkula refer to the lines as "craze cracks" and "cracks."

Figure 1 shows electron micrographs of toughened polystyrene and ABS resins (25). Crack or craze branching at rubber particles is readily apparent. Matsuo refers to the traces as "branched crazes." Once again, crazes can be seen to have propagated on planes which are stress free in the static case. Matsuo comments, "in contrast with the straight crazes observed in PS and PMMA" (20), however ,"branching" is noted "in the toughened PS and ABS."

STRAIN AND STRAIN RATE BEHAVIOR. It is clear from the theories of toughening hitherto proposed that thinking started from consideration of the ABS yield stress, which is lower than the S/AN ultimate strength. The consideration of strain is informative. Bars of S/AN (injection-molded Lustran A21) were pulled in tension at room temperature; crazing began at 1.8–2.0% elongation, and fracture occurred at 2.3–2.4% elongation. Bars of an ABS (injection-molded Cycolac CG: 41% insolubles in acetone, 29% butadiene), were strained similarly; the onset of yielding was seen at 2.5% strain, and blushing was observed to occur at an elongation between 2 and 2.5%. No change in the slope of the stress–strain curve was noted in the 1.8–2.0% region of the ABS $\sigma-\epsilon$ curve. Thus, the blush and yield onset in the ABS were found at an elongation greater than that (29) which initiates crazing in unmodified S/AN.

Notching is known to raise yield stress through lateral constraint effects (29); thus theories of toughening which are based on mechanisms whereby matrix resin yield stress is lowered would predict that notching should suppress the blushing which marks yielding of toughened resins. No such effect would be predicted by the branching theory of toughening. Tensile experiments on ABS bars (Cycolac H ABS resin), made with notches cut by sharp razor edges, showed intensive blushing right to the edge of the razor cut, indicating that blushing is insensitive to sharp notching.

SONIC MODULUS. If crack or craze branching is the operative mechanism in toughening, toughness should be directly related to the difference in sonic speeds in matrix and dispersed phases. Experiments to confirm this effect were undertaken using three commercial ABS resins. These were selected to represent the three main rubber types encountered commercially: an acrylonitrile/butadiene copolymer rubber, a butadiene rubber with grafted styrene/acrylonitrile copolymer, and a block polymer of

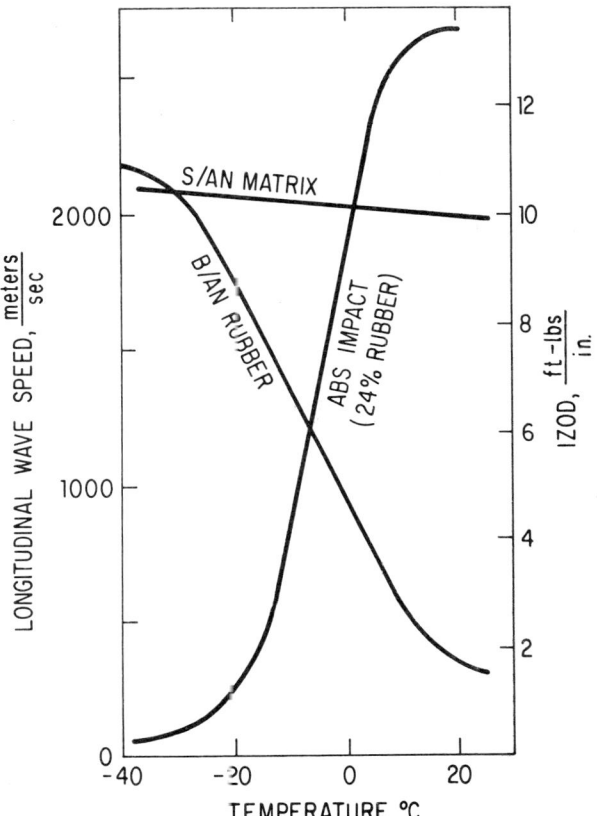

Figure 5. Relation between wave speeds and toughening by a nitrile rubber

butadiene and styrene/acrylonitrile copolymer. Izod impact strengths were measured at various temperatures on molded bars. Sonic speed–temperature data were measured on commercial styrene/acrylonitrile copolymers closely approximating typical ABS matrix copolymer, and on rubber-phase material isolated from the three ABS resins by dissolving the matrix resin in acetone. "Rubber" contents of the three resins were assigned by percentage insoluble in acetone.

The sonic speed used was longitudinal wave speed, measured using a pulse propagation meter (Model 4, H. M. Morgan Co., Cambridge, Mass.). Pulse frequency was 8–10 kHz. Transverse wave velocity was calculated from longitudinal wave velocity using the approximation

$$\frac{V_t}{V_e} = \sqrt{\frac{G}{E}} = \sqrt{\frac{1}{2(1+\nu)}}$$

where V is wave velocity, G is shear modulus, E is Young's (tensile) modulus, ν = Poisson's ratio. Since ν for polymers ranges from 0.3 in the glassy state to 0.48 in the rubbery state, $V_t/V_e = 0.6 \pm 0.02$.

Test results, plotted in Figures 5, 6, and 7, show that impact strengths vary with difference in sonic speed in the way predicted by the crack/craze branching theory.

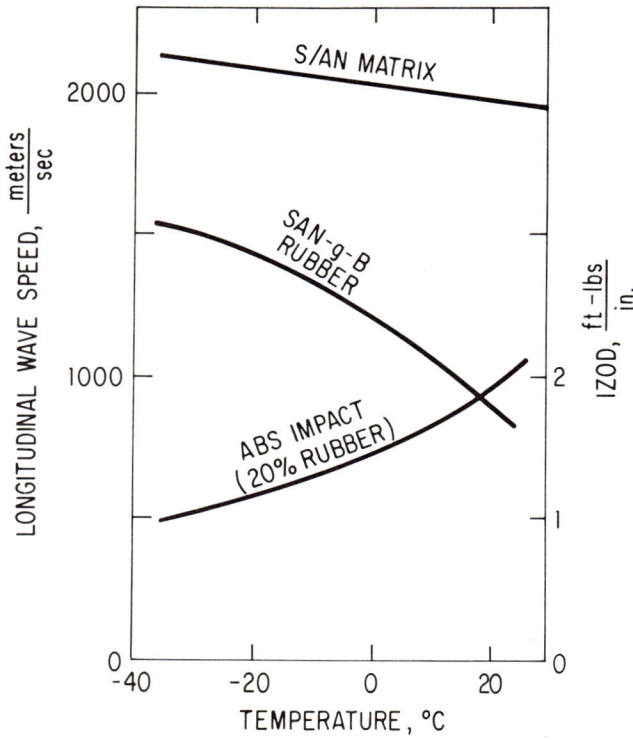

Figure 6. Relation between wave speeds and toughening by a grafted butadiene rubber

RUBBER PARTICLE SIZE AND SHAPE. If rubber particles act as crack or craze branch points along an advancing crack in matrix polymer, impact strength should depend on the frequency with which branch points are encountered. If C = rubber phase volume fraction, N = number of dispersed particles, and d = average particle diameter, $N \sim C \div d^3$. N is maximized as C increases or d decreases. The probability of an advancing crack hitting a particle as it advances an incremental distance is proportional to cross sectional area Nd^2, which equals C/d. Again, C

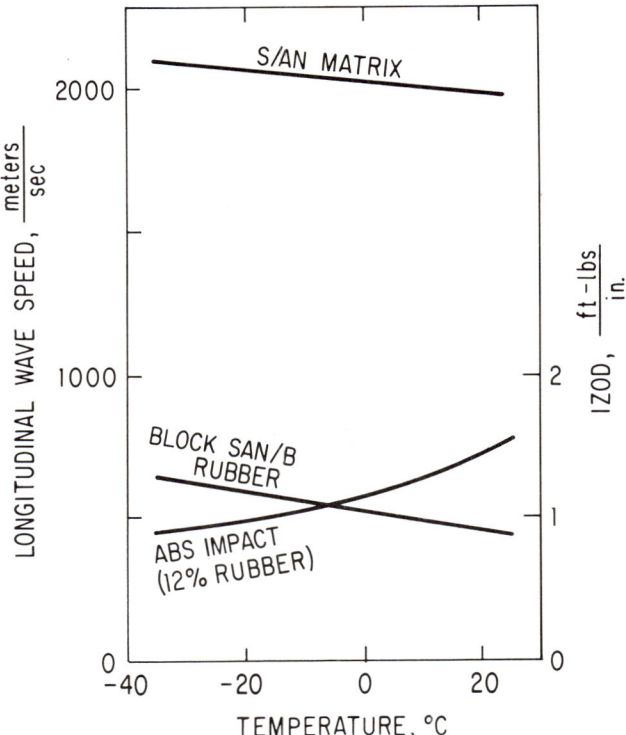

Figure 7. Relation between wave speeds and toughening by a SAN/butadiene block copolymer rubber

should be large and d small. It is difficult to set a lower limit on d on à priori grounds. A separate phase of rubber is needed because branching is a consequence of dynamic elastic wave propagation through a finite space—one molecule won't do—so d must be > 1000 A, say. Now, toughened S/AN art teaches that the preferred rubber particle size is 0.1–1.0μ (3), whereas in toughened polystyrene, the preferred particle size is 1–25μ [Angier et al. state that commercial materials seem to utilize 1–10μ particles in general (1)]. How can this be understood in view of the similarities between polystyrene and S/AN in modulus, strength, elongation, specific heat, etc.? Kambour (18) has measured depth of "craze" material at crack surfaces by interferometric techniques. He finds that the disturbed depth was ~0.7μ in S/AN (Tyril 767, 30% AN) and 0.9–2.8μ in polystyrene. We found 0.5–0.6μ crazes in Lustran A20 S/AN, and ~2μ crazes in Styron 678 polystyrene. The crude correspondence of craze layer thickness and preferred rubber particle size may be explained as follows. If rubber particle size is small compared with thickness of the craze layer [which may or may not be followed closely by a crack (*19*)] or craze tip radius, the perturbation in the elastic strain field introduced by the particle is too small to cause deviation of the craze. Thus, minimum effective particle size should be of the same order as the craze thickness characteristic of the matrix polymer. Experimental evidence on this point indicates that there is indeed a very sharp transition in impact strength at this limit. In Lustran A20 polystyrene/acrylonitrile, we find craze thicknesses of 0.5–0.6μ. By appropriate choice of mixing process conditions (shear rate, etc.), dispersions of 20% polybutadiene/acrylonitrile rubber (Hycar 1043) in poly S/AN (Lustran A20) can be produced with rubber particle sizes just above and below the 0.5–0.6μ level. Impact data on injection-molded bars (Table V) show that there is a sharp transition in behavior at the predicted level. Further, the data of Table V pertinent to samples having elongated rubber particles indicate that in these samples, toughening occurred whenever one particle dimension typically exceeded 0.6μ. Elongated particles gave rise to substantial scatter in impact strength.

Table V. Effect of Rubber Particle Size and Shape in Impact: Poly S/AN + 20 wt % Poly B/AN

Rubber Particles	Izod Impact Strength		
	Average ft lb/in	Std. Deviation ft lb/in	Range ft lb/in
Spheroidal, 0.5–3.0μ	3.0	0.2	3.5–2.8
Spheroidal, 0.3–0.5μ	0.5	0.1	
Spheroidal, 0.5μ	0.5	0.1	
Elongated, 0.2 × 1.0μ	0.9	0.6	2.0–0.4
Very Elongated, 1.0 × >3.0μ	3.3	1.5	5.3–1.5
Elongated, 0.3 × 1.0μ	2.2	1.6	4.8–0.6
Elongated, 0.1 × 1.0μ	1.1	0.7	2.2–0.6

Profusion of branching should be proportional to number of rubber particles greater in size than the minimum discussed above. At a given rubber content, the number of rubber particles varies as the reciprocal of the third power of particle diameter. Thus, number of particles drops rapidly as particle size climbs above the effective minimum. Laboratory tests show that stiffness properties depend on total rubber content irrespective of particle size (provided the specimen dimensions are large compared with particle dimensions); hence, narrow particle size distribution is essential if maximum toughness is to be combined with minimum loss in stiffness properties (modulus, creep).

INTERPHASE ADHESION. Adhesion between phases must be excellent if a crack or craze traveling in the matrix is to propagate into the dispersed phase instead of traveling around the particle as though a void were present. It is felt that the poor adhesion between S/AN and ungrafted polybutadiene rubber (Table IV) explains the poor toughening obtained using this rubber (*12*). On the other hand, nitrile rubber and S/AN-grafted butadiene rubber exhibit excellent adhesion, and toughen well (Figures 5 and 6).

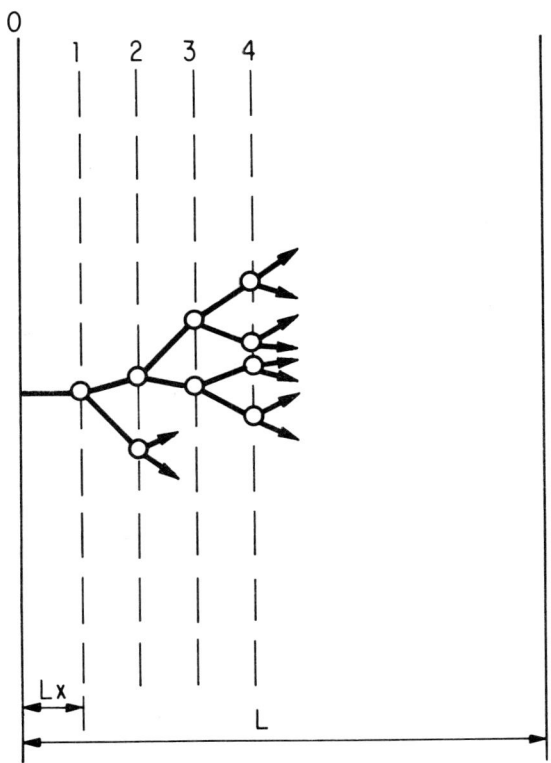

Figure 8. *Craze branching model*

RUBBER CONTENT. In the theories of toughening where the role of rubber particles is (a) to absorb energy directly or (b) to induce matrix yielding through stress concentration or hydrostatic tension effects, energy absorption should increase linearly with the number of rubber particles (proportional to rubber content if particle size is invariant). On the other hand, if dynamic craze/crack branching is the operative mechanism, evidence of an exponential law may be expected. The exponential form of the law may be derived as follows.

Assume a regular, three-dimensional network of rubber particles, spaced on planes which are L_x distance apart (Figure 8), where

$$L_x = \frac{N_x}{L}$$

N_x being number of rubber particles along an X-axis and L being specimen thickness. Total number of rubber balls $= N_x N_y N_z$. The number of cracks leaving each plane $= 2^n$, where $n =$ plane number. In a two-dimensional situation, the total number of cracks across the specimen is

$$\sum_0^{N_x - 1} 2^n$$

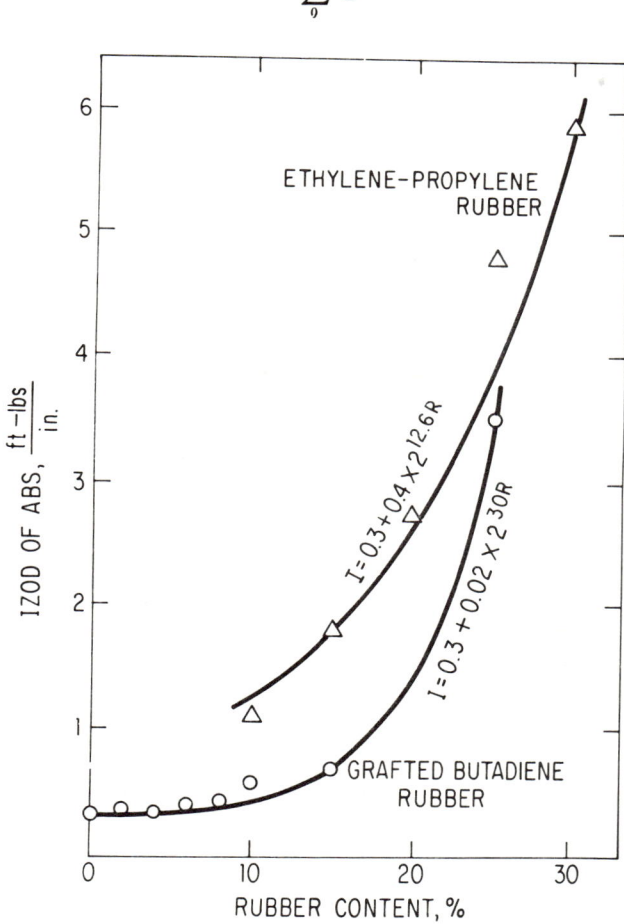

Figure 9. *Effect of rubber content on toughness of rubber-toughened resins*

Length of each crack is of the order of $L_x = \dfrac{L}{N_x}$, $N_x \geqslant 1$.

Per unit of specimen width, total crack/craze area is

$$A_c = \frac{L}{N_x} \sum_0^{N_x - 1} 2^n, \; N_x \geqslant 1 \qquad (8)$$

If we postulate that the energy required to break a specimen is simply surface energy (neglecting kinetic and elastic strain energies), then

$$\text{Impact Energy} \approx 2\gamma A_c \qquad (9)$$

where γ is the effective surface energy associated with a unit area of crazed or cracked surface. Combining Equations 8 and 9, impact strength, to a first approximation, is predicted to be an exponential with base 2. This crude derivation has neglected important processes such as cooperative stress relief, craze recombination, failure to branch at every particle, etc.

Experimental data on the effect of rubber content on impact strength are shown in Figure 9. Data are for specimens containing two different types of rubber. Base 2 exponentials fit the data well; we have been unable to find nonexponential functions which fit as well.

Summary

Existing studies of the fracture of apparently brittle plastics such as polystyrene or styrene/acrylonitrile copolymer indicate that the polymers are actually enormously tough in very thin layers near crack or craze interfaces. Because the energy-absorbing volume is so low in these polymers, their engineering usefulness has been limited. The technology of ABS and impact polystyrene resins shows that incorporation of a dispersed rubber phase of appropriate properties can raise impact toughness by an order of magnitude, apparently by increasing the energy-absorbing volume in the resin. Principal existing theories of the toughening of brittle plastics by rubber dispersion all appear quite limited in ability to explain known phenomena and have substantial difficulties. It is proposed that the mechanism whereby rubber dispersions greatly increase energy-absorbing volume is one which causes cracks and/or crazes to branch dynamically at rubber sites through the Yoffe mechanism (35). This theory explains existing morphological information (25, 33), strain and strain rates effects, need for interphase adhesion, and dependence of toughness on rubber particle size and size distribution, rubber content and temperature.

Acknowledgments

The author thanks H. A. Davis and R. P. Schatz for optical and electron micrographic measurements, and O. J. Cope for assistance in chemical and physical characterization of rubbers and resins.

Literature Cited

(1) Angier, D. J., Fettes, E. M., *Rubber Chem. Technol.* **1965**, 38, 1164.
(2) Basdekis, C. H., "ABS Plastics," p. 57, Reinhold, New York, 1964.
(3) Belgium Patent **646,258**.
(4) Berry, J. P., *J. Mech. Phys. Solids* **1960**, 8, 194.
(5) Berry, J. P., *J. Polymer Sci.* **1961**, 50, 107.
(6) Berry, J. P., *Nature* **1960**, 185, 91.
(7) Bucknall, C. B., Smith, R. R., *Polymer* **1965**, 6, 437.
(8) Bueche, A. M., White, A. V., *J. Appl. Phys.* **1956**, 27, 980, 9.
(9) Cotterell, B., *Appl. Mater. Res.* **1965**, 4 (4), 227.
(10) Cotterell, B., *Intern. J. Fracture Mech.* **1965**, 1, 96.
(11) Dulaney, E. N., Brace, W. F., *J. Appl. Phys.* **1960**, 31, 2233.
(12) Frazer, W. J., *Chem. Ind.* **Aug. 13, 1966**, 33, 1399.
(13) Gent, A. N., Lindley, P. B., *Proc. Roy. Soc. (London) Ser. A* **1959**, 249 195.
(14) Goodier, J. N., *Trans. ASME* **1933**, 55, A39, p. 39.
(15) Griffith, A. A., *Trans. Roy. Soc. A* **1921**, 221, 163.
(16) Griffith, A. A., *Proc. Intern. Congr. Appl. Mech., 1st,* **1924**, p. 55.
(17) Haward, R. N., Mann, J., *Proc. Roy. Soc. (London) Ser. A* **1964**, 282, 120.
(18) Kambour, R. P., *J. Polymer Sci. A2* **1966**, 4, 17.
(19) *Ibid.*, p. 349.
(20) *Ibid.*, p. 359.
(21) Kambour, R. P., *Polymer* **1964**, 5, 143.
(22) Knight, A. C., *J. Polymer Sci. A* **1965**, 3, 1845.
(23) Mann, J., Bird, R. J., Rooney, G., *Makromol. Chem.* **1966**, 90, 207.
(24) Martin, J. R., *Antec Tech. Papers (SPE)* **1966**, XII, Paper XXV-1.
(25) Matsuo, M., *Polymer* **1966**, 7, 421.
(26) McPherson, A. T., Klemin, A., "Engineering Uses of Rubber," p. 74, Reinhold, New York, 1956.
(27) Merz, E. H., Claver, G. C., Baer, M., *J. Polymer Sci.* **1956**, 22, 325.
(28) Newman, S., Strella, S., *J. Appl. Polymer Sci.* **1965**, 9, 2297.
(29) Orowan, E., *Repts. Progr. Phys.* **1949**, XII, 185.
(30) Orowan, E., personal communication, 1966.
(31) Ritchie, P. D., "Physics of Plastics," p. 126, Van Nostrand, Princeton, 1965.
(32) Roberts, D. K., Wells, A. A., *Engineering* **1954**, 178, 820.
(33) Schmitt, J. A., Keskkula, H., *J. Appl. Polymer Sci.* **1960**, III, 132.
(34) Strella, S., *J. Polymer Sci. A2* **1966**, 4, 527.
(35) Yoffe, E. H., *Phil. Mag.* **1951**, 42, 739.

RECEIVED October 31, 1969.

Permeability and Phase Structure of PVC/EVA Systems

HELGE STORSTRÖM[1] and BENGT RÅNBY

Department of Polymer Technology, Royal Institute of Technology, Stockholm, Sweden

> *Thin films of poly(vinyl chloride) (PVC) containing various amounts of an ethylene vinyl acetate (EVA) copolymer (55% ethylene by weight) were prepared from mechanical blends and graft copolymerized blends. The film properties were studied by gas permeability measurements with helium, argon, and carbon dioxide as penetrants and by light transmission measurements. Addition of small amounts of EVA ($\leqslant$ 5% by weight) give negligible changes in permeability and opacity. With larger amounts of EVA added, both permeability and opacity of the films increased sharply. The data indicate that addition of more than 5 wt % EVA increases the heterogeneity of the system, probably because of the development of a continuous EVA phase, thus enclosing (embedding) the PVC particles. This phase structure model is supported by published electron micrographs of thin sections of polymer blends. Permeability measurements have proved valuable in studies of multicomponent polymer systems.*

The technological properties and the commercial application of several polymer blends have been studied extensively. Investigations of the basic principles, however, relating the phase structure of the blends to the properties of the individual components have not been carried out to an extent justified by the industrial value of these materials. Several methods have been used, the most successful being optical and electron microscopy and dynamic-mechanical measurements. Critical factors and difficulties in the morphological studies of polymer blends have been

[1] Present address: AB Casco, S-10061 Stockholm 11, Sweden.

discussed by Holliday et al. (6, 7, 8), Rosen (16), and Scott (17). A combination of different methods appears necessary for a successful approach to these problems. Bergen (4) compared graft copolymers and physical blends and found that differences in dynamic-mechanical properties could be related to the morphology of the systems. Kollinsky and Markert (11) concluded from studies of methyl methacrylate–butyl acrylate copolymer blends that electron microscopy and turbidity measurements were more sensitive methods than torsional vibration experiments for detecting heterogeneities in these polymer systems. Matsuo et al. (12) studied several polymer blends using electron microscopy and a staining technique developed by Kato (10). PVC–acrylonitrile–butadiene copolymer (NBR) blends showed increasing compatibility with increasing AN content of the rubber component. In some blends the NBR appeared to be the dispersing (continuous) phase, even at concentrations as low as 5 wt % NBR in the blend. Matsuo (13) also studied PVC–EVA blends, using the same EM techniques as applied previously (12), and they found that EVA is most likely the dispersing phase.

Permeability measurements for polymer blends prepared by mixing different latices have been reported by Peterson (14). Interpreting transport data for such heterogeneous systems as polymer blends is extremely difficult, however (3, 9, 15). The main purpose of the present investigation is, therefore, to study the applicability of gas permeation measurements to characterize polymer blends and not to evaluate the different theoretical models for the permeation process in heterogeneous polymer systems.

Experimental

Materials. The polymer samples were prepared either by physical blending of PVC with an EVA copolymer or by suspension polymerization of vinyl chloride with the EVA copolymer dissolved in the monomer. The resulting dried polymer samples were blended with additives (organic Ba–Cd stabilizer and Pb–stearate) to the following general composition:

> PVC + EVA   100 parts
>
> Additives   3 parts

Blending was carried out on a steam-heated two-roll mill. PVC and the additives were first milled together until a continuous slab was formed, then the EVA was milled into the blend. The total milling time in all experiments was 15 minutes, and the roll mill temperature was varied (160°, 170°, and 180°C).

The samples prepared by suspension polymerization were mixed with the additives and treated on the two-roll mill the same as the mechanical blends.

PVC, suspension polymerized, had the following properties: K value $= 65$, $M_w = 74.000$, density $d = 1.39$ grams/ml, and refractive index $n = 1.5422$ at 20° C. Permeability at 50°C:

$$P_{He} = 4.5 \times 10^{-10} \text{ (ml, STP)(cm)(cm)}^{-2}\text{(sec)}^{-1}\text{(cm Hg)}^{-1}$$

and

$$P_{CO_2} = 0.32 \times 10^{-10} \text{ (ml, STP)(cm)(cm)}^{-2}\text{(sec)}^{-1}\text{(cm Hg)}^{-1}$$

The EVA copolymer of 55 wt % ethylene and 45 wt % vinyl acetate had the following properties: $(\eta_{sp}/c) = 3.22$ dl/gram in toluene solution, corresponding to a Mooney viscosity of about 41, and $M_w = 170.000$. Density $d = 0.985$ gram/ml, and refractive index $n = 1.4739$ at 20°C. Permeability at 50°C (for films cast from toluene solution):

$$P_{He} = 52 \times 10^{-10} \text{ (ml, STP)(cm)(cm)}^{-2}\text{(sec)}^{-1}\text{(cm Hg)}^{-1}$$

and

$$P_{CO_2} = 211 \times 10^{-10} \text{ (ml, STP)(cm)(cm)}^{-2}\text{(sec)}^{-1}\text{(cm Hg)}^{-1}$$

Grafted samples were prepared by dissolving the EVA copolymer in vinyl chloride, followed by suspension polymerization. In moderately grafted samples, some pure EVA remained. In strongly grafted samples no pure EVA could be extracted.

The films for permeability and light transmission measurements were prepared as follows: sheets were pressed for 5 minutes at 170°C, then pressed further between polished chromium-coated plates, 7×7 cm² with total deviation of less than 0.01 mm. The sheets were heated with contact pressure for 5 minutes, then a pressure of 540 kg/cm² was applied, and heating was stopped. The press temperature decreased to about 70°C in 3 hours with the pressure maintained.

Measurements. PERMEABILITY. Permeability measurements were made using a Barrer-type (2, 18) vacuum apparatus for studying the permeation of helium, argon, and carbon dioxide. The permeability cell was sealed with mercury and thermostated in an aluminum block. For degassing the cells were kept under vacuum overnight ($p = 10^{-5}$ mm Hg) at the highest temperature to be used. The pressure was recorded on one side of the membrane using a differential pressure meter (model 306-2) with a membrane pressure sensor (model 306 B-2A, The Decker Corp., Bala Cynwyd, Pa., USA).

DIFFUSIVITY. Diffusivity values derived from time lag measurements of permeation ($D = l^2/6\tau$) vs. membrane composition of PVC/EVA show the same general tendency as the permeability–composition data. The D values are, however, less reproducible, probably owing to short time lag values, and therefore are not reported here. Solubility values can also be calculated ($S = P/D$), but they are not reported for the same reason. No separate solubility measurements were made.

LIGHT SCATTERING. Light scattering measurements were carried out using a Beckman DK-2A ratio recording spectrophotometer. Transmission was measured at 435.8 nm (mμ) and recorded as turbidity ($= \log (I_o/I)/l$ cm^{-1}), where l is film thickness.

In this work a system of two-component polymer blends with a composition range much wider than those of practical importance was studied to obtain basic information about the transport process in the blends in relation to their morphology. Within the composition region of practical importance, studies were also made of the effect of different variables of sample preparation which influence the heterogeneity of the blends.

In the system chosen for study, the EVA component has a much greater permeability coefficient than PVC. For helium the ratio P_{EVA}/P_{PVC} is 11–12 at 50°C; for argon the ratio is 180; for carbon dioxide it is about 700. This indicates that the greatest resistance against permeation in a blend will be in the PVC component. The data also indicate that measurements of P_{CO_2}/P_{He} as a function of EVA content can give valuable information about the morphology of the blends assuming that their phase structure is heterogeneous. The observed values for the ratio P_{CO_2}/P_{He} at 50°C are 0.07 for PVC and 4 for EVA.

Discussion

Polymers Mixed by Milling. The effect of EVA concentration in the blends on gas permeation and light transmission through the film was studied. The permeability and the diffusion coefficients at 50°C for the penetrants helium, argon, and carbon dioxide are shown in Figures 1, 2,

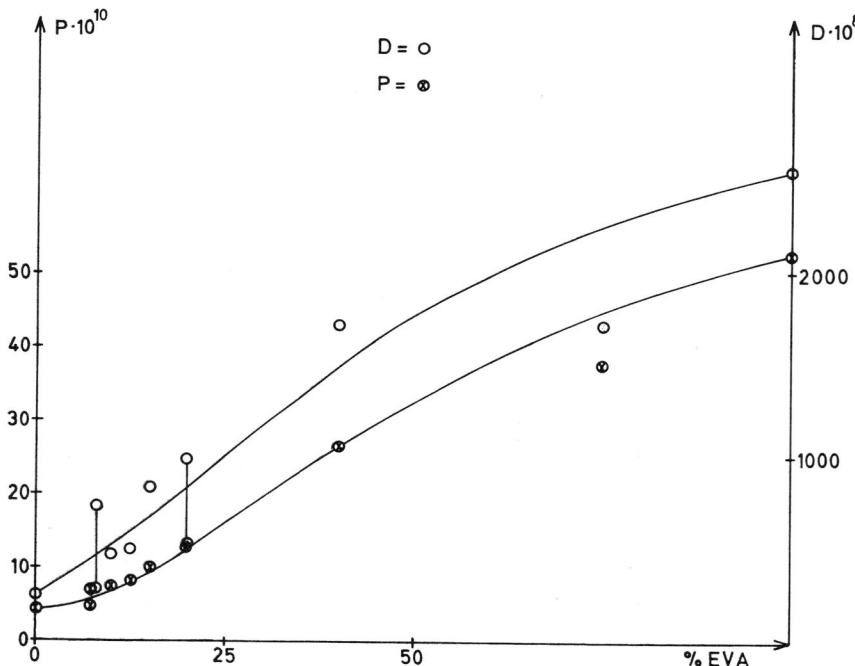

Figure 1. Effect of EVA content in physical blends of PVC/EVA on permeability and diffusion coefficients (P and D) at 50°C for helium. P in (ml STP)-$(cm)(cm)^{-2}(sec)^{-1}(cm\ Hg)^{-1}$, D in $(cm)^2(sec)^{-1}$

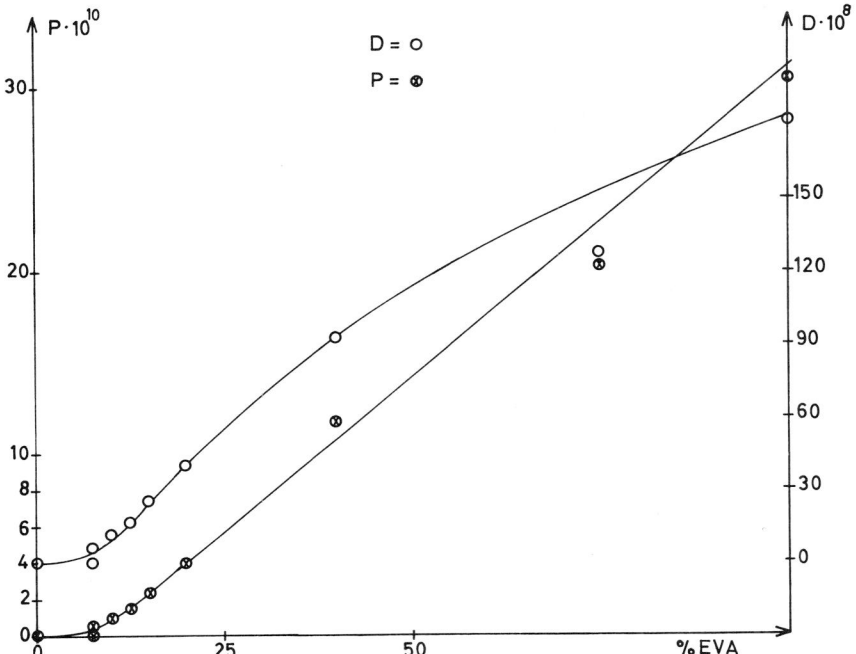

Figure 2. Effect of EVA content in physical blends of PVC/EVA on permeability and diffusion coefficients (P and D) at 50°C for argon

and 3. The ratio P_{CO_2}/P_{He} vs. blend composition for measurements at 50°C is given in Figure 4.

When a more permeable polymer is blended with PVC, the permeability coefficient is expected to increase. The magnitude of the increase, however, will depend on the morphology of the blend. A laminar composite film containing 10% EVA would give an increase in permeability of about 10% compared with a pure PVC film (*15, 18*). The reported measurements for PVC with 10% EVA give an increase of about 400% (Figures 1, 2, and 3).

Considering heterogeneous models for the film structure, we realize that if PVC with its low permeability were the continuous phase, there should only be small increases in permeability with the addition of EVA polymer. Such effects have been observed for a system of butadiene-based polymer modifier added to PVC to increase the impact strength (*1*). Addition of 15% modifier increased the permeability less than 10%. Electron micrographs of this film showed that the butadiene-based modifier was dispersed in the PVC phase.

On the other hand, if PVC particles were dispersed in a matrix of a highly permeable polymer like EVA, we may assume that after a sufficient "wall" thickness of the EVA matrix is reached the PVC phase would not contribute significantly to the permeability. The effect of the PVC particles would then mainly be to give the permeating molecules a more tortuous diffusion path. In this case, the net effect of added EVA could be a sharp increase in permeability. For addition of more than 5% EVA this was indeed observed. The spread in the permeability data in the region with 7–8% EVA in the blends is considerable (Figures 1 and

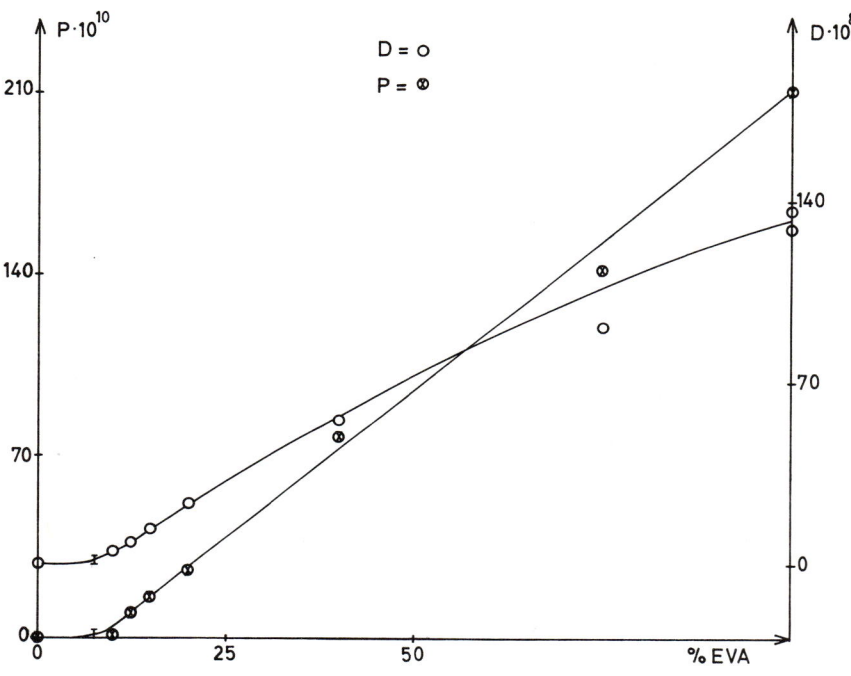

Figure 3. Effect of EVA content in physical blends of PVC/EVA on permeability and diffusion coefficients (P and D) at 50°C for carbon dioxide

2). In the same region there is a sharp increase both in the turbidity of the blends (Figure 5) and in the permeability ratio P_{CO_2}/P_{He} (Figure 4). These results indicate that the formation of a continuous EVA phase is critical in this region. The permeability data show that the EVA component most likely becomes the continuous phase in this region. Wilkes and Marchessault (20) have shown electron micrographs of poly(vinyl acetate) (PVAc) films formed from latex containing poly(vinyl alcohol) (PVA). In their case, 3% PVA was enough to give a coherent PVA phase with honeycomb structure, enclosing the PVAc latex particles. Matsuo et al. (12) have also obtained similar results for PVC–rubber blends using electron microscopy. Therefore, it seems reasonable to as-

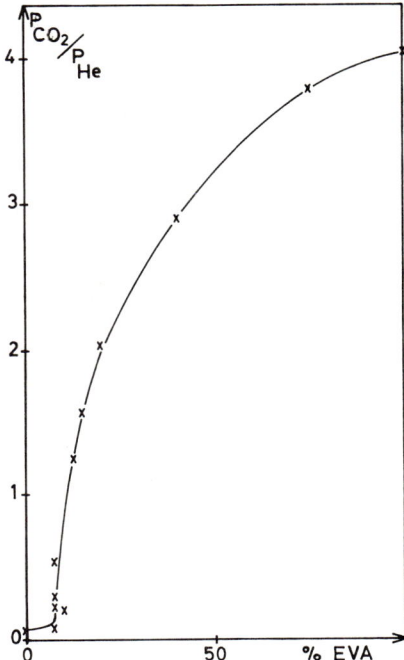

Figure 4. Dependence of the permeability ratio (P_{CO_2}/P_{HE}) at 50°C on the blend composition for physical blends of PVC/EVA

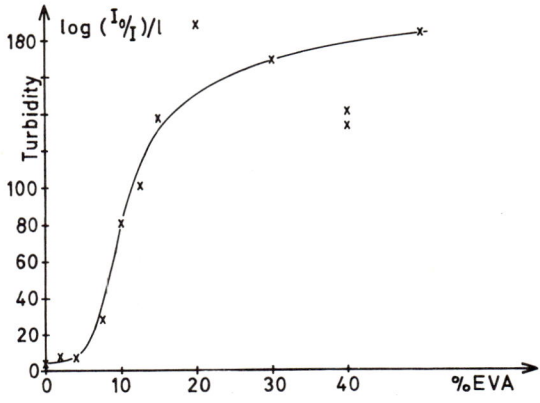

Figure 5. Effect on turbidity of EVA content in physical blends of PVC/EVA

sume in our case that a continuous EVA phase, enclosing the PVC particles, starts to develop at an EVA content of 7–8% in the PVC–EVA blends.

The phase structure of PVC–EVA blends at very low EVA contents (< 5%) is not resolved in these studies. The two polymers are probably not compatible on a molecular level. In any case the solubility should be very low. It seems more likely that the initial amounts of EVA added are accommodated in the PVC component as a dispersed nonhomogeneous phase, filling crevices and voids inside, on the surface, and between the PVC particles. In this way, the EVA polymer could be present in the blends without causing measurable changes in permeability and opacity. When more EVA is added to the blends—e.g., to 7.5% EVA content— and the samples are milled at 160°C for 15 minutes, widely scattered permeability values are obtained. The variation in the measured P_{CO_2}/P_{He} ratios is six-fold for different samples. These results are well in line with the data in Figure 4, which indicate that measurements in the EVA concentration region of 7–8% are difficult to reproduce, owing to small but inevitable variations in the sample properties. It is assumed that a coherent EVA phase begins to form in this region. Mechanical measurements for PVC–EVA blends (*19*) have also indicated that the processing conditions are critical in this concentration region.

Polymers Mixed by Grafting and Milling. MILLING TEMPERATURE. Moderately grafted samples containing 7.5% EVA were milled for 15 minutes at 160°, 170°, and 180°C. Figure 6 shows the effect of the milling temperature on the coefficients for permeation and diffusion and the activation energies E_P and E_D of these processes. Milling at increasing temperatures decreases the particle size of PVC. The apparent morphology as measured on electron micrographs is given in Table I.

Table I. Effect of Milling Temperature on Apparent Particle Size in PVC/EVA Blends with 7.5% EVA

Milling Temperature, °C		Apparent Particle Size, μm
160		0.2–1.0
170	some	≤ 0.1
180	most	≤ 0.1
	few	≥ 0.3

Changing the milling temperature has a pronounced effect on the mechanical properties (*19*) and the physical state of the polymer blend (Table I). Figure 6 shows that the permeability and diffusion coefficients decrease considerably on increasing the milling temperature. This decrease can be correlated with the increased amount of very small PVC particles (< 0.2 μm) as observed by electron microscopy. The effect is

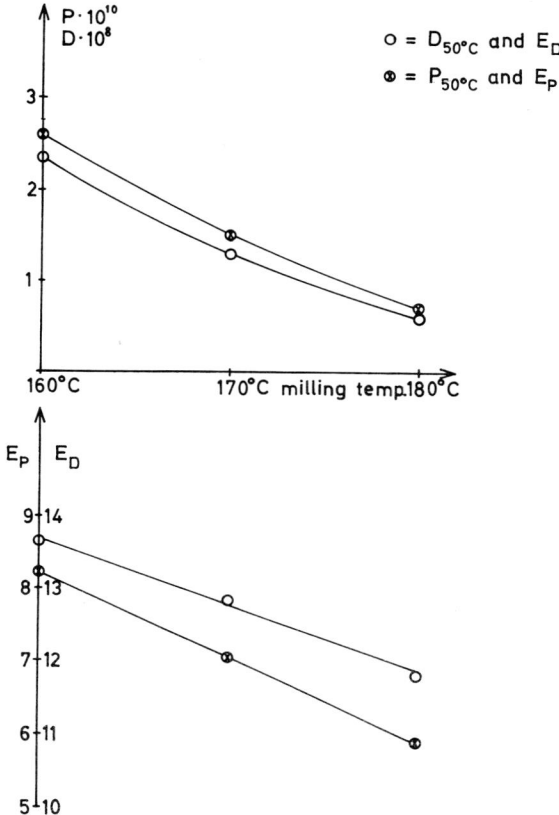

Figure 6. Influence of milling temperature on P, D, E_P and E_D for CO_2 on grafted polymer blends of PVC/EVA

attributed to a more tortuous flow path for the penetrants. It is interesting that the activation energy for permeation is reduced at increasing milling temperatures.

GRAFTING. The effect of grafting was studied using 7.5 and 12.5% EVA in the blends (Table II), which were milled at 160°C for sample preparation. The results from turbidity measurements are shown in Figure 7.

No significant change in turbidity is observed on varying the degree of grafting. Comparison with turbidity values for physical blends indicates great differences in morphology between physical blends and grafted samples. This may arise from improved compatibility of the two

Table II. Effect of Grafting on Permeability Coefficient $(P)^a$ and Permeability Ratio in PVC/EVA Blends at 50°C

EVA Content, %	Grafting	$P_{He} \times 10^{10}$	$P_{CO_2} \times 10^{10}$	P_{CO_2}/P_{He}
7.5	moderate	5.81	2.35	0.40
7.5	strong	6.47	1.71	0.26
12.5	moderate	8.12	6.87	0.85
12.5	strong	7.70	4.86	0.63

^{a} Units for P: (ml, STP) (cm)(cm)$^{-2}$ (sec)$^{-1}$ (cm Hg)$^{-1}$.

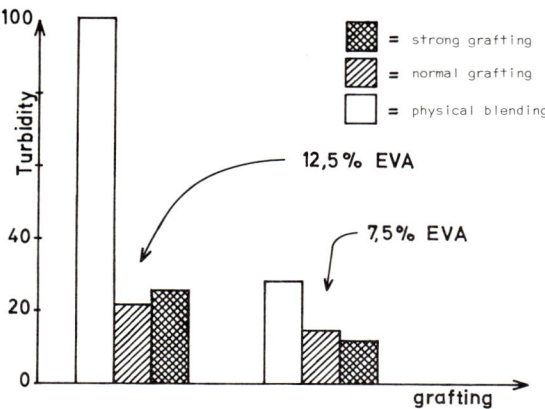

Figure 7. Effect of method of production and grafting of PVC/EVA blends on turbidity for two levels of EVA content

polymer components and fewer voids in the grafted blends. The levels of turbidity of the grafted samples containing 12.5% EVA are approximately the same as in the nongrafted sample containing 7.5% EVA. Grafting the EVA polymer molecules with vinyl chloride is expected to reduce the mobility of the matrix phase of EVA, thus reducing the permeation rate through the matrix. This is supported by comparing Table II with Figures 1 and 3. This effect should be stronger for large penetrant molecules than for small molecules—i.e., P_{CO_2}/P_{He} should decrease with an increased degree of grafting. This agrees with the experimental data (Table II). The results may, however, also be affected by differences in void formation in the resulting film samples. The results of Frazer (5) from impact strength studies of ABS plastics also indicate a reduced mobility of the rubbery component on increasing the degree of grafting. The data available so far are few and scattered. They only indicate the possibilities of using permeability as a tool for phase studies of polymer blends.

PRESSING TEMPERATURE. Films were pressed at 170° and 185°C, and the effect on their permeability was studied (Table III). Pressing at a higher temperature results in only a slight decrease in P values for the grafted samples. For the ungrafted physical blends containing 7.5% EVA, the permeability decreased significantly by pressing, indicating redistribution of the components in the blend and a decrease in the voids to give a greater barrier effect. The P values for blends containing 20% EVA do not depend on the pressing temperature.

Table III. Effect of Pressing Temperature on Permeability at 50°C of PVC/EVA Films

EVA Content, %	Grafting	Penetrant	Permeability, $P \times 10^{10}$	
			170°C	185°C
7.5	no	CO_2	1.23–3.67	0.45
7.5	moderate	CO_2	2.63	2.35
7.5	strong	CO_2	1.86	1.64
20	no	A	4.10	3.85

[a] Units for P: (ml, STP) (cm)(cm)$^{-2}$ (sec)$^{-1}$ (cm Hg)$^{-1}$.

Conclusions

(1) PVC and EVA form incompatible polymer blends as indicated from permeability and opacity measurements.

(2) The PVC phase can accommodate about 5 wt % EVA without measurable changes in permeability or turbidity.

(3) An EVA content of 7–8 wt % is a critical range where a homogeneous EVA phase begins to form, as indicated from the rapidly increased permeability and diffusion coefficients and the increased opacity.

(4) The data obtained indicate that on increasing the EVA content, the EVA gradually encloses the PVC particles and forms the homogeneous matrix in the blends.

Acknowledgments

The authors are greatly indebted to Stiftelsen Svensk Polymerforskning (Swedish Foundation of Polymer Research) for financial support to one of us, (H.S.) and to Fosfatbolaget AB for special samples and useful information about their properties.

Literature Cited

(1) Amagi, Y., Kureha Chemical Ind. Co., Japan, private communication, 1967.
(2) Barrer, R. M., *Trans. Faraday Soc.* **35,** 628 (1939).
(3) Barrer, R. M., "Diffusion in Polymers," J. Crank, G. S. Park, Eds., Chap. 6, p. 165, Academic, London, 1968.
(4) Bergen, R. L., Jr., *Appl. Polymer Symp.* **7,** 41 (1968).
(5) Frazer, J., *Chem. Ind.* **33,** 1397 (1966).
(6) Holliday, L., *Chem. Ind. (London)* **20,** 794 (1963).
(7) Holliday, L., Mann, J., "Advances in Materials," Pergamon, New York, 1965.
(8) Holliday, L., Ed., "Composite Materials," Chaps. VI, VII, Elsevier Material Science Series, Elsevier, Amsterdam, 1966.
(9) Jost, W., "Diffusion in Solids, Liquids, Gases," Academic, New York, 1960.
(10) Kato, K., *J. Polymer Sci. Pt. B,* **4,** 35 (1966).
(11) Kollinsky, F., Markert, G., *Makromol. Chem.* **121,** 117 (1969).
(12) Matsuo, M., Nozaki, Y., Jyo, Y., *Polymer Eng. Sci.* **9** (3), 197 (1969).
(13) Matsuo, M., private communication, 1970.
(14) Peterson, C. M., *J. Appl. Polymer Sci.* **12,** 2649 (1968).
(15) Rogers, C. E., "Physics and Chemistry of the Organic Solid State," D. Fox et al., Eds., Vol. II, Chap. 6, p. 509, Interscience, New York, 1966.
(16) Rosen, S. L., *SPE Tech. Papers* **XII., XIV.-4,** 1 (1966).
(17) Scott, K. W., *Rubber Chem. Technol.* **40** (2), 323 (1967).
(18) Stannett, V., Yasuda, H., "Testing of Polymers," Vol. I, Chap. 13, p. 393, Interscience, New York, 1965.
(19) Trautvetter, W., *Makromol. Chem.* **101,** 214 (1967).
(20) Wilkes, G. L., Marchessault, R. H., *J. Appl. Phys.* **37** (11), 3974 (1966).

Received April 29, 1970.

9

Blends of Chlorinated and Normal PVC

G. AJROLDI, G. GATTA, P. D. GUGELMETTO, R. RETTORE, and G. P. TALAMINI

Montecatini Edison S.p.A., Stabilimento Petrolchimico, Porto Marghera, Italy

Poly(vinyl chloride) (PVC) was chlorinated with or without $CHCl_3$ as swelling agent, to provide samples of CPVC for study. The following properties were determined: thermal stability (dehydrochlorination), Vicat softening point, dynamic modulus, impact strength at various temperatures, and creep at 80° and 100°C. All these properties depend on both chlorine content and type of chlorination process. At constant chlorine content the CPVC samples prepared in the presence of the swelling agent have higher thermal stability, higher Vicat softening point, lower impact strength, and lower creep deformation than the samples prepared without the swelling agent. The logarithm curves of dynamic modulus vs. temperature of the blends of PVC with CPVC of high chlorine content, exhibit a sharp flex at the glass transition temperature of PVC. Chlorination without a swelling agent results in a blend between the PVC and CPVC of various chlorine contents. Moreover PVC and CPVC with high chlorine content are incompatible. When mixed together, heterogeneous blends are obtained.

During the past few years many studies have been made to find an industrial way to obtain postchlorinated poly(vinyl chloride) (CPVC). The processes which, for simplicity and low cost, seem most promising are those in which PVC is chlorinated in the heterogeneous phase. These processes can be carried out either in the presence or absence of a liquid-dispersing phase and with the PVC either swollen by a solvent (swollen process or gel phase) (4, 14) or in the unmodified powder state (unswollen process) (7, 21).

The above-mentioned processes yield CPVC which is suitable for conversion into rigid articles; the homogeneous solution processes, however, yield products used in the fiber and adhesive industries (15).

The CPVC obtained by the various processes (9, 10, 20, 22) differs in many chemical physical properties. According to the most recent studies carried out with infrared (5, 8, 22) and NMR techniques (13, 18, 19), these differences are not justified by the determination of the structure, in terms of the ratio between 1.2 and 1.1 dichloroethylene units.

The differences in the properties of CPVC obtained by the homogeneous and heterogeneous processes (swollen processes) have already been studied (10, 22) and interpreted as a result of differences in the distribution of the chlorine atoms on the macromolecule. However data concerning the properties of the CPVC obtained by the various heterogeneous type processes (swollen and unswollen) are scarce. It is known that heterogeneous processes yield CPVC which is more or less nonhomogeneous regarding chlorine content in the macromolecule (20, 22) and which also contains the structure of the block copolymer between vinyl chloride and postchlorinated vinyl chloride unit (1). A recent study (9) compares the dynamic-mechanical properties of CPVC obtained by swollen and unswollen heterogeneous processes, and in the latter case a blend of PVC and CPVC is undoubtedly obtained.

The present work considers the chemical physical, thermal, and mechanical properties of CPVC obtained with various types of heterogeneous methods and of blends of CPVC with a high chlorine content. In fact, these blends are used widely in industry because their processability is better than that of pure CPVC containing the same amount of chlorine.

Experimental

Polymer Samples. Commercial PVC, Sicron 548 FM, of the type Gs, 6361 112 (Montecatini Edison S.p.A., Milan) was chlorinated by four different heterogeneous processes, designated A, B, C, and D. A, B, and C are characterized by the presence of chloroform as swelling agent. In process D, PVC is chlorinated without swelling agent.

Process A. PVC was kept in suspension in a mixture of chloroform and an aqueous solution of HCl and chlorinated at 50°C with chlorine gas activated by ultraviolet light (4).

Process B. PVC, kept in suspension in a mixture of chloroform and 1-fluorodichloro-2-difluoromonochloroethane ($CFCl_2$–CF_2Cl) with a ratio by weight of 0.94, was chlorinated at 35°C with chlorine gas activated by ultraviolet light (6).

Process C. Chloroform (60 pbw) was absorbed by PVC (100 pbw); a dry smooth powder was obtained again. Chlorination was brought about at 35°C by passing chlorine gas, activated by ultraviolet light, through the powder which was kept moving in a rotating ball (14).

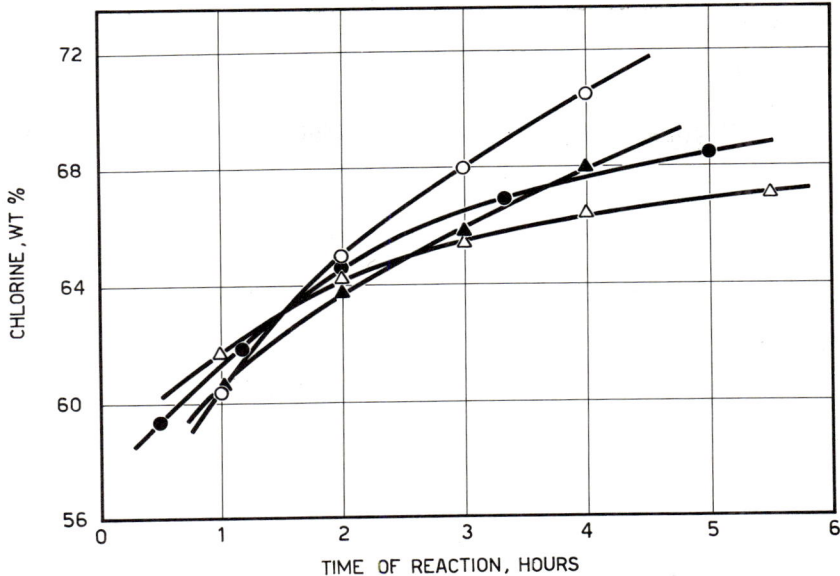

Figure 1. Typical conversion curves

○ *Process A*, ● *Process B*, ▲ *Process C*, △ *Process D*

PROCESS D. PVC, kept in the fluidized bed state, was chlorinated at 50°C, with gas chlorine activated by ultraviolet light.

Typical curves of percent chlorine *vs.* time for the four chlorination processes are shown in Figure 1.

Where not otherwise specified, samples of CPVC were purified by successive treatments with methanol, an aqueous solution of $NaHCO_3$, and water until the sodium salt disappeared. They were dried until a constant weight was attained.

Osmotic measurements (16) were used to check that under these experimental conditions there was no molecular degradation of the polymer—*i.e.*, the degree of polymerization remained constant during the chlorination. The values for the density of the CPVC samples obtained agree with published data (9, 20).

Measuring Methods. Chlorine content was determined by the oxygen flask method (2) on a polymer purified by precipitation from the solution in cyclohexanone. Thermal stability, as HCl evolution, was determined according to ASTM method D-793-49, determining the quantity of HCl evolved by the polymer maintained at 180°C in a nitrogen atmosphere. From the slope of the straight line for the amount of HCl evolved with time, the constant K for the dehydrochlorination rate (DHC) is deduced.

The Vicat softening point was determined according to ASTM D 1525 58 T, with a 5-kg load on a circular specimen with a diameter of 3 cm. The specimens were obtained by sintering the samples of CPVC (without stabilizers) in a molding press, at temperatures near T_g and at pressures varying according to the chlorine content of the CPVC.

PREPARATION OF SPECIMENS FOR MECHANICAL MEASUREMENTS. The specimens for mechanical testing were obtained by blending the powdered polymer with 1.5 phr of Ba–Cd stabilizer, 5 phr of $CaCO_3$, and 0.5 phr of lubricant (stearic acid). The dry blend was processed in a roll mill at 200°C for 5 minutes. The 1-mm sheets obtained were preheated in an oven at 200°C for 5 minutes and molded in a molding press at a pressure of 80–100 kg/cm². The specimens were obtained by milling with suitable equipment.

Table I. Chlorinated PVC samples, Procedures A, B, and C

Sample	Chlorine Content, wt %	Thermal Stability, $K \times 10^3\ hr^{-1}$	Vicat Softening Point, °C	Density grams/ml
Sicron 548 FM	56.6	1.66	87	1.405
A1	60.5	0.87	—	—
A2	61.0	0.80	106	1.470
A3	61.2	0.79	107	—
A4	62.2	0.59	112	—
A5	63.5	0.45	—	—
A6	63.8	0.37	120	—
A7	64.7	—	124	—
A8	67.1	0.19	—	—
A9	68.0	0.15	143	1.570
A10	68.2	—	144	—
A11	70.6	0.095	—	1.605
B1	58.9	1.17	96	1.440
B2	63.7	0.46	119	—
B3	64.4	0.31	—	—
B4	64.6	0.33	124	1.520
B5	66.2	0.25	133	1.545
B6	66.3	0.24	—	—
B7	67.1	—	140	—
B8	67.2	0.17	138	1.555
B9	67.4	0.18	—	—
B10	68.6	0.15	149	1.580
C1	62.0	0.69	111	1.485
C2	63.3	0.46	118	1.500
C3	64.6	0.30	124	—
C4	65.0	—	126	—
C5	65.2	0.31	—	—
C6	65.4	0.24	129	1.530
C7	65.7	—	131	—
C8	67.4	—	140	—

DYNAMIC MECHANICAL MEASUREMENTS. The apparatus here employed was the free oscillation torsion pendulum described previously (3). The storage shear modulus G', the loss modulus G'', and the logarithmic decrement Δ were derived by well-known equations (17).

The apparatus operates with samples of dimensions of 10 × 1 × 0.1 cm at frequencies of 2.8–1.5 cps in the temperature range −180°C to T_g and at frequencies of about 0.2 cps above T_g.

IMPACT STRENGTH. The Izod (ASTM D-256) impact strength measurements were carried out with a Zwick apparatus using specimens of dimensions 6.3 × 1.3 × 0.3 cm and a 0.025-cm bending radius notch. The specimens were preheated to the various temperatures in a oven, with circulating air, for 2 hours. They were removed from the oven in a closed Dewar, and measurements were carried out within a maximum of 3–4 seconds from the time of extraction from the Dewar.

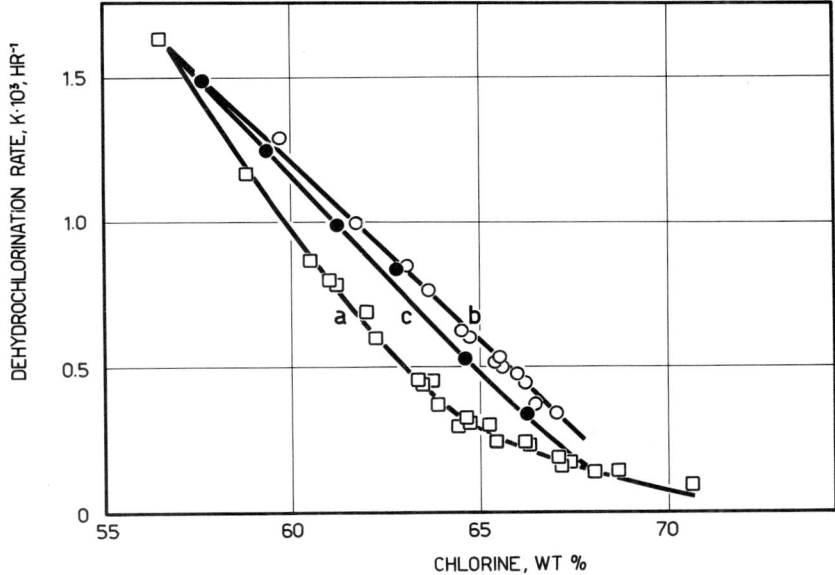

Figure 2. Thermal stability of CPVC at 180°C

Curve a: Processes A, B, and C
Curve b: Process D
Curve c: Blends of PVC and CPVC

CREEP. Creep measurements were carried out in a Frank apparatus preheated to the desired temperature by an air bath, using tensile microspecimens 0.1 cm thick, obtained by milling, as described by ASTM D-1708.

Results and Discussion

Thermal Stability. The thermal stability at 180°C of the CPVC specimens obtained with Processes A, B, and C (Table I, Figure 2, Curve a) increases as the chlorine content increase. No behavioral differences

Table II. Chlorinated PVC Samples, Procedure D

Sample	Chlorine Content, wt %	Thermal Stability, $K \times 10^3$, hr^{-1}	Vicat Softening Point, °C
D1	59.8	1.30	97
D2	61.7	1.00	104
D3	63.0	0.85	109
D4	63.6	0.77	113
D5	64.5	0.62	117
D6	64.7	0.60	119
D7	65.4	0.52	123
D8	65.5	0.53	123
D9	65.6	0.50	124
D10	66.0	0.48	125
D11	66.2	0.44	127
D12	66.4	0.38	130
D13	67.1	—	132
D14	67.1	0.34	133
D15	68.0	—	138

Table III. CPVC (Procedure A)/PVC Blends

Sample	Blend, parts by weight CPVC[a]	Blend, parts by weight PVC[b]	Chlorine Content, wt %	Vicat Softening Point, °C	Thermal Stability, $K \times 10^{-3}$, hr^{-1}
E1	10	90	57.7	88	1.50
E2	25	75	59.4	90	1.26
E3	40	60	61.2	92	1.00
E4	55	45	62.8	98	0.84
E5	70	30	64.6	110	0.53
E6	85	15	66.3	127	0.34

[a] Sample A9, see Table I.
[b] Sample Sicron 548 FM, see Table I.

are observed among the CPVC specimens obtained with the three different heterogeneous processes, all of which use chloroform as a swelling agent. The chlorine content being equal, the CPVC's obtained with the three procedures have the same thermal stability.

The constants K of DHC of the specimens of CPVC obtained with Process D (Table II, Figure 2, Curve b) vary linearly with chlorine content, and their values are always higher (lower thermal stability) than those of Curve a.

The difference between the K values of Curves a and b diminishes and seems to disappear entirely for elevated chlorination values. It can be assumed reasonably that for highly chlorinated CPVC (chlorine content >70%) the values of K become practically equal to those of the CPVC obtained with Processes A, B, and C.

Figure 2, Curve c also shows the values of K of dehydrochlorination relative to six blends prepared by mixing PVC with a sample of CPVC of high chlorine content (68%) obtained by Process A (Table III). Assuming the HCl evolution rate of the two components of the blend to be additive, Curve b, which is linear like Curve c, shows that the blends (chlorine content being equal) behave in about the same way as the CPVC obtained by the unswollen process. The presence of PVC, in any case, reduces the thermal stability of the CPVC.

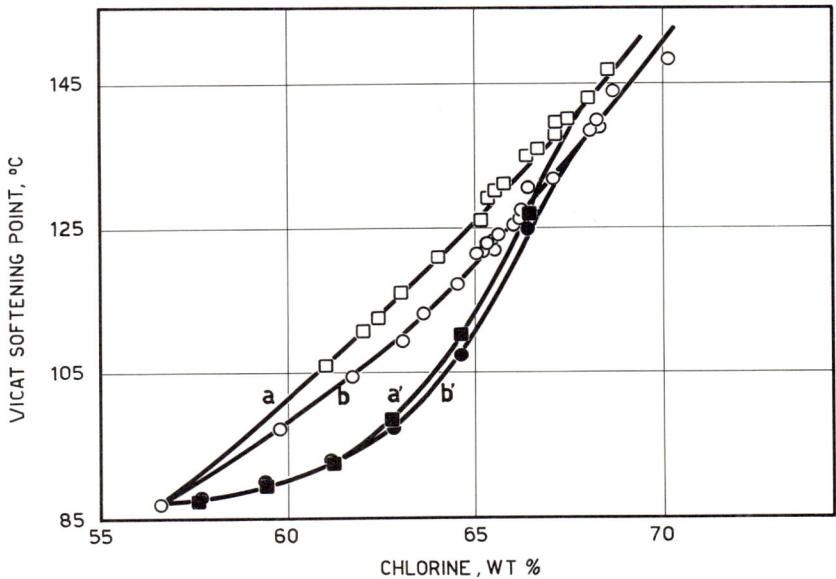

Figure 3. Vicat of CPVC and blends with PVC

Curve a: Processes A, B, and C
Curve b: Process D
Curve a': Blends of PVC and CPVC, type A
Curve b': Blends of PVC and CPVC, type D

This behavior supports the hypothesis formulated by Heidingsfeld (9) that the CPVC obtained by heterogeneous unswollen processes besides having molecules whose chlorine atoms are nonuniformly distributed (block copolymers), might also contain, in a different quantity, unaltered PVC. This product behaves similarly to a mixture of the two components (highly chlorinated CPVC and PVC), contrary to the CPVC specimens obtained with the swollen processes which should have a relatively more uniform distribution of the chlorine content on the macromolecules.

Vicat Softening Point. The Vicat softening point was determined on some CPVC samples prepared by the swollen processes (A, B, C; *see*

Table IV. CPVC (Procedure D)/PVC Blends

Sample	Blend, parts by weight		Chlorine Content, wt %	Vicat Softening Point, °C
	$CPVC^a$	PVC^b		
E7	10	90	57.7	88
E8	25	75	59.4	90
E9	40	60	61.2	93
E10	55	45	62.8	98
E11	70	30	64.6	107
E12	85	15	66.3	125

^{a} Sample D15, see Table II.
^{b} Sample Sicron 548 FM, see Table I.

Table V. Samples for Mechanical Measurements^b

Sample	Blend, parts by weight		Chlorine Content, wt %	Polymers Vicat, °C	Dynamic-Mechanical Properties	Creep	Izod
	CPVC	PVC^a					
C7	—	—	65.7	131	yes	no	no
D7	—	—	65.4	123	yes	no	no
C8	—	—	67.4	140	yes	no	no
E13	85(C8)^c	15	65.7	126	yes	no	no
D13	—	—	67.1	132	yes	yes	yes
E14	85(D13)^d	15	65.5	122	yes	yes	yes
E17^e	85(D13)	15	65.5	—	yes	no	no
E15	90(D13)	10	66.0	—	no	yes	no
E16	95(D13)	5	66.5	—	no	yes	yes
A10	—	—	68.2	144	no	yes	yes
B7	—	—	67.1	140	no	yes	yes
A7	—	—	64.7	124	no	yes	yes

^{a} Sicron 548 FM, Type GS, 6361112, see Table I.
^{b} See preparation of specimens for mechanical measurements in text.
^{c} See Table I.
^{d} See Table II.
^{e} Obtained by coprecipitation of blend E14.

Table I). The results appear in Figure 3 (Curve a), and all lie on a single curve, regardless of the type of process. Curve b shows the Vicat values of some samples obtained with Process D (unswollen process, Table II).

The pattern of the two curves is approximately similar, Curve b being a few degrees lower than Curve a. The differences between the two curves are of course attenuated at low degrees of chlorination, but this phenomenon seems to take place at high degrees of chlorination also, the two curves tending to have equal Vicat values where the chlorine is more than 70%.

Blends of PVC and A-type CPVC with 68% chlorine (Figure 3, Curve a′, Table III) and blends of PVC and D-type CPVC with 68% chlorine (Figure 3, Curve b′, Table IV) display Vicat softening point plots which are quite similar. In particular, the addition of normal PVC to the CPVC greatly diminishes the Vicat and reduces the differences between the Vicat values of the CPVC obtained with Processes A and D. In essence, blends with PVC cancel the differences between the Vicat of the two original CPVC's.

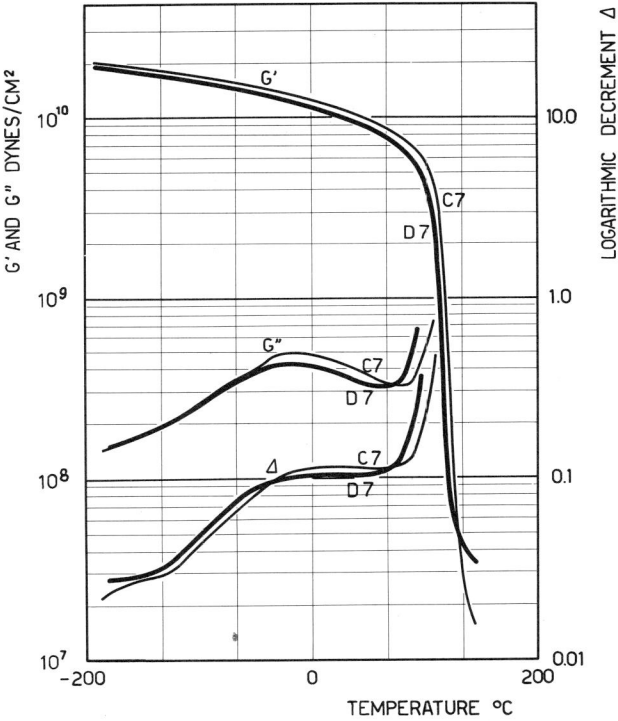

Figure 4. Dynamic-mechanical properties of samples C7 and D7

The behavior of the two series of blends (Curves a' and b') is interesting because it shows that small additions of PVC lower the softening point noticeably. If the solubility of PVC in CPVC were perfect, the mixtures would behave like random copolymers with lower chlorine content, and the points pertaining to the blends would fall on Curves a or b. The observed behavior could be caused by an incompatibility of the two components.

Dynamic Mechanical Measurements. To explain the differences in the physical properties found between the samples obtained with Processes A, B, and C and those obtained with Process D, the dynamic mechanical properties of the various CPVC and of their blends with PVC were studied. The real part of the complex shear modulus G' and the logarithmic decrement Δ for two samples having 65.5% chlorine, obtained by Process C and D, respectively (sample C7 and D7, Table V) are plotted in Figure 4, *vs.* temperature, on a semilogarithmic scale. The glass transition temperature T_g, calculated at $G' = 10^9$ dynes/cm², is 118° and 114°C for sample types C and D, respectively, in good agreement with the data obtained by the Vicat test.

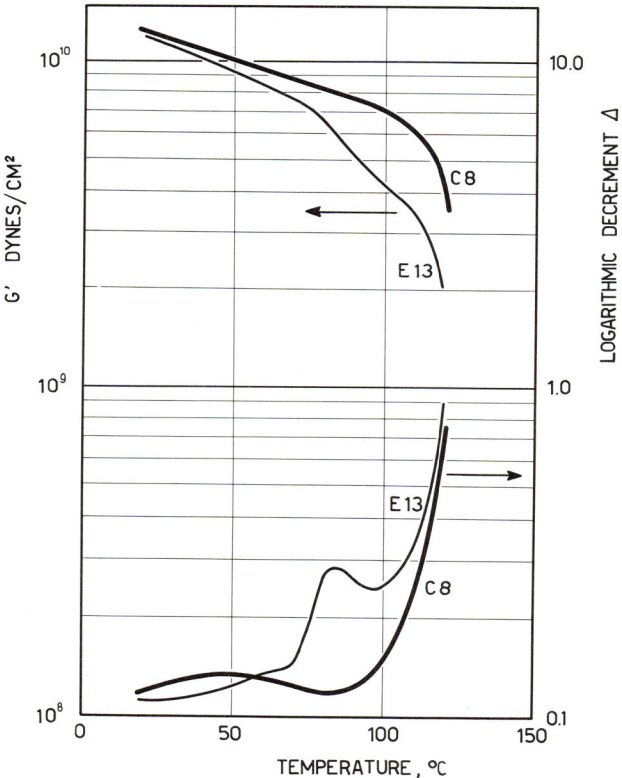

Figure 5. Dynamic-mechanical properties of samples of CPVC C8 and its blends (E13) with 15% pure PVC. C8 = 67.8% chlorine, E13 = 65.7% chlorine.

Between $-120°$ and 60°C a flat and large secondary dispersion is evident in the Δ–T curve. This dispersion is broadened with respect to that of PVC because of the presence of the chlorinated units which stiffen the chain.

The two different chlorination processes do not substantially influence the dynamic-mechanical properties. This contrasts with Heidingsfeld's results (9); he found a well-defined maximum at about 90°C for

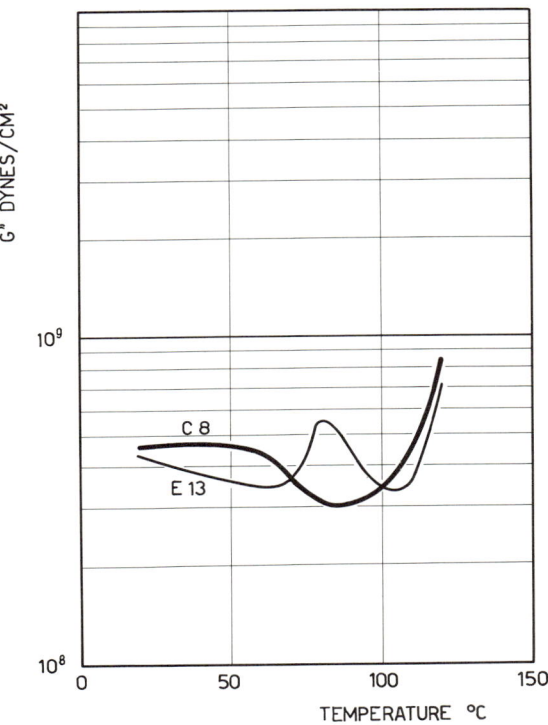

Figure 6. Modulus G" vs. temperature of the same samples as in Figure 5

the sample prepared by chlorination without adding $CHCl_3$, certainly caused by the pure PVC. In this respect we must point out that:

(1) Process D is substantially different from that used by the aforementioned authors.

(2) Some preliminary studies have shown that the degree of heterogeneity of the CPVC obtained without the swelling agent is influenced greatly by the morphological properties of the raw PVC.

A study was also carried out on the dynamic-mechanical properties of two blends of CPVC and PVC (samples E13 and E14, Table V) obtained starting with samples of CPVC, C type (sample C8, Table V) and D type (sample D13, Table V), both having about the same chlorine content.

The samples were mixed with PVC Sicron 548 FM, in the ratio 85/15, to obtain blends having 65.7% (E13) and 65.5% (E14) chlorine, respectively.

The results of the dynamic-mechanical tests carried out on the blends and on the corresponding pure CPVC's are given in Figures 5, 6, 7, and 8.

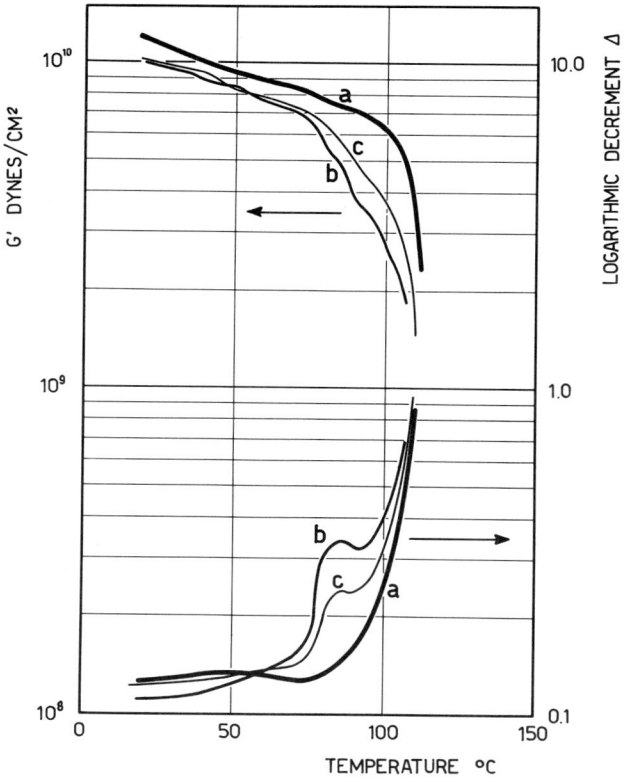

Figure 7. Dynamic-mechanical properties

Curve a: sample of CPVC D13, 67.1% chlorine
Curve b: blends E14 (85% D13–15% pure PVC), 65.5% chlorine
Curve c: blends E17 (obtained by coprecipitation of the E14 blend)

Samples C8 and D13 differ in T_g; in fact, the T_g (extrapolated) of C8 is about 10°C higher than that of D13. For the blends, a clearly defined maximum of the logarithmic decrement Δ is observed at about 85°C—*i.e.*, at the glass transition temperature of the homopolymer PVC. This indicates incompatibility between PVC and CPVC (*11*). The incompatibility observed could be caused by a mixing difficulty arising from the high viscosity in the molten state of the two components. Nevertheless, in the dynamic-mechanical measurements, a maximum of G'' of specimen E14, molded at 180°, 200°, and 220°C for periods ranging from 3 to 15 minutes was always present at about 85°C. Figures 9 and 10 give the results obtained on materials molded at the two extreme conditions and show that G'' maximum remains evident even though its intensity diminishes as the molding time and temperature increase. This may be attributed to better mixing.

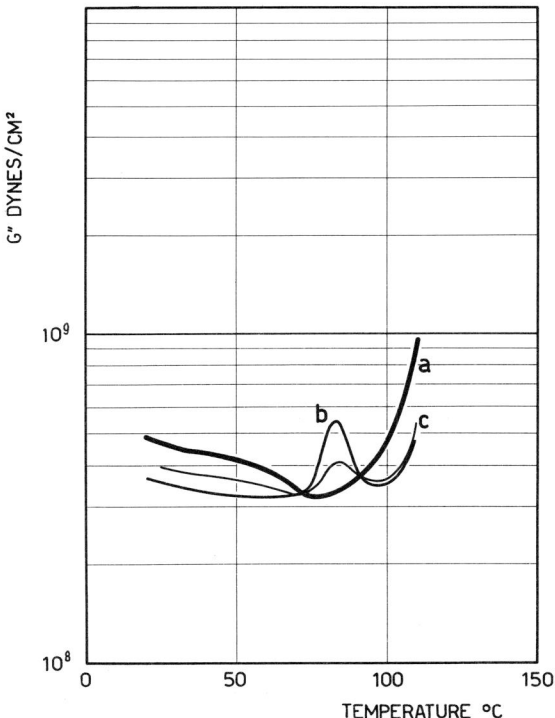

Figure 8. Modulus G'' vs. temperature of the same samples as in Figure 7

As further proof and to improve an already intimate mixture, blend E14 was also obtained by coprecipitation and then molded at 200°C for 3 minutes (sample E17). (Conditions for coprecipitation: from a solution in cyclohexanone 1% concentration, with methanol at a ratio of 7/1, room temperature.) In this case also the maximum Δ at 85°C is still evident (Figures 7 and 8, Curve c). Furthermore, the height of this maximum does not vary with the molding time, which, for a temperature of 200°C, was 15 minutes.

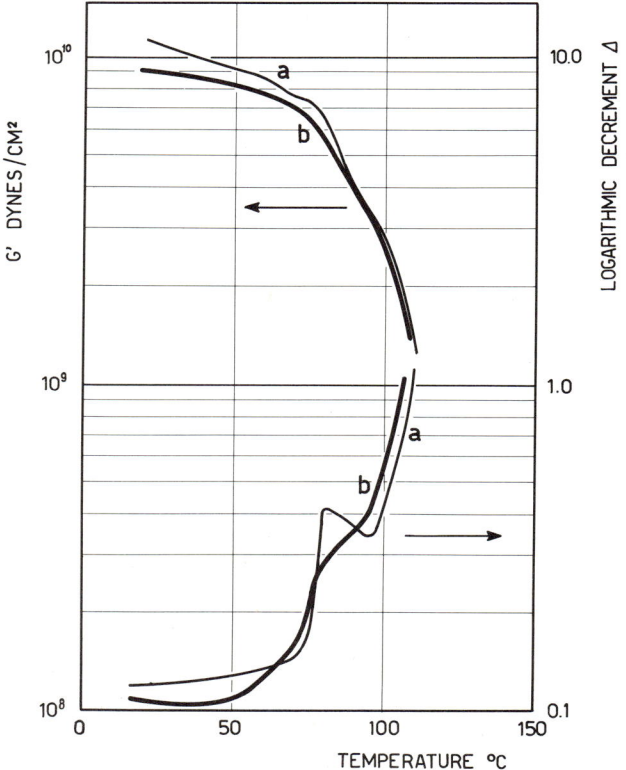

Figure 9. E14 dynamic-mechanical properties

Curve a: E14 mill rolls molded at 180°C for 3 minutes
Curve b: Same as Curve a but at 220°C for 6 minutes

Analogous behavior was observed by Oswald and co-workers (12) for blends of chlorinated polyethylene with a high chlorine content when the difference between the T_g's of the two components was greater than 30°C. In our case the values of ΔT_g are about 45° (E13) and 34°C (E14) in regard to the homopolymer PVC.

The data we found also show that the greater the ΔT_g, the greater the degree of incompatibility. No research has been done to discover the minimum value of ΔT_g for which incompatibility still occurs, but studies are in progress. On the basis of this ascertained incompatibility it is easy to explain the Vicat values obtained for the CPVC and PVC blends.

Creep Properties. The creep properties, determined at 100°C under a constant load of 50 kg/cm², of some CPVC types A, B, and D and of blends (designated E) obtained by mixing 5, 10, and 15% by weight PVC with CPVC type D, are shown in Figure 11 (*see* Table V). Similar results were obtained for the measurements carried out at 80°C under a constant load of 100 kg/cm² (Figure 12).

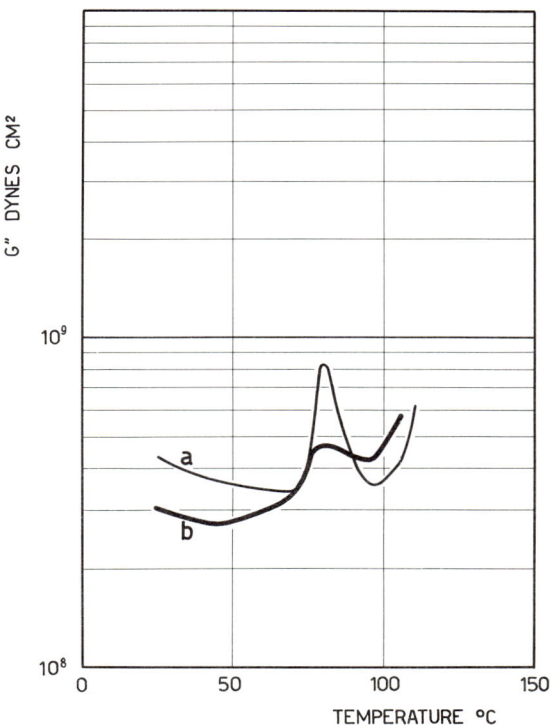

Figure 10. Modulus G" vs. temperature of the same samples as in Figure 9

From these data we have calculated the creep rate $d\epsilon/dt$ at 100 minutes. The results are plotted in Figure 13 against the chlorine content to summarize the results obtained. When the chlorine content is the same, the creep rate is greater for the sample of CPVC obtained by the unswollen process (for example D13 > B7, %Cl = 67.7 for both), and although the Vicat of the blends differs by only a few degrees from that of the original CPVC (sample D13), the creep rate of the blends is

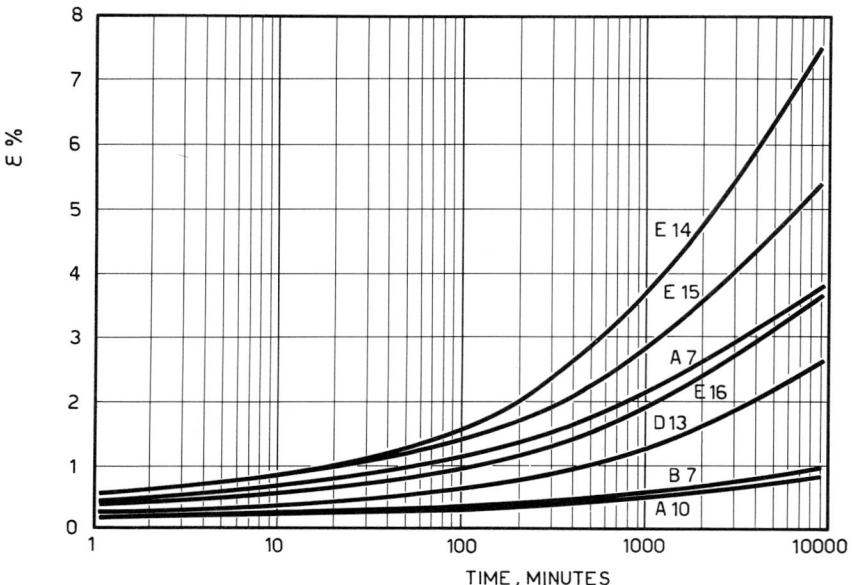

Figure 11. Tensile creep at 100°C, constant load of 50 kg/cm² of some CPVC samples (see *Table V*)

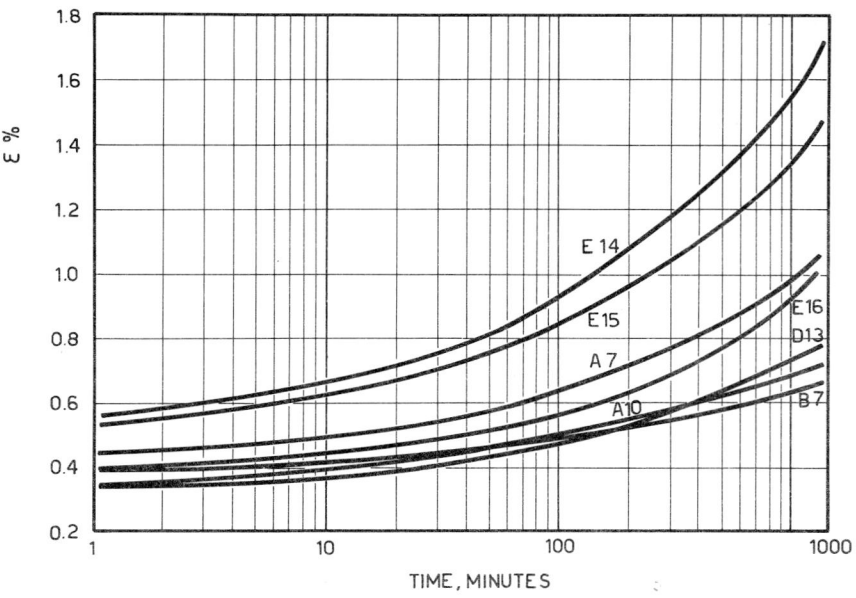

Figure 12. Tensile creep at 80°C, constant load of 100 kg/cm² of some CPVC samples (see *Table V*)

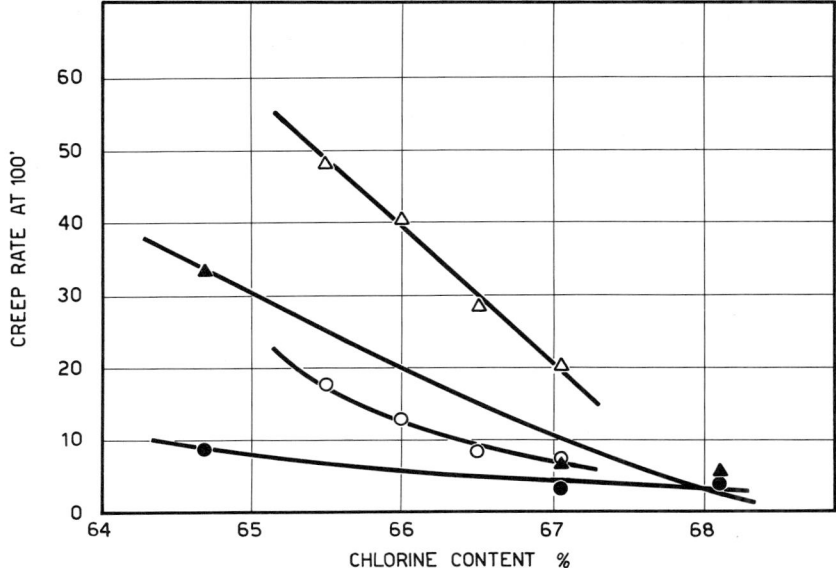

Figure 13. Rate of creep at 100 minutes of some CPVC samples (see Table V)

△: 100°C, 50 kg/cm² and ○: 80°C, 100 kg/cm², CPVC D type samples and its blends with PVC
▲: 100°C, 50 kg/cm² and ●: 80°C, 100 kg/cm², CPVC A type or B type samples

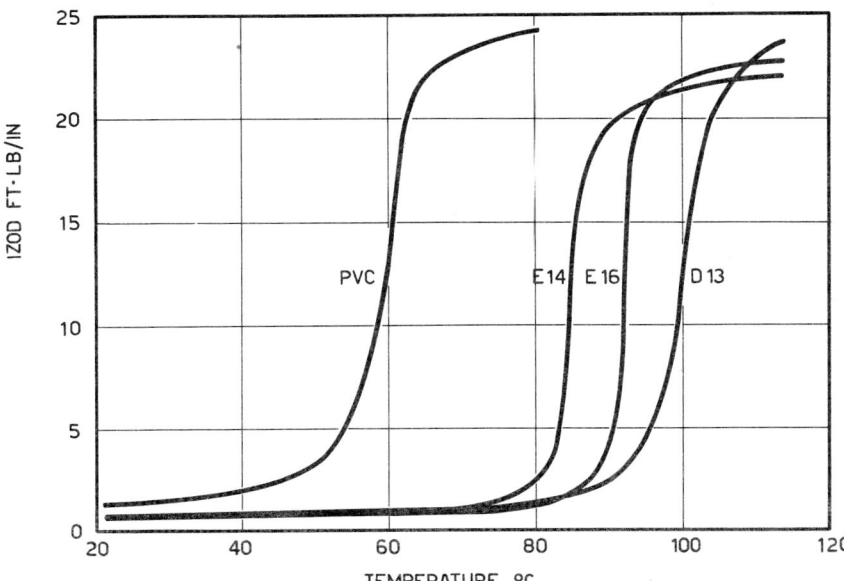

Figure 14. Izod impact strength vs. temperature

greater and increases as the PVC content increases (Figure 13). This agrees well with the result of the dynamic-mechanical measurements. In fact, at these temperatures the PVC phase can no longer withstand heavy loads.

Impact Strength as a Function of Temperature. Impact strength measurements were carried out on a sample of CPVC type D (sample D13, Table V) and on two blends, E16 and E14 (Table V) at temperatures up to almost T_g (Figure 14). The stiffening of the polymeric chain, owing to the introduction of chlorine, moves the brittle-tough transition zone up to higher temperatures, while the transition zone of the blends appears around 93° and 85°C, respectively—*i.e.*, *ca.* 5° and 15°C lower than the corresponding brittle-tough zone of the original CPVC. Recalling the biphase nature of the system, this fact tells us that the added PVC acts as a reinforcer.

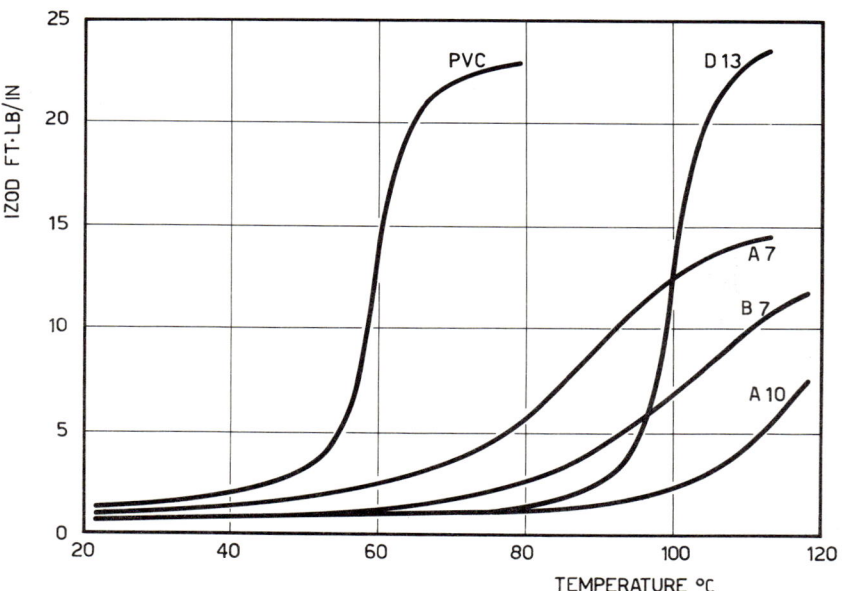

Figure 15. Izod impact strength vs. temperature

Figure 15 gives the same measurements for samples A7, B7, and A10 (Table V) of CPVC all obtained by swollen processes. The higher the chlorine content, the higher the temperature of the transition from brittle to tough behavior, and the higher the chlorine content, the lower the value of the impact strength Izod, beyond the transition zone.

Finally, comparing sample D13 with sample B7 (Figure 15) both having the same chlorine content (67.1%), one (D13) obtained by the unswollen and the other (B7) by the swollen chlorination process, it is observed that the brittle-tough transition zone for sample B7 is higher and less definite, and the impact strength value is lower than for sample D13. The trend of the Izod–T curve for sample D13 is steeper and like that of PVC. This could mean that even if heterogeneity could not be found by dynamic-mechanical measurements, it should exist on a microscopic scale.

Conclusions

All the chlorination processes in which the swelling agent is present give products with similar physical properties, whereas the materials obtained by process without a swelling agent show a lower thermal stability, lower Vicat, poorer creep properties, and little improvement in impact behavior.

The study of the blends has shown a lack of compatibility between PVC and CPVC and a substantial decay of all the physical properties with the exception of the impact strength. This demonstrates the need to specify at well-defined limits the amount of PVC used as an aid in processing.

Literature Cited

(1) Bier, G., *Kunststoffe* **55**, 694 (1965).
(2) Cheng, F. W., *Microchem. J.* **3**, 537 (1959).
(3) Convalle, G., *Proc. Intern. Congr. Plastics, 15th*, Turin, 1963.
(4) Dannis, M. L., Heights, M., Ramp, F. L., U. S. Patent **2,996,489** (1961).
(5) Fredriksen, O., Crowo, J. A., *Makromol. Chem.* **100**, 231 (1967).
(6) Gatta, G., Rettore, R., Italian Patent **821,081** (1968).
(7) Germain, C. M., Canadian Patent **701327** (1965).
(8) Germar, H., *Makromol. Chem.* **86**, 89 (1965).
(9) Heidingsfeld, V., Kuska, V., Zelinger, J., *Angew. Makromol. Chem.* **3**, 141 (1968).
(10) Kaltwasser, H., Klose, W., *Plaste Kautschuk* **13**, 515, 583 (1966).
(11) Nielsen, L. E., "Mechanical Properties of Polymers," Reinhold, New York, 1962.
(12) Oswald, H. J., Kubu, E. T., *SPE Tech. Papers* 9, XVIII—1 (1963).
(13) Petersen, J., Ranby, B., *Makromol. Chem.* **102**, 83 (1967).
(14) Rettore, R., Gatta, G., Italian Patent **759,847** (1967).
(15) Ruebensaal, C. L., FIAT Final Rept. 1071 (March 14, 1947).
(16) Sammartin, P., Montecatini Edison S.p.A., private communication.
(17) Schmieder, K., Wolf, K., *Kolloid Z.* **127**, 66 (1952); **134**, 149 (1953).
(18) Sobajima, S., Tagaci, N., Watase, H., *J. Polymer Sci., Pt. A2*, **6**, 223 (1968).
(19) Svegliado, G., Zilio-Grandi, F., *J. Appl. Polymer Sci.* **13**, 1113 (1969).
(20) Takadono, S., Yoshida, Y., Fukawa, K., *J. Chem. Soc. Japan, Ind. Chem. Sect.* **67**, 1928 (1964).
(21) Taylor, J. H., British Patent **948,372** (1964).
(22) Trauvetter, W., *Kunststoffe* **2**, 54, (1966).

RECEIVED April 29, 1970.

Copolymers

10

Copolymerization in the Presence of Depolymerization Reactions

PAUL WITTMER

Kunststofflaboratorium der Badische Anilin- & Soda-Fabrik AG, Ludwigshafen am Rhein, Germany

> *The copolymerization equation is valid if all propagation steps are irreversible. If reversibility occurs, a more complex equation can be derived. If the equilibrium constants depend on the length of the monomer sequence (penultimate effect), further changes must be introduced into the equations. Where the polymerization is subjected to an equilibrium, α-methylstyrene was chosen as monomer. The polymerization was carried out by radical initiation. With methyl methacrylate as comonomer the equilibrium constants are found to be independent of the sequence length. Between 100° and 150°C the reversibilities of the homopolymerization step of methyl methacrylate and of the alternating steps are taken into account. With acrylonitrile as comonomer the dependence of equilibrium constants on the length of sequence must be considered.*

In general polymerizations proceed in only one direction, so that we can write for a single step of the propagation reaction

$$P_n^{\cdot} + M \xrightarrow{k_p} P_{n+1}^{\cdot} \tag{1}$$

$P_n^{\cdot}$ is a reactive molecule of the polymer with n monomer units in the chain, and it is unimportant whether the polymerization mechanism is radical or ionic. The rate constant of the propagation step is k_p. Under certain conditions, monomer units can be split off the reactive polymer molecule. Then it is necessary to consider also the depolymerization reaction (with the rate constant k_d).

$$P_n^{\cdot} + M \underset{k_d}{\overset{k_p}{\rightleftarrows}} P_{n+1}^{\cdot} \tag{2}$$

According to Equation 2 each step of the propagation is an equilibrium for which Dainton and Ivin (2, 4) have found:

$$T_c = \frac{\Delta H}{\Delta S° + R \ln[M]_{eq}} \quad (3)$$

Equation 3 shows that for a given monomer concentration $[M]_{eq}$ at temperatures above a critical value T_c the rate of the depolymerization step becomes greater than the rate of the polymerization step and dominates the reaction. The critical temperature T_c is called "ceiling temperature" (22, 23). (ΔH is the enthalpy of polymerization, and $\Delta S°$ is the entropy of polymerization at the monomer concentration $[M] = 1$ mole/liter.) The concentration of the monomer at equilibrium $[M]_{eq}$ is identical to the equilibrium constant K, which is defined by the rate constants k_p and k_d.

$$[M]_{eq} = K = \frac{k_d}{k_p} \quad (4)$$

To describe the composition of copolymers which are polymerized at a given mole ratio of the comonomers, the well known Mayo equation (11) is used:

$$\frac{d[M]_1}{d[M]_2} = \frac{[M]_1}{[M]_2} \cdot \frac{r_1[M]_1 + [M]_2}{[M]_1 + r_2[M]_2} \quad (5)$$

This equation gives the composition of a copolymer which is polymerized at a given composition of the monomer mixture. The parameters of copolymerization r_1 and r_2 can be determined experimentally. They are given by the ratio of the rate constants of the homopolymerization (k_{11} and k_{22}) to the rate constants of the alternating step (k_{12} and k_{21}).

$$r_1 = \frac{k_{11}}{k_{12}} \; ; \quad r_2 = \frac{k_{22}}{k_{21}} \quad (6)$$

Equation 5 is valid only for irreversible reactions. For a copolymerization reaction that contains a polymerization–depolymerization equilibrium, Equation 5 is oversimplified. Depolymerization (Equation 2) is not considered in Equation 5. The following derivation of a generally valid equation for a copolymerization will include this retroreaction.

Copolymerization Including Depolymerization Steps

We consider a system with the two monomers, M_1 and M_2. The possible propagation steps are reversible according to Equation 2. The reaction scheme is as follows:

(a) $\mathrm{M_1^{\cdot}} + \mathrm{M_1} \underset{k_{\overline{11}}}{\overset{k_{11}}{\rightleftarrows}} \mathrm{M_1} - \mathrm{M_1^{\cdot}}$

(b) $\mathrm{M_1^{\cdot}} + \mathrm{M_2} \underset{k_{\overline{12}}}{\overset{k_{12}}{\rightleftarrows}} \mathrm{M_1} - \mathrm{M_2^{\cdot}}$

(c) $\mathrm{M_2^{\cdot}} + \mathrm{M_2} \underset{k_{\overline{22}}}{\overset{k_{22}}{\rightleftarrows}} \mathrm{M_2} - \mathrm{M_2^{\cdot}}$

(d) $\mathrm{M_2^{\cdot}} + \mathrm{M_1} \underset{k_{\overline{21}}}{\overset{k_{21}}{\rightleftarrows}} \mathrm{M_2} - \mathrm{M_1^{\cdot}}$

The reactive position in the chains is written as a dot. The constants k_{11}, k_{12}, k_{22}, and k_{21} are the rate constants of the propagation steps; $k_{\overline{11}}$, $k_{\overline{12}}$, $k_{\overline{22}}$, and $k_{\overline{21}}$ are the rate constants of the depolymerization steps. It is assumed that the constants are independent of the last member of the chain.

The changes in monomer concentration $[\mathrm{M_1}]$ and $[\mathrm{M_2}]$ with time are given in the following equations derived from kinetics:

$$\frac{d[\mathrm{M_1}]}{dt} = -k_{11}\sum_{i=1}^{\infty}[\mathrm{M_1^{\cdot}}]_i[\mathrm{M_1}] + k_{\overline{11}}\sum_{i=2}^{\infty}[\mathrm{M_1^{\cdot}}]_i - k_{21}\sum_{i=1}^{\infty}[\mathrm{M_2^{\cdot}}]_i[\mathrm{M_1}] + k_{\overline{21}}[\mathrm{M_1^{\cdot}}]_1 \quad (7a)$$

$$\frac{d[\mathrm{M_2}]}{dt} = -k_{22}\sum_{i=1}^{\infty}[\mathrm{M_2^{\cdot}}]_i[\mathrm{M_2}] + k_{\overline{22}}\sum_{i=2}^{\infty}[\mathrm{M_2^{\cdot}}]_i - k_{12}\sum_{i=1}^{\infty}[\mathrm{M_1^{\cdot}}]_i[\mathrm{M_2}] + k_{\overline{12}}[\mathrm{M_2^{\cdot}}]_1 \quad (7b)$$

Subscripts 1 and 2 within the brackets denote the type of monomer; subscript i outside the brackets denotes chain length.

The Bodenstein principle which states that the radical concentrations are constant, is assumed valid (7).

$$\frac{d\sum_{i=1}^{\infty}[\mathrm{M_1^{\cdot}}]_i}{dt} = \frac{d\sum_{i=1}^{\infty}[\mathrm{M_2^{\cdot}}]_i}{dt} = 0 \quad (8)$$

Thus, the kinetic model is:

$$k_{21}\sum_{i=1}^{\infty}[\mathrm{M_2^{\cdot}}]_i[\mathrm{M_1}] - k_{\overline{21}}[\mathrm{M_1^{\cdot}}]_1 = k_{12}\sum_{i=1}^{\infty}[\mathrm{M_1^{\cdot}}]_i[\mathrm{M_2}] - k_{\overline{12}}[\mathrm{M_2^{\cdot}}]_1 \quad (9)$$

Dividing Equation 7a by 7b and considering Equation 9 we get

$$\frac{d[M]_1}{d[M]_2} = \left(\frac{k_{12} \sum_{i=1}^{\infty} [M_1^{\cdot}]_i [M_2] - k_{\overline{12}}[M_2^{\cdot}]_1}{k_{21} \sum_{i=1}^{\infty} [M_2^{\cdot}]_i [M_1] - k_{\overline{21}}[M_1^{\cdot}]_1}\right) \cdot$$

$$\frac{1 + \dfrac{k_{11}[M_1] - k_{\overline{11}}\left(\sum_{i=2}^{\infty}[M_1^{\cdot}]_i\right)\left(\sum_{i=1}^{\infty}[M_1^{\cdot}]_i\right)^{-1}}{k_{12}[M_2] - k_{\overline{12}}[M_2^{\cdot}]_1\left(\sum_{i=1}^{\infty}[M_1^{\cdot}]_i\right)^{-1}}}{1 + \dfrac{k_{22}[M_2] - k_{\overline{22}}\left(\sum_{i=2}^{\infty}[M_2^{\cdot}]_i\right)\left(\sum_{i=1}^{\infty}[M_2^{\cdot}]_i\right)^{-1}}{k_{21}[M_1] - k_{\overline{21}}[M_1^{\cdot}]_1\left(\sum_{i=1}^{\infty}[M_2^{\cdot}]_i\right)^{-1}}} \quad (10)$$

The term in parentheses is equal to 1 according to Equation 9. The quotient of the summations can be simplified further according to Equation 11.

$$\frac{\sum_{i=2}^{\infty}[M_1^{\cdot}]_i}{\sum_{i=1}^{\infty}[M_1^{\cdot}]_i} = \frac{\sum_{i=1}^{\infty}[M_1^{\cdot}]_i - [M_1^{\cdot}]_1}{\sum_{i=1}^{\infty}[M_1^{\cdot}]_1} = 1 - x_1 \quad (11)$$

where x_1 is the mole fraction of the reactive chains ending with monomer M_1 with a sequential length 1. In general, the mole fraction of the reactive chains with sequential length i is given as:

$$x_i = \frac{[M_1^{\cdot}]_i}{\sum_{i=1}^{\infty}[M_1^{\cdot}]_i} \quad (12a)$$

Accordingly

$$y_i = \frac{[M_2^{\cdot}]_i}{\sum_{i=1}^{\infty}[M_2^{\cdot}]_i} \quad (12b)$$

The rate constants $k_{\overline{11}}$ and $k_{\overline{22}}$ can be replaced by the equilibrium constants K_1 and K_2 according to Equation 4. In addition, the reactivity ratios are introduced according to Equation 6.

$$\frac{d[M_1]}{d[M_2]} = \frac{1 + \dfrac{r_1 \dfrac{[M_1]}{[M_2]} - r_1 \dfrac{K_1}{[M_2]}(1 - x_1)}{1 - \dfrac{k_{\overline{12}}}{k_{12}} \cdot \dfrac{[M_2^{\cdot}]_1}{[M_2]}\left(\sum\limits_{i=1}^{\infty}[M_1^{\cdot}]_i\right)^{-1}}}{1 + \dfrac{r_2 \dfrac{[M_2]}{[M_1]} - r_2 \dfrac{K_2}{[M_1]}(1 - y_1)}{1 - \dfrac{k_{\overline{21}}}{k_{21}} \cdot \dfrac{[M_1^{\cdot}]_1}{[M_1]}\left(\sum\limits_{i=1}^{\infty}[M_2^{\cdot}]_i\right)^{-1}}} \quad (13)$$

The terms with summations in Equation 13 can be transformed as follows:

$$\frac{k_{\overline{12}}}{k_{12}} \cdot \frac{[M_2^{\cdot}]_1}{[M_2]}\left(\sum_{i=1}^{\infty}[M_1^{\cdot}]_i\right)^{-1} = \frac{k_{\overline{12}}}{k_{12}} \cdot \frac{y_1}{[M_2]} \cdot \frac{\sum\limits_{i=1}^{\infty}[M_2]_i}{\sum\limits_{i=1}^{\infty}[M_1^{\cdot}]_i} \quad (14a)$$

$$\frac{k_{\overline{21}}}{k_{21}} \cdot \frac{[M_1^{\cdot}]_1}{M_1}\left(\sum_{i=1}^{\infty}[M_2^{\cdot}]_i\right)^{-1} = \frac{k_{\overline{21}}}{k_{21}} \cdot \frac{x_1}{M_1} \cdot \frac{\sum\limits_{i=1}^{\infty}M_1^{\cdot}{}_i}{\sum\limits_{i=1}^{\infty}M_2^{\cdot}{}_i} \quad (14b)$$

The summations in Equations 14a and 14b can be calculated using Equation 9. After dividing by:

$$\sum_{i=1}^{\infty}[M_2]_i$$

and considering Equations 12a and 12b we get:

$$\frac{\sum\limits_{i=1}^{\infty}[M_1^{\cdot}]_i}{\sum\limits_{i=1}^{\infty}[M_2^{\cdot}]_i} = \frac{k_{21}[M_1] + k_{\overline{12}}\, y_1}{k_{12}[M_2] + k_{\overline{21}}\, x_1} \quad (15)$$

Now we introduce two new parameters (28), q_1 and q_2:

$$q_1 = \frac{k_{\overline{12}}}{k_{21}} \; ; \quad q_2 = \frac{k_{\overline{21}}}{k_{12}} \tag{16}$$

Combining Equations 13, 14, 15, and 16 yields Equation 17:

$$\frac{d[M_1]}{d[M_2]} = \frac{1 + \dfrac{r_1 \dfrac{[M_1]}{[M_2]} - r_1 \dfrac{K_1}{[M_2]}(1-x_1)}{1 - q_1 \dfrac{y_1}{[M_2]} \cdot \dfrac{[M_2] + q_2 x_1}{[M_1] + q_1 y_1}}}{1 + \dfrac{r_2 \dfrac{[M_2]}{[M_1]} - r_2 \dfrac{K_2}{[M_1]}(1-y_1)}{1 - q_2 \dfrac{x_1}{[M_1]} \cdot \dfrac{[M_1] + q_1 y_1}{[M_2] + q_2 x_1}}} \tag{17}$$

The terms x_1 and y_1 in Equation 17 must be determined by an experimentally accessible expression. This is possible by calculating the distribution of the sequential lengths of the reactive chain ends. For monomer M_1 we can postulate the following equations

$$\frac{d[M_1^{\cdot}]_1}{dt} = k_{21}[M_1]\sum_{i=1}^{\infty}[M_2^{\cdot}]_i - k_{\overline{21}}[M_1^{\cdot}]_1 + k_{\overline{11}}[M_1^{\cdot}]_2 - k_{11}[M_1^{\cdot}]_1[M_1] -$$
$$k_{12}[M_1^{\cdot}]_1[M_2] + k_{\overline{12}}[M_2^{\cdot}]_1 \cdot x_1$$

$$\frac{d[M_1^{\cdot}]_2}{dt} = k_{11}[M_1^{\cdot}]_1[M_1] + k_{\overline{11}}[M_1^{\cdot}]_3 - k_{11}[M_1^{\cdot}]_2[M_1] - k_{\overline{11}}[M_1^{\cdot}]_2 -$$
$$k_{12}[M_1^{\cdot}]_2[M_2] + k_{\overline{12}}[M_2^{\cdot}]_1 \cdot x_2 \tag{18}$$

$$\frac{d[M_1^{\cdot}]_3}{dt} = k_{11}[M_1^{\cdot}]_2[M_1] + k_{\overline{11}}[M_1]_4 - k_{11}[M_1^{\cdot}]_3[M_1] - k_{\overline{11}}[M_1]_3 -$$
$$k_{12}[M_1^{\cdot}]_3[M_2] + k_{\overline{12}}[M_2^{\cdot}]_1 \cdot x_3$$

$$\frac{d[M_1^{\cdot}]_j}{dt} = k_{11}[M_1^{\cdot}]_{j-1}[M_1] + k_{\overline{11}}[M_1^{\cdot}]_{j+1} - k_{11}[M_1^{\cdot}]_j[M_1] - k_{\overline{11}}[M_1^{\cdot}]_j -$$
$$k_{12}[M_1^{\cdot}]_j[M_2] + k_{\overline{12}}[M_2^{\cdot}]_1 \cdot x_j$$

The last terms in Equation 18 express the fact that new reactive chain end $\sim(M_1)_{n-1} - M_1^{\cdot}$ is formed by the reversible alternating polymerization step

$$\sim\!\!M_2 - (M_1)_n - M_2^{\cdot} \xrightarrow{k_{\overline{12}}} \sim\!\!M_2 - (M_1)_{n-1} - M_1^{\cdot} + M_2$$

The factor x_i corrects for the fact that each step in the reaction does not lead to a chain end with i units of monomer M_1, but only some do. We can safely assume that the sequential distribution of the chain lengths of the unreactive and reactive sequences is the same. Under all ordinary conditions the concentration of radicals assumes a value such that the rate of disappearance equals their rate of creation (7). This steady state is realized not only for the total concentrations of radicals $\sum_{i=1}^{\infty}[M_1^{\cdot}]_i$ and $\sum_{i=1}^{\infty}[M_2^{\cdot}]_i$ but also for the concentrations of radicals with special chain lengths $[M_1^{\cdot}]_i$ and $[M_2^{\cdot}]_i$ (11). The differential quotients in Equation 18 assume the value 0. In addition we express the first two terms on the right in the first line of the equation system 18 by Equation 9.

$$k_{12}[M_2]\sum_{i=1}^{\infty}[M_1^{\cdot}]_i - k_{\overline{12}}[M_2^{\cdot}]_1 = [M_1^{\cdot}]_1(k_{11}[M_1] + k_{12}[M_2]) -$$
$$k_{\overline{11}}[M_1^{\cdot}]_2 - k_{\overline{12}}[M_2^{\cdot}]_1\, x_1$$

$$k_{11}[M_1^{\cdot}]_1[M_1] = [M_1^{\cdot}]_2(k_{11}[M_1] + k_{12}[M_2]) + k_{\overline{11}}([M_1^{\cdot}]_2 - [M_1^{\cdot}]_3) -$$
$$k_{\overline{12}}[M_2]_1\, x_2$$

$$k_{11}[M_1^{\cdot}]_2[M_1] = [M_1^{\cdot}]_3(k_{11}[M_1] + k_{12}[M_2]) + k_{\overline{11}}([M_1^{\cdot}]_3 - [M_1^{\cdot}]_4) -$$
$$k_{\overline{12}}[M_2^{\cdot}]_1\, x_3 \qquad (19)$$

$$k_{11}[M_1^{\cdot}]_{j-1}[M_1] = [M_1^{\cdot}]_j(k_{11}[M_1] + k_{12}[M_2]) + k_{\overline{11}}([M_1^{\cdot}]_j -$$
$$[M_1^{\cdot}]_{j+1}) - k_{\overline{12}}[M_2^{\cdot}]_1\, x_j$$

Now we divide by $k_{12}[M_2]\sum_{i=1}^{\infty}[M_1^{\cdot}]_i$ and introduce the reactivity ratios. Considering Equation 12a we can write:

$$1 = x_1\left(r_1\frac{[M_1]}{[M_2]} + 1\right) - x_2 r_1\frac{K_1}{[M_2]} + \frac{k_{\overline{12}}}{k_{12}} \cdot \frac{[M_2^{\cdot}]_1}{\sum_{i=1}^{\infty}[M_1^{\cdot}]_i} \cdot \frac{1-x_1}{[M_2]}$$

$$x_1 r_1\frac{[M_1]}{[M_2]} = x_2\left(r_1\frac{[M_1]}{[M_2]} + r_1\frac{K_1}{[M_2]} + 1\right) - x_3 r_1\frac{K_1}{[M_2]} -$$
$$\frac{k_{\overline{12}}}{k_{12}} \cdot \frac{[M_2^{\cdot}]_1}{\sum_{i=1}^{\infty}[M_1^{\cdot}]_i} \cdot \frac{x_2}{[M_2]}$$

$$x_2 r_1 \frac{[M_1]}{[M_2]} = x_3 \left(r_1 \frac{[M_1]}{[M_2]} + r_1 \frac{K_1}{[M_2]} + 1\right) - x_4 r_1 \frac{K_1}{[M_2]} -$$

$$\frac{k_{\overline{12}}}{k_{12}} \cdot \frac{[M_2 \cdot]_1}{\sum_{i=1}^{\infty} [M_1 \cdot]_i} \cdot \frac{x_3}{M_2} \qquad (20)$$

$$x_{j-1} r_1 \frac{[M_1]}{[M_2]} = x_j \left(r_1 \frac{[M_1]}{[M_2]} + r_1 \frac{K_1}{[M_2]} + 1\right) - x_{j+1} r_1 \frac{K_1}{[M_2]} -$$

$$\frac{k_{\overline{12}}}{k_{12}} \cdot \frac{[M_2 \cdot]_1}{\sum_{i=1}^{\infty} [M_1 \cdot]_i} \cdot \frac{x_j}{M_2}$$

The expression:

$$[M_2 \cdot]_1 \left(\sum_{i=1}^{\infty} [M_1 \cdot]_i\right)^{-1}$$

in Equation 20 is given according to Equations 12b and 15:

$$\frac{[M_2 \cdot]_1}{\sum_{i=1}^{\infty} [M_1 \cdot]_i} = y_1 \cdot \frac{k_{12}[M_2] + k_{\overline{21}}\, x_1}{k_{21}[M_1] + k_{\overline{12}}\, y_1} \qquad (21)$$

Introducing q_1 and q_2 into Equation 20 we can write

$$1 = x_1 \left(r_1 \frac{[M_1]}{[M_2]} + 1\right) - x_2 r_1 \frac{K_1}{[M_2]} + q_1 \frac{y_1}{[M_2]} \cdot \frac{[M_2] + q_2 x_1}{[M_1] + q_1 y_1} (1 - x_1)$$

$$x_1 r_1 \frac{[M_1]}{[M_2]} = x_2 \left(r_1 \frac{[M_1]}{[M_2]} + r_1 \frac{K_1}{[M_2]} + 1\right) - x_3 r_1 \frac{K_1}{[M_2]} - q_1 \frac{y_1}{[M_2]} \cdot$$

$$\frac{[M_2] + q_2 x_1}{[M_1] + q_1 y_1} x_2$$

$$x_2 r_1 \frac{[M_1]}{[M_2]} = x_3 \left(r_1 \frac{[M_1]}{[M_2]} + r_1 \frac{K_1}{[M_2]} + 1\right) - x_4 r_1 \frac{K_1}{[M_2]} - q_1 \frac{y_1}{[M_2]} \cdot$$

$$\frac{[M_2] + q_2 x_1}{[M_1] + q_1 y_1} x_3$$

$$x_{j-1}r_1\frac{[M_1]}{[M_2]} = x_j\left(r_1\frac{[M_1]}{[M_2]} + r_1\frac{K_1}{[M_2]} + 1\right) - x_{j+1}r_1\frac{K_1}{[M_2]} -$$

$$q_1\frac{y_1}{[M_2]} \cdot \frac{[M_2] + q_2x_1}{[M_1] + q_1y_1} x_j \tag{22}$$

Here Equation 23 is valid.

$$\sum_{i=1}^{\infty} x_i = 1 \tag{23}$$

Equation 24 expresses the pattern of distribution

$$x_i = \rho \cdot \sigma^i ; \quad \sigma < 1 \tag{24}$$

Inserting Equation 23 into Equation 24 we get:

$$\sum_{1}^{\infty} x^i = \rho \cdot \sigma(1 + \sigma + \sigma^2 + \sigma^3 + \ldots) = \rho\sigma\frac{1}{1-\sigma} = 1 \tag{25}$$

It follows that

$$\rho = \frac{1-\sigma}{\sigma} ; \quad x_1 = (1-\sigma)\sigma^{i-1} \tag{26}$$

Inserting Equation 26 into any line of the system of Equations 22 and solving for σ give in every case

$$\sigma = (1 - x_1) = \frac{1}{2r_1K_1}\left\{r_1([M_1] + K_1) + [M_2] - \psi\right\} -$$

$$\sqrt{\left(\frac{r_1([M_1] + K_1) + [M_2] - \chi}{2r_1K_1}\right)^2 \frac{[M_1]}{K_1}} \tag{27}$$

As an abbreviation the term ψ is introduced

$$\psi = q_1y_1\frac{[M_2] + q_2x_1}{[M_1] + q_1y_1} \tag{28}$$

Thus, Equation 27 is in this case a possible distribution function. It is of the type of the Schulz-Flory (25) distribution function. The expressions ρ and σ express implicitly the probabilities of propagation, depolymerization, and alternating polymerization (chain termination). The validity of the Schulz-Flory distribution function in this example of a polymerization with reversible propagation steps is evident. This type of distribution is always present if the distribution of the chain lengths is the result of competing reactions between randomly occurring propagation and termination steps. If depolymerization steps are involved, the over-all chain length is smaller; the chain length distributed however is determined by the same mathematical law.

The same equations can be derived for the monomer M_2.

$$y_1 = (1 - r_i) r_i^{i-1} \tag{29}$$

$$r_i = 1 - y_1 = \frac{1}{2r_2 K_2} \{r_2([M_2] + K_2) + [M_1] - \phi\} -$$
$$\sqrt{\left(\frac{r_2([M_2] + K_2) + [M_1] - \phi}{2r_2 K_2}\right)^2 - \frac{[M_2]}{K_2}} \tag{30}$$

The term ϕ is introduced as an abbreviation

$$\phi = q_2 x_1 \frac{[M_1] + q_1 y_1}{[M_2] + q_2 x_1} \tag{31}$$

After arranging Equations 27 and 30 and raising to the second power we can write (considering Equations 28 and 31:

$$y_1 = \frac{[M_1]}{q_1 x_1} \cdot \frac{([M_2] + r_1 x_1 [M_2])(1 - x_1) - r_1 x_1 [M_1]}{(q_2 - r_1 K_1)(1 - x_1) + r_1 [M_1]} \tag{32a}$$

$$x_1 = \frac{[M_2]}{q_2 y_1} \cdot \frac{([M_1] + r_2 y_1 [M_1])(1 - y_1) - r_2 y_1 [M_2]}{(q_1 - r_2 K_2)(1 - y_1) + r_2 [M_2]} \tag{32b}$$

Knowing the constants and the monomer concentrations in Equations 32a and 32b, we can calculate the numerical values for x_1 and y_1. This can be done graphically, but it is more practical to use a computer. Using Equation 17 we can then calculate the composition of the polymer formed. The relationship by which the two monomers are copolymerized is not a simple function of the ratio of the two monomer concentrations in the monomer mixture. The result depends on the absolute value of the monomer concentrations also.

Special Case where $K_2 = 0$, $q_1 = 0$, and $q_2 = 0$. In this case only the homopolymerization of the monomer M_1 is reversible. We can write the simpler equation (10, 13, 29):

$$\frac{d[M_1]}{d[M_2]} = \frac{1 + r_1 \frac{[M_1]}{[M_2]} - r_1 \frac{K_1}{[M_2]}(1 - x_1)}{1 + r_2 \frac{[M_2]}{[M_1]}} \tag{33}$$

$$1 - x_1 = \frac{r_1([M_1] + K_1) + [M_2]}{2\,r_1 K_1} - \sqrt{\left(\frac{r_1([M_1] + K_1) + [M_2]}{2\,r_1 K_1}\right)^2 - \frac{[M_1]}{K_1}} \quad (34)$$

Special Case where $K_2 \neq 0$, $q_1 = 0$, **and** $q_2 = 0$. In this case the two homopolymerization reactions are reversible; the alternative steps are irreversible. Here Equation 35 is valid (10, 29):

$$\frac{d[M_1]}{d[M_2]} = \frac{1 + r_1 \frac{[M_1]}{[M_2]} - r_1 \frac{K_1}{[M_2]}(1 - x_1)}{1 + r_2 \frac{[M_2]}{[M_1]} - r_2 \frac{K_2}{[M_2]}(1 - y_1)} \quad (35)$$

x_1 is given by Equation 34, y_1 by Equation 36:

$$(1 - y_1) = \frac{r_2([M_2] + K_2) + [M_1]}{2\,r_2 K_2} - \sqrt{\left(\frac{r_2([M_2] + K_2) + [M_1]}{2\,r_2 K_2}\right)^2 - \frac{[M_2]}{K_2}} \quad (36)$$

Special Case where $K_2 = 0$, $q_1 = 0$, **and** $q_2 = 0$. **Equilibrium Constant** K_1 **Depends on the Chain Length.** According to Szwarc and co-workers (12, 26) the equilibrium constants of the anionic polymerization of α-methylstyrene depend on the degree of polymerization (chain lengths). For the reaction

the equilibrium constant K, defined by Equation 4 is smaller by three orders of magnitude than the one for the reaction of monomers with longer chains. The equilibrium constant for the reaction of monomeric α-methylstyrene with compound II according to the following formulas

$$\underset{\underset{\big|}{\bigcirc}}{\overset{CH_3}{\underset{|}{\theta\ C}}}-CH_2-CH_2-\underset{\underset{\big|}{\bigcirc}\ III}{\overset{CH_3}{\underset{|}{C}}}-CH_2-\underset{\underset{\big|}{\bigcirc}}{\overset{CH_3}{\underset{|}{C}}}-CH_2-\underset{\underset{\big|}{\bigcirc}}{\overset{CH_3}{\underset{|}{C}\ \theta}}$$

are not significantly different from the one for longer chains. The differences in the equilibrium constants depend on differences in the depolymerization constant, $k_{\overline{11}}$ and not the propagation constant k_{11}. The differences in the rate constants are explained by steric and electrostatic influences.

The monomer units in I are linked tail to tail and not in the normal head-to-tail manner. Therefore I is not comparable with a normal dimer. In substance II a dimer lies to the right of the tail-to-tail linkage. In the copolymerization of α-methylstyrene with, for example, methyl methacrylate, the addition of one molecule of α-methylstyrene corresponds to substance I,

$$\sim\sim CH_2-\underset{\underset{|}{CH_3}}{\overset{COOCH_3}{\underset{|}{C}}}-CH_2-\underset{\underset{\big|}{\bigcirc}}{\overset{CH_3}{\underset{|}{C}\cdot}}$$

the addition of two molecules of α-methylstyrene—i.e., a sequence of two monomer units—gives substance II.

$$-CH_2-\underset{\underset{|}{CH_3}}{\overset{COOCH_3}{\underset{|}{C}}}-CH_2-\underset{\underset{\big|}{\bigcirc}}{\overset{CH_3}{\underset{|}{C}}}-CH_2-\underset{\underset{\big|}{\bigcirc}}{\overset{CH_3}{\underset{|}{C\cdot}}}$$

The addition of a monomer unit to a sequence with the length 1 or with a length greater than 1 corresponds to the addition of a monomer to I and II, respectively.

Since the equilibrium constants for the addition of monomer onto the sequence of length 1 or to longer sequences are so different, the following situation must be considered. The living sequences with two monomer units cannot split off a monomer unit, but higher sequences are subject to polymerization–depolymerization equilibria with an equilibrium constant which no longer depends on chain length. We consider the special case in which the equilibrium constant K_2 and the constants

q_1 and q_2 are zero. We get the following equation (13, 29):

$$\frac{d[M_1]}{d[M_2]} = \frac{1 + \frac{[M_1]}{[M_2]} \cdot r_1 \cdot \left(1 - \frac{r_1 \frac{K_1}{[M_2]} \cdot \sigma}{1 + r_1 \frac{[M_1]}{[M_2]}}\right)}{1 + r_2 \frac{[M_2]}{[M_1]}} \tag{37}$$

σ corresponds to Equation 34:

$$\sigma = \frac{1}{2} \cdot \frac{r_1([M_1] + K_1) + [M_2]}{r_1 K_1} - \sqrt{\left(\frac{r_1([M_1] + K_1) + [M_2]}{2 r_1 K_1}\right)^2 - \frac{[M_1]}{K_7}} \tag{38}$$

Copolymerization of α-Methylstyrene

In most vinyl monomer polymerizations measurable monomer concentrations at equilibrium are only apparent at elevated temperatures ($T > 150°C$). However, the corresponding concentrations of α-methylstyrene are measurable at room temperature. This is caused by the enthalpy of polymerization which is, compared with other monomers, relatively low (at a comparable entropy of polymerization).

The existing equilibrium monomer concentrations have been measured by several authors (9, 17, 30, 31) for anionic polymerizations (from the type described by Szwarc (24) at different temperatures. The thermodynamic terms ΔH and $\Delta S°$ can be calculated from these data according to Equation 3. Table I shows the results of these calculations and a value for ΔH which was measured calorimetrically (20). (The values

Table I. Enthalpy (ΔH) and Entropy ($\Delta S°$) for the Polymerization of α-Methylstyrene

Reference	ΔH, kcal/mole	$\Delta S°$, cal/(mole °C)
McCormick (17)	6.96	24.8
Worsfold and Bywater (30,31)	8.02	28.75
Hopff and Lussi (9)	8.50	30.6
Roberts and Jessup (20)	8.42	—

are valid for the transition: monomer in solution → polymer in solution —i.e., ΔH_{ss} or $\Delta S°_{ss}$ defined by the terminology of Dainton and Ivin (3). Only the calorimetrically measured value corresponds to the transition: liquid monomer → solid polymer—i.e., $\Delta H°_{1c}$ and $\Delta S°_{1c}$). The values measured by different authors are in a good agreement except for the values given by McCormick.

From the values of ΔH and $\Delta S°$ one can calculate (Equation 3) the equilibrium concentrations at different temperatures. The results are plotted in Figure 1. Curves 1–3 are in good agreement below 10°C; at elevated temperatures larger deviations are apparent. This stems from the fact that nearly all measurements are carried out at temperatures below 0°C; thus in this region the curves correspond to actually measured values. At higher temperatures the curves are determined by extrapolation. Only McCormick has measured these values directly at elevated temperatures—*i.e.*, up to 60°C (Curve 1). For this reason here we have used McCormick's data to calculate the equilibrium concentrations even though his thermodynamic constants show a larger deviation than the other measurements.

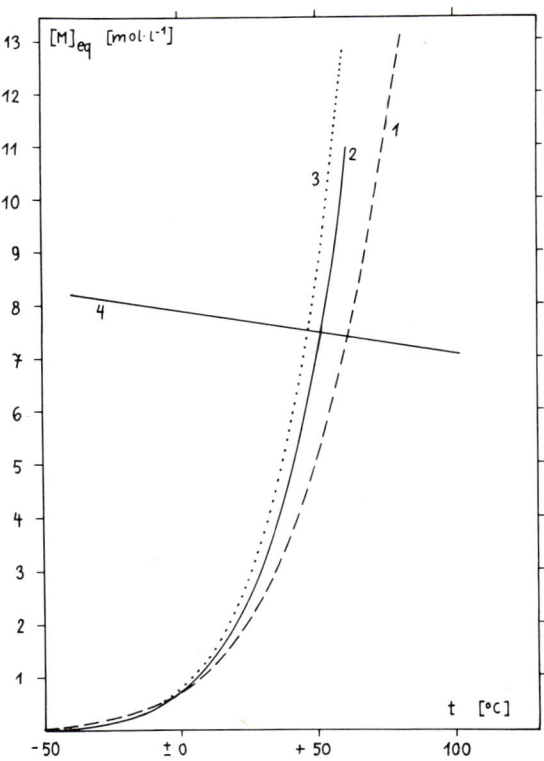

Figure 1. Polymerization of α-methylstyrene. Equilibrium concentrations of the monomer.

Curve 1: Values by McCormick (17)
Curve 2: Values by Worsfold and Bywater (30, 31)
Curve 3: Values by Hopff and Lussi (9)
Curve 4: Monomer concentration of the undiluted monomer (4)

In Figure 1 the fourth plot shows the monomer concentration of the undiluted monomer—i.e., the highest monomer concentration possible (*14*). At temperatures on the right side of the intersection of this curve with the equilibrium concentration–temperature curve, the rate of the depolymerization step is greater than the rate of the polymerization step even if the monomer is undiluted. These temperatures are quite low compared with those in other polymerization reactions. Normally they are 100° to 200°C higher. The calculated ceiling temperature for the undiluted monomer, which has been calculated from the measurements made by McCormick is 61°C. This is in good agreement with the ceiling temperature of 60°C which has been measured for three cationic polymerizations using different initiation (*15*). The values of the equilibrium constants at different temperatures according to McCormick (*17*) are given in Table II.

Table II. Equilibrium Constants for the Polymerization of α-Methylstyrene According to McCormick (*17*)

T, °C	K, mole/liter	T, °C	K, mole/liter
0	0,70	80	12,9
10	1,10	90	16,9
20	1,70	100	22,9
30	2,53	110	27,9
40	3,65	120	35
50	5,1	130	43
60	7,1	140	54
70	9,7	150	67

Measurements on α-Methylstyrene–Methyl Methacrylate at 20°–100°C. All copolymerizations were carried out without solvent. Below 80°C azobisisobutyronitrile was used as initiator. At 100°C the reactions were initiated thermally. At temperatures of 50°, 60°, and 80°C the reactions were carried out in dilatometers. At 20°C small flasks were used, and the reactions were conducted in a temperature-controlled room over a period of days. At 100°C sealed glass tubes were preferred. The reactions were stopped at yields below 5%. The composition of the copolymers was determined by oxygen analysis in the analytical laboratories of BASF. The method for determining oxygen was developed in the Untersuchungslaboratorium of BASF (*18*).

The results are given in Table III. At higher α-methylstyrene concentrations than those given in Table III, the reactions were so slow that no precipitable polymer could be obtained in a reasonable time.

In evaluating these measurements the Mayo equation 5 is used. This is also done in two other publications (*5, 27*). This equation cannot be used in its strictest sense, as noted previously. However, by appropriate

Table III. Copolymerization of α-Methylstyrene (M_1) and Methyl Methacrylate (M_2)

M_1, mole % in Monomer Mixture	M_1 mole % in Polymer				
	20°C	50°C	60°C	80°C	100°C
5	7.7_9	8.1_3	9.1_0	7.7_9	6.9_7
					8.8_8
10	17.2_1	15.4_7	17.7_5	14.9_6	14.6_8
20	29.2_7	26.8_7	25.3_8	25.1_9	21.4_5
30	34.4_1	32.8_9	33.5_0	31.3_3	26.9_3
40	41.5_5	40.5_9	39.7_3	37.3_1	34.0_0
50	45.8_7	45.9_5	44.6_8	42.7_8	37.0_0
60	50.5_7	53.4_1	52.0_5	47.4_2	43.0_9
70	57.9_0	60.4_9	59.6_1	51.8_3	43.4_0
			57.7_5		
80	64.1_0	64.0_9	65.9_9	54.4_9	
90				68.4_1	
95	81.3_7				

Table IV. Copolymerization Parameters from Calculations Using Equation 33

T, °C	r_1		r_2	
20	0.3	(0.25)[a]	0.5	(0.50)[a]
50	0.51	(0.35)	0.55	(0.55)
60	0.6	(0.30)	0.55	(0.55)
80	0.81	(0.15)	0.65	(0.55)
100	1.0	(0.0)	0.7	(0.60)

[a] Values in parentheses were obtained by using the Mayo equation 5.

means of fitting the reactivity ratios, it can be used as a good approximation. The r values have been determined by the usual methods (6, 16). They are given in Table IV (in brackets); the calculated curves are given in Figure 2 together with the measured values. Figure 2 shows that the measured and calculated values agree well. This method of describing the results is a mathematical formality, and the resulting parameters have no real meaning. This can be explained by the following consideration. The r values depend on the temperature according to the following equation (11):

$$r_1 = \frac{A_{11}}{A_{12}} \exp\left(-\frac{E_{11} - E_{12}}{RT}\right) \qquad (37)$$

where A_{11} and A_{12} are the frequency factors of the Arrhenius equation for the homopolymerization step and the alternation step, respectively, and E_{11} and E_{12} are the corresponding activation energies. An analogous

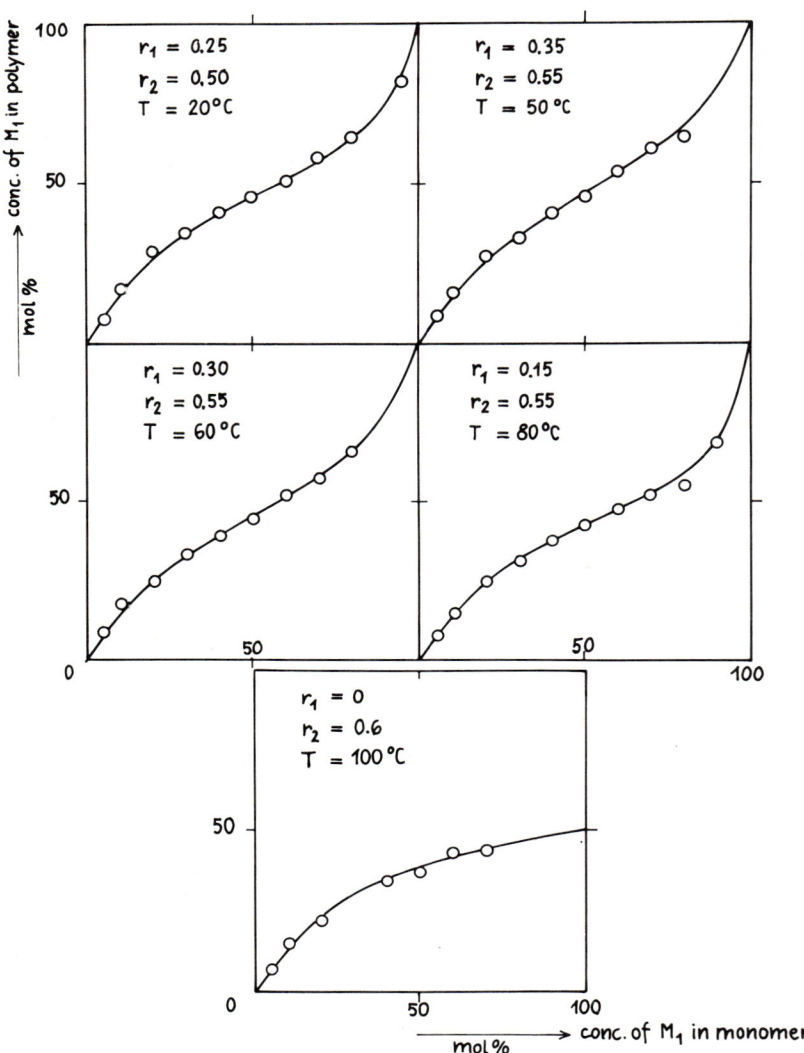

Figure 2. Copolymerization of α-methylstyrene (M_1) and methyl methacrylate (M_2). Curves calculated using Equation 5.

equation is valid for r_2. Generally the activation energies E_{11} and E_{12} are very similar; thus, the r values depend only weakly on the temperature.

According to Equation 37 the logarithm of the r values depends on temperature according to the Arrhenius equation. According to Figure 3 the condition fixed by Equation 37 is fulfilled for r_2 values but not for r_1

values. The temperature dependence of the r_1 values is not in agreement with the conditions of Equation 5.

In the following the reversibility of the polymerization of α-methylstyrene has been taken into account. The copolymerization curves are calculated *via* Equation 33 together with Equation 34. The equilibrium constants necessary for the calculation are taken from Table II. The depolymerization of methyl methacrylate (*1*) can be neglected in the temperature range investigated since the equilibrium constants for this monomer (K_2) are extremely small compared with the value for α-methylstyrene—*e.g.*, (*21*) at 100°C, $K_1 = 22.9$ mole/liter, $K_2 = 0.12$ mole/liter; at 80°C, $K_1 = 12.9$ mole/liter, $K_2 = 0.049$ mole/liter.

Except for small deviations the r_2 values correspond with the values obtained using the Mayo equation 5. The r_1 values have been fitted to the experimental data by trial and error and are tabulated in Table IV. The calculated curves as well as the measured points are plotted in Figure 4. The measured points and the calculated curves are in good agreement at low temperature—*i.e.*, 20°–80°C; however at 100°C deviations become apparent.

The r values, calculated by Equation 33 follow the pattern of an Arrhenius equation, as shown in Figure 5. The differences in the activation energies are given by the following data:

$$E_{11} - E_{12} = 3.45 \text{ kcal/mole} \qquad E_{22} - E_{21} = 0.95 \text{ kcal/mole}$$

These values are comparable with those given in Ref. *11*, p. 180.

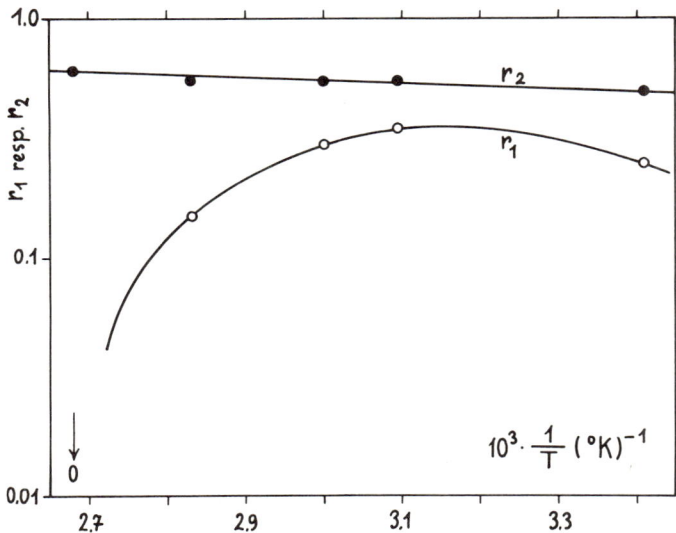

Figure 3. Temperature dependence of the r values used in Equation 5

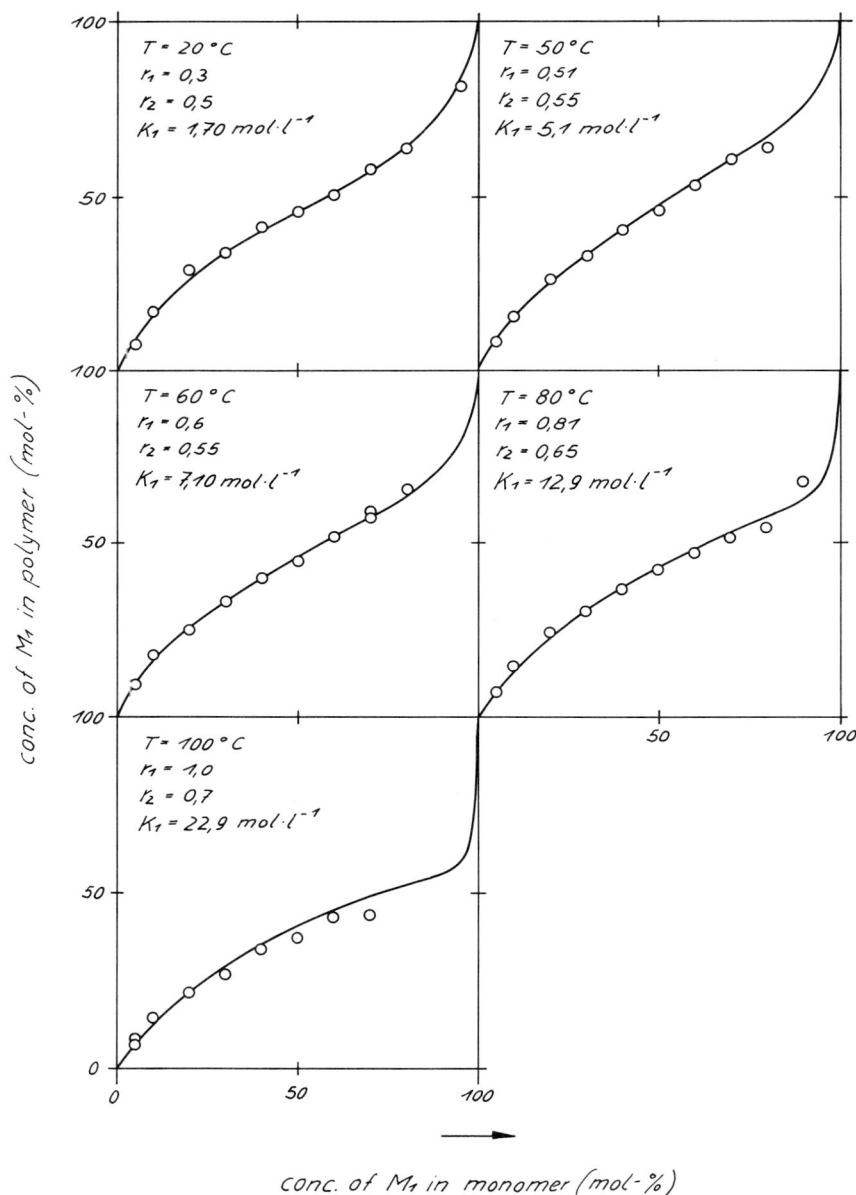

Figure 4. Copolymerization of α-methylstyrene (M_1) and methyl methacrylate (M_2). Curves calculated by Equation 33 for 20°, 50°, 60°, 80°, and 100°C.

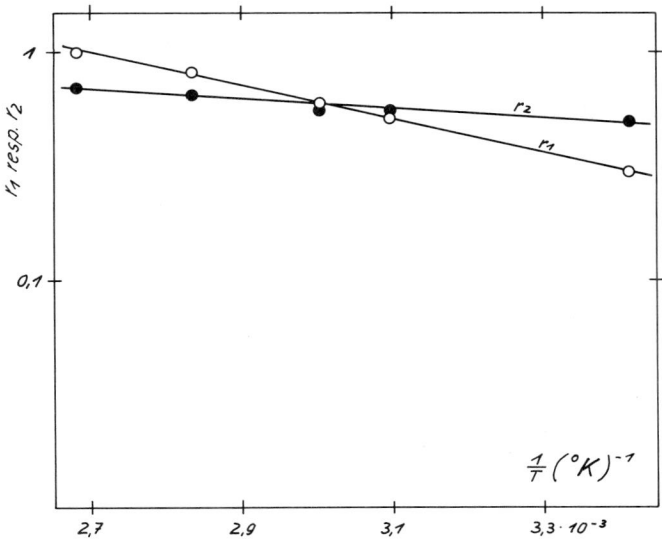

Figure 5. Dependence on temperature of the r values used in Equation 33

One can describe the copolymerization of α-methylstyrene and methyl methacrylate with Equations 5 and 33. Equation 5 reflects only a mathematical approach, whereas Equation 33 takes into account the polymerization–depolymerization equilibrium investigated in the homopolymerization of α-methylstyrene.

In Figure 6 Equation 37 combined with Equation 38 was used to evaluate the copolymerization behavior. It is now assumed that the sequence with two monomer units M_1 cannot depolymerize, but that longer sequences are subject to the polymerization–depolymerization equilibrium.

Using Equation 37 copolymerization curves can be calculated again. A suitable choice of the copolymerization parameters allows a good fit of the curve to the experimental results.

In Figure 7 the r values are plotted according to the Arrhenius equation. The temperature dependence of the r values does not agree with the Arrhenius equation. Therefore, one of the basic assumptions underlaying Equation 37 is false. To explain the experimental copolymerization data, the condition of an equilibrium constant independent of the α-methylstyrene sequence length is best considered.

Measurements on α-Methylstyrene–Methyl Methacrylate at 100°–150°C. All polymerizations were carried out in small sealed glass tubes using thermal initiation. The results are given in Table V and Figure 8.

The copolymerization parameters for the higher temperatures were calculated from the values at lower temperatures using the Arrhenius equation. The copolymerization curves calculated by Equations 33 and 34 do not agree with the experimental results.

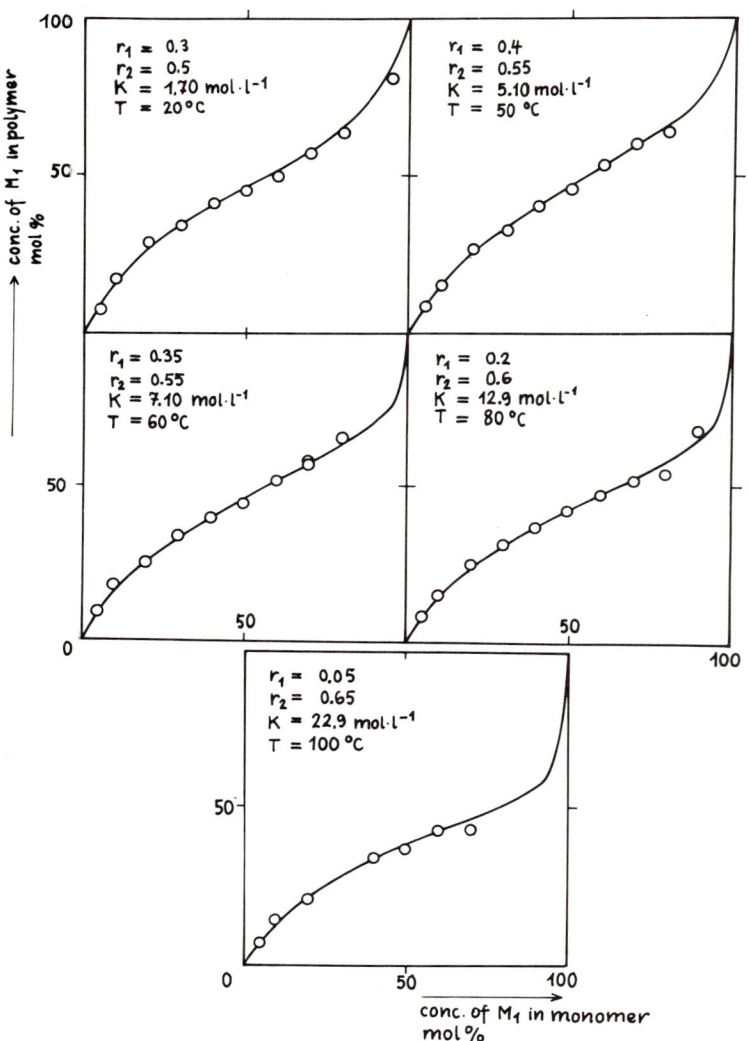

Figure 6. Copolymerization of α-methylstyrene and methyl methacrylate. Curves calculated by Equation 37 for 20°, 50°, 60°, 80°, and 100°C.

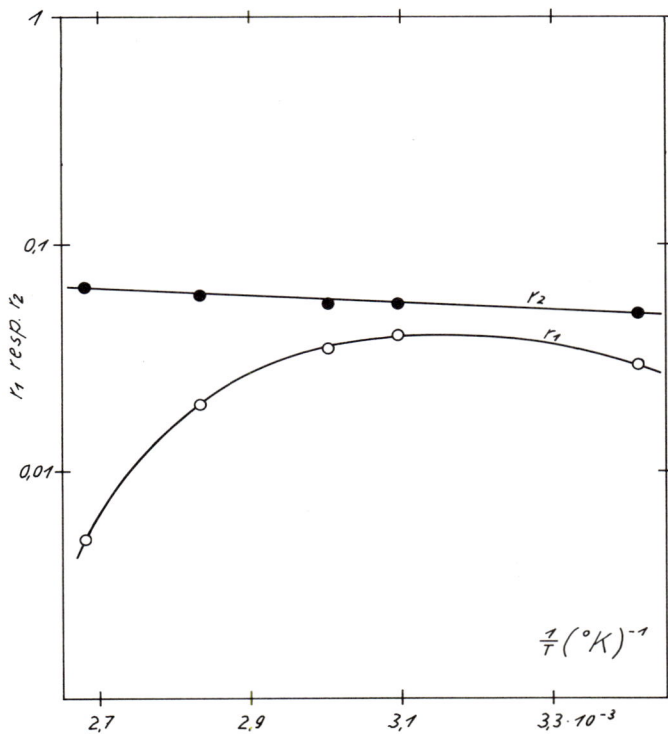

Figure 7. Dependence on temperature of r values used in Equation 37

Table V. Copolymerization of α-Methylstyrene (M_1) and Methyl Methacrylate at 100°–150°C

M_1 mole % in Monomer Mixture	M_1, mole % in Polymer				
	100°C	110°C	120°C	130°C	150°C
5	6.9_7 8.8_8	7.2_4	6.1_6	7.5_2	1.4_9
10	14.6_8	11.3_5	11.0_7	11.0_7	0.6_8
20	21.4_5	21.4_5	13.2_9	13.8_4	3.6_8
30	26.9_3	25.2_0	19.4_6	16.9_2	4.2_3
40	34.0_0	27.5_3	24.3_3	19.1_8	7.0_1
50	37.0_0	31.9_4	28.3_9	23.1_6	7.0_1
60	43.0_9	36.4_1	32.5_2	26.9_3	6.4_4
70	43.4_0	37.3_0	33.1_2	25.4_8	7.2_8
80		40.0_4	35.0_0	32.5_2	7.8_4
90			40.0_3	32.5_2	15.2_4

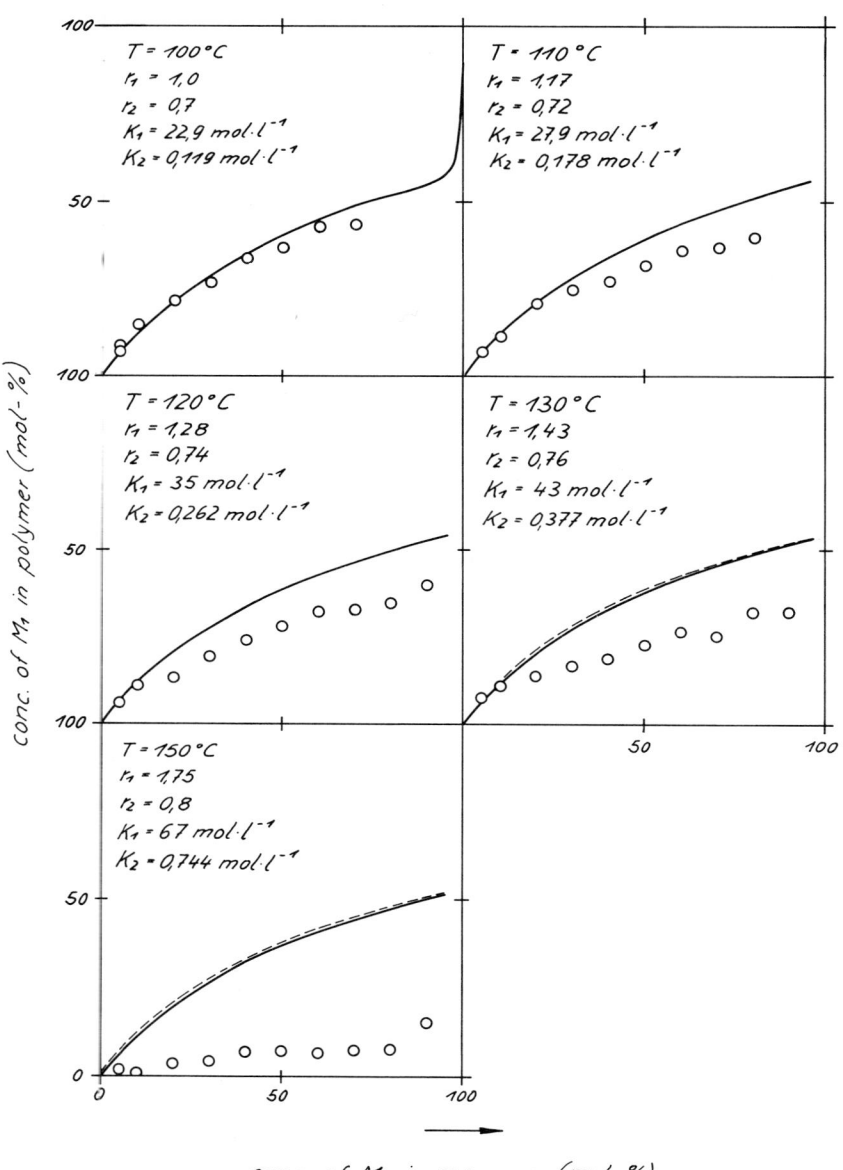

Figure 8. Copolymerization of α-methylstyrene and methyl methacrylate. Curves calculated using Equation 33 for 100°, 110°, 120°, 130°, and 150°C. Dotted curves: reversibility of M_2 taken into account (Equation 35).

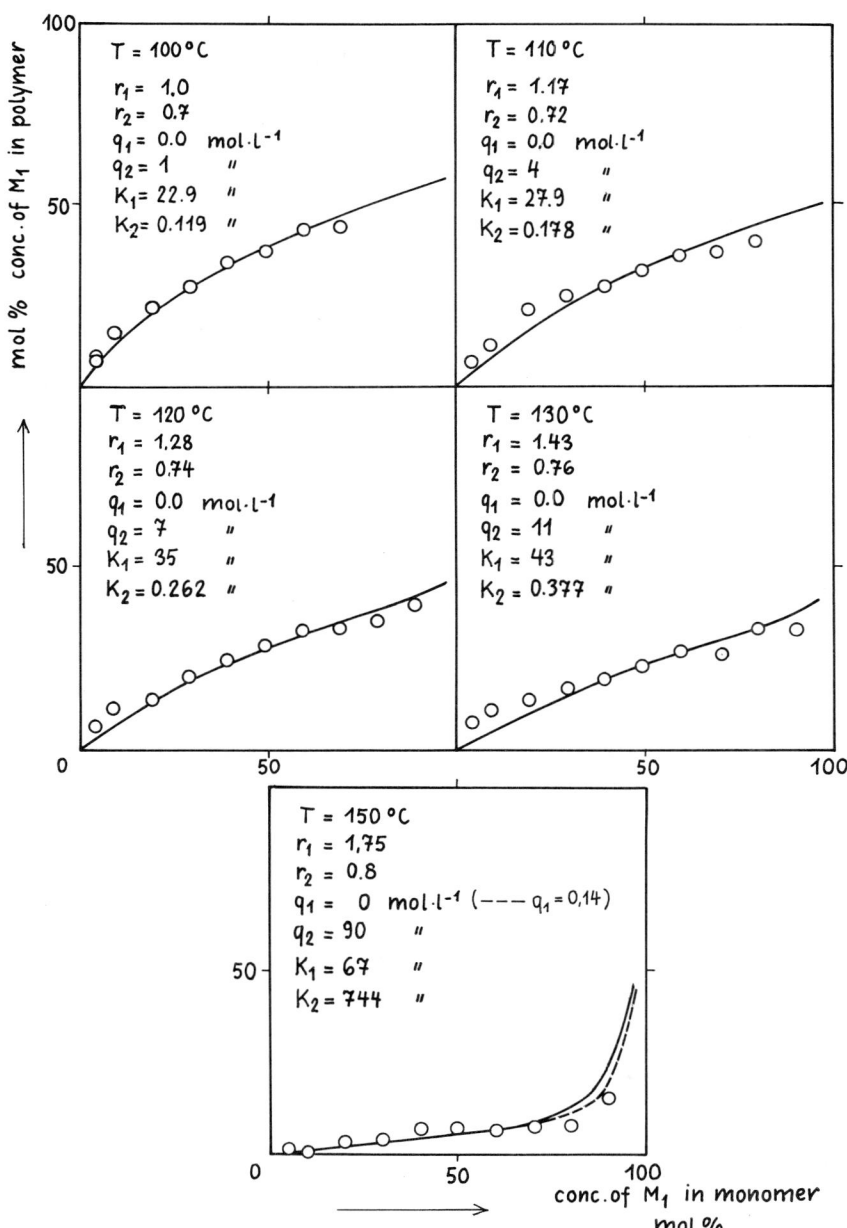

Figure 9. Copolymerization of α-methylstyrene and methyl methacrylate. Curves calculated by Equation 17 for 100°, 110°, 120°, 130°, and 150°C.

If one takes into account the reversibility of monomer M_2 (Equations 35 and 36), relatively little change is observed in the shape of the curves. The equilibrium concentrations of methyl methacrylate used in these calculations were obtained from older measurements (*21*). The experimentally determined concentrations of α-methylstyrene are lower than the calculated ones. For a better fit of the curves to the measured

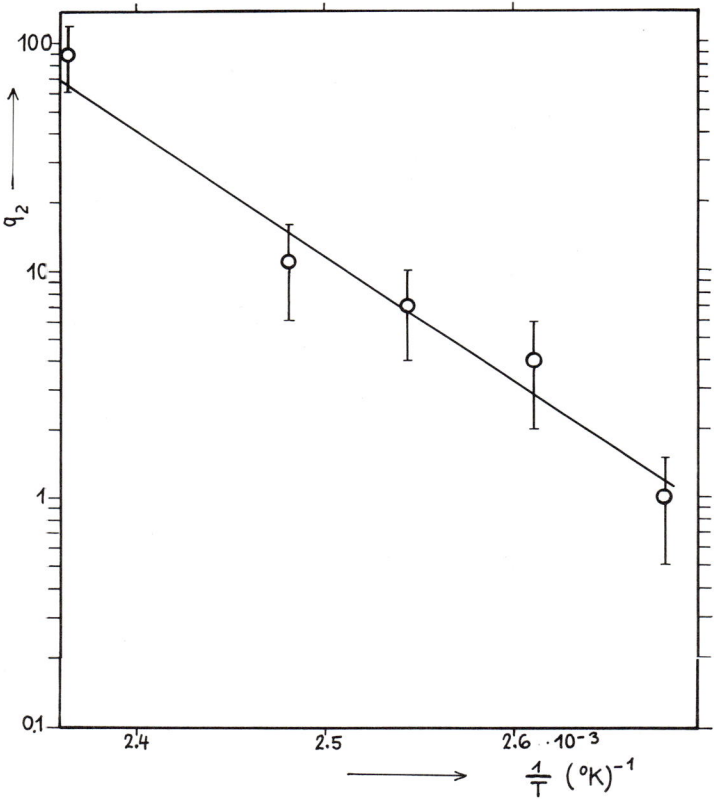

Figure 10. Temperature dependence of q_2 used in Equation 17

values, the reversibility of the alternating steps had to be considered—*i.e.*, Equation 17 in connection with Equations 32a and 32b. To fit the parameters q_1 and q_2 and to calculate the copolymerization curves, a computer was used. With rising temperature q_2 increases; q_1 is zero, and it appears reasonable to assign a small positive value to q_1 only at the highest temperature explored—*i.e.*, 150°C. The results are shown in Figure 9.

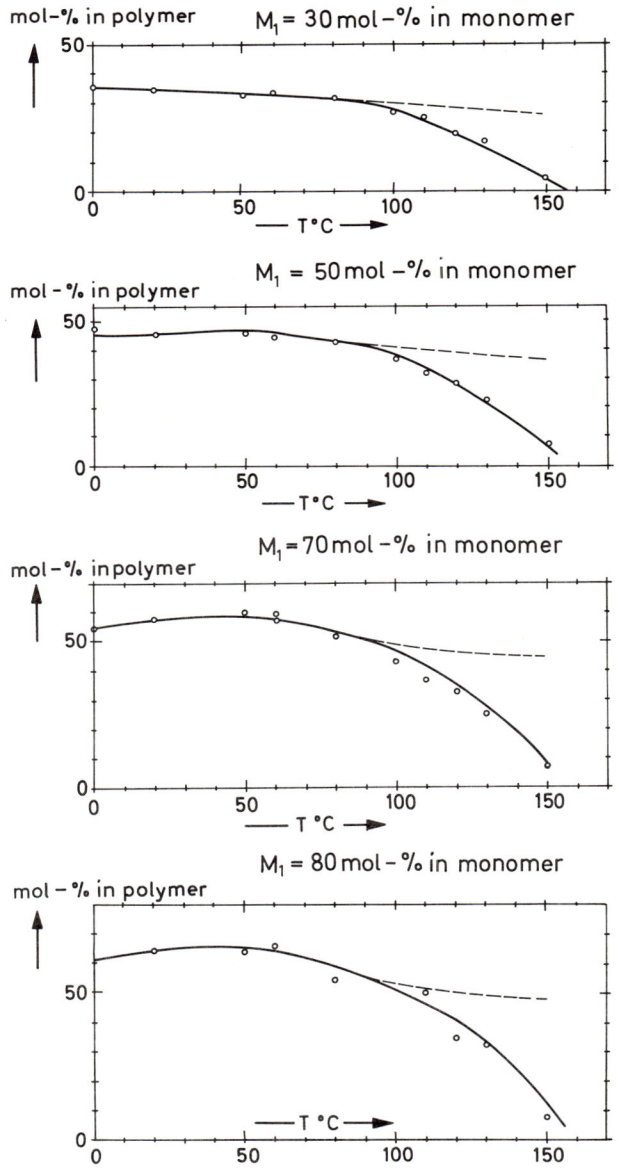

Figure 11. Composition of copolymers as a function of the polymerization temperature. Calculated by Equations 17 and 33. Dotted curves calculated by Equation 33.

The calculated curves agree well with the measured values. At low concentrations of α-methylstyrene (lower left corner of the diagram) there are greater deviations. The experimentally measured concentrations of α-methylstyrene lie above the calculated ones. This difference cannot yet be explained. It is noteworthy that the statistical deviation is higher than at lower temperatures.

The parameter q_2 is by definition a quotient of two rate constants. Therefore, its temperature dependence should follow the Arrhenius equation (Figure 10).

The differences in activation energies are calculated to be

$$\Delta E q_2 = E_{\overline{21}} - E_{12} = 25 \text{ kcal/mole}$$

Figure 11 shows a compilation of the compositions of the polymers which have been polymerized from different monomer mixtures as a function of polymerization temperature. The curves plotted next to the measured points were calculated at temperatures below 100°C by Equation 33 and at temperatures above 100°C by Equation 17. The dotted curves for temperatures above 100°C were calculated by Equation 33. In addition to the measured values taken from Tables III and V, Figure 11 also contains some measured points at 0°C. Polymerization was done in flasks which were stored in a thermally controlled room for a long time (*e.g.*, with 30 mole % α-methylstyrene, 34 days, with 50 and 70 mole %, 166 days). It is apparent that the curve derived by Equation 17 agrees well with the measured points. However, the dotted curve at higher temperatures, calculated by Equation 33 shows significant deviations.

Table VI. Copolymerization of α-Methylstyrene (M_1) and Acrylonitrile (M_2)

M_1 mole % in Monomer Mixture	M_1, mole % in Polymer						
	0°C	20°C	50°C	60°C	70°C	80°C	100°C
5	40.1$_1$	39.0$_9$	36.5$_2$	35.6$_7$	35.2$_6$	34.0$_2$	32.8$_0$
10	45.8$_5$	44.4$_0$	42.3$_9$	42.7$_7$	41.6$_9$	41.2$_7$	39.6$_3$
20	48.8$_8$	48.2$_0$	47.5$_3$	46.7$_9$	46.9$_6$	46.0$_8$	45.4$_3$
30	50.6$_4$	50.5$_5$	49.6$_5$	49.8$_0$	49.9$_6$	49.0$_7$	48.5$_8$
40	51.6$_0$	51.9$_9$	50.6$_3$	51.5$_1$	51.7$_1$	50.8$_3$	50.7$_9$
					51.5$_3$		50.8$_0$
50	52.8$_9$	52.4$_7$	52.3$_6$	53.6$_1$	53.7$_3$	53.1$_2$	52.8$_0$
55.78			53.3$_8$				
60	53.9$_4$	53.6$_9$	54.7$_2$	54.6$_8$	54.8$_7$	54.0$_4$	54.4$_7$
70	55.0$_8$	56.0$_7$	56.4$_2$	57.1$_6$	57.1$_2$	55.9$_6$	56.8$_5$
80	58.1$_5$	59.1$_5$					59.8$_2$
80.29			59.7$_1$	60.1$_2$	60.1$_5$	58.9$_2$	
90	62.5$_9$	64.2$_0$	65.3$_6$	65.5$_2$	65.2$_4$	64.1$_6$	
95	67.2$_7$	68.8$_0$					

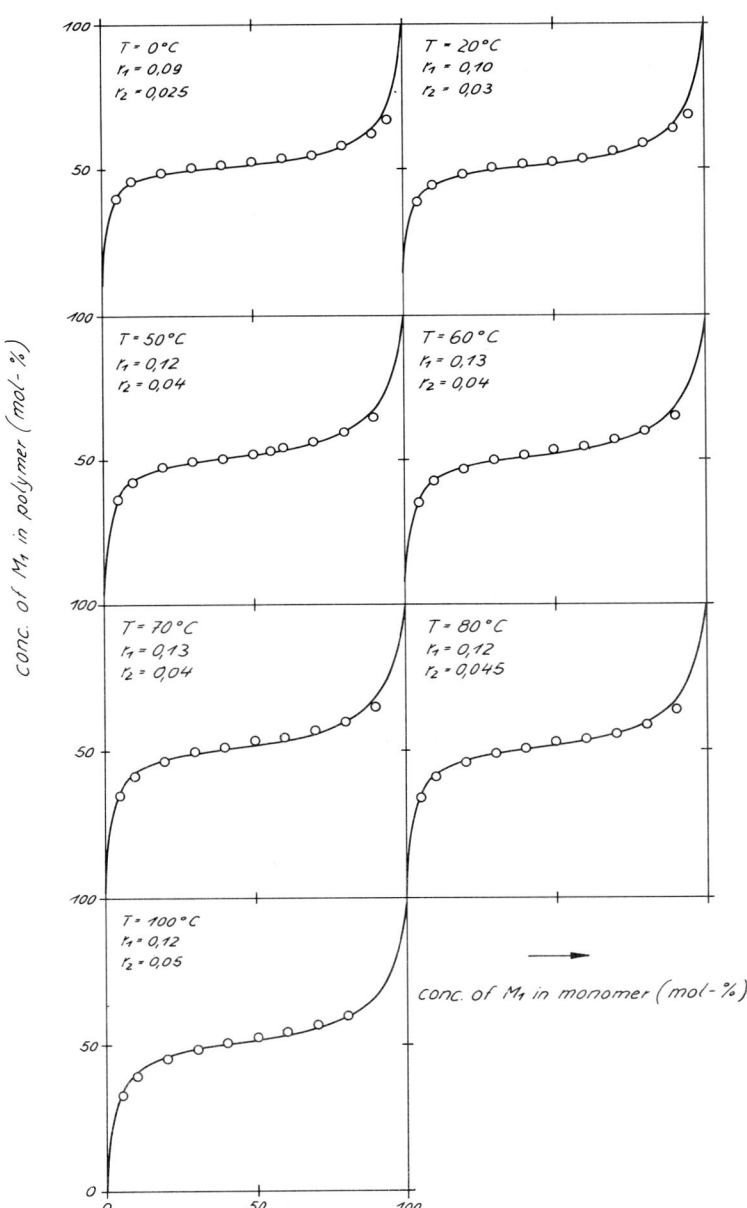

Figure 12. Copolymerization of α-methylstyrene and acrylonitrile. Curves calculated by Equation 5 for 0°, 20°, 50°, 60°, 70°, 80°, and 100°C.

Measurements on α-Methylstyrene–Acrylonitrile. Polymerizations were carried out in dilatometers without solvent. Initiator was azobisisobutyronitrile. At 100°C the reaction was initiated thermally. At 0° and 20°C the reaction was carried out in flasks in a thermally controlled room. Yields were below 5%. The composition of the copolymers was calculated from nitrogen determination (Kjeldahl method). In an older reference (29) polymerizations were carried out at 20°, 50°, 60°, and

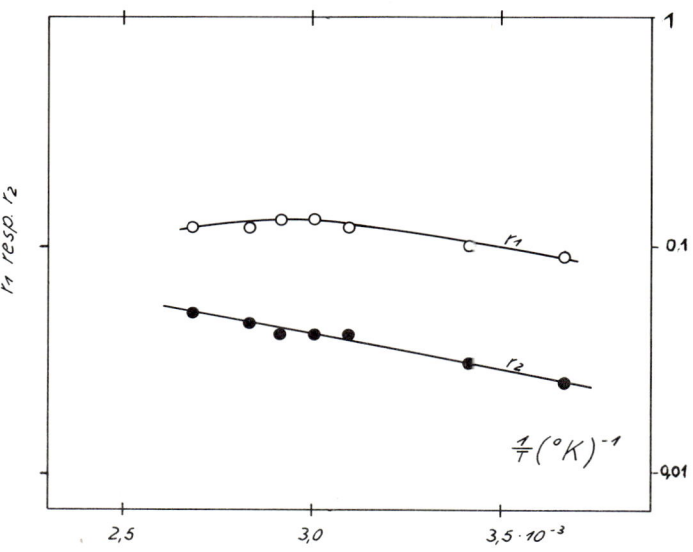

Figure 13. Dependence on temperature of r values used in Equation 5

70°C. The copolymerization parameters, derived from Equation 33 yielded a straight line in an Arrhenius plot. This led to the conclusion that the sequence of two monomer units is subject to the same polymerization–depolymerization equilibrium as in the case for longer sequences. This must be corrected if one considers a wider temperature range. Table VI shows the result of the measurements. Here too it is possible to use the Mayo equation to describe the measurements if the copolymerization parameters at each polymerization temperature are chosen properly (Figure 12). If the logarithm of the measured values is plotted against the reciprocal of absolute temperature, a small deviation from the straight line for r_1 is observed; this is shown in Figure 13.

In the following the reversibility of the polymerization of α-methylstyrene has been considered. The copolymerization curves were calcu-

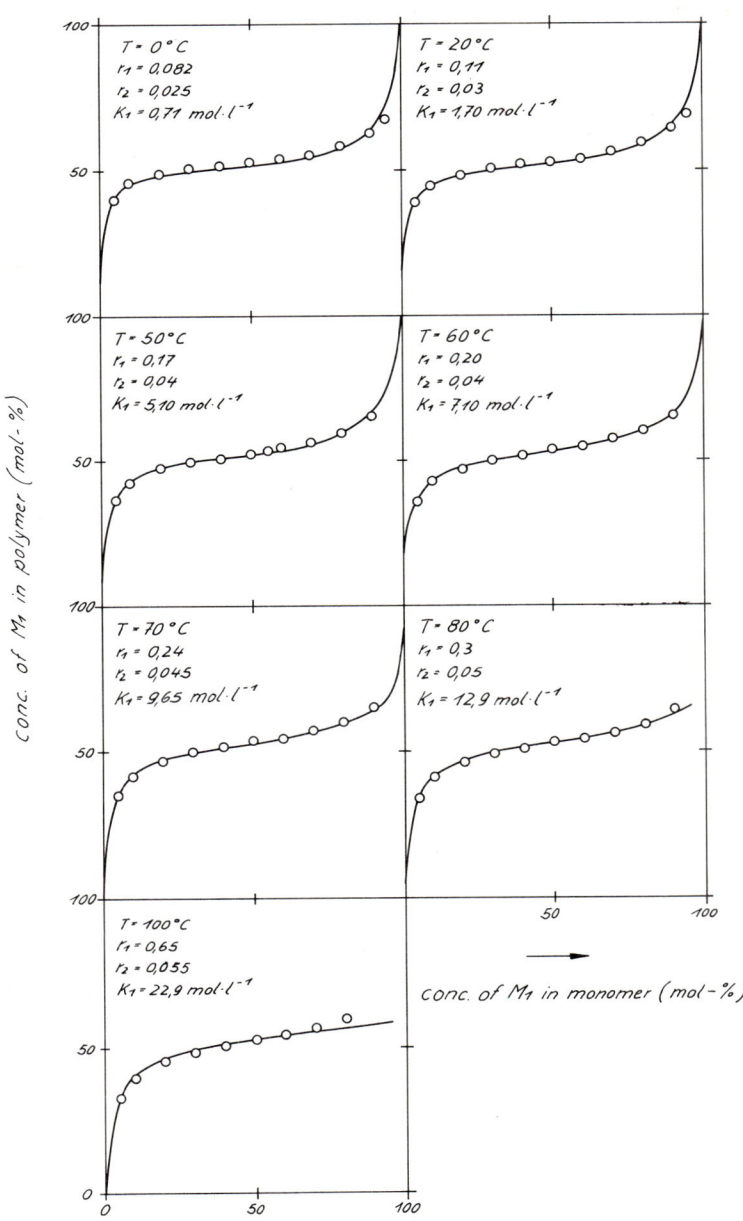

Figure 14. Copolymerization of α-methylstyrene and acrylonitrile. Curves calculated by Equation 33 for 0°, 20°, 50°, 60°, 70°, 80°, and 100°C.

lated by Equations 33 and 34. As Figure 14 shows a proper choice of the copolymerization parameters yields a good description of the measured points. Only at the highest polymerization temperature investigated (100°C) is there a deviation to higher values at the highest concentrations of α-methylstyrene. Even in choosing higher numerical values for r_1 it is impossible to get better agreement in the upper part of the curve without losing good agreement in the lower part.

Figure 15 shows that the copolymerization parameter r_1 used in Equation 33 does not follow the Arrhenius equation. In a previous publication (29) measurements were carried out only in the temperature

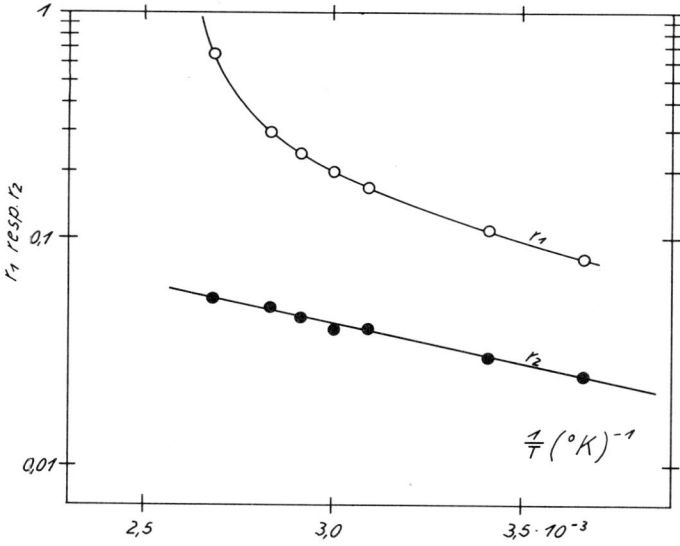

Figure 15. Dependence on temperature of r values used in Equation 33

range 20°–70°C. In this range r_1 still follows the Arrhenius law. Measurements in a wider temperature range, however, demonstrate that this is a purely formal agreement without any real meaning. Figure 16 shows data calculated by Equation 37. This calculation includes the condition that sequences of two monomer units do not depolymerize. Here too, good agreement is reached between the measured points and calculated curves. The r values in Figure 16 follow the Arrhenius equation as shown in Figure 17. The differences in activation energies are calculated to be

$$E_{11} - E_{12} = 1.87 \text{ kcal/mole}$$
$$E_{12} - E_{21} = 1.46 \text{ kcal/mole}$$

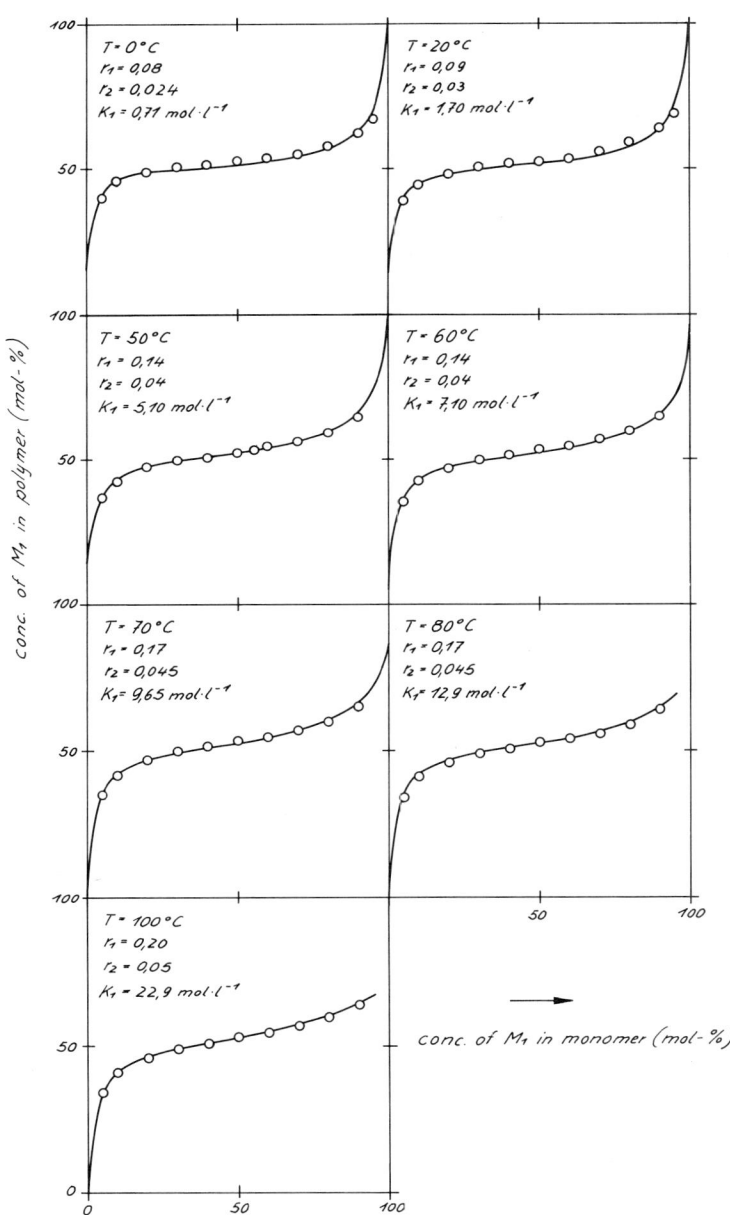

Figure 16. Copolymerization of α-methylstyrene and acrylonitrile. Curves calculated by Equation 37 for 0°, 20°, 50°, 60°, 70°, 80°, and 100°C.

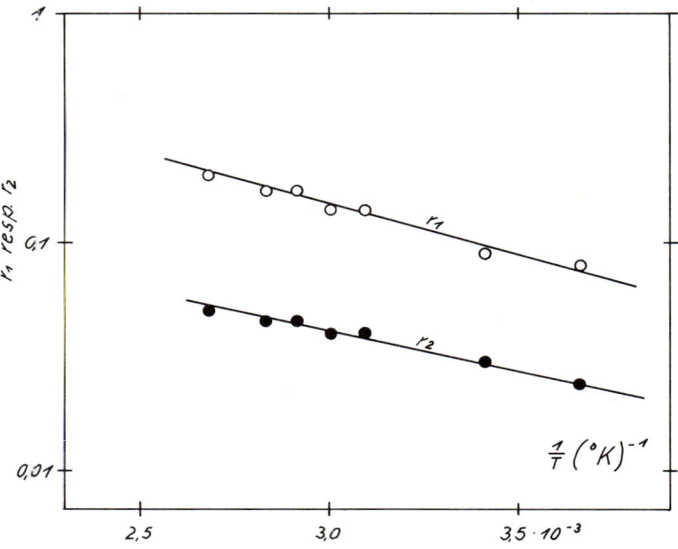

Figure 17. Dependence of r values used in Equation 37 on temperature

Table VII. Copolymerization Parameters Used To Describe the Copolymerization of α-Methylstyrene (M_1) and Acrylonitrile (M_2)

T, °C	Equation 5; Mayo Equation		Equation 33; Sequences of two Monomer Units are Depolymerizable		Equation 37; Sequences of two Monomer Units are not Depolymerizable	
	r_1	r_2	r_1	r_2	r_1	r_2
0	0.09	0.025	0.082	0.025	0.08	0.024
20	0.10	0.03	0.11	0.03	0.09	0.03
50	0.12	0.04	0.17	0.04	0.14	0.04
60	0.13	0.04	0.20	0.04	0.14	0.04
70	0.13	0.04	0.24	0.045	0.17	0.045
80	0.12	0.045	0.30	0.05	0.17	0.045
100	0.12	0.05	0.65	0.055	0.20	0.05

Table VII gives all copolymerization parameters r_1 and r_2 calculated by the different equations.

Summary

The copolymerization behavior of the system α-methylstyrene–acrylonitrile can best be described under the condition that sequences of two monomer units do not depolymerize. This has also been noted by others (19). Hence, a penultimate effect is important in this reaction (8). The r_1 value is smaller than the one in the system α-methylstyrene–methyl methacrylate (Table IV). From this r_1 value it is thought that only occasionally larger sequences than two α-methylstyrene molecules are incorporated in the copolymers. Thus, depolymerization of such longer sequences is also a rare occurrence. Therefore, the r values found by using the Mayo equation 5 are not much different from these calculated by Equation 37. This also explains the relatively small deviation from the Arrhenius law in Figure 13.

It is not possible to predict which mechanism is involved in a certain copolymerization. In the system α-methylstyrene–methyl methacrylate depolymerization of sequences of two monomer units seemed to occur as well as depolymerization of α-methylstyrene from longer sequences. In the system α-methylstyrene–acrylonitrile the sequence of two monomer units of α-methylstyrene is stable and does not depolymerize. The reversibility of the polymerizations of α-methylstyrene and methyl methacrylate can be explained by sterically induced strain in the polymer chain (13). In the copolymer α-methylstyrene–methyl methacrylate this strain involves the whole polymer chain whereas in the α-methylstyrene–acrylonitrile system the strain is broken by the acrylonitrile sequences and is built up again in the α-methylstyrene. This explains the difference in the depolymerization tendencies of sequences of two units of α-methylstyrene and longer sequences in this system.

Acknowledgment

I thank Badische Anilin- & Soda-Fabrik AG for permission to publish this work. I also thank Agathe Dörr for carrying out the experimental work, Dipl.-Math. F. Hafner for numerical calculations, and W. Merz for carrying out the oxygen determinations.

Literature Cited

(1) Bywater, S., *Trans. Faraday Soc.* **51**, 1267 (1955).
(2) Dainton, F. S., Ivin, K. J., *Nature* **162**, 705 (1948).
(3) Dainton, F. S., Ivin, K. J., *Trans. Faraday Soc.* **46**, 331 (1950).
(4) Dainton, F. S., Ivin, K. J., *Quart. Rev. (London)* **12**, 61 (1958).
(5) Doak, K. W., Deahl, M. A., Christmas, I. H., "Abstracts of Papers," 137th Meeting, ACS, April 1960, p. 151.
(6) Fineman, M., Ross, S. D., *J. Polymer Sci.* **5**, 259 (1950).
(7) Flory, P. J., "Principles of Polymer Chemistry," pp. 113, 179, Cornell University Press, Ithaca, N. Y., 1943.
(8) Ham, G. E., *J. Polymer Sci.* **45**, 169, 183 (1960).
(9) Hopff, H., Lüssi, H., *Makromol. Chem.* **62**, 31 (1963).
(10) Johnsen, U., Kolbe, K., *Makromol. Chem.* **116**, 173 (1968).

(11) Küchler, L., "Polymerisationskinetik," pp. 69, 161, 174, Berlin-Göttingen-Heidelberg, 1951.
(12) Lee, C. L., Smid, J., Szwarc, M., *J. Am. Chem. Soc.* **85,** 912 (1963).
(13) Lowry, G. G., *J. Polymer Sci.* **57,** 463 (1960).
(14) Luce, E. N., "Styrene, Its Polymers, Copolymers, and Derivatives," R. H. Boundy, R. F. Boyer, Eds., p. 696, New York, 1952.
(15) Mathieson, A. R., *J. Chem. Soc.* **1960,** 2773.
(16) Mayo, F. R., Lewis, F. M., *J. Am. Chem. Soc.* **66,** 1594 (1944).
(17) McCormick, H. W., *J. Polymer Sci.* **25,** 488 (1957).
(18) Merz, W., *Anal. Chim. Acta* **50,** 305 (1970).
(19) O'Driscoll, K. F., Gasparro, F. P., *J. Macromol. Sci.* **A1** (4), 643 (1967).
(20) Roberts, D. E., Jessup, R. S., *J. Res. Natl. Bur. Std.* **46,** 11 (1951).
(21) Schulz, G. V., Wittmer, P., *Z. Physik. Chem. (Frankfurt)* **41,** 274 (1964).
(22) Snow, R. D., Frey, F. E., *Ind. Eng. Chem.* **30,** 176 (1938).
(23) Snow, R. D., Frey, F. E., *J. Am. Chem. Soc.* **65,** 2417 (1943).
(24) Szwarc, M., *Nature* **178,** 1168 (1956).
(25) Vollmert, B., "Grundriss der Makromolekularen Chemie," p. 271, Berlin-Göttingen-Heidelberg, 1962.
(26) Vrancken, A., Smid, J., Szwarc, M., *Trans. Faraday Soc.* **58,** 2036 (1962).
(27) Walling, C., Briggs, E. R., Wolfstirn, K. B., *J. Am. Chem. Soc.* **70,** 154 (1948).
(28) Walling, C., *J. Polymer Sci.* **16,** 315 (1955).
(29) Wittmer, P., *Makromol. Chem.* **103,** 188 (1967).
(30) Worsfold, D. J., Bywater, S., *J. Polymer Sci.* **26,** 299 (1957).
(31) Worsfold, D. J., Bywater, S., *Can. J. Chem.* **36,** 1141 (1958).

RECEIVED February 25, 1970.

11

Demixing Phenomena in Copolymers

FRITZ KOLLINSKY and GERHARD MARKERT

Röhm & Haas GmbH, 61 Darmstadt, West Germany

> *Mixtures of chemically homogeneous methyl methacrylate–n-butyl acrylate copolymers appear homogeneous optically (clear) and in torsional oscillation experiments (one damping maximum) for $\Delta MMA \leq 10$–25 mole %. There is no compatibilizing effect, and the compatibility range depends upon chemical composition and molecular weight. An analogy exists for "natural" copolymers whose chemical composition shows a broad, continuous frequency distribution. Inhomogeneity results from separation processes which lead to microphases of different size and composition. At a given over-all composition the extent of grafting is a decisive parameter for heterogeneity. For microphase sizes above the resolving power of torsional oscillation experiments compositions are obtainable from locations of the damping maxima. Splitting of damping curves of homogeneous systems into several maxima by a principle of equivalence reflects a separation process.*

The structure of copolymers, even those derived from comonomers having r parameters near 1 is characterized by a broad chemical distribution (Figure 1). This distribution (3) can be utilized as a measure of chemical heterogeneity. Of the various product properties which depend upon heterogeneity, compatibility was chosen for our investigation. Positive proof of the existence of compatibility is problematic especially because no clear definition exists. Even if a monophase system—*i.e.,* homogeneity—is observed, this result may depend on the test method utilized. Only in those cases where multiphase systems are found one can offer definite proof of inhomogeneity. Under these circumstances we limit ourselves to the use of the term inhomogeneity and only in special cases refer to homogeneity.

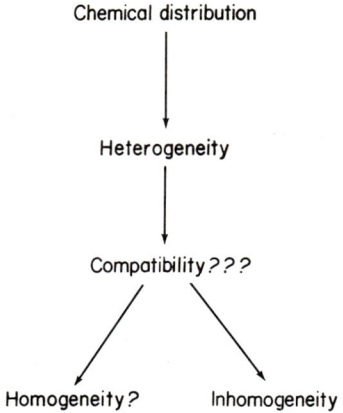

Figure 1. Schematic showing broad chemical distribution structure of copolymers

Composition of Copolymers

In analyzing a system of equal molar parts of methyl methacrylate (MMA) and butyl acrylate (BA) with $r_1 = 2.37$ and $r_2 = 0.34$ (2), it is possible to calculate the composition of a "natural" copolymer—i.e., a polymer found at 100% conversion. This is illustrated in Figure 2. The upper part (a) shows the usual copolymerization diagram, that is the dependence of the instantaneous copolymer composition (p_{MMA}) on the monomer composition (m_{MMA}). The lower part (b) shows the general distribution (H_a) of the chemical composition of the copolymer. That part of the curve to the left of the initial composition holds for a definite natural MMA–BA copolymer. The lined area represents the distribution of the system under study. Comparing the composition of the instantaneously formed copolymers at the beginning (70 mole % MMA) and at the end (0 mole % MMA) of the copolymerization the existence of heterogeneity becomes obvious. One would expect these copolymers to be turbid. However, as shown later, deviations occur. Do these deviations thus signify true monophase systems or do other effects simulate this homogeneity?

Experimental

All polymers were prepared with ethyl azodiisobutyrate in ethyl acetate at ca. 77°C and a concentration of ca. 50% unless otherwise specified. Chemically homogeneous copolymers were prepared with a maximum 3 mole % scatter in composition. Following stabilization they were isolated by partial distillation and reprecipitation.

All tests except electron optical studies were carried out on 0.5-mm thick films prepared from homogeneous 20% ethyl acetate solutions by evaporation and drying for 48 hours at 80°C.

Three methods were used to characterize the films:

(1) Visual (clear/turbid).

(2) Torsion pendulum according to Schmieder and Wolf (6) (1 or more damping maxima).

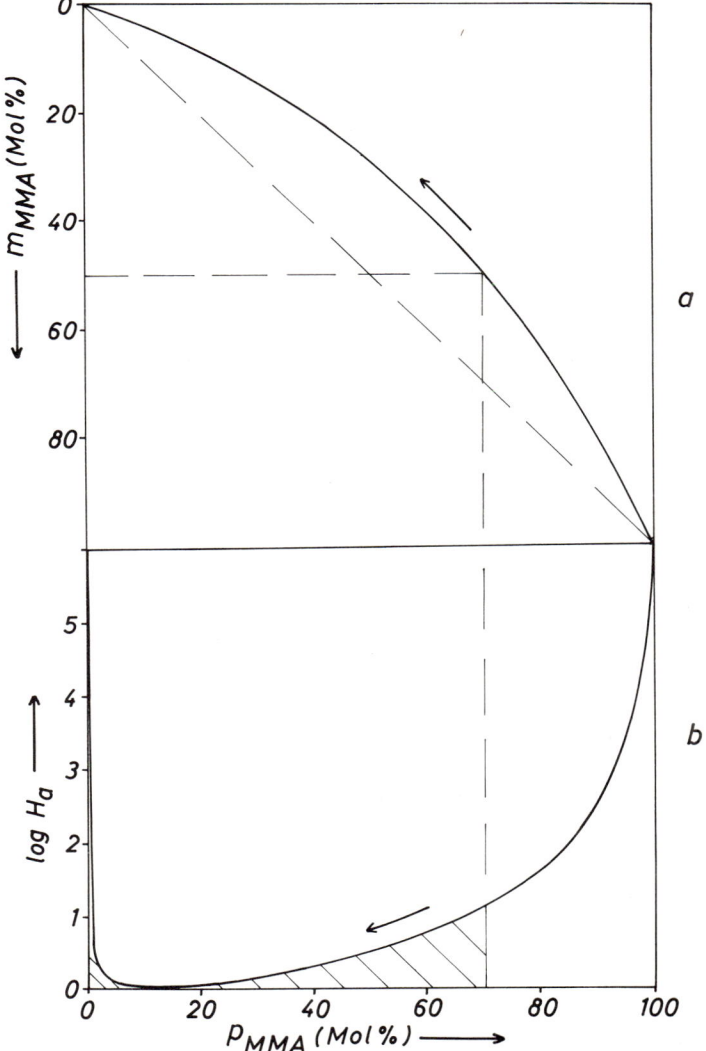

Figure 2. *Copolymerization diagram (a) and chemical distribution (b)*

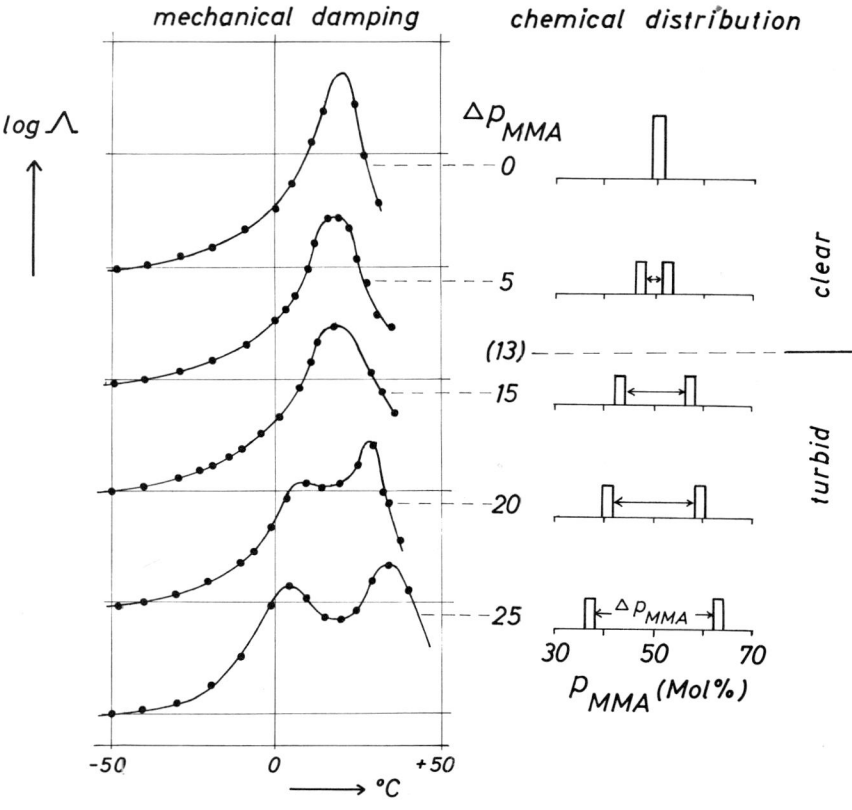

Figure 3. Binary mixtures of copolymers

(3) Electron microscopy (homogeneous—monophase/inhomogeneous—multiphase).

Each method gives unambiguous results only if resolving power and selectivity are not exceeded. Thus, as previously mentioned, recognition of inhomogeneities depends on the method used. Reproducibility of the degree of inhomogeneity found is excellent, as preparation conditions can be kept sufficiently constant. This even though no thermodynamical equilibrium could have been said to be established.

Binary Mixtures of Copolymers

It is possible to consider a natural equimolar MMA–BA copolymer as consisting of a multitude of adjacent chemically homogeneous copolymers with a composition from 70 to 0 mole % of MMA. Investigations of model mixtures composed of only two chemically homogeneous copolymers should give a first approximation.

If equal parts of two chemically homogeneous copolymers with different MMA contents are mixed so that mixtures having a constant gross composition of 50 mole % of MMA are obtained (Figure 3), a sharp transition from clear to turbid occurs with increasing difference in the MMA content, designated Δp_{MMA}. This transition is reproducible within ± 1 mole % of p_{MMA}. In this particular case the Δp_{MMA} at the transition point is 13. This marks the homogeneity span where mixtures, even those mixtures of several chemically homogeneous copolymers are homogeneous. On the other hand, electron microscopic results reveal multiphase systems to exist at Δp_{MMA} of 10; however, at the turbidity limit the degree of dispersion decreases abruptly. Unexpectedly, torsion pendulum experiments in the present case are less sensitive than visual tests; this probably arises from insufficient distinction of damping maxima. The diagram shows that from a Δp_{MMA} of 0 to 15 there is only a broadening of the damping maximum Λ. Only at a Δp_{MMA} of 20 does splitting into two maxima occur. In the homogeneous system the damping maximum is characteristic of the gross composition of the mixture. In the inhomogeneous system the damping maxima reflect the gross composition of the microphases. The torsion pendulum experiment thus provides quantitative information on composition ranges in the process of phase separation. As shown in Figure 4 the homogeneity range Δp_{MMA}

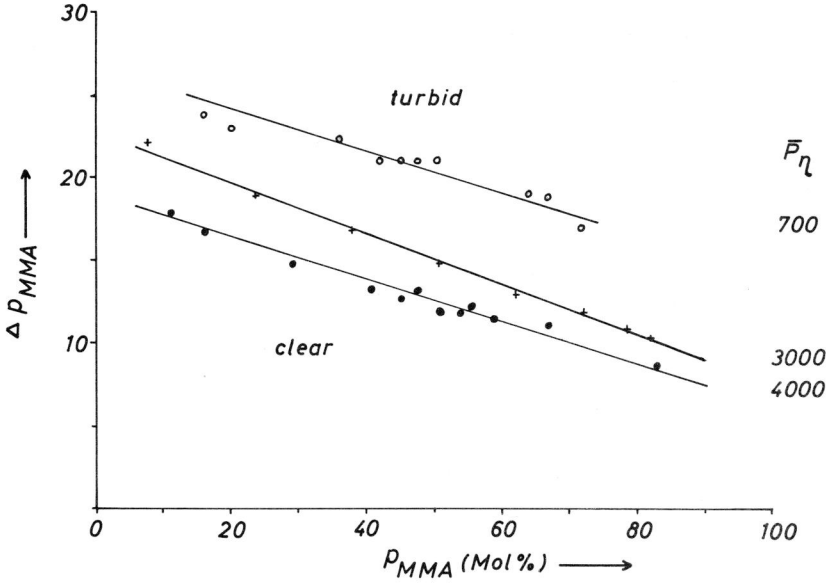

Figure 4. Dependence of homogeneity range on gross composition of p_{MMA} and molecular weight

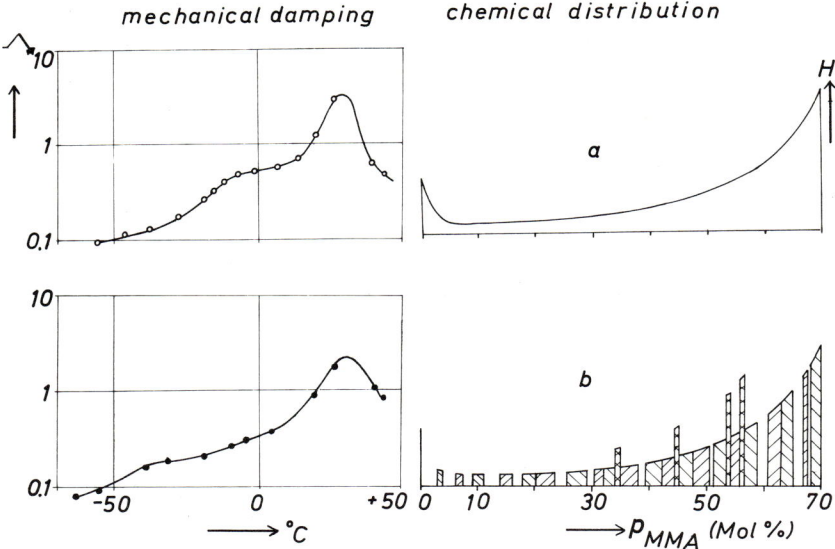

Figure 5. Results of torsion pendulum experiments on (a) a natural copolymer and (b) a simulated copolymer (22 compounds)

depends on the gross composition of p_{MMA} and on molecular weight. Increasing MMA content as well as increasing molecular weight results in a decrease of compatibility.

Another important result was the finding that inhomogeneous binary mixtures of chemically uniform copolymers did not "homogenize" upon adding a third copolymer capable of forming clear films with each of the two mixture components. As Molau (4) already found in the styrene–acrylonitrile system there is no "homogenization."

The conclusions thus far are:

(1) Mixtures of chemically homogeneous copolymers yield homogeneous films at Δp_{MMA} below 25.

(2) The range Δp_{MMA} depends both on absolute composition and molecular weight.

(3) There is no "homogenization" by third copolymers compatible with each of the mixture components.

(4) On the basis of the foregoing it follows that a natural MMA–BA copolymer with a Δp_{MMA} of 70 should be inhomogeneous. However, in some cases the opposite is found. Why?

Natural Copolymers

To see whether or not knowledge of the behavior of binary mixtures of chemically homogeneous copolymers can be applied to natural copolymers, a model of such systems was investigated. This study confirmed that a natural copolymer consists of a mixture of many chemically homogeneous copolymers displaying a continuous change of chemical composition.

The results of torsion pendulum experiments on an inhomogeneous natural copolymer with a MMA–BA ratio of 50–50 mole % and on a natural copolymer imitated by mixing 22 chemically homogeneous copolymers is shown in Figure 5 together with the corresponding chemical distribution. The damping functions in both experiments have the same basic form, while the temperatures at which the main maxima occur are identical. Differences in the shape of the curve below $-10°C$ are not the result of holes or overlaps of the imitated distribution necessitated by experimental difficulties. This is shown by comparing the damping function of the imitated distribution with that of a four-component mixture having a completely uniform chemical distribution (Figure 6). Thus, the proposed composition of a natural copolymer is basically confirmed. A more detailed analysis of the damping function confirms that corresponding to its turbid appearance this natural copolymer is multiphase (Figure 7). The "gross composition" corresponding to the main maximum is located between that of a homogeneous natural copolymer ($T_{A,max}$ 20°, identical with a chemically homogeneous copolymer) and that of chemically homogeneous copolymer containing 68–70 mole % MMA, which represents those parts of the natural copolymer richest in MMA.

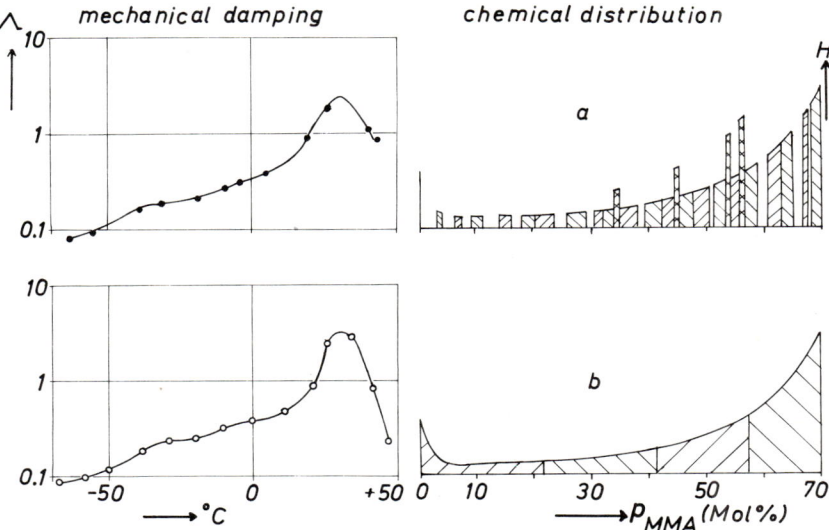

Figure 6. Simulated copolymers: (a) 22 compounds, (b) four compounds

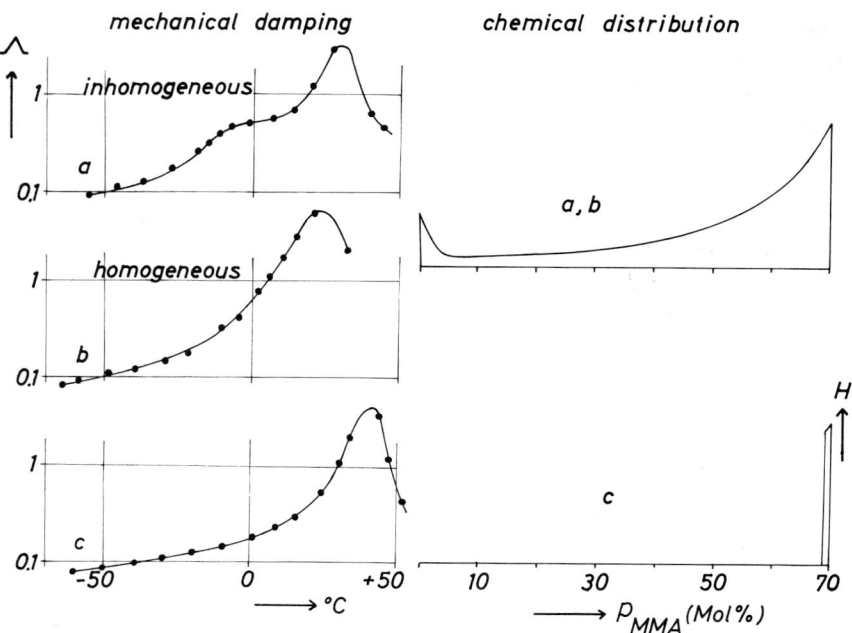

Figure 7. Confirmation of multiphase nature of natural copolymers (a,b) and chemically homogeneous copolymer (c)

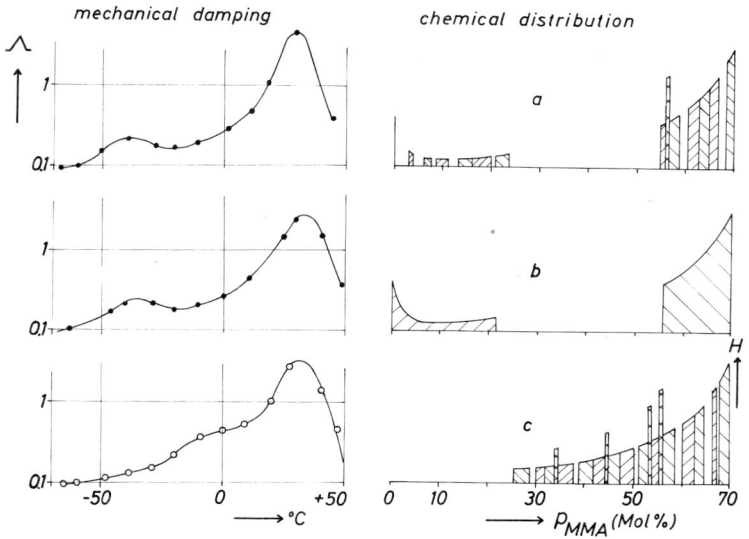

Figure 8. Simulated copolymers with "holes" in their distribution (a,b,c)

Using the homogeneity conditions found for the binary systems where Δp_{MMA} was below 25, the position of the main maximum can now be explained. Fractions with the highest MMA concentration of *ca.* 70 mole % are homogeneously miscible with fractions containing more than 55 mole % MMA. This also holds for mixtures having MMA concentrations for example between 60 and 43 mole % MMA and so on. This means homogeneous microphases with "gross compositions" scattered practically over the entire composition range result from demixing occurring mainly in the region with the most frequent compositions.

Investigation of simulated natural copolymers having holes in the distribution supports this view as shown in Figure 8. If the ranges

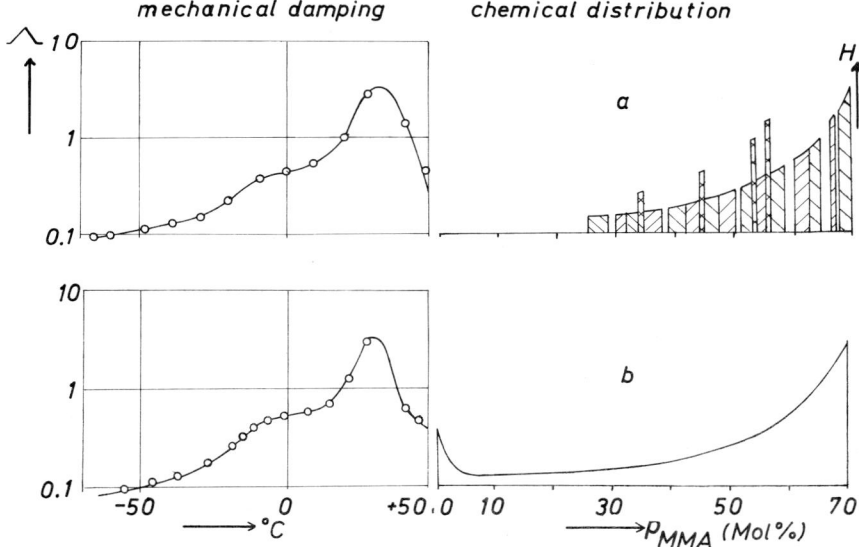

Figure 9. Comparison of damping function for (a) simulated copolymer with a "hole" in the distribution and (b) a natural copolymer

richest in BA or on the other hand the central ranges are omitted from the distribution, the corresponding parts of the damping function will diminish. The effect is independent of whether either the four- or the 22-component system is being investigated.

A second important result is that the damping function of a distribution having the hole within the range richest in BA is largely similar to that of the inhomogeneous natural copolymer (Figure 9). Since the conversion is practically 100%, the only explanation is that the parts richest in BA are so finely dispersed that they are below the resolving

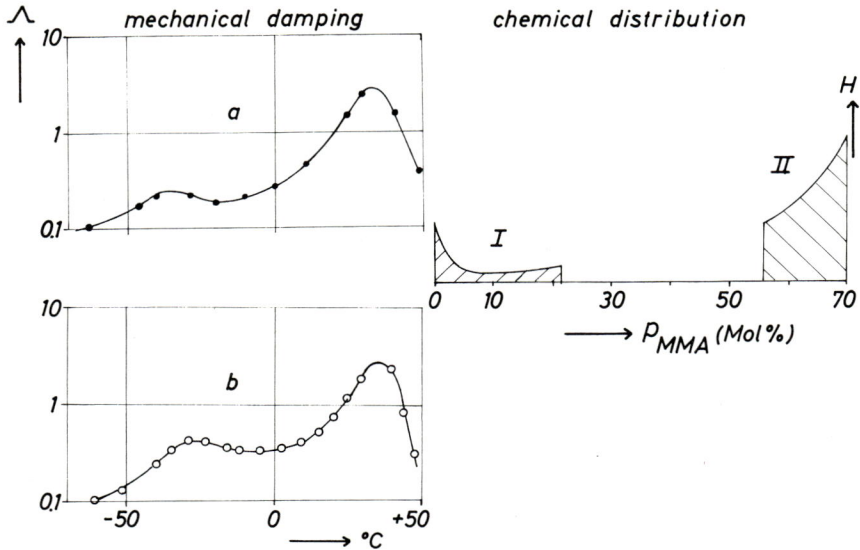

Figure 10. Simulated copolymers having a "hole." (a): Mixture of I and II. (b): I polymerized in the presence of II.

power of the torsion pendulum experiment. A possible mechanism for this becomes probable as follows. If one tries to obtain a distribution with a central hole by polymerizing the BA-rich part I in the presence of the MMA-rich polymer II (Figure 10), damping increases in the temperature range $-10°$ to $-30°C$ as compared with an actual mixture of Polymer I and II. At the same time the resulting polymer is clearer.

Riess (5) has shown that block copolymers of the type X–Y inhibit the demixing of polymer blends consisting of X and Y. In our case the resulting graft copolymers having an average grafting degree about 1 are comparable in effect with block polymers. The formation of graft copolymers is generally favored by transfer grafting of those parts richest in BA onto the BA elements of MMA-rich polymer previously formed and existing in high concentration. Thus, the inhibition of demixing simulates in the central sector the existence of a composition formed by finely dispersed fractions composed of copolymers rich and poor in BA. Furthermore, this could be the reason why no mechanical absorption typical for BA-richest copolymer fractions is found in torsion pendulum experiments done on natural copolymers.

If graft copolymers determine the homogeneity or inhomogeneity of natural copolymers, any influence on the grafting process should change the resulting characteristics of the polymer. Among the factors influencing grafting are molecular weight, solvent, and initiator.

Molecular Weight. Increasing molecular weight leads to an increasing tendency toward inhomogeneity, as we have already seen in the binary system. At the same time transfer grafting is reduced because of the reduced amount of initiator. Copolymers displaying a degree of polymerization up to $P_\eta = 600$ form a clear film and reveal only one broadened damping maximum (Figure 11). Copolymers with higher degrees of polymerization P_η, for example 2000, form turbid films which have a definite secondary maximum.

Solvent. Solvents also may be significant in grafting. In a poor solvent such as 2-propanol, rather than in a good solvent like ethyl acetate, polymerization yields clear films up to $P_\eta = 900$. Polymers formed in 2-propanol are solvated preferentially by the residual monomer. As a result, transfer grafting and the length of graft branches is increased. In contrast the influence of the solvent on film formation is comparatively small. Even though conditions for grafting are better in 2-propanol, the resulting turbid films are mechanically more inhomogeneous than would be the case for films prepared using ethyl acetate (Figure 12). Differences in the structure of the graft polymers should be the cause of the fact that more pronounced demixing is observed above the molecular weight limit of homogeneity as for ethyl acetate.

Initiator. During polymerization in ethyl acetate exchanging ethyl azodiisobutyrate for *tert*-butyl perpivalate the visual homogeneity limit rises from $P_\eta = 600$ to *ca.* 4000. In accordance with the visual result

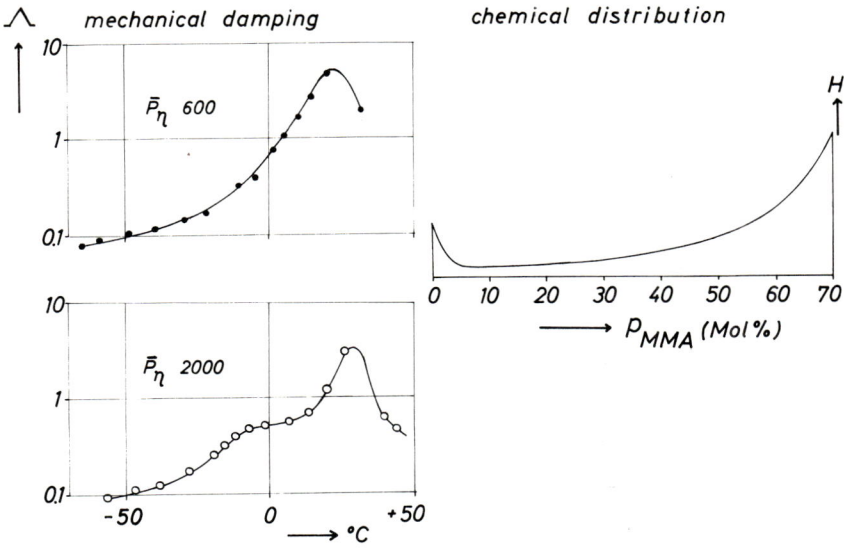

Figure 11. Damping maximum for natural copolymers

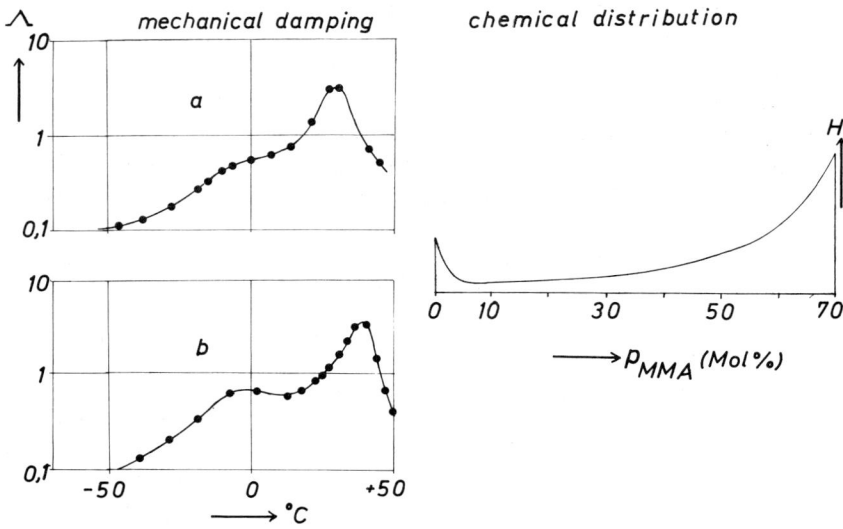

Figure 12. Natural copolymers polymerized in (a) ethyl acetate and (b) 2-propanol

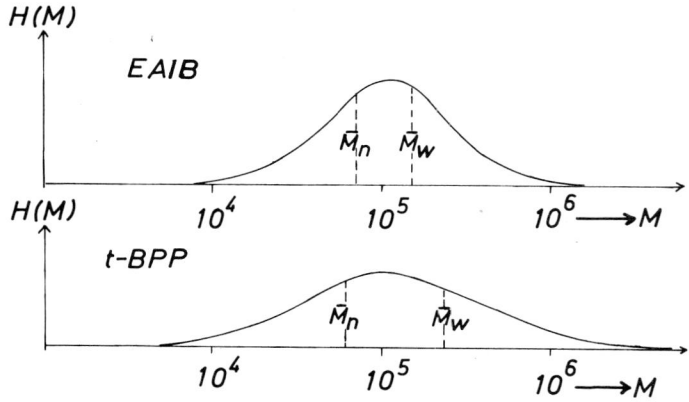

initiator	$\bar{M}_n$	$\bar{M}_w$	$\frac{\bar{M}_w}{\bar{M}_n} - 1$
EAIB	70,000	150,000	1.15
t-BPP	60,000	230,000	2.84

Figure 13. Molecular weight distribution

only one (although considerably broadened) damping maximum is found. As might be expected, the initiator choice has a considerable influence. As generally known, grafting, caused by either polymer or initiator radicals, results in an expansion of the molecular weight distribution toward higher molecular weights (Figure 13). Polydispersity calculated from gel chromatography experiments shows significantly increased grafted fractions when *tert*-butyl perpivalate was used. Other factors such as polymerization temperature, concentration, and transfer agents also influence inhomogeneity. In these cases, however, several parameters must be varied to keep the molecular weight constant. Hence, it becomes difficult to separate the effects these parameters contribute.

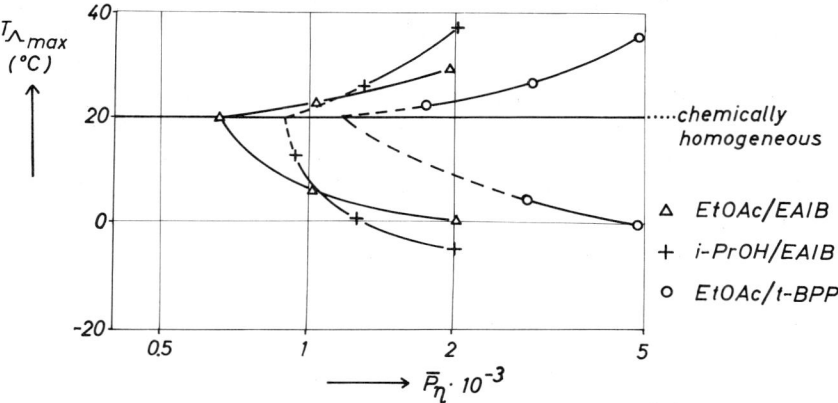

Figure 14. *Phase separation in natural copolymers*

A quantitative consideration of the demixing effects described earlier reveals a general pattern. Depending on the conditions of polymerization different temperature differences are found between main and secondary maxima of the damping curve. This is shown in Figure 14 in which the temperature of the damping maximum $T_{\Lambda,\max}$ is plotted vs. the average degree of polymerization P_η. Parameters are three different conditions of polymerization. For the chemically homogeneous copolymer the temperature of the damping maximum was found to be 20°C. Any deviation from this indicates demixing, regardless of the number of maxima found. The peak formed by the two branches of the curve corresponds in each case to the homogeneity borderline of a particular polymerization system. This limit and the shape of the branches are characterized by type and quantity of the graft polymers formed. With increasing molecular weight the damping function of a homogeneous system is split into several maxima. This is in accordance with a principle of equivalence and reflects the demixing process.

Our results of investigations in the MMA–BA system should be of general relevance at least to the acrylate–methacrylate system.

Summary

Mixtures of chemically homogeneous methyl methacrylate–n-butyl acrylate copolymers are homogeneous for composition ranges Δp_{MMA} below 25 mole %. The copolymers appear optically clear and show one maximum in torsion pendulum experiments. There are no homogenization effects. The value of Δp_{MMA} depends both on absolute position of the chemical composition and molecular weight. Similar conditions apply to "natural" copolymers with a broad uninterrupted frequency distribution of chemical composition. Inhomogeneities result from demixing and lead to microphases with different degrees of dispersion and composition. In addition for a given gross composition the degree of graft reactions which is influenced by solvents, initiators, etc., is a decisive parameter for homogeneity. If the microphases are beyond the resolving power and the selectivity of the torsion pendulum experiment, the positions of the damping maxima reflect the gross composition of these phases.

Acknowledgment

We thank W. Geymayer for the electron microscopic plates, G. Schreyer and J. Hennig for carrying out the torsion pendulum measurements, W. Wunderlich for determining molecular weights, and H. Knöll for technical assistance.

Literature Cited

(1) Gerrens, H., *Chem. Ing. Tech.* **39**, 1053 (1967).
(2) Kretz, R., *Angew. Makromol. Chem.* **11**, 41 (1970).
(3) Markert, G., *Makromol. Chem.* **109**, 112 (1967).
(4) Molau, G. E., *J. Polymer Sci. Pt. B*, **3**, 1007 (1965).
(5) Riess, G., Kohler, J., Tournut, C., Banderet, A., *Makromol. Chem.* **101**, 58 (1967).
(6) Schmieder, K., Wolf, K., *Kolloid-Z.* **127**, 65 (1952).

RECEIVED February 9, 1970.

12

Dynamic and Stress–Optical Properties of Polyblends of Butadiene–Styrene Copolymers Differing in Composition

GERARD KRAUS and K. W. ROLLMANN

Phillips Petroleum Co., Research and Development Dept., Bartlesville, Okla. 74003

> *Dynamic viscoelastic and stress–optical measurements are reported for blends of crosslinked random copolymers of butadiene and styrene prepared by anionic polymerization. Binary blends in which the components differ in composition by at least 20 percentage units give 2 resolvable loss maxima, indicative of a two-phase domain structure. Multiple transitions are also observed in multicomponent blends. All blends display an elevation of the stress–optical coefficient relative to simple copolymers of equivalent over-all composition. This elevation is shown to be consistent with a multiphase structure in which the domains have different elastic moduli. The different moduli arise from increased reactivity of the peroxide crosslinking agent used toward components of higher butadiene content.*

When two chemically different polymers are mixed, the usual result is a two-phase polyblend. This is true also when the compositional moities are part of the same polymer chain such as, for instance, in a block polymer. The criterion for the formation of a single phase is a negative free energy of mixing, but this condition is rarely realized because the small entropy of mixing is usually insufficient to overcome the positive enthalpy of mixing. The incompatibility of polymers in blends has important effects on their physical properties, which may be desirable or not, depending on the contemplated application.

In many copolymerizations compositional differences arise either intermolecularly or intramolecularly (or both) as a result of the kinetics of the copolymerization. This is true in the anionic batch copolymerization of styrene and butadiene which allows the preparation of copolymers of almost any degree of randomness (3). Whereas it is easy to distinguish between outright block polymers and random copolymers by their characteristic mechanical behavior (1, 4, 5), subtle differences are likely to arise from compositional heterogeneity in random copolymers, more specifically copolymers free of long sequences of either comonomer in which composition nevertheless varies along the chain. In this connection we wished to establish some limits of compatibility between molecules or molecular sequences differing in monomer ratio only. As a first step a series of polyblends was investigated which had been prepared exclusively from uniformly random butadiene–styrene copolymers. (The term "uniformly random" is used to denote a copolymer for which composition is independent of conversion, precluding the possibility of any sort of compositional heterogeneity on a scale of more than a few monomer units). In this report we describe the results of both dynamic and stress–birefringence measurements on these polyblends. Microheterogeneity is clearly detectable in binary blends of components differing in composition by 20 percentage units and may be present in blends of even narrower composition distribution. The results are analyzed by alternative one-parameter equivalent mechanical model treatments, one of which is shown to be moderately successful in describing both the dynamic and stress–optical properties of the blends in terms of the properties of the components.

Experimental

Polymers. The polymers used in the blending experiments were prepared by anionic polymerization using an alkyllithium initiator and a chemical randomizing agent to control monomer sequence, in the manner described by Hsieh and Wofford (3). Randomness was checked in each case by measuring the styrene content as a function of conversion. Table I gives descriptive data for these polymers.

Table I. Polymer Characterization

Polymer	Charge Ratio B/S	Styrene Anal., %	Microstructure (B)			GPC	
			cis, %	trans, %	vinyl, %	$M_w/1000$	$M_n/1000$
A	9:1	9.7	45.0	39.9	15.1	197	163
B	8:2	19.6	35.9	49.0	15.1	182	153
C	7:3	29.5	35.5	49.4	15.1	138	119
D	6:4	39.6	35.8	50.0	14.2	134	115
E	5:5	49.3	31.8	53.0	15.2	121	105
F	10:0	0.0	48.6	42.5	8.9	260	203

The styrene contents were confirmed analytically by ultraviolet absorption spectroscopy (7). The microstructure of the contained butadiene units was determined by infrared analysis (10). Molecular weights were determined by the gel permeation technique using a commercial Waters Associates chromatograph. The solvent was tetrahydrofuran. Since separate calibrations against light scattering molecular weights were available only for polybutadiene (of the same type as Polymer F) and for 75:25 butadiene:styrene, the molecular weights of the copolymers were obtained from a universal calibration curve of $[\eta]M$ vs. eluent volume, as proposed by Benoit and associates. (2).

Polymer Blending and Sample Preparation. Two methods were used to prepare polymer blends. Solution blending was accomplished by mixing toluene solutions of the component polymers (5% solids) with good agitation and coagulating the mixture in a large excess of 2-propanol under rapid stirring. The coagulated blends were then dried at 30°C *in vacuo*. Crosslinking agent (Di-Cup 40C) was added on a two-roll laboratory mill. The amount used was 0.75%, corresponding to 0.3% of active dicumyl peroxide.

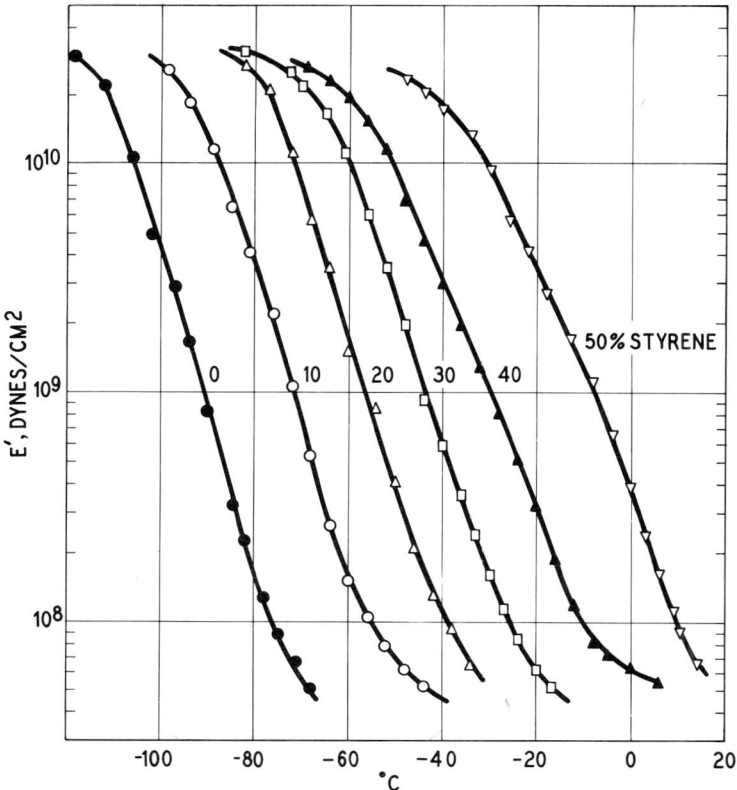

Figure 1. Storage moduli of random copolymers

In preparing dry blends the peroxide was added to each component separately, and the resulting "masterbatches" were blended for 5 minutes on a roll mill at 70°C. All samples were press-cured for 30 minutes at 153°C into films approximately 0.2 mm thick.

Dynamic Measurements. A Vibron direct reading viscoelastometer (Toyo Measuring Instruments Co., Ltd., Tokyo, Japan) was used to determine viscoelastic properties. This instrument and its operation have been described in detail by Takayanagi and Yoshino (*11, 14*). All polymers and blends were examined at a standard frequency of 110 cps. In a few selected examples measurements were also made at 3.5 cps.

Stress–Birefringence. The apparatus used to determine simultaneously stress and birefringence was of conventional design, employing compensation for the optical measurement. All measurements were carried out at 24°C. The sample chamber was a 2-liter Dewar with optically flat windows. The upper end of the sample strip was connected to a stress transducer, while the lower end could be manipulated from the outside to impose any desired strain. The actual strain was measured through the windows with a cathetometer, using gage marks stamped on the sample for reference. To measure the stress-optical coefficient, successively increasing strains were imposed on the sample. At each strain the stress was recorded continuously for 5 minutes, at which time the compensator (Babinet type, Gaertner Instrument Co.) reading was made. Plots of birefringence *vs.* stress showed good linearity. From these the stress–optical coefficients were calculated by the method of least squares.

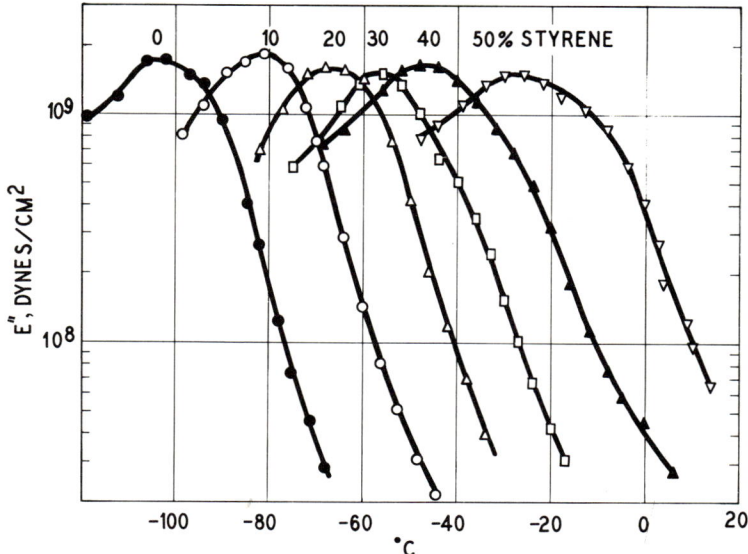

Figure 2. Loss moduli of random copolymers

Results

Dynamic Measurements. Figures 1 and 2 show the dynamic storage and loss moduli E' and E'' at 110 cps for the six component polymers, A to F. The position of the loss maximum is plotted as T_{max}^{-1} vs. styrene content in Figure 3. A good straight line is obtained which passes through the expected values for pure polybutadiene (15% vinyl) and polystyrene

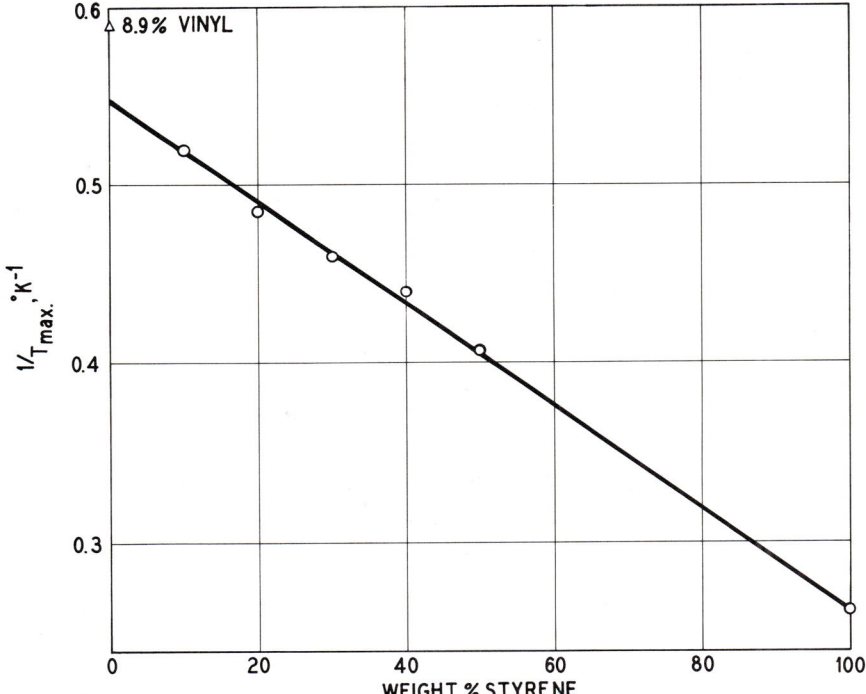

Figure 3. Position of loss maximum as function of copolymer composition (circles for 15% vinyl in butadiene portion)

—i.e., −91° and 108°C, respectively. These values lie slightly above the dilatometric glass transitions, as would be expected for 110-cps data.

Figure 4 shows data for a mechanical blend of equal weights of Polymers A and E. The two-phase nature of this blend is immediately obvious from the observation of separate transitions for the component polymers which are almost unshifted from their positions in Figures 1 and 2. The corresponding solution blend is shown in Figure 5. The differences between the dry blend and the solution blend are very small, indicating that the incompatibility is not the result of inadequate dry blending technique. The same was found also for other blends of narrower composition distribution. For this reason further discussion will be confined to dry blends only.

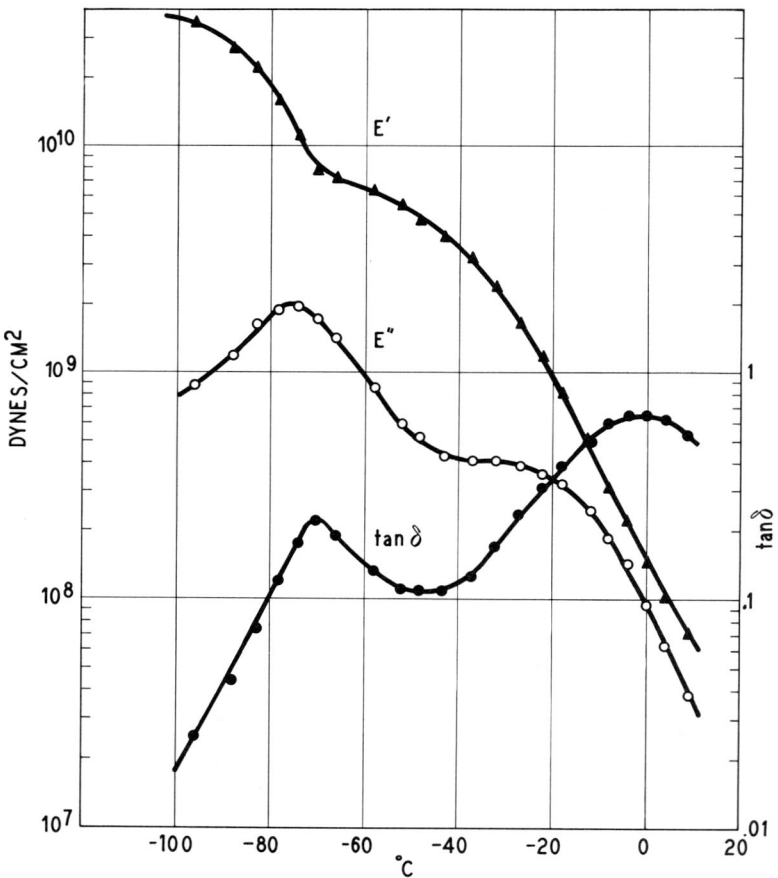

Figure 4. Dynamic properties of 1:1 mechanical blend of Polymers A(10% styrene) and E(50% styrene)

Figure 6 shows a blend of equal weights of Polymers B and D. Separate loss peaks are no longer resolved, but a distinct broadening of the dispersion region is observed which suggests that the blend is two-phase. A ternary blend of equal weights of Polymers B, C, and D (Fig-

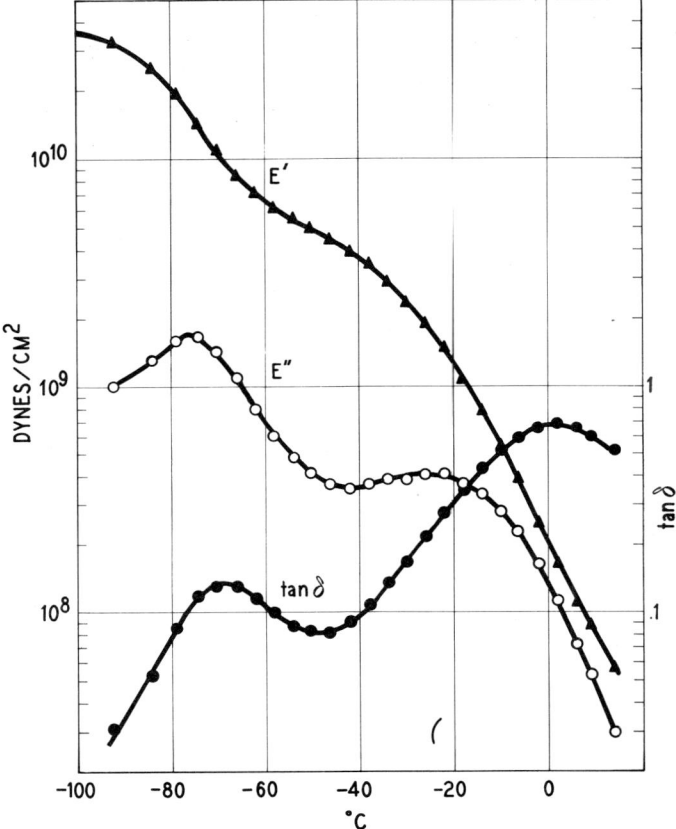

Figure 5. Dynamic properties of 1:1 solution blend of Polymers A and E

ure 7) shows a narrower dispersion region. The dynamic properties of this blend would be difficult to distinguish with certainty from those of Polymer C, which has the same over-all composition. On the other hand, a five-way blend of equal amounts of A, B, C, D, and E (Figure 8) displays a broad dispersion region with an interesting characteristic: the principal peaks in E'' and tan δ are displaced from each other to an unusually high degree. Obviously the maximum in E'' is weighted most heavily by the components of lower T_g, whereas the maximum in tan δ is weighted predominantly by the components of higher T_g or higher styrene content. The behavior of the blends of Figures 4 to 8, all of which contain 30% styrene over-all, is typical also of blends of other average compositions, as well as of blends of unequal amounts of the components. Results of all blending experiments are summarized in Table II.

Comparison of the positions of the loss maxima leads to some interesting observations. In all binary blends, which differ by more than 20 percentage units in component composition, the E'' maxima occur within a few degrees of their position in the components. The maxima in tan δ

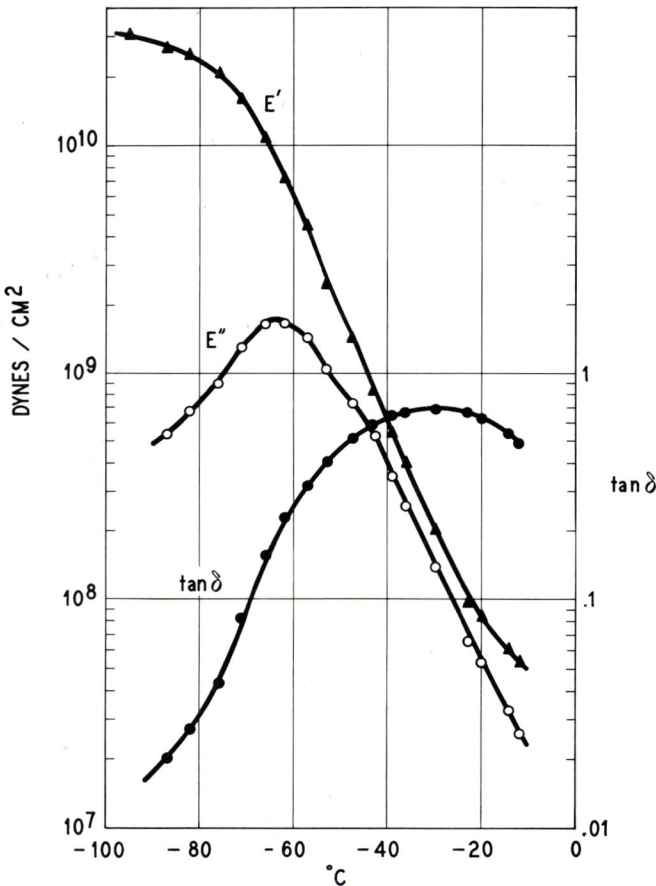

Figure 6. Dynamic properties of 1:1 blend of Polymers B(20% styrene) and D(40% styrene)

tend to be affected more by blending, but the behavior is not entirely consistent. In binary blends both maxima in the loss tangent occur generally at lower temperatures than they do in the components. The shift is of the order of 5–10° for the lower maximum, but varies from zero to 35° for the upper maximum. In Blend 3, representing the narrowest composition distribution tested, the maxima in E'' and tan δ occur nearly at the positions indicated for a compositionally homogeneous copolymer of the same styrene content. This blend may well consist of a single phase. This is not so for the other multicomponent blends, 4 and 5. These show severe broadening of the dispersion region and two or more resolvable loss maxima. In Blend 5 (Figure 8) there is obviously significant interaction of components or partial miscibility.

An interesting aspect of broad composition distribution is the reversal with temperature of the dynamic properties relative to a narrow

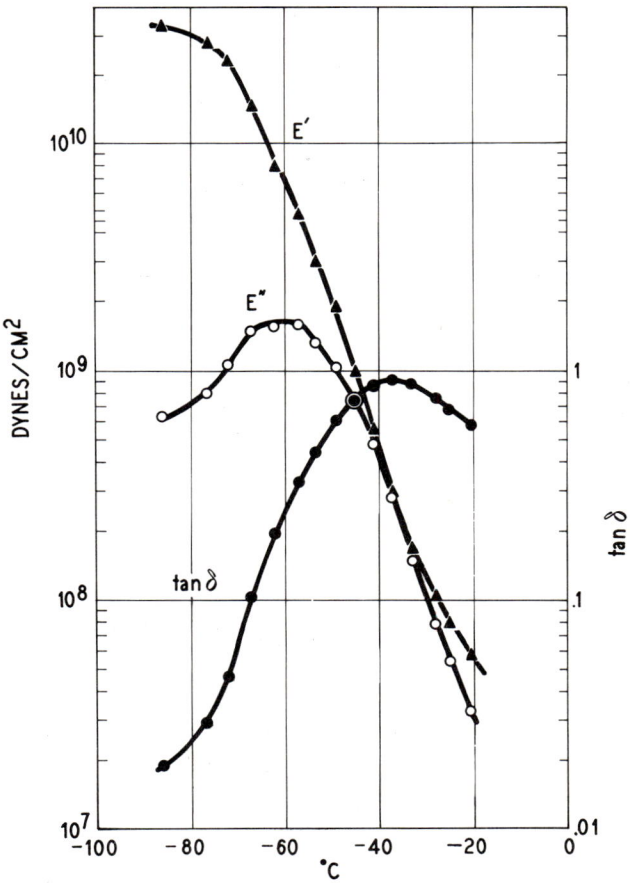

Figure 7. Dynamic properties of 1:1:1 blend of Polymers B(20% styrene), C(30% styrene), and D(40% styrene)

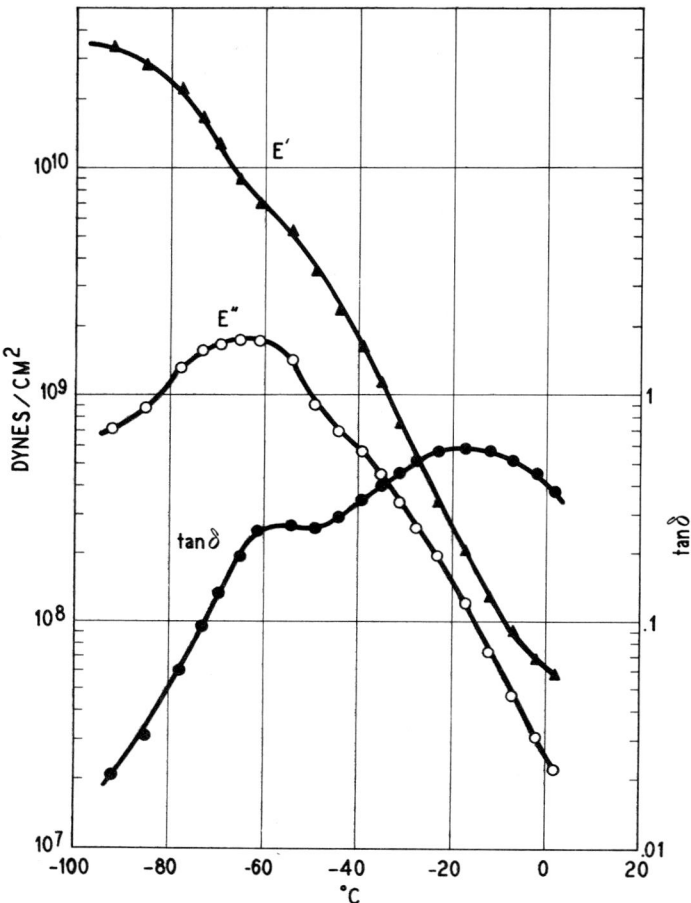

Figure 8. Dynamic properties of five-component blend (Polymers A to E)

distribution polymer of the same styrene content. As an illustration we list E' and tan δ for Blend 5 (Figure 8) and Polymer C at two temperatures:

	E' $dynes/cm^2$		tan δ	
	$-30°C$	$-70°C$	$-30°C$	$-70°C$
Blend 5 (30% styrene, broad composition distribution)	7.5×10^8	1.3×10^{10}	0.48	0.125
Polymer C (30% styrene, uniform composition)	1.6×10^8	3.0×10^{10}	0.98	0.035

These reversals are a direct consequence of the multiphase structure and occur throughout the entire series of blends studied. Similar reversals would be expected with frequency at fixed temperature.

Stress Birefringence. Confirmation of the multiphase nature of the polyblends described may be obtained by stress–optical measurements. Since this technique apparently has not been used before to demonstrate incompatibility in polyblends, a brief description is given of the rationale behind the method.

The stress–optical coefficient, K_σ, of an elastomer network is a constant, independent of extension ratio and crosslink density. It is directly proportional to the difference between the longitudinal and transverse polarizabilities of the statistical chain segment ($b_1 - b_2$):

$$K_\sigma = 2\pi (n^2 + 2)^2 (b_1 - b_2)/45\, nkT \tag{1}$$

Here n is the average refractive index, k is Boltzman's constant, and T is absolute temperature (13). If a polyblend were to form a homogeneous network, the stress would be distributed equally between network chains of different composition. Assuming that the size of the statistical segments of the component polymers remains unaffected by the mixing process, the stress–optical coefficient would simply be additive by composition. Since the stress–optical coefficient of butadiene–styrene copolymers, at constant vinyl content, is a linear function of composition (Figure 9), a homogeneous blend of such polymers would be expected to exhibit the same stress–optical coefficient as a copolymer of the same styrene content. Actually, all blends examined show an elevation of K_σ which increases with the breadth of the composition distribution (Table III). Such an elevation can be justified if the blends have a two- or multiphase domain structure in which the phases differ in modulus. If we consider the domains to be coupled either in series or in parallel (the true situation will be intermediate), then it is easily shown that

$$K_\sigma = K_{\sigma 1} V_1 + K_{\sigma 2} V_2 \ldots \text{(series)} \tag{2}$$

or

$$K_\sigma = (K_{\sigma 1} V_1 E_1 + K_{\sigma 2} V_2 E_2)/(V_1 E_1 + V_2 E_2) \ldots \text{(parallel)} \tag{3}$$

assuming linear elasticity. Here V_1, V_2 are volume fractions, and E_1, E_2 are Young's moduli. Extension to more than 2 components is obvious. Of the two equations only the second is capable of accounting for an elevation in K_σ. Predictably, it overestimates the stress-optical coefficient since simple parallel coupling is an obvious exaggeration. Values calculated by Equations 2 and 3 are shown in the fifth and sixth columns of Table III.

Table II. Loss

Blend No.	A (9:1)	B (8:2)	C (7:3)	D (6:4)	E (5:5)	F (10:0)
1	50	—	—	—	50	—
2	—	50	—	50	—	—
3	—	33.3	33.3	33.3	—	—
4	33.3	—	33.3	—	33.3	—
5	20	20	20	20	20	—
6	—	66.7	—	—	33.3	—
7	—	—	—	75	—	25
8	33.3	—	—	66.7	—	—
9	—	—	—	—	60	40
10	—	50	—	—	—	50
11	—	—	—	50	—	50
12	66.7	—	—	33.3	—	—
13	—	—	50	—	50	—
14	25	—	—	—	75	—

^a Column 9 identifies corresponding loss maxima of the components (C) with those observed in the blend (B).

The elastic moduli of the component polymers were obtained by extrapolating Mooney–Rivlin plots (8, 9) of the stress to zero strain:

Polymer	A	B	C	D	E	F
$E \times 10^{-6}$, dynes/cm^2	26.7	24.9	18.0	13.2	12.3	30.0

The variation in modulus appears to be caused by increased reactivity of the dicumyl peroxide crosslinking agent toward butadiene-rich polymers.

The last column in Table III shows a weighted mean, using empirical weight factors of 0.73 and 0.27 for parallel and series coupling, respectively. These values fit the observed stress–optical coefficients within experimental error. The elevation of the stress-optical coefficient, together

Maxima in Blends[a]

Over-all % Styrene	Col. 9[a]	E" Maxima, °C	tan δ Maxima, °C
30	C	−81; −28	−69; +5
	B	−76; −30	−70; −2
30	C	−67; −46	−50; −17
	B	−64; −48[b]	−30[c]
30	C	−67; −57; −46	−50; −33; −17
	B	−62; − ; −48[b]	− ; −37; −
30	C	−81; −57; −28	−69; −33; +5
	B	−73; −52[b]; −19	−70[b]; −37[b]; −7
30	C	−81; −67; −57; −46; −28	−69; −50; −33; −17; +5
	B	−65; −44[b]; −20[b]	−57; − ; −17
30	C	−67; −28	−50; +5
	B	−70; −23	−65; −3
30	C	−104; −46	−86; −17
	B	−93; −59	−92; −29
30	C	−81; −46	−69; −17
	B	−77; −54	−72; −20
30	C	−104; −28	−86; +5
	B	−93; −30	−89; +6
10	C	−104; −67	−86; −50
	B	−93; −70	−92; −70
20	C	−104; −46	−86; −17
	B	−97; −41	−89; −39
20	C	−81; −46	−69; −17
	B	−82; −52	−70; −52
40	C	−57; −28	−33; +5
	B	−54; −30[b]	−34[b]; −4
40	C	−81; −28	−69; +5
	B	−84; −32	−79; +1

[b] Shoulder.
[c] Single broad maximum.

with its successful (if empirical) description by a calculation based on a heterogeneous model, seems to us ample confirmation of the multiphase nature of these blends.

Discussion

In the present section an attempt is made to describe both stress–optical and dynamic properties of blends in terms of those of the components, using a more detailed, but simple equivalent mechanical model. We consider a (binary) blend as consisting of domains of Polymer 2 dispersed in Polymer 1. These domains are thought to be coupled in series–parallel fashion. The simple analysis of the stress–optical coeffi-

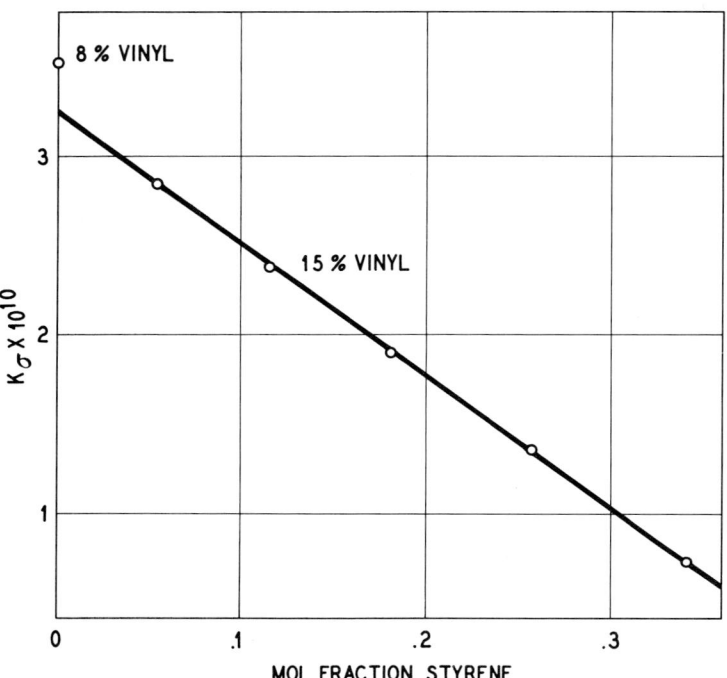

Figure 9. Stress-optical coefficients for component polymers

Table III. Stress-Optical Coefficients for Blends

Blend No.[a]	Over-all Styrene Content, %	$K_\sigma \times 10^{10}$		$K_\sigma \times 10^{10}$ (calc.)		
		Obs. cm²/dyne	By Additivity cm²/dyne	Series	Parallel	Weighted Mean
1	30	2.05	1.90	1.82	2.20	2.10
2	30	1.98	1.90	1.88	2.01	1.98
3	30	2.02	1.90	1.89	1.98	1.96
4	30	2.04	1.90	1.84	2.09	2.03
5	30	2.01	1.90	1.86	2.10	2.03
6	30	2.01	1.90	1.84	2.07	2.01
7	30	2.12	2.00	1.93	2.29	2.19
8	30	2.00	1.90	1.87	2.10	2.03
9	30	2.41	2.04	1.91	2.51	2.35
10	10	3.12	2.98	2.96	3.13	3.09
11	20	2.79	2.55	2.48	2.88	2.77
12	20	2.52	2.38	2.36	2.55	2.50
13	40	1.41	1.36	1.32	1.44	1.41
14	40	1.51	1.36	1.29	1.64	1.55

[a] For composition of blends *see* Table II.

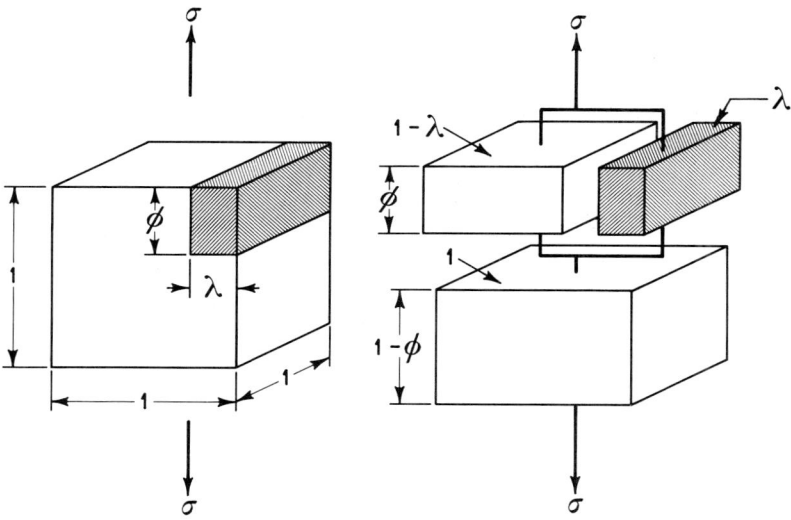

Figure 10. The Takayanagi model

cients outlined above suggests that the model should be capable of emphasizing parallel coupling, particularly in systems of approximately equal component volumes. Physically, such a system resembles an interlocking network of two phases rather than a dispersion of one phase in the other.

One model which incorporates the desired features and which has been used with some success to describe the behavior of polyblends is that of Takayanagi and associates (*12*). This model is illustrated in Figure 10. The parameters λ and ϕ represent the state of mixing, and their product equals the volume fraction of the disperse phase. An equally plausible arrangement for coupling the elements of the model would be a series combination of "1" and "2," in parallel with "1." Takayanagi rejects this alternative on the grounds that it leads to poorer fits of experimental data. The ambiguity in the arrangement of the series and parallel elements is an inherent difficulty with all such models. Another difficulty is conceptual. The Takayanagi model is anisotropic in the sense that it gives different mixing laws depending on how the cube of Figure 10 is oriented with respect to the stress. This would be appropriate for describing the properties of fibers or laminates, but for isotropic blends the apparent difficulty can only be overcome by letting $\lambda = \phi$, which does not produce good fits to experimental data. A model which removes this difficulty without introducing additional parameters is shown in Figure 11. It seems particularly suitable for describing the idea of interpenetrating networks. Its geometric parameters a and b are uniquely related to V_2 by the equation

$$V_2 = a^2 (3b - 2a) \tag{4}$$

Two ways of arranging the various elements are shown in Figure 11; still others are possible. However, only the two cases shown, which represent

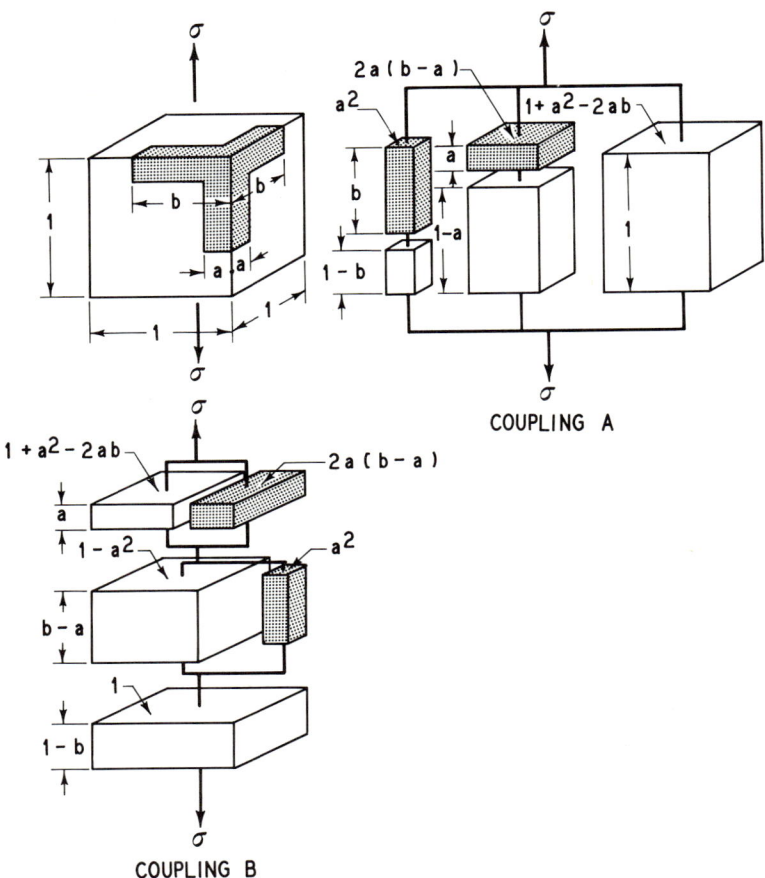

Figure 11. The "isotropic" model

the extremes series-parallel (coupling A) and parallel-series (coupling B), have been evaluated. While the "isotropic model" may be mentally more satisfying, it must be recognized that all such models merely furnish phenomenological descriptions by providing essentially empirical sets of combination rules. The advantage of one or the other model can be demonstrated only by its ability to represent data.

The equations describing the moduli are as follows:

Takayanagi model:

$$E = \left[\frac{\varphi}{\lambda E_2 + (1 - \lambda)E_1} + \frac{1 - \varphi}{E_1}\right]^{-1} \quad (5)$$

Isotropic model (coupling A):

$$E = \frac{a^2}{(1 - b)/E_1 + b/E_2} + [(1 - a)^2 + 2a(1 - b)]E_1 + \frac{2a(b - a)}{(1 - a)/E_1 + a/E_2} \quad (6)$$

Isotropic model (coupling B):

$$E = \left\{\frac{a}{a(2b - a)E_2 + (1 - a)^2 E_1} + \frac{b - a}{a^2 E_2 + (1 - a^2)E_1} + \frac{1 - b}{E_1}\right\}^{-1} \quad (7)$$

In all three equations E_1 and E_2 are now the complex moduli; the storage and loss moduli for the blend are obtained by direct substitution into these equations and separation of the real and imaginary parts to obtain separate mixture rules for each. Analytical expressions have been obtained for these, but they are lengthy and cumbersome. All the calculations described, therefore, were carried out by computer. The substitution of complex moduli into the solution of the equivalent purely elastic problem is justified by the correspondence principle of viscoelastic stress analysis (6).

Table IV. Parameters for Equations 5 and 6

Blend No.[a]	Equation 5		Equation 6	
	λ	φ	a	b
1	.490	.990	.490	1.000
2	.555	.887	.505	.980
6	.395	.820	.390	.969
7	.900	.290	.335	1.000
8	.347	.990	.394	1.000
9	.889	.470	.445	1.000
10	.559	.880	.495	1.000
11	.527	.920	.490	1.000
12	.405	.800	.390	.969
13	.604	.800	.509	.961
14	.267	.980	.335	1.000

[a] For composition of blends see Table II.

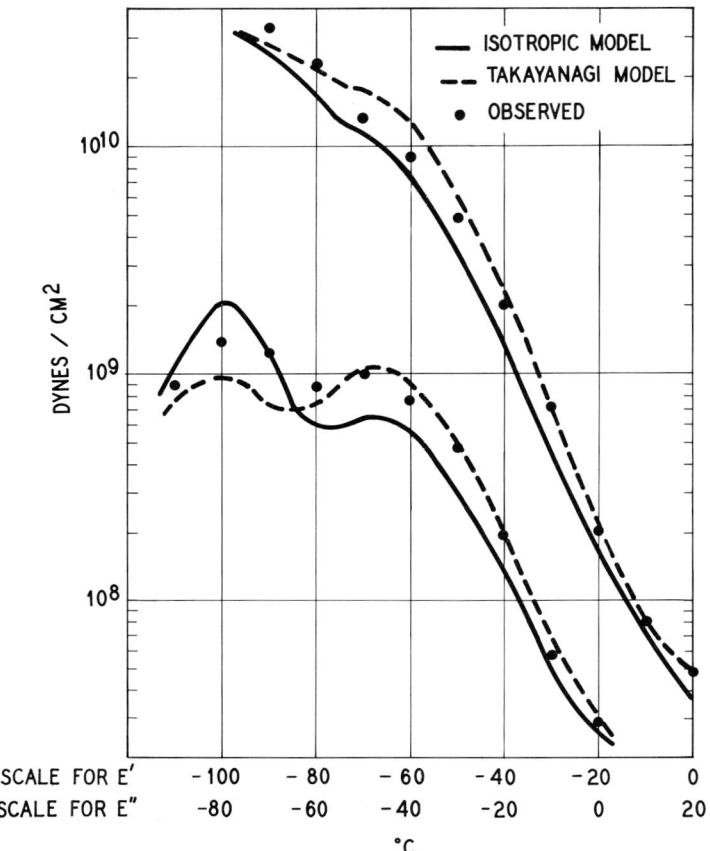

Figure 12. Data on Blend 8 fitted by equivalent models

Expressions for the stress–optical coefficient based on the above models are written easily. One assumes that the birefringences contributed by the various elements of the model are additive by volume and are given by the product of the stress in the element and the stress–optical coefficient of the appropriate component. The resulting equations are:

Takayanagi model:

$$K_\sigma = (1 - \varphi)K_{\sigma_1} + \frac{\varphi\lambda K_{\sigma_2}E_2 + \varphi(1 - \lambda)K_{\sigma_1}E_1}{E_2\lambda + E_1(1 - \lambda)} \tag{8}$$

Isotropic model (coupling A):

$$K_\sigma = K_{\sigma_2}\left[\frac{a^2b}{(1-b)/E_1 + b/E_2} + \frac{2a^2(b-a)}{(1-a)/E_1 + a/E_2}\right] +$$

$$K_{\sigma_1}\left[\{(1-a)^2 + 2a(1-b)\}E_1 + \frac{a^2(1-b)}{(1-b)/E_1 + b/E_2} + \right.$$

$$\left.\frac{2a(b-a)(1-a)}{a/E_2 + (1-a)/E_1}\right] \quad (9)$$

The corresponding expression for coupling B of the isotropic model is not given as it was not used in the final analysis of the stress–optical data.

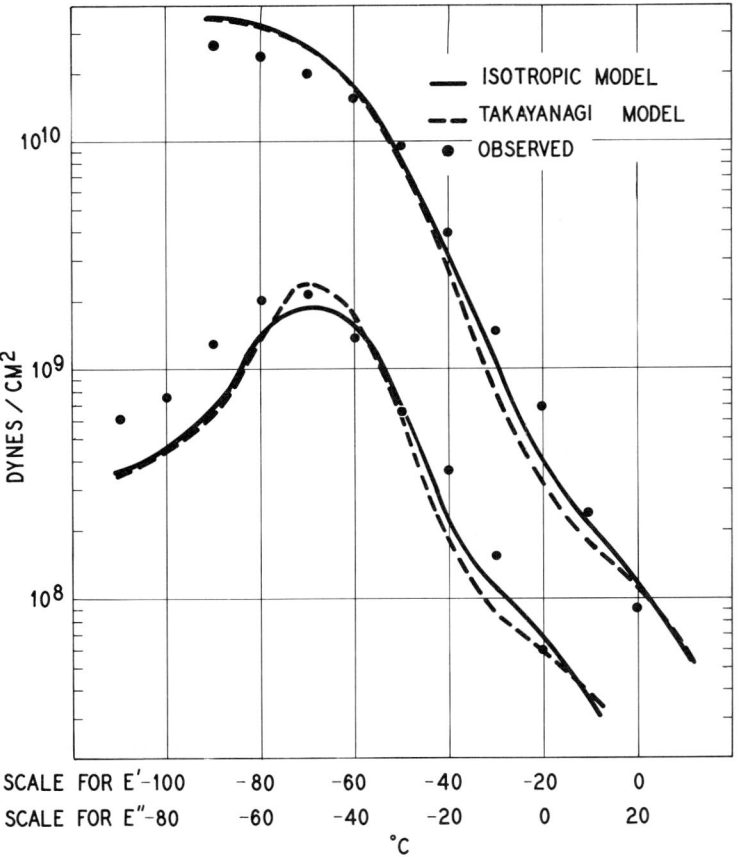

Figure 13. Data on Blend 13 fitted by equivalent models

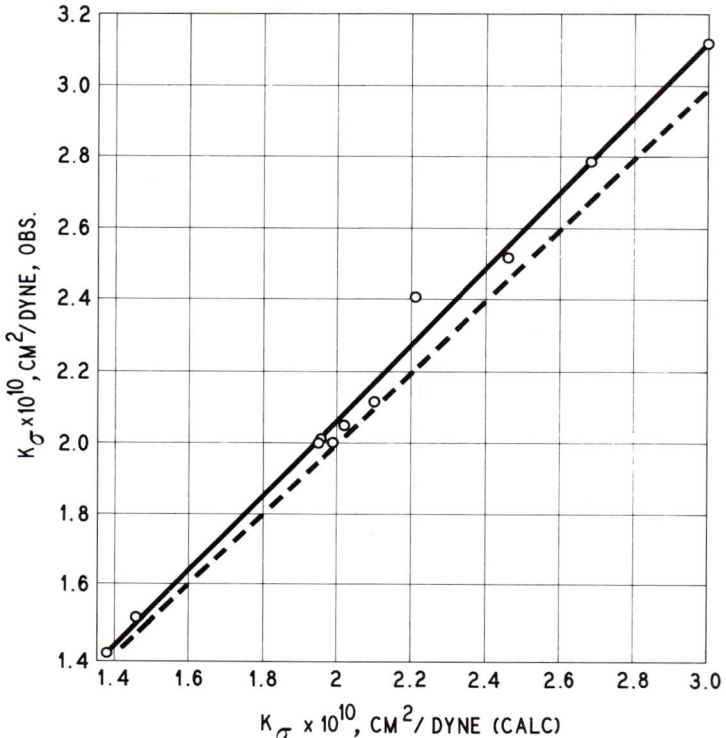

Figure 14. Observed and calculated stress-optical coefficients (calculation by Equation 9 with parameters producing best fit to dynamic data)

Both Equations 8 and 9 assume linear elements. The calculations have also been performed (by computer) for neo-Hookean elements. This leads to a slight dependence of K_σ on strain. The predicted dependence is so small, however, that it would be difficult to detect experimentally.

Application of Equivalent Models to Dynamic Data. Equations 5, 6, and 7 were applied to the data on all binary blends in this study. Equation 7 showed no advantage over the simpler Takayanagi Equation 5. Equation 6 produced slightly better fits than 5 in most cases. Table IV gives the parameters for the Takayanagi and Isotropic (coupling A) models fitting the data most closely. In each case the component present in excess by volume was chosen as the continuous phase. It is interesting that the isotropic model consistently gives the best fit with $b \approx 1$—*i.e.,* the situation idealizing an interlaced network of the two phases. In no case could an exact fit be obtained; examples of the quality of fit produced to the experimental data are shown in Figures 12 and 13.

Analysis of Stress–Optical Data. The slight, if indeed real, improvement of the isotropic model over the Takayanagi model would be of little consequence were it not for a more pronounced difference between the two models in their ability to describe the stress-optical data. When the parameters obtained from the dynamic data (Table IV) are substituted into Equations 8 and 9, Equation 8 produces results which are uniformly too low. Equation 9 also underestimates the magnitude of K_σ but only by an average 7% (Figure 14). For most blends the discrepancy is less than 5%, and all calculated values show the characteristic elevation of the birefringence attributed to the multiphase structure.

It is difficult to assess to what extent the imperfect fits obtained by the equivalent models are the result of partial miscibility. This is because the models do predict minor shifts in the loss maxima, which are in the same direction as the shifts expected from partial miscibility. Thus the major cause of the deviations may well be the inability of the models to describe a complex morphology with a single adjustable parameter.

Conclusions

In the molecular weight range of 120,000–200,000 butadiene–styrene random copolymers form multiphase blends when they differ by 20 or more percentage units in composition. The precise limits of compatibility have not been determined but may lie substantially below this figure. The multiphase structure of these polyblends is indicated by both dynamic and stress–optical properties. As a result of this structure polyblends of broad compositional distribution exhibit a temperature reversal of their dynamic properties relative to narrow distribution polymers of the same average composition: above a certain temperature a broad distribution blend will have higher storage moduli and loss tangents, while at low temperatures the behavior will be reversed. Another reversal occurs in the glassy region.

Acknowledgment

The authors are indebted to R. J. Sonnenfeld for preparing the polymers used in the blending studies and to N. W. Tschoegl for several helpful discussions.

Literature Cited

(1) Childers, C. W., Kraus, G., *Rubber Chem. Technol.* **1967**, 40, 1183.
(2) Grubisic, Z., Rempp, P., Benoit, H., *J. Polymer Sci.*, B **1967**, 5, 753.
(3) Hsieh, H. L., Wofford, C. F., *J. Polymer Sci.*, A-1 **1969**, 449, 461.
(4) Kraus, G., Childers, C. W., Gruver, J. T., *J. Appl. Polymer Sci.* **1967**, 11, 1581.

(5) Kraus, G., Gruver, J. T., *J. Appl. Polymer Sci.* **1967**, 11, 2121.
(6) Lee, E. H., *Quart. Appl. Math.* **1955**, 12, 183.
(7) Meehan, E. J., *J. Polymer Sci.* **1946**, 1, 175.
(8) Mooney, M., *J. Appl. Phys.* **1940**, 11, 582.
(9) Rivlin, R. S., Saunders, D. W., *Phil. Trans. Roy. Soc.*, A243 **1951**, 251.
(10) Silas, R. S., Yates, J., Thornton, V., *Anal. Chem.* **1959**, 31, 529.
(11) Takayanagi, M., *Proc. Intern. Congr. Rheology, 4th,* **1965**, Part 1, 161–187.
(12) Takayanagi, M., Uemura, S., Minami, S., *J. Polymer Sci., C* **1964**, 5, 113.
(13) Treloar, L. R. G., "The Physics of Rubber Elasticity," p. 138, Oxford University Press, London, 1949.
(14) Yoshino, M., Takayanagi, M., *J. Japan Soc. Test. Mat.* **1959**, 8, 330.

RECEIVED December 31, 1969.

Terpolymerization of Cyclopentene, Sulfur Dioxide, and Acrylonitrile

YUYA YAMASHITA, SHOUJI IWATSUKI, and KOZO SAKAI

Department of Synthetic Chemistry, Faculty of Engineering,
Nagoya University, Nagoya, Japan

> *Spontaneous copolymerization of cyclopentene (CPT) with sulfur dioxide (SO_2) suggests the participation of a charge transfer complex in the initiation and propagation step of the copolymerization. The ESR spectrum together with chain transfer and kinetic studies showed the presence of long lived SO_2 radical. Terpolymerization with acrylonitrile (AN) was analyzed as a binary copolymerization between $CPT–SO_2$ complex and free AN, and the dilution effect proved this mechanism. Moderately high polymers showed enhanced thermal stability, corresponding to the increase of AN content in the terpolymer.*

Extensive studies have been reported on the copolymerization of various olefins and sulfur dioxide to yield polysulfones (3, 7). Attention was paid to the significant alternating tendency irrespective of monomer feed ratios and to the presence of ceiling temperature. The mechanism of such peculiar alternating copolymerization has been a matter of considerable discussion, although polymerization through a charge transfer complex suggested by Barb (1) was favored by Furukawa (6) and by us (15). Recently Zutty (16) discovered spontaneous copolymerization of norbornene and sulfur dioxide, suggesting propagation through biradical coupling, and Frazer (4) reported on the unusual initiation of 1,5-cyclooctadiene and sulfur dioxide by bubbling oxygen.

As a development of our studies on charge transfer complexes and polymerization, we reported on the spontaneous copolymerization of cyclopentene and sulfur dioxide (11), and kinetic evidence for the participation of the charge transfer complex in the copolymerization was presented. This paper discusses the terpolymerization of cyclopentene, sulfur dioxide, and acrylonitrile to give further evidence for the charge transfer complex mechanism. The thermal stability of polysulfones is also discussed.

Experimental

Materials. Cyclopentene (CPT) was prepared by dehydration of Commercial acrylonitrile (AN) was washed with dilute sulfuric acid and fractionally distilled after adding lithium aluminum hydride to remove traces of hydroperoxide, which is an active initiator for the present system.

Commercial sulfur dioxide (SO_2) was used without purification. Commercial acrylonitrile (AN) was washed with dilute sulfuric acid and water, dried over calcium hydride, and fractionally distilled.

Toluene was washed repeatedly with concentrated sulfuric acid until the acid was colorless, and after washing with water it was dried over calcium hydride and fractionally distilled. N,N-Dimethylformamide (DMF) was dried over phosphorus pentoxide and distilled carefully.

Polymerization. A weighed mixture of monomer and solvent in a 30-ml ampoule was cooled in a dry ice bath and flushed with nitrogen, sealed and set without stirring in a bath. After polymerization, excess petroleum ether was added to precipitate the copolymer, which was dried under reduced pressure.

Characterization of the Copolymer. The SO_2 content in the copolymer was calculated from the sulfur content determined by the Schoniger method. The reduced viscosity of the copolymer was determined at 30°C with an Ostwald viscometer. DMF containing 0.1% LiCl was used as solvent at 0.2 gram/100 ml. The melting point of the terpolymer was measured by a hot-stage microscope.

Results and Discussion

CPT–SO_2 System. Bulk copolymerization of CPT and SO_2 takes takes place spontaneously at a remarkable rate even at −15°C. In comparison with the thermal initiation of p-dioxene and maleic anhydride which proceeds through a similar charge transfer complex at room temperature (13), the interaction between CPT and SO_2 seems more pronounced, giving the propagating species at a lower temperature.

Dilution with toluene slowed the copolymerization rate, and kinetic measurements were carried out in toluene at 0°–30°C. As reported previously (11), the over-all activation energy of the spontaneous copolymerization of CPT and SO_2 was calculated to be 16.5 kcal/mole from the Arrhenius plot of the initial rate vs. polymerization temperature. Dependence of the intial rate of copolymerization upon monomer concentration was checked at various monomer concentrations and found to be quite high (11); this could not be explained without participation of the monomer in the initiation step.

$$R_p = k_p [\text{CPT}]^3 [\text{SO}_2]^2$$

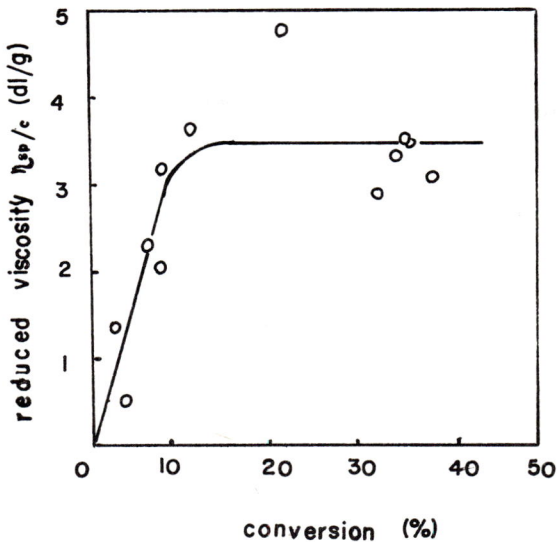

Figure 1. Relationship between conversion and reduced viscosity. CPT, 0.68 gram; SO_2, 0.64 gram; toluene, 1.25 ml at 30°C. Reduced viscosity, 0.2 gram/100 ml at 30°C in DMF + 0.1% LiCl.

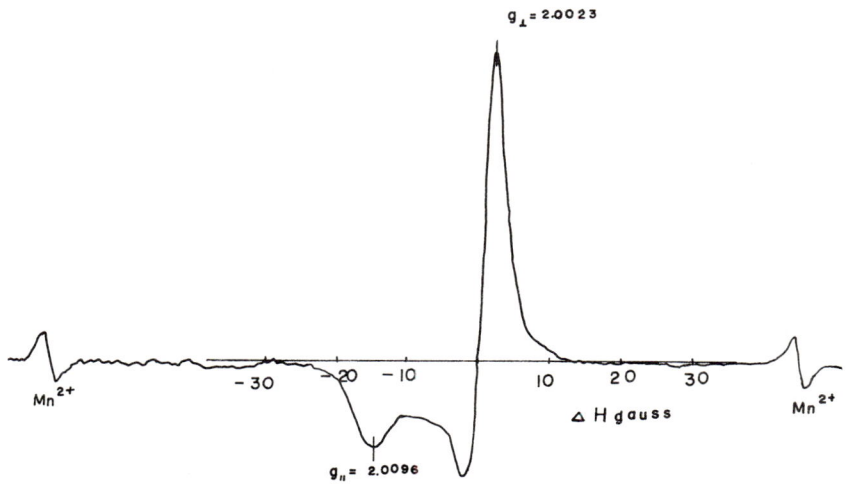

Figure 2. ESR spectrum of $CPT-SO_2$ system measured at the temperature of the polymerization system in toluene

A yellow color developed when CPT and SO_2 were mixed, and the equilibrium constant of the molecular complex formation was measured as $K = 0.0353$ (40°C, in hexane) (2). This complex might be the intermediate in this alternating copolymerization, and it might participate both in the initiation and propagation steps of this spontaneous copolymerization.

The following scheme is proposed to explain the high dependence of the polymerization rate on monomer concentration, assuming initiation from the complex and CPT and propagation through the complex.

$$CPT + SO_2 \underset{}{\overset{K}{\rightleftarrows}} \text{complex}$$

$$\text{complex} + CPT \overset{k_i}{\rightarrow} R\cdot$$

$$R\cdot + \text{complex} \overset{k_p}{\rightarrow} R\cdot$$

Assuming a non-steady state polymerization involving a long lived radical, the following equation is derived (9):

$$R_i = k_i [CPT] [\text{complex}], [\text{complex}] = K [CPT] [SO_2]$$

$$R_p = k_p [R\cdot] [\text{complex}] = k_i k_p K^2 [CPT]^3 [SO_2]^2$$

The validity of the non-steady state assumption is shown in Figure 1, where the molecular weight of the copolymerization mixture increases with conversion at the beginning of copolymerization. The long lived radical is observed from the ESR spectrum in Figure 2, measured at room temperature for polymerization in toluene; this agrees with the SO_2 radical similar to the norbornene–SO_2 system (16).

The nature of the propagating radical was examined from chain transfer studies. The effect of some chain transfer agents on the reduced viscosity of the copolymer is shown in Table I. The fact that carbon tetrabromide is more effective than n-butylmercaptan as a chain transfer agent and that 1,4-dioxane decreases the reduced viscosity of the copolymer shows the presence of an acceptor radical (10) in agreement with the ESR study. Thus, copolymerization between cyclopentene and SO_2 is explained by regarding the charge transfer complex as a new monomer, and propagation proceeds only through the SO_2 radical.

The copolymer produced during the polymerization precipitated as a white powder, soluble in DMF, chloroform, and liquid SO_2. Elementary analysis proved the exact alternating composition irrespective of monomer feed ratio. On heating, they decomposed at 220°–230°C without melting.

Table I. Effect of Chain Transfer Agent on the Reduced Viscosity of the Copolymer[a]

Chain Transfer Agent[b]	Conversion, %	$\eta_{sp/c}$
—	3.20	1.25
CBr_4, 0.005	2.18	0.335
0.048	1.20	0.191
0.16	6.17	0.143
n-BuSH 0.03	5.49	0.420
0.06	6.46	0.346
Dioxane 1.7	50.9	1.56
4.0	52.0	1.03

[a] Solvent = toluene (2.5 ml) for all but the last two runs.
[b] Molar ratio of chain transfer agent to monomer; CPT, 0.68 gram; SO_2, 0.64 gram; temp., 30°C.

CPT–SO_2–AN System. Terpolymerizations were carried out to clarify the nature of the intermediate complex and the propagating radical in the CPT–SO_2 copolymerization system (*11*). Vinyl monomers are grouped by the electron density of their double bonds into three classes; donor, acceptor, and indifferent. Strong donors such as isobutyl vinyl ether polymerizes cationically by adding CPT–SO_2 at 30°C. Moderate donors such as styrene, which can give polysulfone with SO_2, can copolymerize with CPT–SO_2 to give a random copolymer of (CPT–SO_2) and (2-St–SO_2). Acceptors such as maleic anhydride are expected to yield a random copolymer from our previous experiment on the terpolymerization of butadiene–SO_2–maleic anhydride system (*8*). Indifferent monomers such as acrylonitrile, methyl acrylate, methyl methacrylate, and vinylidene chloride are expected to form a random copolymer with the CPT–SO_2 system because their double bonds have intermediate electron density and they copolymerize randomly with another charge transfer complex between p-dioxene and maleic anhydride. However, vinylidene chloride showed very low reactivity and methyl methacrylate, which was reported to polymerize by SO_2 (*5*), did give a block copolymer consisting of CPT–SO_2 and MMA blocks (probably because of steric hindrance).

The terpolymerization of CPT–SO_2 and acrylonitrile is shown in Table II. It was necessary to accelerate the polymerization by adding azobisisobutyronitrile (AIBN) as initiator. The nature of the propagating species may not be different with a different initiator. Polymerization ceased at a low conversion at 40°C in toluene. The terpolymer composition calculated from elemental analysis of C, H, N, and S showed an equimolar ratio of CPT and SO_2. The terpolymers are white powders, soluble in DMF, can be cast into transparent film different from the CPT–SO_2 copolymer, and showed melting temperature without decompo-

Table II. Terpolymerization of CPT–SO$_2$–AN at 40°C in Toluene[a]

CPT, grams	SO$_2$, grams	AN, grams	Toluene, ml	Time, hrs	Conversion, %	Terpolymer Composition, AN/CPT–SO$_2$[b]	Mp^c, °C
2.04	1.94	1.06	4.5	3.8	12.8	0.385	139 ~ 153
0.68	0.64	0.60	2.0	2.2	0.64	0.535	—
1.36	1.28	1.08	4.0	6	—	0.574	193 ~ 200
1.36	1.28	2.13	3.5	2.5	2.5	1.07	138 ~ 145
1.36	1.28	3.20	2.9	1.5	2.5	1.65	188 ~ 199
1.36	1.28	4.24	2.9	2	2.0	2.91	—
0.68	0.64	3.18	2.0	2.5	1.0	4.81	—

[a] AIBN (5 mg) added as initiator for all runs except the second.
[b] Molar ratio of AN to CPT–SO$_2$.
[c] Microscopic melting temperature.

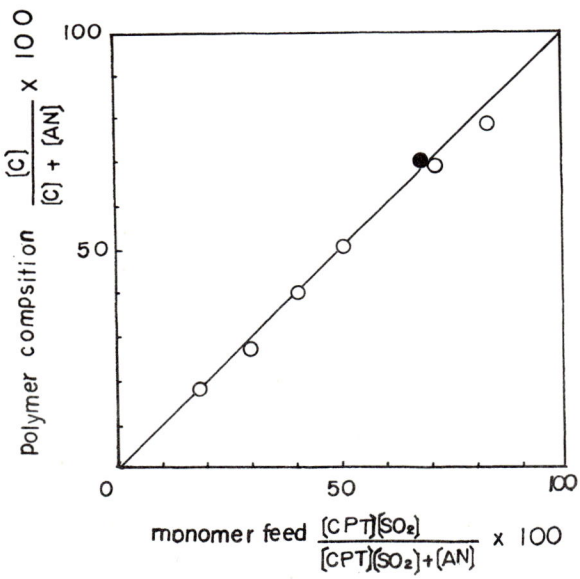

Figure 3. Copolymer composition diagram of CPT–SO$_2$–AN expressed as a binary copolymer
○ AIBN initiated
● Thermally initiated

sition. The terpolymers richer in AN were insoluble in chloroform, and those poorer in AN were soluble in chloroform. Fractional extraction by chloroform proved the homogeneity of the terpolymer composition, showing the randomness of the terpolymerization.

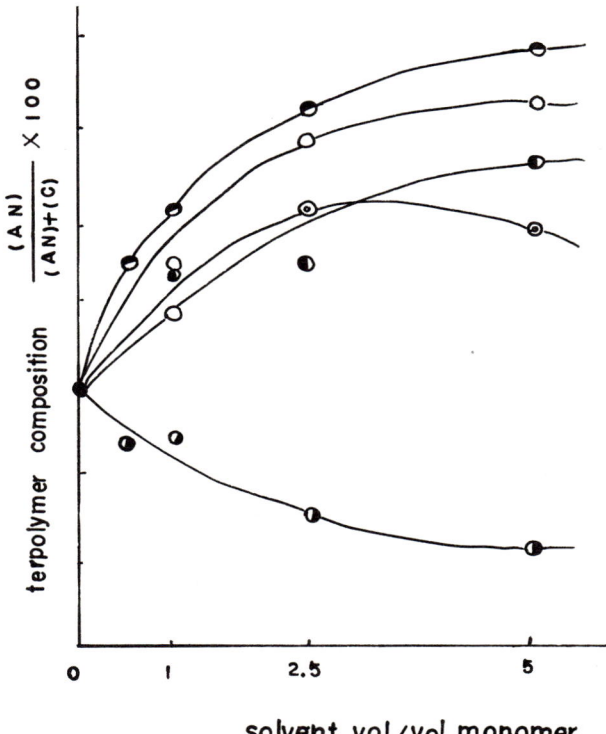

Figure 4. Effect of dilution on terpolymer composition of CPT–SO_2–AN system at 40°C. AIBN, 5 mg; CPT, 1.0×10^{-2} mole; SO_2, 1.0×10^{-2} mole; AN, 1.6×10^{-2} mole.

- ● Bulk
- ⊙ Toluene
- ○ Benzene
- ◐ Chloroform
- ◕ Carbon tetrachloride
- ◑ Acetone

Considering the terpolymerization of CPT–SO_2–AN system as a binary copolymerization of CPT–SO_2 complex and free acrylonitrile, the copolymerization equation can be derived as follows, assuming a fast equilibrium.

$$\mathrm{CPT} + \mathrm{SO}_2 \underset{}{\overset{K}{\rightleftarrows}} \text{complex}$$

$$[\text{complex}] = K\,[\mathrm{CPT}]\,[\mathrm{SO}_2]$$

$$\frac{d\,[\text{complex}]}{d\,\mathrm{AN}} = \frac{[\mathrm{CPT}]\,[\mathrm{SO}_2]}{\mathrm{AN}} \left\{ \frac{r_1\,K\,[\mathrm{CPT}]\,[\mathrm{SO}_2] + [\mathrm{AN}]}{[\mathrm{CPT}]\,[\mathrm{SO}_2] + r_2/K\,[\mathrm{AN}]} \right\}$$

Table III. Terpolymerization of CPT–SO$_2$–AN

CPT, grams	SO$_2$, grams	AN, grams	Solvent, ml	Temp, °C
0.45	0.67	0.40	DMF (1.4)	21
0.47	0.47	0.41	toluene (1.4)	21
0.56	3.00	0.34	—	21
0.44	0.51	0.40	DMF (1.4)	0
0.44	2.59	0.41	—	0
0.45	3.16	0.43	—	0
0.45	0.54	0.41	—	0
2.66	3.13	2.50	—	0
2.65	3.02	2.45	—	0

^{a} Microscopic melting temperature.

From the Finemann-Ross plot of the data in Table II, monomer reactivity ratios are determined. $r_1K = 0.654$, $r_2/K = 1.09$. The copolymer composition diagram is shown in Figure 3, which includes both spontaneous and catalytic polymerization to prove the similarity of both mechanisms.

By introducing the equilibrium constant of the complex formation (2), r_1(complex) = 18.5 and r_2(AN) = 0.0385 is calculated. This corresponds to the Q, e value of the CPT–SO$_2$ complex ($Q = 7.69$ and $e = 0.61$), which suggests an acceptor radical as a propagating species in agreement with the ESR and chain transfer studies.

Binary copolymerization of the CPT–AN system was carried out at 40°C using AIBN as initiator. From the Finemann-Ross plot of the copolymer composition and monomer feed ratios, the monomer reactivity ratios were determined. r_1' (CPT) = 0, r_2' (AN) = 1.97. The increase of the relative reactivity of CPT by forming a complex with SO$_2$ is calculated as follows.

$$\frac{k_{AN \cdot C}}{k_{AN \cdot CPT}} = \frac{k_{AN \cdot C}}{k_{AN \cdot AN}} \frac{k_{AN \cdot AN}}{k_{AN \cdot CPT}} = \frac{1}{r_2 (AN)} \cdot r_2' (AN) = \frac{1.97}{0.0385} = 51.3$$

Thus, CPT is activated 50-fold by complex formation with SO$_2$, and hence the complex is the reactive species even though it is very dilute. These results are comparable with previous studies on vinyl ether–maleic anhydride system (14).

Further evidence for the participation of the charge transfer complex in these terpolymerization systems was obtained by dilution experiments (12). The effect of dilution with various solvents on the AN content of the terpolymer is shown in Figure 4. Except for chloroform, the AN content of the copolymer increases with dilution. This suggests a higher order dependence of monomer consumption on monomer concentration

under Ultraviolet Irradiation

AIBN, mg	Time, hrs	Conversion, %	Terpolymer Composition wt % AN	mp, °C^a	$\eta_{sp/c}{}^b$
31	11	86.3	28.4	186–191	0.09
31	11	88.9	17.0	210–230	0.23
33	11	61.4	6.2	198–210	0.22
5	26.5	56.0	27.7	183–200	0.19
5	26.5	65.2	8.9	199–215	0.30
—	34.5	66.3	10.4	210–217	0.29
—	6	33.8	10.6	193–209	0.29
1	7	56.5	14.8	217–227	0.47
1	7	44.5	8.6	200–216	0.51

^{b} Reduced viscosity at 30°C in DMF with 0.1% LiCl at 0.2 gram/100 ml added.

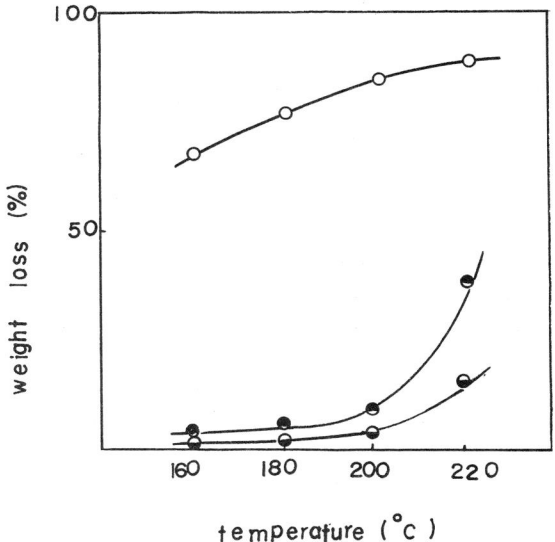

Figure 5. Thermal degradation of polysulfone terpolymers (weight loss after heating one hour under air)

○ CPT–SO$_2$
◐ CPT–SO$_2$–AN (AN, 6.2 wt %)
◕ CPT–SO$_2$–AN (AN, 27.7 wt %)

of CPT–SO$_2$ than AN, which agrees with the complex mechanism showing a second-order dependence of the complex concentration. Such a phenomenon strongly favors the complex mechanism.

The decrease of AN content with dilution in the case of (14) chloroform is difficult to explain. If the K value differs little with solvent, the reactivity of the complex may increase on dilution with chloroform, and in fact terpolymerization is accelearted greatly in chloroform. This unusual effect was observed in other systems (12).

Thus, the alternating copolymerization of the olefin–SO_2 system is shown to proceed through the intermediate charge transfer complex, and by regarding the complex as a new activated monomer, terpolymerization with AN should yield a random terpolymer. Because olefin polysulfone is thermally unstable, terpolymerization might improve the thermal stability of the polysulfone. Table III shows some polymerization experiments at lower temperatures to prepare high conversion samples with moderately high molecular weight. For such purposes polymerization at lower temperatures by a photosensitized process gave fair results. Thermal degradation of these samples is shown in Figure 5, and enhanced thermal stability corresponding to AN content of the terpolymer is observed. This is additional evidence for the randomness of the terpolymer in support of the charge transfer complex mechanism.

Literature Cited

(1) Barb, W. G., *Proc. Roy. Soc.* **A212,** 66 (1952).
(2) Booth, D., Dainton, F. S., Ivin, K. J., *Trans. Faraday Soc.* **55,** 1293 (1959).
(3) Fettes, E. M., Davis, F. O., "High Polymers," Vol. XIII, pp. 225–270, Interscience, New York, 1962.
(4) Frazer, A. H., O'Neill, W. P., *J. Am. Chem. Soc.* **85,** 2613 (1963).
(5) Ghosh, P., O'Driscoll, K. F., *Polymer Letters* **4,** 519 (1966).
(6) Ito, I., Hayashi, H., Saegusa, T., Furukawa, J., *Kogyo Kagaku Zasshi* **65,** 703, 1634 (1962).
(7) Ivin, K. J., Rose, J. B., *Advan. Macromol. Chem.* **1,** 335–401 (1968).
(8) Iwatsuki, S., Amano, S., Yamashita, Y., *Kogyo Kagaku Zasshi* **70,** 2027 (1967).
(9) Iwatsuki, S., Kokubo, T., Yamashita, Y., *J. Polymer Sci., Pt. A-1* **6,** 2441 (1968).
(10) Iwatsuki, S., Nisio, K., Yamashita, Y., *Kogyo Kagaku Zasshi* **70,** 384 (1967).
(11) Iwatsuki, S., Okada, T., Yamashita, Y., *J. Polymer Sci., Pt. A-1* **6,** 2451 (1968).
(12) Iwatsuki, S., Yamashita, Y., *J. Polymer Sci., Pt. A-1* **5,** 1753 (1967).
(13) Kokubo, T., Iwatsuki, S., Yamashita, Y., *Makromol. Chem.* **123,** 256 (1969).
(14) Kokubo, T., Iwatsuki, S., Yamashita, Y., *Macromolecules* **1,** 482 (1968).
(15) Yamashita, Y., Iwatsuki, S., Kokubo, T., *J. Polymer Sci.* **C23,** 753 (1968).
(16) Zutty, N. L., Wilson, C. W., Potter, G. H., Priest, D. C., Witworth, C. J., *J. Polymer Sci.* **A 3,** 2781 (1965).

RECEIVED January 23, 1970.

14

High Impact Polystyrene by Prepolymerization in a Water-in-Oil Emulsion Followed by Suspension Polymerization

RUDOLF B. DE JONG

Chemische Werke Hüls A.G., 4370 Marl, Germany

During the suspension polymerization of solutions of rubber in styrene under the conditions of a "normal" suspension polymerization, stabilization of the droplets of the prepolymerization mixtures in water by the suspending agent impedes the inversion of the rubber phase by restricting agitation within that mixture. This difficulty may be overcome by performing the prepolymerization in a water-in-oil emulsion. W/o emulsions can be made with standard suspending agents as emulsifiers by adding the aqueous phase with stirring to the rubber solution. Phase ratio w:o, type and concentration of the emulsifier, rate of agitation and polymerization are important reaction parameters during emulsification and polymerization. The transition from the w/o emulsion polymerization to a standard suspension polymerization occurs automatically with increasing styrene conversion.

Most impact polystyrenes on the market today are made by polymerizing solutions of rubber in styrene. The styrene solution of rubber is prepolymerized in bulk with stirring to 12–40% conversion. Polymerization is then completed either in bulk in the absence of agitation or in aqueous suspension with agitation. The degree of agitation during the prepolymerization step has a profound effect upon the properties of the final product (5, 6). A high rate of agitation leads to a product with a low gel content and small rubber particles, whereas a product with a large proportion of gel and with large rubber particles is obtained if the agitation during the prepolymerization step is only moderate. The rubber particle size and the gel content are two important structural parameters determining the physical properties of high impact polystyrene.

Prepolymerization may be carried out in the presence of water (7), which aids in heat transfer. At the same time, however, because of the low viscosity of water compared with the polymeric phase, agitation within the organic phase is less intensive and therefore must be accelerated or prolonged to arrive at a rubber particle size and gel content equivalent to that obtained during bulk agitated prepolymerization.

The rubber particle size in the final product increases several fold if the prepolymerization is carried out in the presence of a dilute aqueous solution of an alkane sulfonate or polyvinyl alcohol in place of pure water. The addition of a surface-active agent converts the coarsely dispersed oil–water mixture—obtained as above in the presence of pure water—into an oil-in-water emulsion. In this case even prolonged stirring during prepolymerization does not decrease the rubber particle size appreciably in the final product. The stabilization of the droplets of the organic phase in water by the emulsifier obviously impedes or prevents agitation within the polymeric phase. Figure 1 shows the influence of these three prepolymerization methods (under otherwise equal reaction conditions) on the dispersion of rubber particles in polystyrene.

As explained above the rubber particle size in the final product is a measure for the rate of agitation—under otherwise equal reaction conditions—within the rubber–polystyrene–styrene solution during prepolymerization. Figure 1 shows that agitation is least effective if the organic

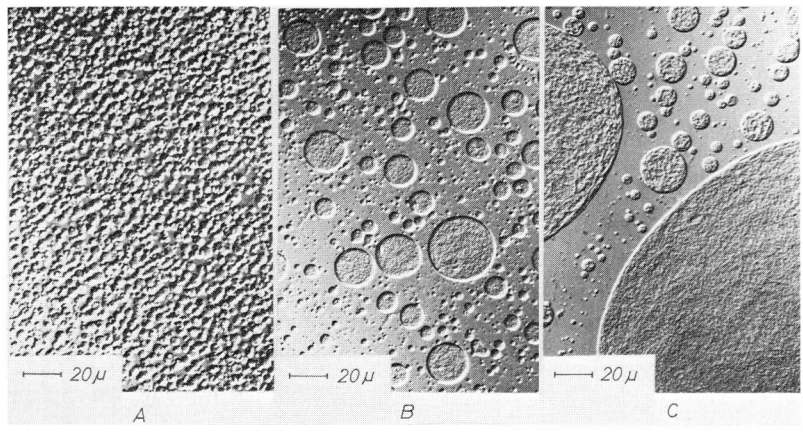

Figure 1. Interference phase contrast micrographs of rubber particles in polystyrene. Prepared by prepolymerization (A) in bulk, (B) in the presence of water, and (C) in an o/w emulsion.

phase is present as an oil-in-water emulsion (o/w). An optimum in agitation efficiency is obtained only in those cases where the organic phase is the continuous phase—*i.e.* in the presence of water in a w/o emulsion. Heat transfer during prepolymerization in a w/o emulsion should be better, and agitation within the polymeric phase should be as good as during the prepolymerization in bulk.

Experimental

The following investigations were performed with a solution of 6% polybutadiene in styrene. The polybutadiene used was Buna CB-10 (Stereokautschuk-Werke Hüls GmbH & Co. KG) which has a cis content of about 95% and a Mooney viscosity ML-4 of 45–50. The screening procedure of the potential emulsifiers was as follows. The compound to be tested was either dissolved in water or in the rubber solution depending on which was the better solvent. The aqueous phase was then added dropwise to the rubber solution with agitation. The stirrer speed chosen was as high as possible but below the speed at which air was beaten into the mixture. Agitation was stopped a short time after the aqueous phase was added. Immediately afterwards, the rate of droplet coalescence was determined. In those cases where the droplets failed to coalesce or did so only slowly—*i.e.* when an emulsion was obtained—the emulsion type was determined by the dye solubility method (*1*). A mixture of the oil-soluble red Sudan III and the water-soluble malachite green was added to the emulsion. A green color indicated an o/w emulsion and a red color a w/o emulsion.

Prepolymerization was then performed with this w/o emulsifier. The polymerization was carried out in a 2-liter cylindrical glass vessel fitted with a four-stage cross-blade agitator operating at 500 rpm. The stirring motor was connected to a watt meter. The watt meter indicated major changes in the viscosity of the prepolymerization mixture such as the demulsification of a w/o emulsion—*i.e.*, the transition of the highly viscous organic phase to the low viscosity of the aqueous phase. Prepolymerizations were performed at 60°–100°C.

The final polymerizations—*i.e.*, the prepolymerization in a w/o emulsion followed by suspension polymerization were carried out in a 40-liter stainless steel reactor. The stirrer speed was varied between 200 and 420 rpm. Prepolymerizations were performed at 60°–100°C, the ensuing suspension polymerizations at up to 140°C.

Emulsifiers for the w/o Emulsion Polymerization

The compounds tested as potential w/o emulsifiers for the rubber solution during this investigation can be divided into three groups: (a) finely divided solids, (b) low molecular weight emulsifiers, and (c) high molecular weight compounds. Finely divided solids, such as calcium phosphate, are used in large quantities as dispersing agents during the suspension polymerization of styrene. Finely divided solids modified by the controlled adsorption of various amphiphilic compounds have been used as emulsifiers for the w/o emulsion polymerization of styrene (8). All attempts to polymerize the rubber solution in a w/o emulsion with the aid of these emulsifiers were unsuccessful.

Emulsifiers suitable for preparing w/o emulsions have HLB numbers in the range of 2–7 (HLB = hydrophile–lipophile balance). The commercial emulsifiers offered in this range are fatty acid (stearic acid, oleic acid, etc.) esters of polyhydric alcohols (glycol, glycerol, sorbitol, etc.). Other typical w/o emulsifiers are the fatty acid salts of multivalent metals (calcium, magnesium, zinc, etc.). Some emulsifiers produced w/o emulsions with pure styrene. However, none of the emulsifiers, which belong to this group and which were examined by us under the conditions described above, produced w/o emulsions with the rubber solution. After the essential part of this work had been completed, a patent (3) appeared in which the prepolymerization of a solution of rubber in styrene in a w/o emulsion was described. Zinc stearate and other compounds with HLB numbers below 10 are claimed as emulsifiers.

A number of water-soluble high molecular weight compounds are used as protective colloids or suspending agents during the suspension polymerization of styrene. Many of these dispersants depress surface and interfacial tensions to a moderate degree. The applicability of the following dispersants as emulsifiers for the w/o emulsification of the rubber solution was investigated: poly(vinyl alcohol) (PVA) (*MW ca.* 80,000, degree of hydrolysis 87%), hydroxyethylcellulose (HEC) (apparent viscosity of a 2% aqueous solution: 300 cp at a shear rate of 1 sec^{-1}), poly(vinyl pyrrolidone) (PVP) and the 1:1 copolymer of styrene and maleic anhydride hydrolized and neutralized with aqueous sodium hydroxide (SMNa). The following additional compounds were examined: a high viscosity hydroxyethylcellulose (apparent viscosity of a 1% aqueous solution: 3,000 cp at a shear rate of 1 sec^{-1}), a high viscosity SMNa and the sodium salt of the azeotropic styrene acrylic acid copolymer (SANa). The high viscosity SMNa copolymer was obtained by neutralizing a 1:1 styrene–maleic anhydride copolymer with a dilute aqueous solution of sodium hydroxide in the presence of a small quantity of polyhydric alcohol (as a crosslinking agent).

Using a phase volume ratio w:o of 1:1 and the rubber solution as the oil phase, w/o emulsions were successfully made with PVA, HEC, SMNa, and SANa. No emulsions were obtained with PVP. It was possible to increase the stability of the w/o emulsions, at the same level of emulsifier concentration, by using the high viscosity HEC or the high viscosity SMNa in place of the corresponding low viscosity products. The SANa copolymer gave the most stable emulsions. It seems worth mentioning that these emulsifiers are not composed of large blocks of hydrophilic and lipophilic groups like the styrene–poly(ethylene oxide) graft copolymers, for instance, which are reported to be good w/o emulsifiers (2). On the contrary, the hydrophilic and lipophilic groups of the emulsifying suspending agents are quite small and are arranged in the polymer chain in an alternating or random manner. It is significant that the efficiency of the emulsifier increases with increasing emulsifier viscosity. It is not yet clear whether the viscosity of the aqueous phase, or the viscosity at the water/oil interface, or both, are responsible for this effect.

Formation and Stability of w/o Emulsions as a Function of the Process Conditions

The formation of a w/o emulsion with the rubber solution as the oil phase depended, like the production of w/o emulsions with low molecular weight oils, to a large extent on the process conditions. It cannot be ruled out, therefore, that a few of the products which were examined as emulsifying agents during these investigations and which were discarded as ineffective, would have yielded positive results if they had been tested under different conditions.

To produce w/o emulsions the aqueous phase had to be added to the rubber solution. The reverse order of addition yielded o/w emulsions; at a phase volume ratio w:o of 1:1 it was impossible to change an o/w emulsion into a w/o emulsion even when good w/o emulsifiers, which displayed only a moderate tendency to produce o/w emulsions, were used. The tendency to form w/o emulsions and the stability of these emulsions increased with the following factors: (the tendency to form o/w emulsions decreased):

(a) with decreasing water content and increasing emulsifier concentration (SANa was an apparent exception; emulsions made with 0.1% of this emulsifier appeared to be more stable than those with 1%);

(b) with decreasing viscosity of the oil phase (an increase in the rubber content of the oil phase favored the production of o/w emulsions);

(c) with an increasing rate of agitation (similar observations were made during the emulsification of low molecular weight oils (4).

Demulsification of w/o Emulsions

Of great importance to the polymerization is the smooth resolution of the w/o emulsions at the end of the prepolymerization. The transition of w/o emulsion polymerization to suspension polymerization (type o/w) must be complete and must occur without great effort. It has already been pointed out that the stability of w/o emulsions increased with decreasing viscosity of the oil phase and that the tendency towards formation of o/w emulsions increased with increasing viscosity of the oil phase. Therefore, it was expected that the w/o emulsions would become unstable during the polymerization—i.e., with increasing styrene conversion. Demulsification was not only caused by increasing viscosity of the oil phase, it also depended on the type and quantity of the emulsifier used to prepare the particular emulsion. By increasing the emulsifier concentration, demulsification was delayed. Usually demulsification occurred without outside inducement. However, only a few w/o emulsions prepared with SANa resolved, either partially or not at all. These w/o emulsions separated only a small quantity of the aqueous phase; the entire organic phase remained as a w/o emulsion. There are probably two reasons why the remaining w/o emulsions did not disintegrate further. First, by discarding water the w/o emulsion should have acquired an increased stability. Secondly, agitation—probably necessary for the demulsification of a highly viscous w/o emulsion—within the w/o emulsion is reduced greatly owing to the fact that the remaining w/o emulsion merely slips through the separated aqueous phase when agitated. These w/o emulsions disintegrated readily when a typical o/w emulsifier (e.g., an alkane sulfonate) was added. This result may indicate that only those w/o emulsions demulsify which are prepared with an emulsifier having a distinct tendency to produce o/w emulsions.

Development of a Polymerization Recipe for Producing High Impact Polystyrene

The production of a high impact polystyrene with optimum gel content and rubber particle size required the careful balancing of a number of factors during the w/o emulsion polymerization: phase volume ratio w:o, type and quantity of emulsifier, rate of polymerization, rate of agitation, and styrene conversion during the w/o emulsion polymerization. The phase ratio w:o chosen for the prepolymerizations was the one normally used for a suspension polymerization—i.e., a phase volume ratio w:o of 0.8:1 to 1:1. Of the emulsifiers mentioned above (not counting mixtures of these emulsifiers) the high viscosity HEC seemed to provide the best over-all combination of properties required for the w/o emulsion polymerization, demulsification, and the suspension polymerization. The required amount of this emulsifier ranged between 0.2 and 0.5% of the aqueous phase, depending on the phase ratio w:o, rate of agitation, and rate of polymerization used and on the desired styrene conversion during the w/o emulsion polymerization. Using an agitator speed of 300 rpm and a rate of polymerization of 5–10% per hour, the w/o emulsions demulsified at a styrene conversion of about 30%. An optimum rubber

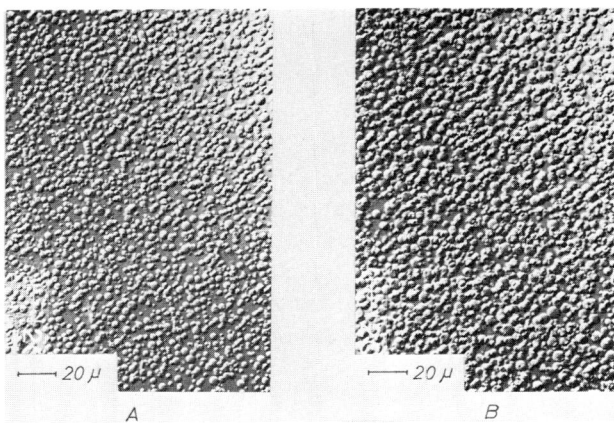

Figure 2. Interference phase contrast micrographs of the rubber dispersion in high impact polystyrene. Prepared by prepolymerization (A) in a w/o emulsion and (B) in bulk.

particle size and gel content in the final product was obtained by regulating and controlling the polymerization rate during the w/o emulsion polymerization or the conversion during the w/o emulsion polymerization, while keeping the other factors within the above limits. Suspension polymerization was started immediately after demulsification. To limit water occlusions in the final polymer beads to about 1%, the rate of styrene conversion during the first hour after demulsification was not allowed to exceed 10% per hour. This guaranteed a certain agitation within the globules of the organic phase, which apparently was necessary to bring about complete demulsification. The reaction conditions used during the remainder of the suspension polymerization were those of the combined bulk–suspension polymerization process during the suspension polymerization step. Figure 2 shows the rubber dispersion in two high impact polystyrenes, which were prepared on the one hand by prepolymerization in a w/o emulsion and on the other hand by prepolymerization in bulk under otherwise equal reaction conditions. The rubber particle size in both products is about equal. This shows that the agitation conditions within the polymer mixture during the w/o emulsion polymerization must have corresponded closely to those of the bulk agitated prepolymerization.

Conclusions

High impact polystyrene can be made by prepolymerization in a w/o emulsion with ensuing suspension polymerization. The processes which take place within the polymer mixture during the w/o emulsion polymerization correspond to those of the bulk agitated polymerization. Outwardly the conditions used for the w/o emulsion polymerization correspond to those of a "normal" suspension polymerization. A number of suspending agents are suitable emulsifying agents for w/o emulsions. The efficiency of a particular emulsifying–suspending agent increases with increasing viscosity of that emulsifier. A successful w/o emulsion polymerization requires the careful balancing of the following factors: phase ratio w:o, viscosity of the rubber solution, type and concentration of the emulsifier, rate of agitation, and rate of polymerization. Demulsification of the w/o emulsion and transition to the suspension polymerization occurs automatically with increasing styrene conversion. The reaction conditions necessary during the suspension polymerization correspond to those of the combined bulk–suspension polymerization process during the suspension polymerization step.

Literature Cited

(1) Becher, P., "Emulsions: Theory and Practice," Reinhold, New York, 1965.
(2) Bartl, H., Bonin, von H., *Makromol. Chem.* **57,** 74 (1962).
(3) Dainippon Celluloid Co. Ltd., Japanese Patent **18,710** (Aug. 14, 1968).
(4) Rodger, W. A., Trice, V. G., Rushton, J. H., *Chem. Eng. Progr.* **52,** 515 (1956).
(5) Ruffing, N. R., Kozakiewicz, B. A., Cave, B. B , Amos, J. L., U. S. Patent **3,243,481** (March 29, 1966).
(6) Schroeder, C. W., Lunk, H. E., Doyle, M. E., Belgian Patent **619,901** (Jan. 7, 1963).
(7) Stein, A., Walter, R. L., U. S. Patent **2,862,907** (Dec. 2, 1958).
(8) Wenning, H., *Makromol. Chem.* **20,** 196 (1956).

RECEIVED February 25, 1970.

15

Comparison of Methyl Methacrylate–Butadiene–Styrene with Acrylonitrile–Butadiene–Styrene Graft Copolymers

RUDOLPH D. DEANIN and ISMAIL S. RABINOVIC[1]

Departments of Chemistry and Plastics, Lowell Technological Institute, Lowell, Mass. 01854

ANTONIO LLOMPART

Chemical Laboratory, DeBell & Richardson, Inc., Hazardville, Conn. 06036

Methyl methacrylate was substituted for acrylonitrile in conventional graft copolymerization of acrylonitrile and styrene onto a polybutadiene backbone. Methyl methacrylate favored more homogeneous polymerization, better melt processability, lighter color, and greater stability toward ultraviolet light aging, while acrylonitrile favored greater strength and higher heat deflection temperature. Ultraviolet light discoloration appeared to be caused by oxidation of C=C bonds to C=O and C—OH groups. This was catalyzed by acrylonitrile units and inhibited by methyl methacrylate units in the graft copolymer.

Copolymerization of styrene and acrylonitrile in the presence of polybutadiene produces a graft copolymer polyblend in which the polybutadiene contributes ductility and impact resistance while the styrene–acrylonitrile forms a continuous matrix which contributes rigidity, strength, and reasonably high heat deflection temperature. This combination of useful properties makes acrylonitrile–butadiene–styrene (ABS) polymers one of the fastest growing of our present commercial plastics (9).

One major deficiency of these materials is their tendency to discolor, particularly under exposure to ultraviolet light during long term outdoor

[1] Present address: Rohm & Haas Co., Bristol, Pa.

Table I. Emulsion Polymerizations

Ingredients and Conditions	Order of Addition (8)	Polymerization Recipes		
		ABS	AMBS	MBS
Polymer Components				
Firestone FR-S-2004 59.2% polybutadiene latex	1	50.7	50.7	50.7
(polybutadiene equivalent)		30	30	30
styrene	7	45	45	45
acrylonitrile	8	25	12.5	—
methyl methacrylate	9	—	12.5	25
Dispersion Medium				
deionized distilled water (total)	12	180	180	180
Hercules Dresinate 731 rosin soap	4	2.8	2.8	2.8
Vanderbilt Darvan 1 polymerized Na alkyl naphthalene sulfonate	5	0.125	0.125	0.125
NaOH	2	0.146	0.146	0.146
Initiator System				
cumene hydroperoxide	10	0.65	0.65	0.65
$FeSO_4 \cdot 7H_2O$	13	0.09	0.09	0.09
$Na_4P_2O_7$	3	0.5	0.5	0.5
dextrose	6	1	1	1
Chain-Transfer Agent				
n-dodecylmercaptan	11	0.1	0.1	0.1
Polymerization Conditions				
temperature, °C		65	65	65
time, hours		6	6	6
Stabilizers				
Shell Ionol di-*tert*-butyl-*p*-cresol		1.2	1.2	1.2
Monsanto Santonox R 4,4'-thiobis (6-*tert*-butyl-*m*-cresol)		1	1	1
Conversion, %		96	95	95

Table II. Bulk Polymerizations

Ingredients	Polymerization Recipes		
	ABS	AMBS	MBS
Firestone Diene 35 NFA polybutadiene	30	30	30
Styrene	45	45	45
Acrylonitrile	25	12.5	—
Methyl methacrylate	—	12.5	25

use (4, 7). Acrylonitrile is known for its marked tendency toward discoloration (5, 6, 12), so its presence in the polymer could be considered a mixed blessing. Conversely, poly(methyl methacrylate) is widely recognized as the most light-stable of all commercial polymers, retaining its good color throughout the most severe ultraviolet and outdoor aging (*10*). Its contribution to rigidity, strength, and heat deflection temperature, while probably not as great as acrylonitrile, should be considerable.

This study was therefore undertaken to prepare and evaluate acrylonitrile–butadiene–styrene (ABS) and methyl methacrylate–butadiene–styrene (MBS) polymers under similar conditions to determine whether replacement of acrylonitrile by methyl methacrylate could improve color stability during ultraviolet light aging, without detracting seriously from the good mechanical and thermal-mechanical properties of conventional ABS plastics. For purposes of control, the study also included briefer evaluation of commercial ABS, MBS, and acrylonitrile–butyl acrylate–styrene plastics.

Experimental

Commercial ABS is prepared primarily by free-radical emulsion copolymerization of styrene and acrylonitrile in the presence of polybutadiene latex (*3*). This method was therefore adapted for the preparation of ABS and MBS terpolymers and an intermediate AMBS tetrapolymer under similar conditions (Table I). Polymerizations were charged into 12-ounce crown-cap bottles, sparged and flushed with nitrogen, and

Table III. Milling of Emulsion Polymers on a 6" × 13" Two-Roll Differential Speed Mill

Polymer	Temperature, °C	Time, min	Behavior
Marbon Cycolac H-1000 ABS	166	2–3	good
Experimental ABS	166	2–3	good
Experimental AMBS	166	15	rough
Experimental MBS	146	2–3	good
Mazzucchelli Sicoflex MBS 1.2	176	4	good
Mazzucchelli Sicoflex MBS 1.5	176	4	good

Table IV. Molding of Emulsion Polymers: Compression-Molding of 6" × 6" × 1/8" Sheets

Polymer	Temperature, °C	Preheat, min	Hot Press, min	Cold Press, min
Marbon Cycolac H-1000 ABS	190	2	1.5	1.5
Experimental ABS	190	2	1.5	1.5
Experimental AMBS	190	2	1.5	1.5
Experimental MBS	140	2	1.5	1.5
Mazzucchelli Sicoflex MBS 1.2	190	2	1.5	1.5
Mazzucchelli Sicoflex MBS 1.5	190	2	1.5	1.5

Table V. Properties of Experimental Polymers

Property	ABS	AMBS	MBS
Melt index, 5000-gram load/200°C, gram/10'	0.02	0.33	0.17
Melt index, 4320-gram load/272°C, gram/10'	0.03	1.19	0.63
Barcol hardness	36	32	34
Rockwell L hardness	81	77	74
Flexural modulus, 10^5 psig	2.6	2.5	2.7
Flexural yield strength, 10^3 psig	8.5	7.6	7.4
Izod notched impact strength, fpi	7.4	5.1	5.9
Heat deflection temperature/264 psig, °C	80	75	68
Ultraviolet light, Fadeometer hours to initial discoloration	<20	<20	100

Table VI. Properties of Acrylic Polymers

Property	Experimental MBS	Commercial MBS 1.2	Commercial MBS 1.5	Commercial AN–BA–ST
Manufacturer	LTI	Mazzucchelli		BASF
Trade name	MBS	Sicoflex		Luran S
Melt index/200°C	0.17	0.60	1.03	
Melt index/272°C	0.63	Degradation		
Barcol hardness	34	70	63	
Rockwell L hardness	74	94	87	
Flexural modulus	2.7	3.4	3.2	
Flexural yield strength	7.4	11.8	9.9	
Impact strength	5.9	1.1	2.9	
Heat deflection temp.	68	78	75	
Ultraviolet light	100	60	60	20

tumbled end-over-end at 42 rpm in a constant-temperature water bath. After completion of reaction, latexes were steam-stripped, filtered, stabilized, and coagulated with 0.2% H_2SO_4 in a Waring blender. The coagulated polymers were vacuum-filtered, washed with distilled water, and dried overnight in a 63°C circulating-air oven.

Bulk polymerization was also studied briefly (Table II), adapting a technique useful in production of impact styrene (1). Solid polybutadiene was dissolved in styrene, mixed with acrylonitrile and/or methyl methacrylate, charged into a 12-ounce crown-cap bottle, flushed with nitrogen, and tumbled end-over-end at 42 rpm in a constant-temperature water bath for 16 hours at room temperature, 48 hours at 80°C, and 48 hours at 90°C, then heated in an oven 24 hours at 110°C and 24 hours at 150°C. ABS polymerization produced a grainy, inhomogeneous, light-orange product, while AMBS and MBS polymerizations produced clear homogeneous white products.

Emulsion polymers and commercial controls were milled (Table III) and molded (Table IV) to produce test samples, and properties were evaluated following standard ASTM procedures (Tables V–VII). During accelerated ultraviolet light aging, the polymers were also examined by infrared spectroscopy of films cast from o-dichlorobenzene solution (Table VIII).

Discussion of Results

Comparison of experimental ABS, AMBS, and MBS polymers (Table VIII) indicated that replacement of acrylonitrile by methyl methacrylate did produce marked improvement in color stability during accelerated

Table VII. Properties of ABS Polymers

Property	LTI Experimental Polymer	Commercial Marbon Cycolac H-1000
Melt index/200°C	0.02	1.14
Melt index/272°C	0.03	1.25
Barcol hardness	36	35
Rockwell L hardness	81	78
Flexural modulus	2.6	2.3
Flexural yield strength	8.5	7.5
Impact strength	7.4	7.0
Heat deflection temperature	80	72
Ultraviolet light	<20	20

ultraviolet light aging. Infrared spectroscopy suggested that the mechanism of discoloration in ABS was a rapid conversion of C═C bonds (2900, 2840, 970, and 910 cm^{-1}) into C═O (1720 cm^{-1}), C—OH (3380 cm^{-1}), and CO$_2$H groups (broadening of 3380 and 1720 cm^{-1} bands) during aging; this apparently was inhibited by methyl methacrylate in MBS but not in the presence of acrylonitrile in AMBS. Earlier studies have similarly pointed to polybutadiene as the component responsible for aging (2, 4, 11).

With respect to effects on other properties (Table V) replacement of acrylonitrile by methyl methacrylate also improved processability (melt index) and had no clear effect on rigidity (hardness and flexural modulus) but did lower strength (flexural yield and impact) and heat deflection temperature somewhat, probably owing to the loss of strong hydrogen bonding of acrylonitrile units. Incidentally, the heterogeneity of ABS and homogeneity of AMBS and MBS in bulk polymerizations

Table VIII. Fadeometer Accelerated

Polymer	Measurement
Commercial ABS	Color
Experimental ABS	Color
	IR 3380 cm^{-1}
	2900
	2840
	1720
	970
	910
Experimental AMBS	Color
	IR 3380 cm^{-1}
	2900
	2840
	1720
	970
	910
Experimental MBS	Color
	IR 3380 cm^{-1}
	2900
	2840
	1720
	970
	910
Commercial MBS 1.2	Color
Commercial MBS 1.5	Color
Commercial AN/BA/ST	Color

[a] Infrared absorption bands are described as b = broad, m = medium, s = strong, w = weak, vw = very weak.

suggested that even partial replacement of acrylonitrile by methyl methacrylate might permit the development of commercial bulk polymerization processes, which could be as desirable here as in impact styrene production.

Comparison of experimental and commercial acrylic polymers (Table VI) indicated that the present commercial MBS materials were much more rigid, brittle, and higher softening than the experimental MBS, either because they had lower polybutadiene content or perhaps because they were optimized with respect to clarity rather than impact strength. Thermal degradation of commercial MBS during the high temperature melt index test was characteristic of the depolymerization of poly(methyl methacrylate) homopolymer chains, suggesting that the commercial material contained such a fraction. Incidentally, the blue tint of the com-

Ultraviolet Light Aging

Hours in Fadeometer According to ASTM D1499-59T[a]

0	20	40	60	100	200
—	sl.ylw.	sl.ylw.	sl.ylw.	sl.ylw.	
—	dk.br.	v.dk.br.	v.dk.br.	dk.br.ylw.	dk.br.ylw.
—	mb	mb		mb	mb
s	s	m		m	m
w	—	—		—	—
m	m	s		s	s
s	m	vw		—	—
s	m	vw		vvw	—
—	dk.br.	dk.br.	dk.br.	br.ylw.	br.ylw.
—	sb	sb		sb	sb
s	s	m		m	m
m	—	—		—	—
s	s	s		sb	sb
s	m	vw		—	—
s	m	vw		vw	—
—	—	—	—	sl.ylw.	sl.ylw.
—	vw		sb	sb	sb
s	s		m	m	m
m	m		—	—	—
s	s		sb	sb	sb
s	s		w	w	—
s	s		w	w	—
—	—	—	hazy spots	hazy spots	
—	—	—	hazy spots	hazy spots	
—	sl.ylw.	sl.ylw.	sl.ylw.	sl.ylw.	

mercial MBS polymers suggested that stabilizers and/or brighteners had been added, to help methyl methacrylate protect the polybutadiene backbone against premature discoloration. Both commercial MBS and commercial acrylonitrile–butyl acrylate–styrene polymers discolored slightly during accelerated ultraviolet light aging, suggesting that the presence of either butadiene or acrylonitrile units could be a factor in discoloration. Even styrene–acrylic ester copolymers discolor eventually because of the styrene, so all of these systems contain nonacrylic portions which probably contribute to discoloration.

Comparison of experimental and commercial ABS polymers (Table VII) indicated that the experimental polymer was much harder to process and higher in mechanical and thermal-mechanical properties, probably because of much higher molecular weight, which could be controlled by increasing the chain-transfer agent. The ultraviolet light stability of the commercial polymer may have been caused by greater homogeneity by increment addition of monomer during polymerization, different polymerization recipe, greater purity of the final polymer, or addition of better stabilizers during manufacture; in practice, commercial ABS is very difficult to stabilize thoroughly against prolonged ultraviolet light and outdoor aging.

Conclusions

Replacement of acrylonitrile by methyl methacrylate in ABS can improve processing and ultraviolet light stability, without sacrificing rigidity but with some decrease in strength and heat deflection temperature. It may also permit development of production by bulk polymerization. The mechanism of ultraviolet light stability appears to be the inhibiting action of methyl methacrylate which serves to retard the ultraviolet light-catalyzed atmospheric oxidation of C=C bonds in the polybutadiene backbone.

Acknowledgment

This work was part of a thesis submitted by I. S. Rabinovic in partial fulfillment of the requirements for the degree of Master of Science and was supported in part by a grant from the National Aeronautics and Space Administration.

Literature Cited

(1) Amos, J. L., McCurdy, J. L., McIntire, O. R., U. S. Patent **2,694,692** (1954).
(2) Boyle, D. J., Gesner, B. D., *J. Appl. Polymer Sci.* **12**, 1193 (1968).
(3) Calvert, W. C., U. S. Patent **2,908,661** (1959).
(4) Gesner, B. D., *J. Appl. Polymer Sci.* **9**, 3701 (1965).
(5) Golub, M. A., "Chemical Reactions of Polymers," E. M. Fettes, Ed., p. 111, Interscience, 1964.
(6) Grassie, N., "Chemical Reactions of Polymers," pp. 621–6, Interscience, 1964.
(7) Kelleher, P. G., Boyle, D. J., Miner, R. J., *Am. Chem. Soc. Div. Org. Coatings Plastics Chem.* **29** (2), 508 (Sept. 1969).
(8) Kolthoff, I. M., Medalia, A. I., *J. Polymer Sci.* **5** (4), 391 (1950).
(9) "Modern Plastics Encyclopedia for 1968-69," p. 118 (1968).
(10) "Modern Plastics Encyclopedia for 1968-69," p. 136 (1968).
(11) Shimada, J., Kabuki, K., *J. Appl. Polymer Sci.* **12**, 655, 671 (1968).
(12) Smets, G. J., "Chemical Reactions of Polymers," p. 89, Interscience, 1964.

RECEIVED October 16, 1969.

16

New Transparent Impact-Resistant Polymer Blends from Resins and Rubbers Having Minimized Compositional Heterogeneities

I. Blends of Resins with Diene Copolymer Elastomers

R. G. BAUER, R. M. PIERSON, W. C. MAST, N. C. BLETSO, and L. SHEPHERD

The Goodyear Tire & Rubber Co., Research Division, Akron, Ohio 44316

> *Transparent thermoplastic impact polymers were prepared from selected diene elastomers and resins. By matching rubber and resin refractive indices and adjusting compatibilities with grafting procedures, transparent impact polyblends were prepared having useful engineering properties. In copolymerization reactions if one of the comonomer reactivity ratios exceeds unity while the other is less than unity, no critical (azeotropic) composition exists. When this was the case, constant composition (homogeneous) copolymers were prepared by proportioning a mixture of monomers whose composition was the same as that of the copolymer being formed, or alternatively, the more reactive comonomer was incrementally fed at a rate to prepare the constant composition copolymer. Graft copolymerizations were used partially to compatibilize the resin and rubber components of these two-phase systems, making transparent polymers having high impact resistance.*

For many years elastomers have been used to modify rigid plastics, primarily to improve their toughness without significantly lowering their rigidity. These polyblends utilize diene type elastomers, whereby tough, opaque, or translucent resins are produced. The first polyblends reported were those obtained by mechanical mixing (in the melt) of two individual polymers. For the most part, these mechanical blends have been replaced by graft copolymers, whereby the grafted comonomer(s) tend to stabilize the heterophase structure. This heterophase structure has been observed in all of the impact polyblend systems.

Transparency in Polyblends

Where the two phases are completely compatible, a homogeneous polyblend results which behaves like a plasticized resin (one phase). If two polymers are compatible, the mixture is transparent rather than opaque. If the two phases are incompatible, the product is usually opaque and rather friable. When the two phases are partially compatibilized at their interfaces, the polyblend system may then assume a hard, impact-resistant character. However, incompatible or partially compatible mixtures may be transparent if the individual components are transparent and if both components have nearly the same refractive indices. Furthermore, if the particle size of the dispersed phase is much less than the wavelength of visible light (requiring a particle size of 0.1μ or less), the blends may be transparent.

There are other parameters which must be controlled to produce and maintain optical clarity in polyblends. It is known that in a random copolymerization, compositional heterogeneity can result from the high conversion copolymerization of monomers to polymer (12). This heterogeneity produces a range of polymers which are not necessarily compatible and which may have differing refractive indices. Compositional heterogeneity can thus lead to opacity or translucency in a copolymer, ultimately contributing to the haze of the polyblend. While partial incompatibility of the two phases (resin and rubber) appears to be a requirement for impact modification, essentially homogeneous composition resin and graft rubber components, having nearly matched refractive indices are necessary to produce polyblends having optical properties approaching those of the individual components—i.e., they are transparent.

It was the object of this investigation to use the aforementioned considerations to produce impact-resistant polyblend systems with the optical characteristics of homopolymer systems.

Experimental

Monomer Purification. All polymers were prepared from either column purified or distilled monomers. The acrylate and methacrylate esters, styrene, and vinyl nitrile type monomers were purified by passing them through Rohm and Haas Amberlyst exchange resins (salt forms), while the diene monomers were either distilled directly from cylinders and condensed in a dry ice trap or alternatively caustic washed to remove the inhibitor.

Polymer Preparation. Polymerization reactions were most often conventional emulsion type, although solution, bulk, or suspension techniques may also be used. The diene elastomers were graft copolymerized and either investigated in that form or blended with resins having matching refractive indices and suitable solubility parameters. No attempt was made to measure the actual graft level, and all references to graft level indicate only monomer charge levels.

Where copolymer compositional heterogeneity was a problem, monomer charges and feed ratios were adjusted to produce constant composition homogeneous polymers. While the required composition for the monomer charge and feed ratios may be determined by a random process, it is much more convenient to use an integrated form of the copolymer equation to calculate these quantities (2).

Physical Measurements. INTRINSIC VISCOSITIES. Intrinsic viscosities were obtained using dilute solution viscometry (Cannon-Ubbelohde viscometers).

REFRACTIVE INDEX MEASUREMENTS. Refractive indices were measured at 23°C on molded polymer sheets, using an Abbe type refractometer. Suitable contact liquids were employed as required.

TRANSMISSION AND HAZE MEASUREMENTS. Light transmission measurements were made with a Gardner color difference meter and the haze measurements, using a Hazemeter attachment on the Gardner colorimeter. Test pieces were $3 \times 3 \times 1/8$ inch sheets having glossy surfaces obtained by molding on chrome plates. All measurements were made at 23°C.

INSTRON CAPILLARY RHEOMETRY. Melt flow rheology was measured according to procedures first proposed by Merz and Colwell (10) and supplied by the Instron Engineering Corp.

TENSILE PROPERTIES. Tensile properties were obtained using both Instron and Baldwin-type tensiometers. Samples were elongated at 0.2 and 2.0 inches/min.

FLEXURAL PROPERTIES. Both flexural modulus and flexural strength values were obtained. These values were measured at 23°C and also over a range of temperatures for the MBAS polymer (see Figure 4). In the flexural tests, a molded bar is tested as a simple beam, the bar resting on two supports, and the load is applied midway between. The test is continued until rupture or 5% strain, whichever occurs first. The test fixture is mounted in a universal tester, and the tester is placed in an appropriate temperature environment.

HEAT DISTORTION TEMPERATURE. Heat distortion temperatures were obtained in flexure, using a load of 264 psig on molded bars $5 \times 1/2 \times 1/2$ inches. The heat distortion temperatures were taken as the temperatures at which test specimens had deformed 0.010 inch, where a heating rate of 2°C per minute was applied.

IMPACT STRENGTH. Impact strengths were measured using the notched Izod technique at both 23° and −40°C. Samples were conditioned at these temperatures after notching, prior to testing.

HARDNESS MEASUREMENTS. Hardness was measured on two scales, the Rockwell R scale, and the Shore D scale. The Rockwell test measures nonrecoverable deformation after a 1/2-inch sphere has been pressed into a molded sheet of the plastic and the load released. In the Shore

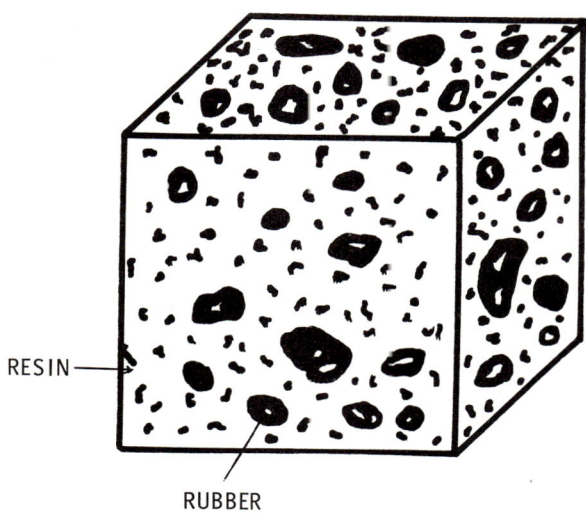

Figure 1. Two phases. Small graft rubber particles suspended in a hard matrix.

durometer type D test a hardness of 100 corresponds to a 10-lb load on a spherical tip of 0.004 inch in radius. This test is a short, 2-sec test. Both measurements were made at 23°C.

Results and Discussion

To obtain two-phase polymer systems with acceptable physical and optical properties, due consideration must be given to the method of preparation. These polymer systems consist of two distinct phases, small graft rubber particles being suspended in a relatively hard matrix of a resinous copolymer (Figure 1). Frequently, the two phases form interpenetrating networks where neither phase is clearly the dispersed phase (Figure 2). Various methods may be used to prepare these two-phase polymers, and depending on their method of preparation, the resulting materials can be classified as either mechanical polyblends or graft copolymers.

It is known that most mixtures of elastomers and resinous copolymers are rather polydisperse, their incompatibility giving rise to poor interfacial adhesion. Preferably, a resinous monomer or monomers are grafted to the rubber, and they act as a link between the rubber and the resin phases. As Rosen (13) has pointed out, one means of predicting the affinities of polymer pairs from measurable properties is through the use of the solubility parameter, and this has proved useful in this study. The solubility parameter concept was originally derived from the thermodynamics of nonelectrolyte solutions. Hildebrand proposed the concept that the square root of the cohesive energy density was a numerical value identifying the solvency behavior of specific solvents (7). Brydson (4) has elaborated on the use of the solubility parameter in plastics technology, and his approach is useful in making first approximations about polymer–polymer interactions. The method, of course, cannot be used to predict the specific behavior of any particular polyblend system. For example, Hansen (6) has concluded that although polymers having similar solubility parameters tend to be compatible, other factors affect this phenomenon, such as polymer crystallinity and molecular weight. Burrell's (5) excellent critique of the solubility parameter approach considers

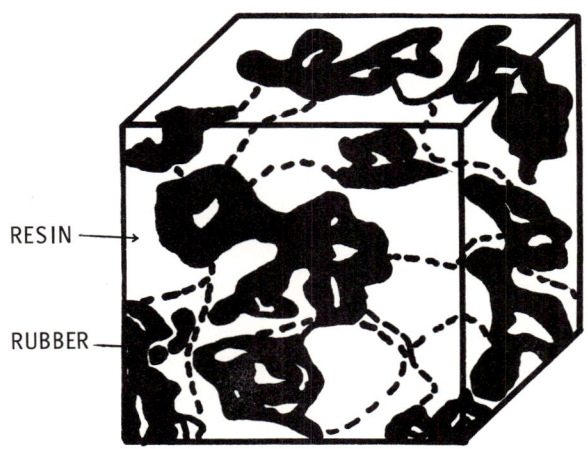

Figure 2. Interpenetrating network structure

many of these factors in more detail. These investigators recognized that the solubility parameter concept must be divided into consideration of dispersion forces (London), permanent dipole–permanent dipole, and hydrogen bonding energies of cohesion.

Considering the number of variables possible in preparing two-phase polymers, it isn't surprising that many candidates can be prepared. However, the overriding consideration of transparency greatly limits the number of possible permutations. An important consideration is that although many high conversion copolymers are heterogeneous and as a result appear hazy or opaque to transmitted light, constant composition copolymers can be prepared which are transparent.

From comonomer reactivity ratios, Wall (15) observed that certain critical monomer compositions were possible which he designated co-

Table I. Optical Haze of a Copolymer System

Copolymers	Percent Haze[b]
[c]Bulk charged copolymer of MMA–α-MeSt (90–10)[a]	14.7
Constant composition copolymer of MMA–α-MeSt (90–10)[c]	3.2

[a] Wt % ratios.
[b] Measured using a Gardner color difference meter on 1/8-inch thick moldings.
[c] The azeotropic composition for MMA/α-MeSt (M_1/M_2) is approximately 59.3 wt % MMA and 40.7 wt % α-MeSt using published reactivity ratios of $r_1 = 0.50 \pm 0.03$ and $r_2 = 0.14 \pm 0.01$.

polymerization azeotropes. From kinetics of chain propagation in copolymerization, the critical concentration is

$$(f_1)c = (1 - r_2)/(2 - r_1 - r_2)$$

The value of this critical concentration, $(f_1)_c$, lies within the permissible range $0 < f_1 < 1$ only if both r_1 and r_2 are greater than unity or if both are less than unity. The case of $r_1 > 1$ and $r_2 > 1$ is unknown. If one of the reactivity ratios exceeds unity while the other is less than unity, no critical composition exists.

In practice, there are effective means for nearly eliminating the spread of chemical compositions (14). The essential condition for the formation of a chemically uniform copolymer is that the monomers must be maintained in the same ratio throughout polymerization. There are two procedures for maintaining this condition. First, it can be achieved by making a continuous addition of either the more reactive monomer alone or certain mixtures of the comonomers as the reaction proceeds. The second method of producing chemically homogeneous copolymers is to proportion a mixture of monomers whose composition is the same as that of the copolymer being formed.

A significant decrease in the spread of compositions for copolymer produced by monomer proportionation compared with that produced by bulk charging of monomers can be seen by comparing the optical haze values obtained on a 90–10 methyl methacrylate–α-methylstyrene (MMA–α-MeSt) copolymer system (Table I).

The initial charge ratio and incremental monomer additions required to produce a constant composition 90–10 methyl methacrylate–α-methylstyrene are shown in Figure 3.

Using these techniques to prepare the resin and rubber components of polyblends, transparent impact polymer blends can be obtained which exhibit mechanical properties comparable with ABS materials. Like ABS, they have a fairly flat modulus curve, which varies gradually over a wide temperature range (Figure 4). Impact properties are outstanding for these transparent impact polymers even at temperatures as low as $-40°C$.

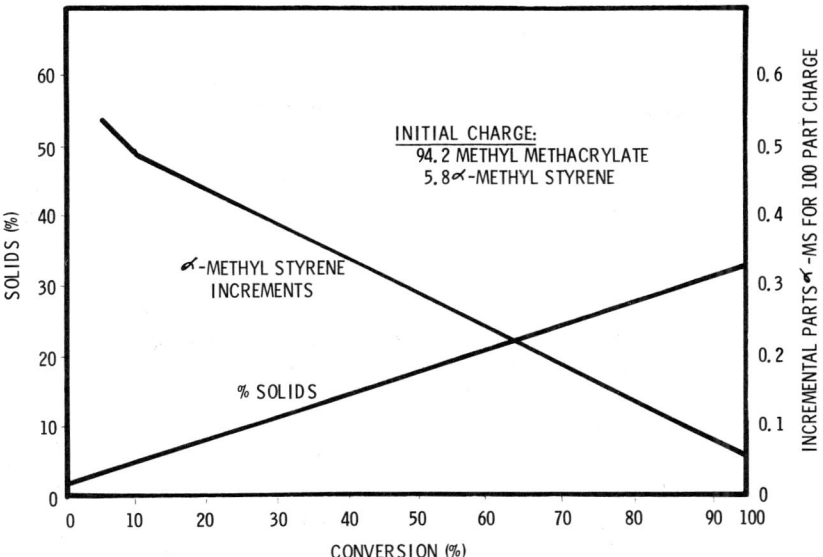

Figure 3. Methyl methacrylate–α-methylstyrene 90–10 constant composition resin

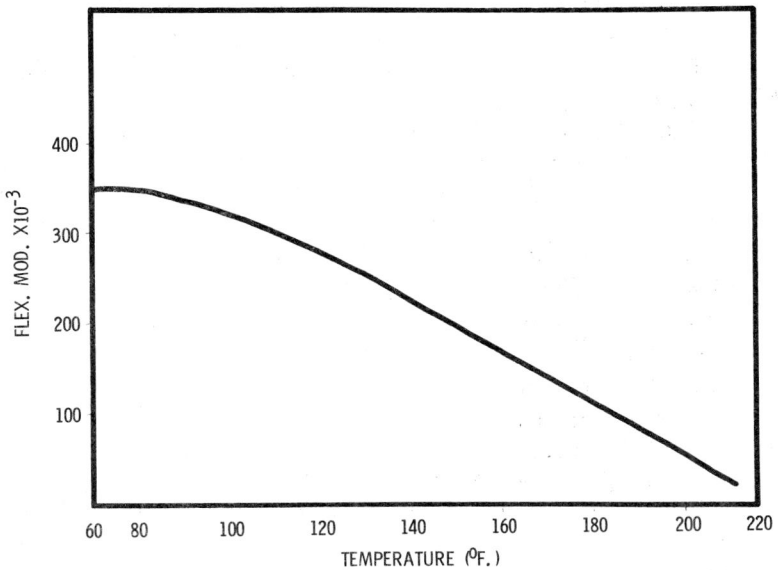

Figure 4. Flexural modulus vs. temperature

**Table II. Effect of Graft Resin on Blend Optical Properties
(Resin–Backbone Rubber = 2.5–1.0)**

Graft Resin, %	Transmission, %	Haze, %	Refractive Indicies	
			Graft Rubber	Resin
0	87.2	38.0	1.5044	1.5064
10	90.6	11.2	1.5046	1.5064
15	89.8	10.5	1.5039	1.5064
22	89.5	11.5	1.5034	1.5064
27	88.6	6.0	1.5028	1.5064

**Table III. The Effect of Graft Resin on Polyblend Mechanical
Properties (Resin–Backbone Rubber Ratio = 2.5–1.0)**

Graft, %	Notched Izod Impact Strength (ft.-lbs/in.)	Heat Distortion Temp. (°C) at 264 psig	Shore D Hardness	Flexural Modulus psig × 10^{-5}	Tensile Strength, psig
0	0.45	73	71	158	1960
10	3.20	73	75	171	3290
15	1.25	75	77	183	3560
22	0.75	77	79	218	4020
27	0.60	77	80	292	4180

The physical properties of these polymers, as well as their transparency, is at least in part owed to their structure. Some authors have indicated that this type of structure represents a polymeric oil-in-oil (POO) emulsion (*11*).

The importance of the graft "handle" on a 62/38 butadiene–methyl methacrylate rubber can be illustrated by its effect on the optical properties of the polyblend. From Table II it can be seen that the reduction in percent haze is dramatic for an increase of methyl methacrylate graft from 0 to 27% by weight, while there is no apparent change in the light transmission. The blend resin in this polyblend system was an 88–12 methyl methacrylate–styrene copolymer, and the total resin to backbone rubber ratio was kept at 2.5–1.0. The measured refractive indices are included for each component (the graft rubber and the blend resin). The difference in refractive index amounts to no more than 0.004 unit for any of the components.

The effect of the graft resin on the polyblend mechanical properties for this same system (2.5–1.0 resin–backbone rubber) can be seen in Table III.

It was observed that the notched Izod impact strength went through a maximum at a rather low graft level, whereas the other measured mechanical properties appeared to increase up to the maximum graft level (27%). Merz and co-workers (9) observed that both the tensile energy absorption and the impact strength of a two-phase polymer system (styrene–butadiene elastomer in polystyrene) passed through a maximum as the concentration of an added crosslinking agent (divinyl benzene) was increased. Since it has been shown that the gel content of a SBR–polystyrene graft copolymer was higher than that of the parent SBR elastomer (3), it is reasonable to assume that the gel content of a butadiene–methyl methacrylate elastomer would similarly increase as a result of the graft reaction. The 52–48 butadiene–methyl methacrylate copolymer rubber used in this work was initially gel free; however, following the grafting reaction, gel contents were found to be in the range 79–83%, using 22 parts of methyl methacrylate per 78 parts of rubber backbone. While this doesn't give a quantitative measure of grafting, it is a strong indication that grafting is significant. Other explanations, such as the possible increase in size of the dispersed graft elastomer phase might also be used to explain this phenomenon.

From such studies it is possible to optimize individual properties, such as impact strength, hardness, and flexural and tensile properties, and, within limits, to adjust optical clarity to the maximum attainable at the preferred levels of performance parameters.

Table IV. Typical Physical Properties of MBAS

Property	Value	ASTM Test Method
Tensile strength, lb/in^2		
23°C	5100–6000	D638
−40°C	8500–8700	
Flexural strength, lb/in^2	7700–8400	D790
Impact strength, ft-lb/in		
23°C ($\frac{1}{8}'' \times \frac{1}{2}''$ notched bar-Izod)	2.5–7.4	D256
($\frac{1}{2}'' \times \frac{1}{2}''$ notched bar-Izod)	4.1–8.3	
−40°C ($\frac{1}{8}'' \times \frac{1}{2}''$ notched bar-Izod)	0.5–1.4	
Flexural Modulus, lb/in^2	265,000–350,000	D790
Elongation, %	60–95	D638
Hardness (Rockwell)	R85–106	D785
Heat Distortion Temp, °C		
(unannealed) 264 lb/in^2	72–80	D648
Specific Gravity	1.02–1.20	D792
Water absorption, %	0.2–0.5	D570
Optical properties		
Transmission, %	85–88	D791
Haze, %	3–5	D1003

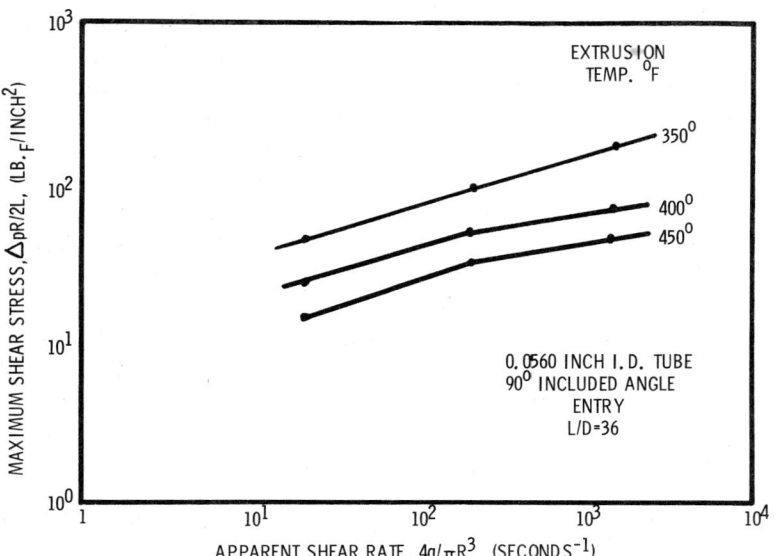

Figure 5. Melt viscosity characteristics

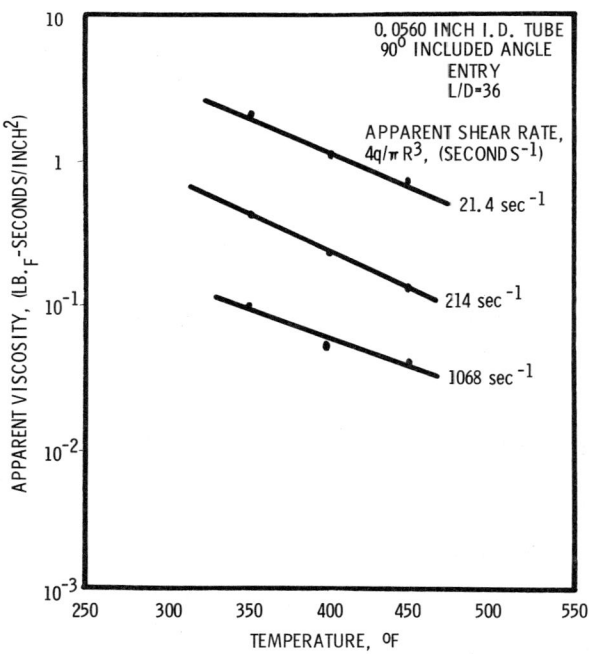

Figure 6. Melt viscosity characteristics

Typical mechanical properties for transparent injection-molded polymers, designated MBAS, having a compositional range of from 11–18%, 1,3-butadiene, 34–39% styrene, and 23–25% each of acrylonitrile and methyl methacrylate are given in Table IV. This polymer is closely akin to the polyblend described previously, differing by containing a butadiene–styrene elastomer backbone, with a terpolymer resin graft consist-

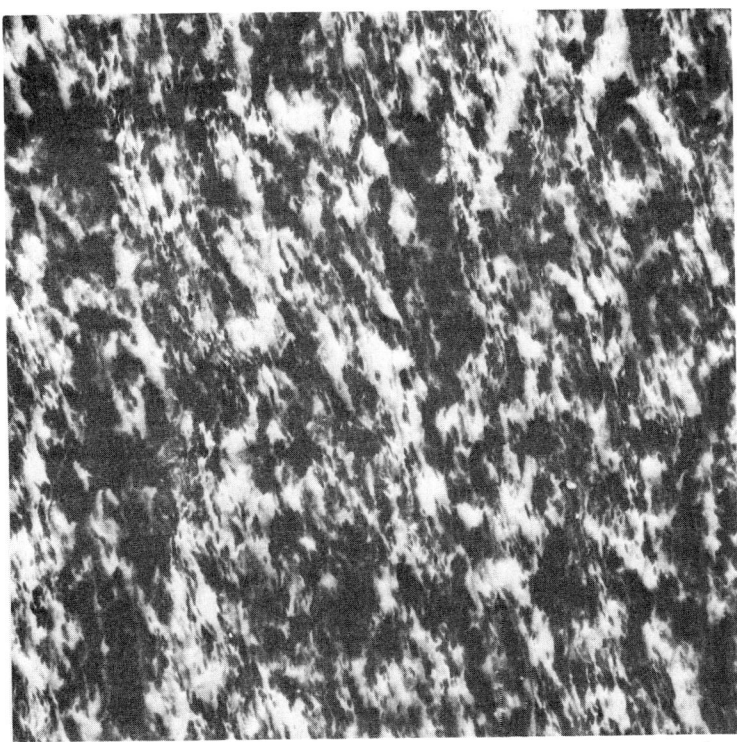

Figure 7. Thin section, osmium tetroxide stained. 1 mm = 800 A, ×12,500.

ing of methyl methacrylate, styrene, and acrylonitrile. To obtain optimum flow behavior, the intrinsic viscosity of the resin fraction (in chloroform at 30°C) was kept in the 0.5–0.7 range.

Melt flow rheology measurements were obtained on the MBAS polymer using an Instron capillary rheometer. The data reported were obtained using an 0.056-inch capillary, 90° included angle, with an L/D of 36. In Figure 5 the maximum shear stress (lb/in^2) is plotted vs. the apparent shear rate (sec^{-1}). The apparent viscosity ($lb\text{-}sec/in^2$) vs. temperature (°F) over the range of 350°–450°F is shown in Figure 6. This MBAS polymer was injection molded (Van Dorn production machine) into high gloss parts at high and low injection rates.

In an effort to interpret the morphology of these transparent polyblends, electron photomicrographs were made from microtomed cross sections of molded sheets. Such a photomicrograph is shown in Figure 7.

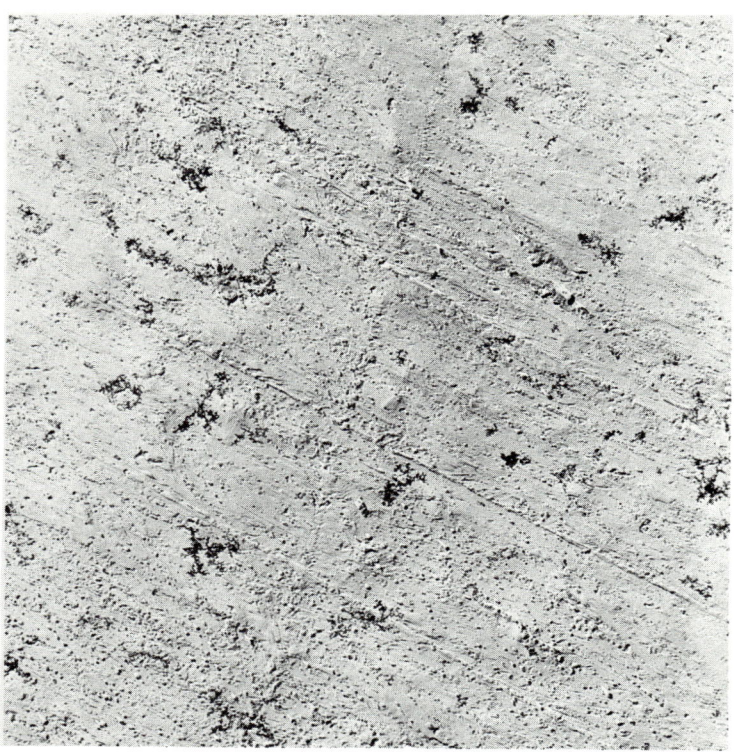

Figure 8. Flat surface replica, germanium shadowed 4:1 angle.
1 mm = 2000 A, ×5,000

The staining method of Kato (8) was used to examine the two-phase character of these blends since they are essentially indistinguishable to the electron beam (or to visible light—*viz.*, using phase contrast microscopy). The method of Kato involves treating the polymer with osmium tetroxide, which oxidizes the rubber and hence produces a dark image for the elastomer fraction. In these two-phase systems the relationship of polymer morphology to impact strength is still an unresolved

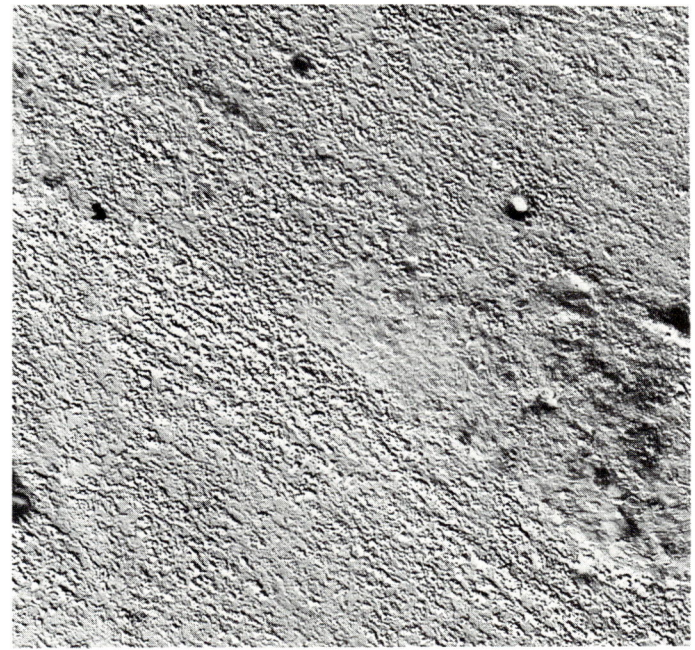

*Figure 9. Flat surface replica, germanium shadowed 4:1 angle.
1 mm = 2000 A, ×5,000*

problem in polymer science, and this photomicrograph is included to indicate the structure present in a system which has both high impact strength and transparency.

Since the transparency of a plastic depends on its surface, in addition to its composition, it is desirable to have melt flow properties which produce high gloss surfaces in molding operations. Evidence for the high gloss of an injection molded piece can be seen in Figure 8, where a flat surface replicate (germanium shadowed, 4:1 angle, 10,000 × magnification) is shown. By comparison, a replicate surface of a matte finish molding at the same magnification is shown in Figure 9.

Conclusion

By combining the concepts of copolymer homogeneity, matching refractive indices, and partial compatibilization *via* grafting, impact resistant polyblend systems can be produced from numerous monomer combinations that approach optical clarity.

Acknowledgment

Electron photomicrographs and surface replicate photographs were skillfully obtained by M. E. Testa of our microscopy department. His assistance and interpretations are gratefully acknowledged, as well as the permission of The Goodyear Tire & Rubber Co. to publish these findings.

Literature Cited

(1) Bauer, R. G., Bletso, N. C., Mast, W. C., Pierson, R. M., U. S. Patent **3,475,516** (Oct. 28, 1969).
(2) Billmeyer, F. W., Jr., "Textbook of Polymer Science," Chap. 11, Interscience, New York, 1962.
(3) Blanchette, J. A., Nielsen, L. E., *J. Polymer Sci.* **22**, 317 (1956).
(4) Brydson, J. A., *Plastics (London)* **26**, 107 (1961).
(5) Burrell, H., *Am. Chem. Soc., Div. Org. Coatings Plastics Chem., Preprints* **28** (1), 682–708 (1968).
(6) Hansen, C. M., *Ind. Eng. Chem., Prod. Res. Develop.* **8**, 2 (1969).
(7) Hildebrand, J. H., Scott, R. L., "The Solubility of Nonelectrolytes," 3rd ed., ACS Monograph Series, No. 17, p. 129, Reinhold, New York, 1950.
(8) Kato, K., *J. Polymer Sci.* **48**, 35 (1966).
(9) Merz, E. H., Claver, G. C., Baer, M., *J. Polymer Sci.* **22**, 337 (1956).
(10) Merz, E. H., Colwell, R. E., *ASTM Bull.* **232**, 63 (1958).
(11) Molau, G. E., *J. Polymer Sci.* **3A**, 1267 (1965).
(12) Nielsen, L. E., *J. Am. Chem. Soc.* **75**, 1435 (1953).
(13) Rosen, S. L., *Polymer Eng. Sci.* **7**, 115 (1967).
(14) Thomas, C. M., Hinds, J. R., *British Plastics* **31**, 522 (1958).
(15) Wall, F. T., *J. Am. Chem. Soc.* **66**, 2050 (1944).
(16) Young, L. J., *J. Polymer Sci.* **54**, 411 (1961).

RECEIVED January 23, 1970.

17

New Transparent Impact-Resistant Polymer Blends from Resins and Rubbers Having Minimized Compositional Heterogeneities

II. Blends of Resins with Acrylic Copolymer Elastomers

R. G. BAUER, R. M. PIERSON, W. C. MAST, N. C. BLETSO, and L. SHEPHERD

The Goodyear Tire & Rubber Co., Research Division, Akron, Ohio 44316

> *Transparent thermoplastic impact polymers were prepared from selected acrylic elastomers and resins. By matching rubber and resin refractive indices and adjusting compatibilities with grafting procedures, transparent saturated impact polyblends were prepared having useful engineering properties. Since these polyblends were completely saturated acrylic polymers, they were expected to show exceptional resistance to oxidative and photolytic decomposition. Accelerated aging tests (xenon arc weatherometer) indicated that they retained a significant level of physical strength, color stability, and freedom from surface deterioration even in the absence of added light stabilizers, tints, or pigments. These thermoplastic polyblends were fabricated into a variety of objects by conventional processing procedures.*

Acrylic polymers have long been recognized as plastics standards for optical clarity and good outdoor aging. While poly(methyl methacrylate) is inherently much more shatter resistant than glass, it has only a modest impact resistance compared with high impact polystyrene, ABS, and other rubber modified polyblends. Most of these high impact resistant polyblends, however, have somewhat limited outdoor aging capabilities and are, for the most part, opaque to translucent.

Many of the current concepts of the mechanical behavior of glassy polymers and composites were summarized at the 1967 Applied Polymer

Symposia on Polymer Modification of Rubbers and Plastics (*11*). Of particular interest to this investigation were the discussions relative to the stress responses of typical organic polymeric glasses. These considerations and currently proposed mechanisms of aging processes for plastic materials, such as those reported at the 1967 Applied Polymer Symposia on Weatherability of Plastic Materials (*10*) were useful in implementing this investigation. The concepts discussed in the first paper in this series (*3*) are fundamental to the development of polymers discussed in this paper. Of additional concern are chemical reactions to prepare graft copolymers on saturated acrylic backbones by chain transfer. The objective of this endeavor, then, was to determine whether transparent impact-resistant polymers could be prepared, combining both the optical clarity and good outdoor aging of the acrylics with the impact resistance of the diene modified polyblends. To date, only transparent unsaturated rubber modified polyblends and opaque saturated acrylic rubber modified polyblends have been reported (*12*).

Experimental

Polymer Preparation. All polymers were prepared from either purified or distilled monomers. The acrylate and methacrylate esters, styrene, and vinyl nitrile type monomers were purified by passing them over Rohm and Haas Amberlyst exchange resins (salt forms), or they were distilled according to appropriate procedures.

Polymerization reactions were principally conventional emulsion types, although solution, bulk, or suspension techniques may also be used.

The acrylic elastomers were graft copolymerized and either investigated in that form or blended with resins having matching refractive indices and suitable solubility parameters. Compositional homogeneity was maintained using techniques similar to those outlined in the first paper (*3*).

Physical Measurements. MOLECULAR WEIGHT. Intrinsic viscosities were determined using dilute solution viscometry (Cannon-Ubbelohde viscometers). For the poly(methyl methacrylate) polymer the following empirical expressions were used to obtain molecular weights (*4*):

$$[\eta]_{25°C}^{CHCl_3} = 4.8 \times 10^{-5} M_v^{0.8}$$

$$[\eta]_{30°C}^{C6H6} = 5.2 \times 10^{-5} M_v^{0.76}$$

GLASS TRANSITION (T_g) AND SECOND-ORDER TRANSITIONS. T_g and viscoelastic responses of the elastomers and polyblends were measured using a torsion wire type apparatus consisting of a modification of the Gehman twist technique for elastomers and also using a duPont thermal analyzer.

OTHER PHYSICAL MEASUREMENTS Other physical measurements were made using the procedures outlined in the previous paper (*3*).

Results and Discussion

Many of the same considerations necessary for developing the polyblends containing diene elastomers (3) are relevant to the preparation of transparent high impact resistant, saturated, acrylic polyblends. Considering first the acrylic rubber backbone, the monomers which most effectively impart rubber-like character are the acrylic esters—*e.g.*, methyl, ethyl, isopropyl, *n*-butyl, isobutyl, 2-ethylhexyl, and *n*-octyl acrylates. These esters and their copolymers with the methacrylate esters, styrene, and styrene derivatives, and vinyl nitrile type derivatives can be used to prepare rubbery backbones. In many cases, the gum strength of these rubbers is somewhat lower than that of a typical diene elastomer. However, by grafting these elastomers, high strength rubbers were produced. The gel structure can be adjusted by introducing low levels of difunctional comonomers, such as divinylbenzene.

Grafting had the same beneficial effect in producing the necessary interdispersion of phases as that observed for a diene rubber polyblend. The feasibility of grafting a monomer to a saturated backbone was considered adequate for this particular system, based on reported studies of similar systems (6, 7, 8).

In this type of system, the backbone rubber acts as a chain transfer site for the monomer, and an initiator is chosen which preferably attacks the α-hydrogen position on the vinyl polymer backbone. Efficiency of the grafting reaction based solely on the chain transfer mechanism depends on several competing reactions (5).

(1) Competition between monomer and backbone for the initiator radicals. Under some conditions, the initiating fragment attacks the backbone directly, resulting in the formation of a macroradical capable of initiating graft copolymerization.

(2) Competition between monomer, solvent, and backbone for the growing polymer radicals. To obtain grafts with linear branches and to suppress homopolymerization, the chain transfer step to the backbone polymer must be the favored process.

(3) Competition between the various terminating processes for the initially formed polymer radical.

(4) Competition between the various termination processes for the growing graft species.

The following experimental observations were made with a methyl methacrylate graft on a 2-ethylhexyl acrylate–styrene copolymer system:

(1) The addition of a good chain transfer agent (such as a mercaptan) to the system during the graft reaction reduced the grafting efficiency.

(2) Grafting efficiency was temperature dependent, increasing with temperature in the range of 40°–70°C.

Table I. Chemical Grafting of Methyl Methacrylate to an 81–19 2-Ethylhexyl Acrylate–Styrene Rubber Backbone[a]

Initiator	Apparent Graft, $\dfrac{W - W_o}{W_o}$
Benzoyl peroxide	0.65
Dicumyl peroxide	0.56

[a] Eight hours treatment at 60°C with initiator level at 1.0 wt % and the backbone rubber/monomer charge ratio at ½.

Table II. Typical Physical Properties of Injection-Molded Transparent Saturated Acrylic Polyblends[a]

Physical Property	Value[b]	ASTM Test Method
Tensile Strength, lbs/in^2		
23°C	5000–5800	D638
−40°C	7600–8500	
Flexural strength, lbs/in^2	6500–7200	
Impact strength, ft lbs/in.		D256
23°C (⅛" × ½" × 2½" notched bar)	0.8–1.5	
−40°C (⅛" × ½" × 2½" notched bar)	0.4–0.6	
Flexural modulus, lbs/in^2	210,000–260,000	D790
Elongation, %	5.0–25.0	D638
Hardness (Rockwell)	R70–85	D785
Heat distortion temp., °C unannealed, 264 psig	70–75	D648
Optical Properties		
Transmission, %	89.0–92.0	D791
Haze, %	4.0–6.0	D1003

[a] Composition: 20.25–27.0% 2-ethylhexyl acrylate, 4.75–6.33% styrene, and 66.67–75.00% methyl methacrylate.
[b] The range obtained depends upon molding conditions and compositions.

(3) Grafting efficiency increased with increasing initiator concentration.

(4) Benzoyl peroxide, dicumyl peroxide, *tert*-butyl peroxypivalate and potassium persulfate were efficient as graft copolymerization initiators. Azobisisobutylronitrile (AIBN), however, was not an efficient initiator.

Because of the complexity of the graft reaction, these conclusions are only qualitative, but they do enable one to make appropriate changes in the polymerization conditions to facilitate the reaction.

Two techniques were used to measure the apparent covalent grafting: a Soxhlet extraction and an elution extraction. The Soxhlet procedure is well known. The elution procedure involved placing the graft copolymer in a nylon pouch (tight weave nylon) and placing the pouch in a brown bottle containing a suitable solvent for the graft comonomers. The solvent was changed several times, and the residual undissolved polymer was used as a measure of the grafting efficiency. The solubility of the rubber backbone in the elution solvent was known, and hence any increase of insoluble weight was attributed to graft copolymer. The apparent graft was calculated using the following expression:

$$\text{apparent graft} = \frac{W - W_o}{W_o}$$

where W_o = the weight of the insoluble rubber backbone after elution, and W = the weight of the insoluble graft rubber after elution.

Table III. Injection Molding Conditions (Van Dorn 6 oz. Screw-Type Machine)

Cylinder temp., °F	
rear	420–440
center	440–470
front	450–490
Nozzle temp., °F	450–490
Mold temp., °F	135–160
Cycle time, sec	
inj. fwd.	10
mold shut	45
Back pressure, psig	125–500
Injection rate	Low to High
Screw, rpm	48
Holding pressure, psig	1200–1400
Current, amps	6.2–6.8

For both techniques some ungrafted polymer will be entrapped in the insoluble graft copolymer, leading to an inaccuracy in the measure of apparent grafting; however, the two techniques agreed within experimental error. The apparent grafting produced by 1.0 wt % of initiator is shown in Table I for two initiators, using the elution extraction procedure.

Typical physical properties for an injection-molded transparent acrylic polyblend resin are given in Table II. The injection molding conditions used are given in Table III. Tensile, flexural, and impact properties are within the range reported for typical ABS and high impact polystyrene resins. Optical properties approach those of the acrylics [*i.e.*, poly(methyl methacrylate)]. The strength properties are on the low side of those reported in the first paper for the transparent diene

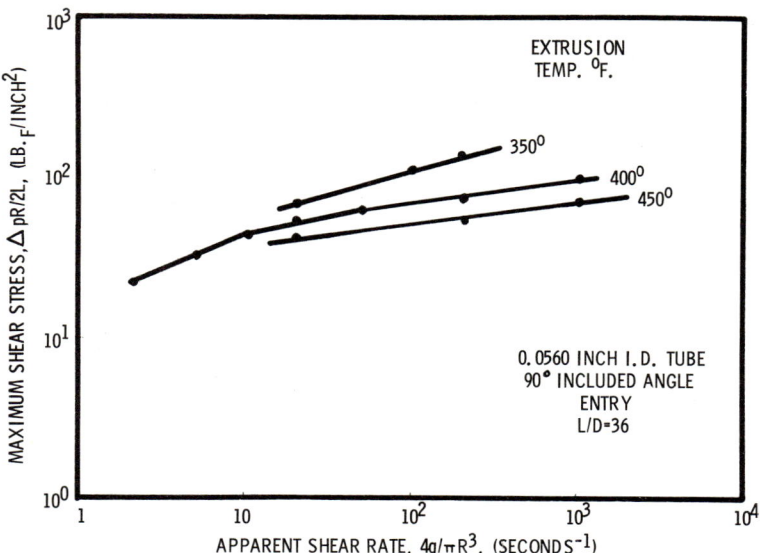

Figure 1. Melt viscosity characteristics

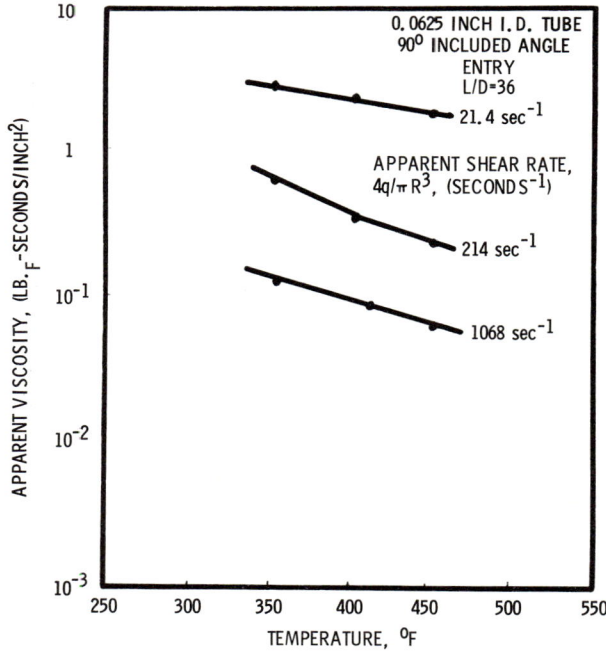

Figure 2. Melt viscosity characteristics

modified polyblends (3), while the optical properties are very similar, percent light transmission being slightly higher. Instron melt flow rheology is shown in Figures 1 and 2. In Figure 1 the maximum shear stress is shown as a function of the apparent shear rate for three extrusion temperatures (350°, 400°, and 450°F). The apparent viscosity is plotted as a function of temperature for three shear rates in Figure 2. Capillary rheometry indicated that this polymer gave the smoothest extrudate at the lower temperature range (325°–400°F) and lower shear rates, having a tendency to show melt fracture at the higher temperature range (400°–500°F). The latter type of flow behavior is characteristic of many thermoplastic resins and is commonly observed with poly(methyl methacrylate) homopolymer (13, 14, 15, 16).

Table IV. Oxygen Absorption of Elastomers

Elastomer	Hours Required to Absorb 1% Oxygen at 212°F
Typical diene elastomer with no antioxidant	2
Typical diene elastomer, stabilized with 1.0 phr of antioxidant	300
Acrylic elastomer with no antioxidant	780
Acrylic elastomer, containing only 0.1 phr Wingstay V and 0.1 phr of DLTDP	~1000

Since a primary goal in this investigation was to examine the aging properties of these materials, a preliminary study of the resistance of the elastomer backbone to oxidation was considered appropriate. A measure of the oxygen absorption of various elastomers in an oxygen atmosphere at an elevated temperature is considered indicative of their relative resistance to oxidation. In Table IV the relative times required for the absorption of 1% oxygen at 212°F are compared for an unstabilized and a stabilized diene elastomer as well as an unstabilized and a stabilized acrylic rubber. The unstabilized acrylic rubber's resistance to oxidation is as good as, or better than, that of a stabilized diene elastomer. Injection-molded test pieces of one of these transparent saturated impact acrylic were exposed to 2000 hours of accelerated aging in an Atlas 6000 watt xenon-arc weatherometer. Physical property retention, appearance, and color were used to measure the resistance of this polymer to accelerated weathering conditions.

Physical property retention is indicated in Figure 3, where tensile strength and notched Izod impact strength were measured at 400-hour intervals. It is apparent that physical strength was maintained satisfactorily over the 2000-hour period, and physical appearance was essentially unchanged—i.e., there was no evidence of crazing, erosion, or

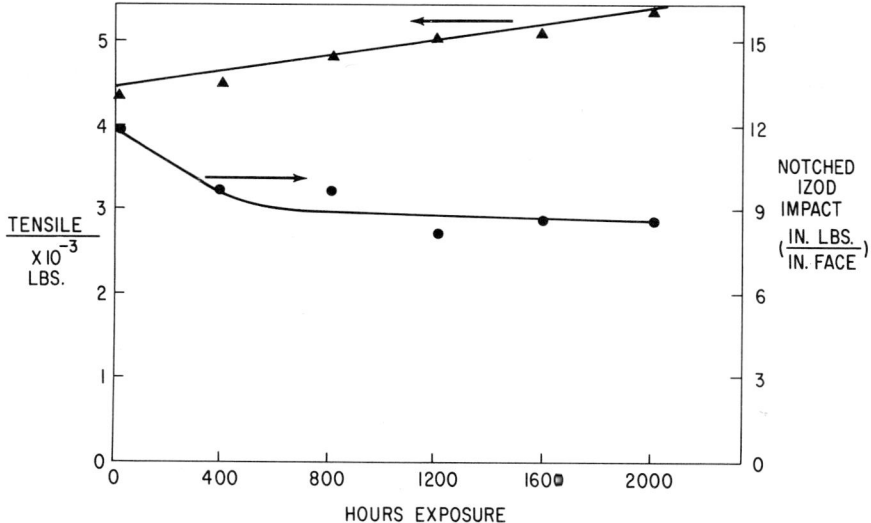

Figure 3. Weatherometer aging of transparent saturated acrylic

discoloration. This polymer contained no light absorber or colorants and only a low level (0.2 phr) of antioxidant. The specimens were essentially colorless (water-white) after the exposure period.

Outdoor aging is currently being conducted on these polymers, and while accelerated aging may have limited value in predicting long term outdoor exposure, the accelerated aging studies of Jordan et al. (9) with poly(methyl methacrylate) copolymers and homopolymers indicated that a 2000-hour carbon-arc weatherometer period was a realistic exposure to screen the relative changes to be expected in about a two-year period in an outdoor environment—e.g., Florida or Arizona.

Conclusion

A variety of comonomers (2) could be used to produce saturated rubber–resin polyblends having optical properties approaching homopolymers and physical properties similar to ABS but with improved resistance to photoxidation.

Acknowledgments

The authors acknowledge the Rubber Chemicals Research section, and particularly W. S. Hollingshead and C. A. McGlothlin for the oxygen absorption measurements. In addition, we thank The Goodyear Tire & Rubber Co. for permission to publish this work.

Literature Cited

(1) Alfrey, T., Jr., Bandel, D., "Copolymerization," Vol. 8 of "High Polymers," pp. 159–160, Interscience, New York, 1952.
(2) Bauer, R. G., Bletso, N. C., Mast, W. C., Pierson, R. M., U. S. Patent **3,475,516** (Oct. 28, 1969).
(3) Bauer, R. G., Pierson, R. M., Mast, W. C., Bletso, N. C., Shepherd, L., ADVAN. CHEM. SER. **99**, 237 (1970).
(4) Bandrup, J., Immergut, E. H., Eds., "Polymer Handbook," Sect. IV, pp. 26–27, Interscience, New York, 1966.
(5) Burlant, W. J., Hoffman, A. S., "Block and Graft Polymers," pp. 16–20, Reinhold, New York, 1960.
(6) Carlin, R. B., Shakespeare, J., *J. Am. Chem. Soc.* **68**, 876 (1964).
(7) Fox, T. G., Gratch, S., *Ann. N. Y. Acad. Sci.* **57**, 367 (1953).
(8) Hayes, R. A., *J. Polymer Sci.* **11**, 531 (1953).
(9) Jordan, J. M., McIlroy, R. E., Pierce, E. M., *Appl. Polymer Symp.* **4**, 205–218 (1967).
(10) Kamal, M. R., Ed., *Appl. Polymer Symp.* **4** (1967).
(11) Keskkula, H., Ed., *Appl. Polymer Symp.* **7** (1967).
(12) *Modern Plastics,* p. 168 (Aug. 1968).
(13) Tordella, J. P., *J. Appl. Phys.* **27**, 454 (1956).
(14) Tordella, J. P., *Trans. Soc. Rheol.* **1**, 203 (1957).
(15) Tordella, J. P., *Rheol. Acta* **1**, 2/3, 216 (1961).
(16) Tordella, J. P., *J. Appl. Polymer Sci.* **7**, 215 (1963).

RECEIVED January 23, 1970.

18

Grafting in an Aqueous Suspension of Vinyl Chloride on Ethylene–Propylene Copolymers

F. SEVERINI, E. MARIANI, and A. PAGLIARI

Montecatini Edison S.p.A., Centro Ricerche Milano, Via Giuseppe Colombo 81, Milan, Italy

E. CERRI

Montecatini Edison S.p.A., Centro Chimico di Ricerca Applicata, Porto Marghera (Venezia), Italy

> *The radical polymerization of vinyl chloride in the presence of an ethylene–propylene rubber yields products consisting of PVC homopolymer, free rubber, and a rubber–PVC graft copolymer. The methods used to separate the components and to study the behavior of the grafting reaction vs. time, temperature, and initiator are described. The physical-mechanical properties of the crude products containing 9% rubber were studied; results indicate that with respect to a PVC having an intrinsic viscosity close to that of the homopolymer in the crude reaction products the above products show comparable values of thermal stability and of resistance to aging and are characterized by much higher values of impact strength and by higher fluidity.*

The brittleness of PVC [poly(vinyl chloride)] at temperatures close to or lower than 0°C limits its possible applications. High impact strength material, without any marked impairment of the other physical-mechanical properties, is prepared according to known techniques by mixing the conventional polymer with varying amounts of elastomeric materials or with elastomer-based materials partially compatible with PVC—e.g., BN rubbers (*19*), certain types of ABS (*10*) or MBS (*8*), chlorinated polyethylenes (*6*), grafted EVA copolymers (*7*), etc. Some of these products and especially those with butadiene-based modifiers are quite susceptible to weathering by atmospheric agents; hence, their use for outdoor applications is limited.

Materials endowed with good mechanical properties, particularly high impact strength, and with a high resistance to aging may be obtained by using saturated or slightly unsaturated olefin rubbers, which are made compatible with the matrix by grafting with PVC. A simple method for preparing a modified PVC of this type consists in the radical polymerization of excess monomer in the presence of an elastomer dissolved in the monomer (2, 12, 13, 14, 20).

In previous work (13) we outlined the preparation of the crude reaction products by the polymerization of vinyl chloride in the presence of ethylene–propylene saturated elastomers, the mechanical properties of the products obtained, and some elastomer-PVC mechanical mixtures.

Crude reaction products having a total rubber content of 9% yield substances characterized by high impact strength without any remarkable variation in the other mechanical properties.

Mechanical mixtures of elastomer with PVC having the same composition as that of other PVC's modified by grafting give materials characterized by a lower impact strength and by macroscopic phenomena of incompatibility, such as the surface migration of rubber. Evidently, the rubber fraction present as a graft copolymer guarantees the apparent homogeneity of the system with an over-all improvement of the compatibility of the components, and of the resilience of the PVC matrix, in which it is dispersed with the unmodified rubber.

In this paper we summarize the results obtained from a study of the reaction of formation of the graft copolymer and the physical-mechanical properties of the crude polymerization products, comprising useful tests to define both processability and resistance of these products to external agents.

Experimental

Materials. Ethylene–propylene copolymer, purified by Kumagawa (9) acetone extraction for 180 hours had a composition (determined by infrared) of $C_2 = 54.5$, $C_3 = 45.5$ wt %, and an intrinsic viscosity determined in toluene at 30°C of 1.38×10^2 cc/gram. Poly(vinyl alcohol) was Elvanol 50-42 from du Pont. The vinyl chloride monomer of Montecatini Edison was 99.99% pure. Initiators used were:

(1) Benzoyl peroxide (Bz_2O_2) purified by crystallizing the Carlo Erba commercial product (99% pure).

(2) *tert*-Butyl perpivalate (TBPV), a commercial product of Noury and van der Lande (75% pure).

(3) Azobisisobutyronitrile (AIBN) purum of Fluka.

(4) 2,4-Dichlorobenzoyl peroxide purified by crystallizing the commercial product of Noury and van der Lande (99% pure).

(5) Acetyl cyclohexane sulfonyl peroxide (ACS) with 30% water.

The solvents used were acetone, dimethylformamide, n-heptane, tetrahydrofuran purified by fractionation of the commercial products, and Celite, K-535 of Mascia e Brunelli S.p.A.

Grafting. The grafting reaction was carried out in a 2-liter stainless steel reactor equipped with blade stirrer and heated to the desired temperature with a thermostatting bath. The water/rubber + monomer ratios are indicated in the text.

PROCEDURE A. The autoclave was initially charged with about one-third of the total water used, which contained the entire amount of poly(vinyl alcohol) used as suspending agent and the crumbed copolymer; after eliminating air by repeated washings with pure nitrogen, the monomer was introduced, and the mass was left to stand for 4 hours at room temperature.

Stirring (300 rpm) was started, and the mass was heated and kept at the desired temperature for 2 hours. During this preliminary treatment, the elastomer was swollen and partially dissolved by the monomer. Finally, the remaining water containing the initiator in suspension was introduced, and the reaction was allowed to proceed until the monomer conversion reached the desired value. The polymer, in the form of pearls, was separated from the suspending liquid by centrifugation, washed thoroughly with water, and dried in an oven at reduced pressure.

PROCEDURE B. A variant of the method consists in precontacting monomer and rubber in the absence of water and subsequently dispersing the pseudo-solution obtained in an aqueous solution of poly(vinyl alcohol) containing the initiator in suspension.

Fractionation of Raw Polymerizates. The raw products were fractionated by the Desreux chromatographic method (4) by introducing a series of solvents through a polymer dispersion on Celite.

To determine the behavior of the grafting reaction, the products were fractionated by separating their three fundamental components: free rubber, PVC, and graft copolymer. The conditions are given in Table I.

A more complete fractionation, which also allows the isolation of some fractions forming the graft product, may be carried out according to the scheme described in Table VI.

The crude polymerizate (2 grams) dried to constant weight and contacted with excess tetrahydrofuran (60 grams) for 12 hours at room temperature yielded a solution containing small amounts of non-perma-

Table I. Fractionation Conditions

Solvent	Temperature, °C	Dissolved Fraction
n-Heptane	60	free rubber or with PVC % ≤1
N,N-Dimethylformamide	30	PVC homopolymer
Tetrahydrofuran	60	graft copolymer

Table II. Polymerization of Vinyl Chloride in the Presence of EPR[a]

Time, Minutes	Conversion to PVC, %	EPR in Crude Product, %	EPR Grafted from 100 Grams of Starting Rubber	Graft Copolymer from 100 Grams of Starting Rubber	VC in Grafted Copolymer, %
80	21.4	27.4	23.2	28.1	16.8
120	34.7	18.4	33.7	40.2	20.8
200	75.3	10.6	47.2	61.3	30.0
720	91.6	9.0	61.1	94.2	34.0

[a]Procedure A; $T = 70°C$; $Bz_2 O_2 = 0.38\%$; Elvanol 50–42 = 1.2%; $R = 2.4$.

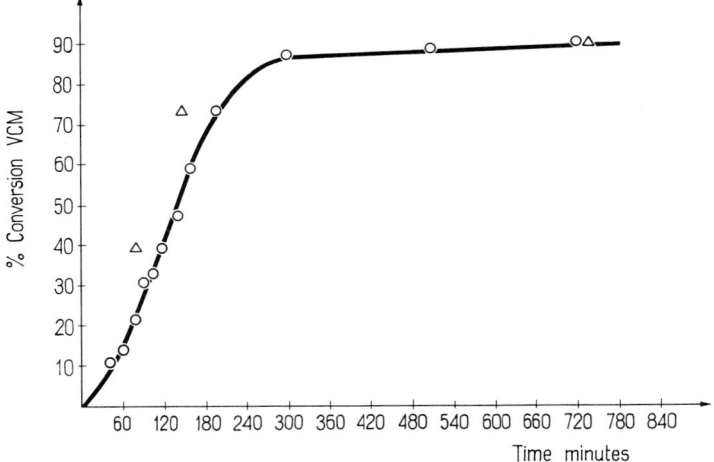

Figure 1. Percent monomer conversion to polymer as a function of reaction time. $T = 70°C$; $Bz_2O_2 = 0.38\%$; *Elvanol 50-42* = 1.2%; $R = 2.4$.

○: Procedure A
△: Procedure B

nent gels ($\leqslant 1\%$) soluble in the other solvents used for fractionation. This solution was poured on a measured amount (120 cc) of Celite K-535; the mass was mixed until the solution was distributed homogeneously on all the Celite used. Tetrahydrofuran was evaporated completely at room temperature and at reduced pressure. The dry residue, which was homogeneized further was placed in a chromatographic column, which may be thermostatted at the desired temperature.

The solution of the extracted component was concentrated at low volume, then poured into excess methanol, and the precipitate was recovered quantitatively by filtering on gooch.

The heptanic extract was recovered by evaporating the solution in a calibrated flask until we obtained a solvent-free film adhering to the walls.

Determination of Chlorine in the Polymers. The chlorine content was determined by heating 0.1–1 gram of the substance to be analyzed to 700°C with a large excess of Eschka mixture. After cooling, the mass was treated with 6N nitric acid, and the chlorine present in the solution as silver chloride was precipitated, filtered, and dissolved again in an aqueous solution of ammonia. From this solution, silver was precipitated as mercaptobenzothiazolate, which allows determination of the chlorine content in the material under examination (15).

Results and Discussion

In a series of runs carried out according to Procedure A using a ratio water/rubber + monomer = 2.4 in the presence of 0.38% of Bz_2O_2 the behavior of the monomer–polymer conversion vs. time at 70°C (Figure 1) shows evidences of autoacceleration, which characterizes the homopolymerization of vinyl chloride (3, 5).

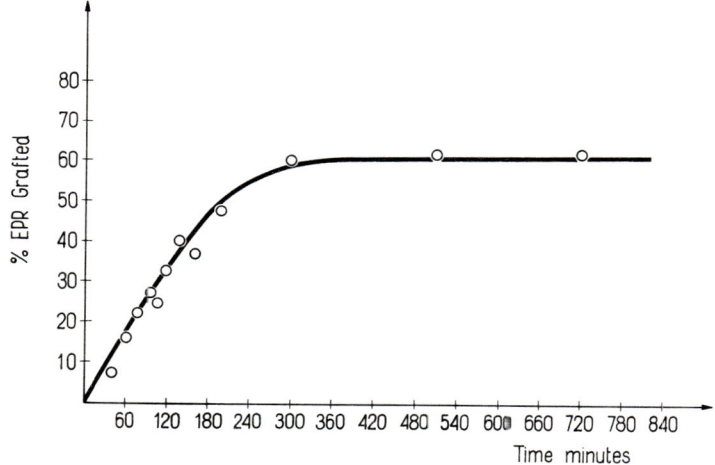

Figure 2. EPR grafted vs. reaction time for polymerization in aqueous suspension. Procedure A; T = 70°C; Bz_2O_2 = 0.38%; Elvanol 50-42 = 1.2%; R = 2.4.

Some runs carried out under the same conditions but according to Procedure B gave the results represented by the points outside the curve, which indicate a higher conversion per hour. Such behavior might be explained by the fact that in the absence of water both swelling and dissolution of the elastomer into monomer are more rapid and complete; hence, the subsequent dispersion of the reacting mass in water is favored.

Figure 2 shows the variation with time of the amount of grafted rubber measured on the basis of the amount of n-heptane-extractable rubber at 60°C. The behavior is asymptotic, and for ethylene–propylene copolymers the yield of grafted rubber never exceeds 60%.

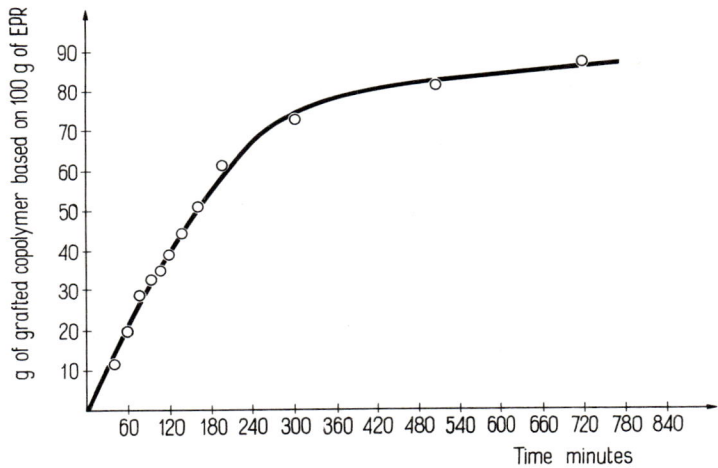

Figure 3. *Grafted copolymer formed as a function of polymerization time. Procedure A as described in Figure 2.*

A comparison between plots 1 and 2 shows that the amount of modified rubber increases with increasing monomer conversion, reaching its highest value nearly five hours after the reaction starts—i.e., when 90% of the monomer is converted into polymer. This behavior might indicate that the amount of reacting rubber depends on the concentration of the monomer present; however, it is also a function of the diffusion of the monomer in the particle consisting of excess PVC until reaching the elastomer.

Monomer diffusion in the rubbery phase of PVC-rich reaction products is difficult, and this was demonstrated by polymerizing vinyl chloride (200 grams) at 70°C in the presence of a crude polymerizate (360 grams) containing 9% total rubber suspended in an aqueous solution of poly(vinyl alcohol) (1.2% with respect to the polymer + monomer weight), so as to obtain a ratio of water/crude + monomer = 2.4 and by using benzoyl peroxide (0.38% with respect to the reacting monomer).

Notwithstanding the excess monomer with respect to the free rubber representing 40% of the total amount of elastomer present in the crude product, at the end of the reaction the amount of PVC in the grafted product increases from 30 to 40%; however, no significant variations in the amount of free rubber are observed; the increase of the amount of grafted PVC is probably caused by branching of the previously grafted chains accessible to the monomer. According to this assumption, we may also interpret the results reported in Figure 3; this figure shows how the amount of graft copolymer present in the crude polymerizate (calculated with respect to 100 grams of reacting rubber) increases even after the first five hours of reaction, when the amount of free rubber is constant. The data were obtained by determining the tetrahydrofuran-extractable fraction at 60°C.

The data in Table III concern materials prepared at 70°C by Procedure B and indicate that with respect to the corresponding runs carried out at 70°C by Procedure A (see Table II) significant variations of the non-extractable rubber content may be observed. Table IV summarizes some results obtained by studying crude polymerizates prepared at 65°C according to Procedure A; with respect to the runs carried out at 70°C (Table III) reduced amounts of suspender and initiator were used. The data suggest that under these conditions both the monomer conversion and conversion of the initial rubber to grafted rubber are slower.

Table III. Polymerization of Vinyl Chloride in the Presence of EPR[a]

Time, Minutes	Conversion to PVC, %	EPR in Crude Product, %	EPR Grafted from 100 Grams of Starting Rubber	Graft Copolymer from 100 Grams of Starting Rubber	VC in Grafted Copolymer, %
80	39.7	18.4	32.0	40.2	28.6
150	75.3	10.6	50.9	68.0	29.7
720	91.6	9.0	60.0	86.0	36.0

[a] Procedure B; $T = 70°C$; $Bz_2O_2 = 0.38\%$; Elvanol 50–42 = 1.2%; $R = 2.4$.

Table IV. Polymerization of Vinyl Chloride in the Presence of EPR[a]

Time, Minutes	Conversion to PVC, %	EPR in Crude Product, %	EPR Grafted from 100 Grams of Starting Rubber	Graft Copolymer from 100 Grams of Starting Rubber	VC in Grafted Copolymer, %
80	8.0	53.3	8.5	11.5	22.2
150	20.8	30.0	17.6	19.7	25.3
390	75.3	10.6	45.3	56.6	25.6
960	89.6	8.8	54.5	85.5	33.0

[a] Procedure A; $T = 65°C$; $Bz_2O_2 = 0.19\%$; Elvanol 50–42 = 0.6%; $R = 1.4$.

The influence of temperature and of initiator type on the amount of grafted rubber present in the reaction crude products were studied by a series of tests; the results are summarized in Table V. In all runs, the monomer conversion was allowed to progress to about 80–90%.

The PVC content in the graft copolymers increases considerably when the reaction temperature decreases. This is shown by the case of crude products prepared at 50° and at 70°C which yield grafted product containing 46 and 29% of PVC.

Table V. Polymerization of Vinyl Chloride in the Presence of EPR with Different Initiators[a]

Initiator	Temperature, °C	Conversion to PVC, %	$[\eta]$ PVC Homopolymer	EPR in Crude Product, %	EPR Grafted from 100 Grams of Starting Rubber	VC in Grafted Copolymer, %
ACS	50	84.5	1.31	9.8	58	46
TBPV	50	82.8	1.31	10.2	57.8	46
DCBP	50	86	1.30	10.1	57.0	47
Bz_2O_2	65	83.3	0.86	8.7	47.1	33
Bz_2O_2	70	91.6	0.74	9.0	61.1	28.4
AIBN	70	88.7	0.71	9.0	62.2	29

[a] Procedure A; initiator = 1.56×10^{-3} mole %; Elvanol 50–42 = 1.2%; R = 2.4.
[b] Intrinsic viscosity measured in cyclohexanone at 30°C and expressed as cc/gram $\times 10^2$.

These results agree with the lower efficiency of the transfer reactions at low temperatures, which on the one hand influences the number of grafted chains and on the other hand favors the formation of PVC chains either free or grafted, characterized by molecular weights that increase on decreasing the reaction temperature.

The amount of grafted rubber is influenced only slightly by the type of initiator used; this suggests that grafting probably takes place on the active centers formed on the rubber by transfer reactions between the growing chains of PVC and the elastomer chains.

In particular, crude polymerizates prepared in the presence of AIBN as initiator, which yield resonance stabilized radicals (17) that are unable to extract hydrogen from hydrocarbon supports (1, 18) show the same content of non-extractable rubber as that of the polymerizates prepared in the presence of active radicals in the hydrogen extraction from hydrocarbon polymers, such as those derived from the decomposition of benzoyl peroxide.

Table VI shows the results obtained from the fractionation, according to the non-simplified scheme, of a crude polymerizate containing

Table VI. Fractionation of a Crude Reaction Product Containing 8.8% EPR

Solvent	Temperature, °C	Dissolved Fraction from 100 Grams Crude Product	VC in Dissolved Fraction, %	$[\eta]$
n-Heptane	30	4.0	—	1.4[a]
n-Heptane	70	1.3	5.7	—
Acetone	30	33.8	99	0.65[b]
N,N-Dimethylformamide	30	54.9	99	1.16[b]
Cyclohexanone	25	1.17	70.5	—
Cyclohexanone	60	1.85	62.0	—
Tetrahydrofuran	40	3.47	15.4	—

[a] Intrinsic viscosity measured in toluene at 30°C and expressed as cc/gram × 10².
[b] Intrinsic viscosity measured in cyclohexanone at 30°C.

Table VII. Main Characteristics of the Crude Polymerizate and Conventional PVC

	PVC	PVC–EPR
Polymerization	suspension	suspension
Polymerization temperature, °C	58	65
$[\eta]$	1.05	0.98[a]
Free rubber, %	—	4
Grafted copolymer, %	—	7.8
PVC homopolymer, %	100	88.2
Bulk density, grams/cc	0.44	0.66

[a] Intrinsic viscosity of PVC present in the crude product measured in cyclohexanone at 30°C.

8.8% total rubber, prepared at 65°C according to the procedure indicated in Table IV.

The components were separated according to the procedures indicated in the experimental part starting from 12 grams of crude product.

The method of fractionation allows effective separation of the different components of the crude polymerizates; the results are valid enough even if some limitations exist concerning the fraction eluted with dimethylformamide, which might dissolve grafted copolymers with a PVC content > 70%.

As may be deduced from the data reported, the amount of monomer converted into graft copolymer is very low—about 3% of the mass; the grafted product on the average contains 30% PVC divided into fractions with PVC contents varying from 15 to 70%.

Characterization of the Crude Polymerizate. PHYSICAL-MECHANICAL PROPERTIES. A preliminary characterization from physical-mechanical and rheological point of view was carried out on the crude polymerizate (Tables VII and VIII).

A comparison was accomplished with a conventional poly(vinyl chloride) (Table VII) having a molecular weight comparable with that of the homopolymer present in the crude grafting product.

Table VIII. Particle Size Distribution of a Crude PVC–EPR Obtained by Method A at 65°C According to the Procedure Indicated in Table IV[a]

% Residue on:		
	60 mesh	4.8
	80 mesh	6.0
	100 mesh	17.0
	140 mesh	30.0
	200 mesh	34.0
	the bottom	11.2

[a] S.A.F. = 164.

A simple compounding recipe was used to minimize the influence on mechanical and rheological properties of the polymer under examination. A commercial powdered Ba–Cd stabilizer (2%), an antioxidant (0.1%) especially suited to prevent the degradation of the elastomeric phase, and 0.5% of a stearic-based lubricating mixture were used.

From the compounds, roll-milled at 185°C for 10 minutes, the test specimens were machined into proper size and shape from compression molded sheets. All tests were carried out according to ASTM methods, and the data are presented in Table IX.

By examining the characteristics one may get some preliminary information on the physical, mechanical, and rheological behavior of the polymeric mixture, which shall be called hereinafter PVC/EPR.

The presence of the elastomer phase determines a slight decrease of the Vicat softening point with respect to PVC; also the heat distortion temperature shows the same trend, although the values are less differentiated.

The impact strength at room temperature or below is considerably improved; actually good results are obtained also at $-10°C$.

Stiffness, as it results from the elastic modulus, the tensile yield strength, and hardness are reduced with respect to that of a homopolymer by about 25%.

Since the polymeric mixture might be of some interest in high impact PVC, we thought it necessary to investigate some characteristics more thoroughly, in particular resilience vs. temperature, resistance to weathering, tensile strength under constant stress, and processability.

The impact strength of a given plastic material basically depends on temperature and on processing conditions. In a high-impact compound, as a function of the type and the amount of modifier used, a range of temperature in which the impact strength undergoes a sharp change (brittle–tough transition) is normally observed.

Table IX. Physical-Mechanical Properties

Properties	Test Method	Units	PVC	PVC/EPR
Specific gravity	ASTM D 792	grams/cc	1.39	1.32
Softening point (Vicat)	ASTM D 1525 (5 kg, oil)	°C	80	75
HDT	ASTM D 648 (264 psi)	°C	70	68
Izod impact strength	ASTM D 256 (with notch)	kg cm/cm		
at 23°C			5	85
at 0°C	ASTM D 758		3	56
at −10°C	ASTM D 758		3	55 ÷ 17
Tensile strength	ASTM D 638			
at yield	($v=20$ mm/min)	kg/cm^2	565	440
at break	($v=20$ mm/min)	kg/cm^2	395	330
ultimate elongation	($v=20$ mm/min)	%	14	20
elastic modulus	($v=5$ mm/min)	kg/cm^2	31.000	24.500
Flexural elastic modulus	ASTM D 790	kg/cm^2	32.000	26.000
Rockwell hardness	ASTM D 785	L scale	91	66
Melt viscosity ($T=180$°C; $\gamma=10$ sec^{-1})	See Ref 16	psig × sec^{-1}	4.7	2.1
Flammability	ASTM D 635		self-extinguish	self-extinguish

On the basis of these considerations, PVC/EPR was processed in two different ways (10 and 20 minutes by roll-milling at 185°C) and evaluated for impact strength at temperatures lower than room temperature. The results of the test, reported in Figure 4, show the excellent impact properties of PVC/EPR in the temperature range of interest. The fairly limited influence of processing conditions on shift of the brittle–tough transition proves that the modifying elastomer has good resistance to the thermomechanical stresses.

The resistance of PVC/EPR to accelerated weathering was evaluated in comparison with several high impact compounds—*i.e.*, PVC/EPR and PVC/CPE and with conventional PVC.

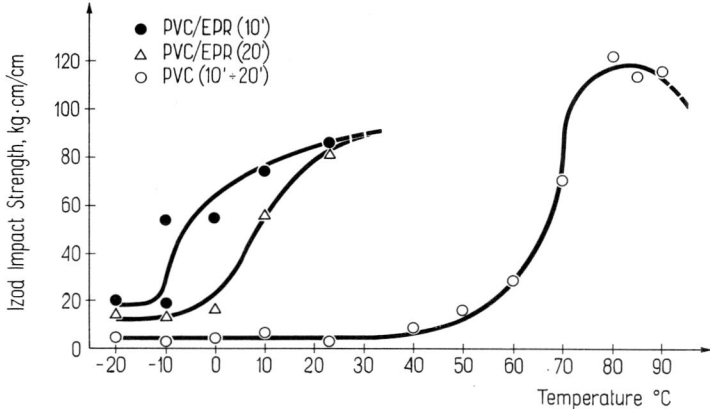

Figure 4. Variation of impact strength as a function of temperature for a crude PVC/EPR and a PVC homopolymer

After 1000 hours exposure in the Atlas Weatherometer, the specimens of different materials show apparently the same alterations of the surface (color changes, degradation spots, faults, loss of gloss, etc.). On the contrary, the retention of the mechanical properties of PVC/EPR was found to be higher than that of plastics aged under the same conditions. The outdoor weathering resistance was also examined by exposing (according to ASTM D 1435-65) the specimens for 1 year in a zone near Venice; the results allowed us to make analogous remarks as above.

The resistance to tensile creep was determined at 23°C using, in analogy with the measurement of impact strength, specimens processed in two different ways: runs were held for 10.000 minutes, and the stress adopted was 250 kg/cm^2. The results plotted in Figure 5 suggest that PVC/EPR in both cases exhibits higher sensitivity to creep compared with PVC homopolymer. The behavior, however, is similar to that observed in compounds of modified PVC having an impact strength at about the same level.

Thermal stability was studied thermogravimetrically on unmodified powder sheets of the compounds previously mentioned. Figure 6 shows instantaneous rates of dehydrochlorination between 165° and 230°C, calculated from the curves of the dynamic thermogravimetric diagrams of the polymers under examination. There are no appreciable differences up to 200°C, whereas at higher temperatures PVC/EPR releases an amount of HCl lower than PVC. This is true even if HCl computation for the two polymers is referred to equal chlorine amounts.

The powder samples of polymer were also subjected to isothermal runs under vacuum for 30 minutes in the same temperature range, showing no appreciable differences in color development.

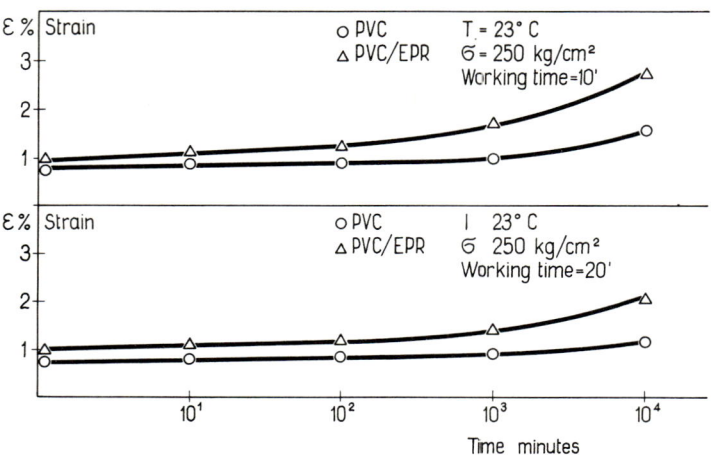

Figure 5. Creep curves for a crude PVC/EPR and a PVC homopolymer

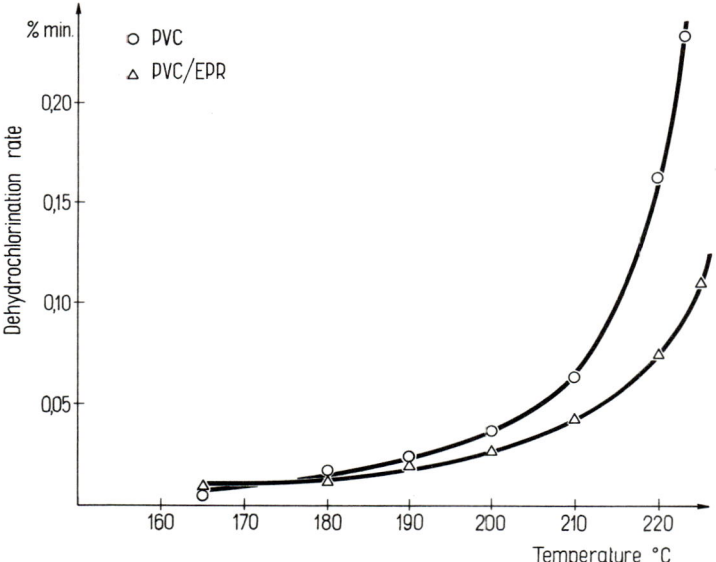

Figure 6. Dehydrochlorination rate vs. temperature for a crude PVC/EPR and a PVC homopolymer

The technological thermal stability—*i.e.*, the proceeding of color development in thermooxidative surroundings—was evaluated in an air circulating oven at 190°C, removing the samples at 5-minute intervals. The results obtained were essentially the same for PVC/EPR and for PVC.

In conclusion, the thermal stability of PVC/EPR is at least equal to that of a PVC having a comparable molecular weight, and no negative effect may be attributed to the presence of the elastomer.

PROCESSABILITY. The term is used generally to describe the flow characteristics of plastic materials during fabrication.

To evaluate extensively the processability of PVC/EPR and to compare it with that of PVC having roughly the same molecular weight, we carried out a set of measurements by a capillary extrusion rheometer and Brabender plastograph. Other runs were carried out by an industrial size extruder.

Table X. Rheologic Measurements by Capillary Rheometer

Operating Temperature, °C	Flow Rate, sec^{-1}	Viscosity of Examined Material, psig/sec	
		PVC	PVC/EPR
180	3	9.7	4.5
	10	4.7	2.1
	30	2.35	1.1
	100	1.0	0.51
	300	0.38	0.23
	1000	0.12	0.087
	1500	0.081	0.061
200	3	4.05	2.5
	10	2.05	1.3
	30	1.1	0.67
	100	0.51	0.32
	300	0.235	0.15
	1000	0.091	0.062
	1500	0.065	0.046

The compounds we examined in extrusion rheometer and in plastograph had the same recipe as that adopted for determining the physical-mechanical characteristics. However, extrusions were carried out with rigid compounds having an industrial recipe suitable for pipes and sections. Gelation of the dry blend, obtained by turbomixer (10 minutes at 1800 rpm, max. temp. 100°C) was carried out by two-roll mill for 10 minutes at 180°C; the blank sheets were mechanically granulated.

The flow properties by extrusion rheometer MCR type (11) were determined at 180° and 200°C in the shear rate range of 1.76 to 1760 sec^{-1}.

Typical results are reported in Table X; data are not correct, owing to the effect of entry, of reservation, and to the non-Newtonian behavior of the medium.

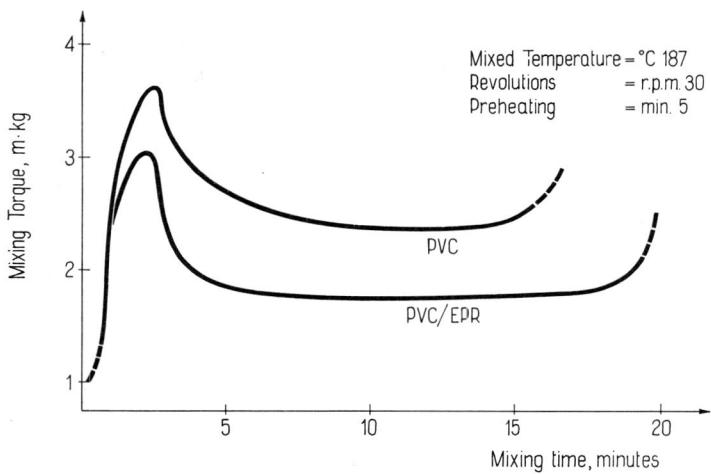

Figure 7. Variation of consistency as a function of time measured by a Brabender plastograph for a crude PVC/EPR and a PVC homopolymer

Table XI. Rheologic Measurements by Brabender Plastograph[a]

Characteristics	Units	Materials	
		PVC	PVC/EPR
Consistency	kg/cm	2.4±0.1	1.8±0.1
Time of decomposition	min	14±2	18±2
Stock temperature of the melt	°C	197±1	189±1

[a] Roller type measuring head: 30 cc. Operating conditions: temperature of the thermostatted bowl = 187°C. Speed of the rotors = 30 rpm. Load applied to the arm = 12 kg.

The melt viscosity of PVC/EPR is lower than that of corresponding homopolymer in the entire shear rate range and temperatures investigated. In particular, one may observe that PVC might exhibit the same fluidity of PVC/EPR but only by operating at a temperature about 20°C higher.

The use of Brabender torque rheometer enabled us to evaluate the behavior of melt viscosity *vs.* time at low shear stresses; the usual operating conditions were adopted—*i.e.*, granules (28 grams) were introduced in the bowl, the temperature was fixed at 187°C, and each run was started after 5 minutes of conditioning.

The flow curves of the two polymers are very similar except for the different consistency degree and the time at which the thermal degradation starts (Figure 7).

The better long term stability of PVC/EPR can be readily explained with its higher fluidity and lower effective test temperature, as measured directly on the sheared polymer melt (Table XI).

The single screw extruder was fitted with a standard equipment (see Table XII); the operating conditions and the results obtained are presented in Table XIII.

Table XII. Single Screw Extruder Main Characteristics

1. Five position control board with thermocouple as sensing element
2. Screw
    diameter D 45 mm
    length L/D 17
    compression ratio   2.2/1
    pitch/diameter ratio   1/1
    number of screw channels   2
3. Die   pipe   rolling shutter
    dimensions   25 mm (diameter)   50 x 16 mm
    thickness   2.5 mm   1.1 mm
    land   20:1   50:1
4. Vacuum operating calibrator
    length   280 mm   500 mm
    sucking area   1   2

Table XIII. Extrusion Conditions and Results

Material	Article	Temperatures, °C (hopper-die)	Stock temperature, °C	Screw speed rpm	Extrusion rate kg/hr	Specific input kwh/kg	Consistency of melt	Gloss	Roughness
PVC/EPR	tube	160-160-160-195-195	198	20	11.3	0.17	good–very good	poor	almost absent
PVC		150-150-150-190-185	201	20	11.0	0.22	good	good	absent
PVC/EPR	Rolling shutter	150-150-150-177-190	194	20[a]	10	0.22	good	poor[a]	present[a]
PVC		150-150-150-190-196	204	20	10.8	0.25	good	good	almost absent

[a] At 30 rpm the gloss and the surface roughness improve markedly.

PVC/EPR exhibits good processability as shown by fluidity, homogeneity, hot tenacity of the melt, specific power absorption, and dynamic thermal stability.

Generally speaking, the appearance of the pipes and profiles was fairly good; it was still noted that the extrudate appearance improves considerably, extrusion temperatures being the same, by operating with screw rotation speeds shifted toward high values.

Conclusions

The polymerization reaction in aqueous suspension of vinyl chloride in the presence of an ethylene–propylene saturated elastomer occurs with the formation of poly(vinyl chloride) homopolymer and rubber–poly(vinyl chloride) grafted copolymers. The first grafting reaction proceeds as far as diffusion of the monomer inside the particles in suspension is possible; afterwards, some chain branching of grafted PVC is possible. Under our experimental conditions the amount of grafted rubber does not exceed 60% of the initial rubber and is little influenced by the type of initiator used.

In comparison with a PVC having an intrinsic viscosity close to that of the homopolymer present in the raw product, compounds based on some reaction raw products, containing 9% rubber, show quite similar thermal stability and weathering resistance; they are characterized further by better resilience and better fluidity.

With regard to conventional PVC we observe that the presence of the elastomeric phase causes a decrease in the Vicat temperature and a higher sensitivity of the PVC/EPR materials in the creep tests.

Acknowledgment

We thank P. D. Gugelmetto for making measurements of the impact strength.

Literature Cited

(1) Allen, P. W., Merret, F. M., *J. Polymer Sci.* **22**, 193 (1956).
(2) Barkhuff, R. A., U. S. Patent **3,408,424** (Dec. 30, 1963).
(3) Bengough, W. I., Norrish, R. G. W., *Proc. Roy. Soc. (London)* **A200**, 301 (1950).
(4) Desreux, V., *Rec. Trav. Chim. Pays Bas* **68**, 769 (1969).
(5) Farber, E., Marvin, K., *Polymer Eng. Sci.* **8**, 11 (1968).
(6) Frey, H. H., *Kunststoffe* **49**, 50 (1959).
(7) Göbel, W., *Kautschuk Gummi* **22**, 116 (1969).
(8) Kenji, Jato, *Japan Plastics Age* **7**, 52 (1969).
(9) Kumagawa, M., Suto, K., *Biochem. Z.* **VIII**, 212–347 (1908).
(10) Martin, J. R., *Rubber Plastics Age* **1966**, 1321.

(11) Merz, E. H., Calwell, R. E., *ASTM Bull.* **232**, 63 (1958).
(12) Natta, G., Severini, F., Beati, E., *J. Polymer Sci.* **34**, 685 (1959).
(13) Natta, G., Severini, F., Pegoraro, M., Beati, E., Aurello, G., Toffano, S., *Chim. Ind. (Milan)* **47**, 960 (1965).
(14) Natta, G., Beati, E., Severini, F., Toffano, S., Italian Patent **12620** (June 17, 1963).
(15) Nebbia, L., Ubaldini, I., *Ann. Chim. Appl.* **51**, 181 (1951).
(16) Pezzin, G., Zinelli, G., *Mater. Plastiche Elastomeri* **32**, 363 (1966).
(17) Pottie, R. F., Lossing, F. P., *J. Am. Chem. Soc.* **83**, 4737 (1961).
(18) Severini, F., Pegoraro, M., Tavazzani, C., Aurello, G., *J. Polymer Sci., Pt. C* **16**, 2805 (1967).
(19) Sharp, J. J., *British Plastics* **1959**, 431.
(20) Schneider, H., Hagel, K. O., Heinrich, E., Glosauer, O., German Patent **1,495,322** (Sept. 7, 1963).

RECEIVED February 9, 1970.

19

Polymeric Systems as Poly(vinyl chloride) Modifiers

J. M. MICHEL

Pechiney-Saint Gobain, Centre de Recherches d'Aubervilliers, France

Systems obtained by polymerizing butadiene–acrylonitrile monomeric mixtures on poly(vinyl chloride) (PVC) latex, using radiochemical and chemical initiators, were studied by fractionation using a precipitation technique. Results show new interesting features which do not appear with mixtures of (PVC) homopolymer and butadiene–acrylonitrile copolymers. Examination of the influence of different physical and chemical parameters (conditions of solubilization, addition of solvent, or polymer chain length modifying agent) enable one to advance a hypothesis on the structure of the copolymers. Results are consistent with a true grafting structure, despite the great difficulties encountered with such complex, oxygen-sensitive systems. Rubber grafted PVC can be compounded easily with PVC homopolymer to get high impact, transparent, and non-whitening resins suitable for numerous applications.

The wide range of physical, chemical, and mechanical properties of poly(vinyl chloride) (PVC) does not always satisfy the demands of some applications. It is therefore necessary to provide one or more of the additional characteristics desired by modifying the resin formula without degrading the other properties.

Since the impact strength characteristics of PVC are insufficient in numerous applications, various solutions have been proposed. All involve blending PVC with an additional resin, but in most cases this causes the disappearance of properties as essential as transparence. For a long time, Pechiney-Saint Gobain has carried out research to produce an addition polymer satisfactory as a reinforcing agent but equally transparent in mixtures with PVC and which does not introduce defects normally encountered with this type of addition (*e.g.*, a white stain on

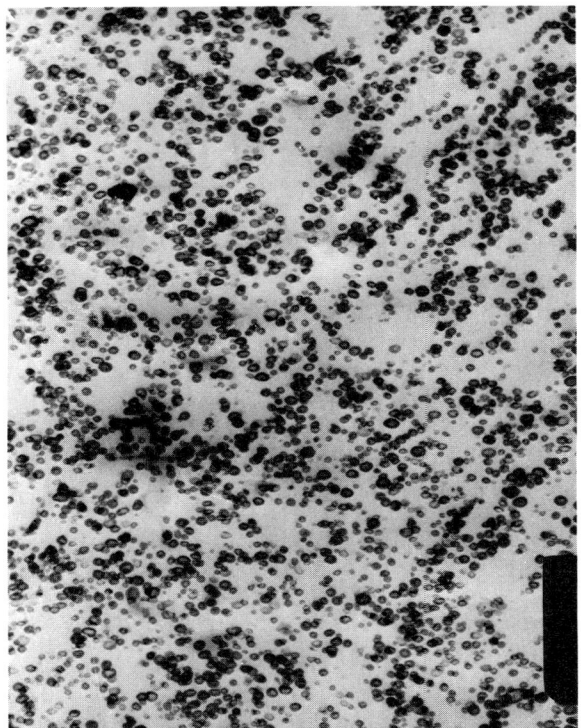

Figure 1. Two-phase system of Lucovyl dispersed in PVC. × 20,000.

accidental bending). These properties are obtained by modifying the macromolecular structure, the chemical nature of constituents, the size of reinforcing agent particles dispersed inside the PVC matrix, their internal cohesion, their refractive index, etc.

The first work was carried out by radiochemical polymerization of a mixture of dienic and acrylic monomers on a PVC latex. A parallel study by chemical means leads to equivalent products and the development of a reinforcing resin which satisfied the requirements listed above. At present a resin of this type is manufactured and sold by Pechiney-Saint Gobain under the name Lucovyl H 4010 (10).

Lucovyl H 4010 is a copolymer obtained by chemical polymerization of a mixture of (principally) butadiene and acrylonitrile on a PVC latex. The mean particle diameter of the resin obtained is very small (less than 0.1 μm). Its refractive index is close to that of PVC, and its compatibility with PVC is excellent. This polymer disperses well in the PVC matrix to give a two-phase system as shown on the electron micrograph in Figure 1 [the granules of H 4010 are shadowed by the technique of Kato (8)].

Lucovyl H 4010 is used in all applications of PVC where the material must be transparent, rigid, tough, and must satisfy food legislation.

During this work we have studied the structure of products obtained from the polymerization of a mixture of butadiene and an acrylic monomer on a PVC latex. More particularly we have studied polymers by fractionation in solution and we describe this here for the specific case of systems of PVC–butadiene acrylonitrile.

Experimental

The butadiene and butadiene–acrylic monomer systems polymerize when irradiated on PVC or vinyl chloride copolymer latex. The structure of the polymer obtained may be grafted if it can be proved that the copolymer properties are different from the blend properties. To elucidate the structure we studied a copolymer obtained by polymerizing butadiene–acrylonitrile on a PVC homopolymer lattice. Owing to practical reasons and to exclude the secondary effect of catalytic residues we used γ radiation. However, we shall observe in a particular case the properties of peroxide-initiated graft copolymer.

Preparation of Samples. PVC LATEX. Polymerization was performed using chemical initiation (persulfate–metabisulfite) and radiochemical initiation (γ-irradiation from ^{60}Co: mean dose rate 10,000 rads/hour). Emulsifier was potassium stearate.

BD–AN. Polymerization was performed chemically (cumyl hydroperoxide/glucose/pyrophosphate/ferrous sulfate) and by radiochemical initiation (*see above*). Monomer ratios were: VC/BD/AN = 5/3/1.

Experimental Parameters. We studied only temperature polymerization for the PVC backbone but the following parameters for grafting: temperature, addition of organic PVC swelling agent, addition of a transfer agent (*tert*-dodecylmercaptan = TDM), and addition of crosslinking agent (divinylbenzene).

Recovery of Polymer. After polymerization the latex was coagulated by adding methanol (with 1% phenyl-β-naphthylamine (PBNA)). The precipitate was washed with methanol (in which 1% PBNA was dissolved), vacuum dried at low temperature, and kept under vacuum in a light-protected vessel.

Method. We sought a simple, fairly rapid method which could be used for many samples and precise enough to yield interesting information. Thus, we chose a fractionation technique in solution. Fractionation conditions were not those ideally recognized in the literature for molecular weight repartition analysis (5), but the chemical nature of the polymers is very different, and slight coprecipitation is not detrimental to the conclusions.

A quantity of 25 cc of a 1% polymer solution was maintained at 25° ± 0.1°C, in a glass vessel directly adjustable on a centrifuge. The

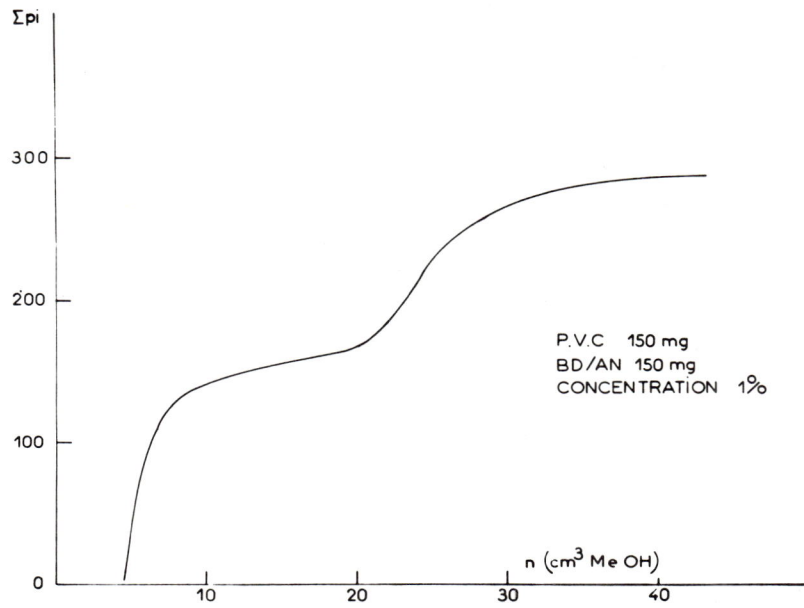

Figure 2. Precipitation curve of mixture of PVC and BD-AN copolymer

precipitation was added slowly with agitation. When turbidity appeared, the solution was centrifuged at constant speed (600 rpm). Generally, solutions were very cloudy; solvent was added by definite increments. The sediment was separated, vacuum dried without additional washing, weighed, and analyzed.

To obtain a good separation of polymer, the precipitation ranges of BD–AN copolymer and PVC homopolymer must be very different. We chose dichloroethane–methanol system (Figure 2). Visible dissolution requires warming at 70°C for 10 minutes, but some copolymers are totally soluble at 20°C giving cloudy solutions.

The reproducibility of curves $P_i = f(n)$ is usually good (Figure 3) for the same polymer and for two polymers prepared under the same conditions (where P_i = weight of fraction i, n = cc of methanol for 25 cc of initial solution). Differences in curves may be related to accidental factors such as temperature increase and variation of rotation speed during centrifuging. Examples of precipitation are given in Tables I, II, and III.

Conditions of recovery of the polymer from the latex must be controlled strictly to avoid modifying the copolymer structure (mainly crosslinking and consequent gel formation by oxidation). A partially soluble polymer can be studied by selective extraction, but experience shows that extraction conditions favor crosslinking and/or degradation. Thus results obtained by extraction must be accepted very cautiously.

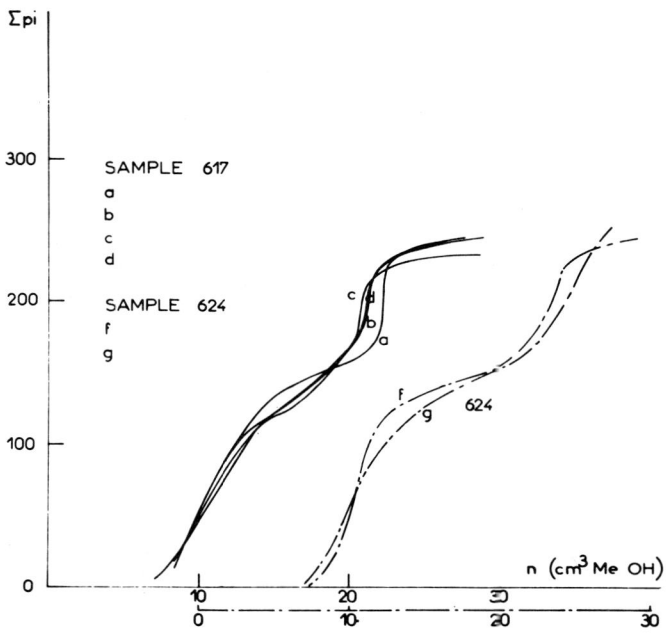

Figure 3. Reproducibility of curves for different samples

Table I. Fractionation of Sample 624[a] (250 mg)

Solution Prepared at 70°C

n, MeOH, cc		Weight, mg	
n_i	$\Sigma\, n_i$	W_i	$\Sigma\, W_i$
7.5	7.5	6	6
0.6	8.1	10	16
1.9	10.	49	65
2	12.	30	95
2	14.	25	120
2	16.	12	132
2	18.	11	143
2	20.	8	151
2	22.	8	159
2	24	37	196
2	26	42	238
Residue		11	249

[a] Sample 624-PVC/BD–AN, 5/3–1. Polymerization by γ-initiation.

Table II. Fractionation of Sample 617ᵃ (250 mg)

Solution Prepared at 70°C					Solution Prepared at 25°C			
n, MeOH, cc		Weight, mg		PVC,	n, MeOH, cc		Weight, mg	
n_i	$\Sigma\, n_i$	W_i	$\Sigma\, W_i$	%	n_i	$\Sigma\, n_i$	W_i	$\Sigma\, W_i$
7.50	7.50	6	6		15	15	7	7
0.50	8	4	10		4	19	8	15
0.50	8.50	9	19		2	21	3	18
0.50	9	8	27	72.5	1	22	17	35
0.50	9.50	10	37		0.5	22.5	38	73
0.50	10	11	48		0.5	23	60	133
1	11	12	60	69	0.5	23.5	62	195
1	12	12	72		0.5	24	28	223
1	13	19	91	66.7	2	26	19	242
1	14	16	107					
1	15	13	120					
2	17	13	133					
1	18	8	144					
1	19	14	155	63.5				
1	20	12	167					
1	21	14	181					
1	22	32	213	21.8				
1	23	20	233					
3	26	4	237					

ᵃ Sample 617-PVC/BD–AN, 5/3–1. Polymerization by γ-initiation with two parts of dichlorethane. PVC in the final resin-53.2%.

Results

Fractionation Curves of Graft Copolymers and Parent Polymers. Figure 4 compares a graft polymer sample (No. 628) and a 60/40 mixture of PVC and copolymer butadiene–acrylonitrile 3/1 prepared at 60°C using a persulfate–TDM catalyst.

The shape of the graft copolymer curve is similar to that of the mixture—*i.e.*, there are two precipitation ranges (A and B), but the n values are different. The A fraction precipitates at higher n values than PVC; the B fraction precipitates at lower n values than copolymer BD–AN, but no conclusions can be deduced because molecular weights for BD–AN chains are unknown.

Analysis of graft copolymer fractions shows that they always contain PVC and BD–AN copolymer in sufficient quantity to eliminate the hypothesis of mechanical coprecipitation (for a mixture we find *ca.* 95% homopolymer). The amount of PVC in the first fraction is higher (*ca.* 70–80%) than in the last.

INFLUENCE OF TEMPERATURE. At a constant polymerization temperature of 60°C the grafting temperature greatly affects the shape of the

Table III. Fractionation of Sample 606[a] (250 mg)

Solution Prepared at 70 °C					Solution Prepared at 25 °C				
n, MeOH, cc		Weight, mg		PVC,	n, MeOH, cc		Weight, mg		PVC,
n_i	$\Sigma\, n_i$	W_i	$\Sigma\, W_i$	%	n_i	$\Sigma\, n_i$	W_i	$\Sigma\, W_i$	%
		0	0		8	8	9		
7	7	75	75[b]	68.6	2	12	6.5	15.5	
0.5	7.5	24	99	63.2	2	16	51.5	67.	
0.9	8.4	18	117		2	18	24.5	91.5	53.5
1.1	9.5	20	137	67.5	2	20	35.5	127.	
0.8	10.3	7.5	144.5		2	22	82.5	210.5	51.4
1.35	11.65	9.	153.5		2	26	20	230.5	51
1.35	13.	6.5	160						
2.5	15.5	8.	168	60.6	Residue		9.5	240	
3.5	19.	54	222[b]	24.4					
1.	20.	6	228	15.2					
1.5	21.5	9	237						
3.5	25.	4	241						
15.	40	4	245						
Residue		8	253						

[a] Sample 606 = PVC/BD–AN, 5/3–1. Polymerization by γ-initiation with one part of benzene. PVC = 52.4%.
[b] Precipitation occurs abruptly.

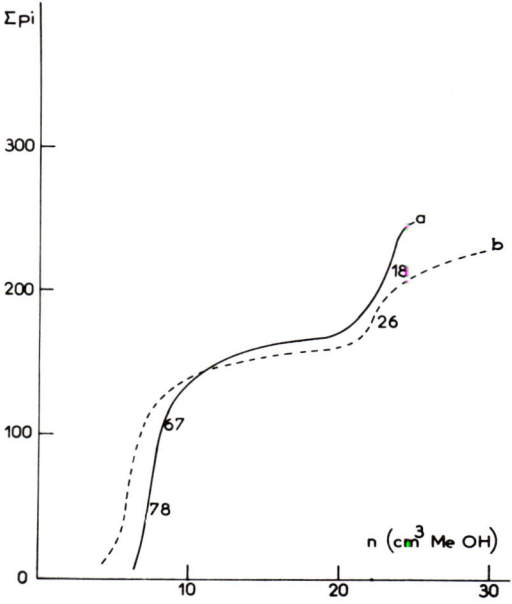

Figure 4. Fractionation curves for (a) sample 628 and (b) 60/40 mixture of PVC and BD-AN copolymer

fractionation curve (Figure 5). At 30°C the maximum precipitation occurs at high n values (18–20, very near the precipitation threshold of BD–AN copolymer). At 70°C the precipitation curve is similar to the PVC homopolymer curve. The shape of the curve changes regularly between 30° and 70°C.

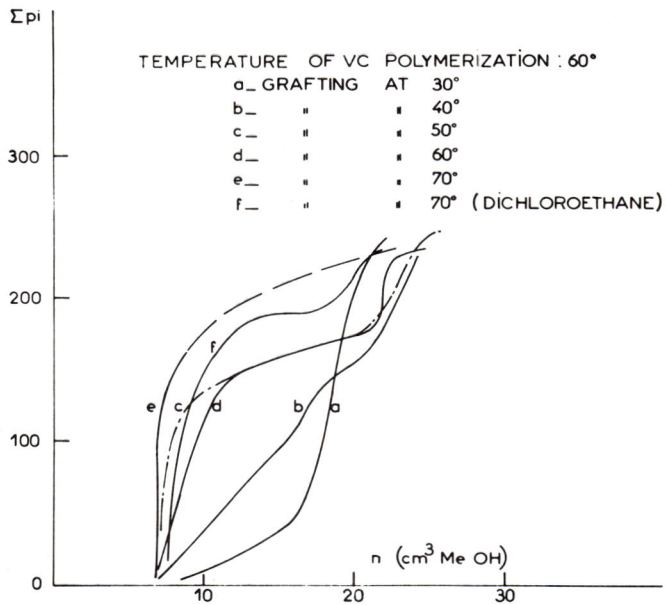

Figure 5. *Effect of grafting temperature on fractionation curve for PVC–BD-AN copolymers*

The effect of the polymerization temperature on the fractionation curve is shown in Figure 6. Vinyl chloride was polymerized at 0°, 30°, 60°C, and the grafting was done at 30°, 60°, and 70°C. Polymers prepared at 0° and grafted at 70° and 30°C are not soluble in dichloroethane, whereas samples prepared at 0° and grafted at 50° are soluble at 20°C. We cannot explain this difference.

The molecular weight of PVC does not appear to exert any noticeable effect, except for the sample grafted at 50°C. This sample, (polymerized at 0°, grafted at 50°C, dissolved at 20°C) was fractionated; the precipitate appears for high values of n ($\sim$22 in a narrow n range).

INFLUENCE OF ADDITIVES. *Organic Swelling Agent.* Initially we thought that adding a swelling agent would render monomer penetration easier in the micelle core of PVC because butadiene and acrylonitrile themselves do not swell PVC. Two organic liquids were studied—benzene and dichloroethane.

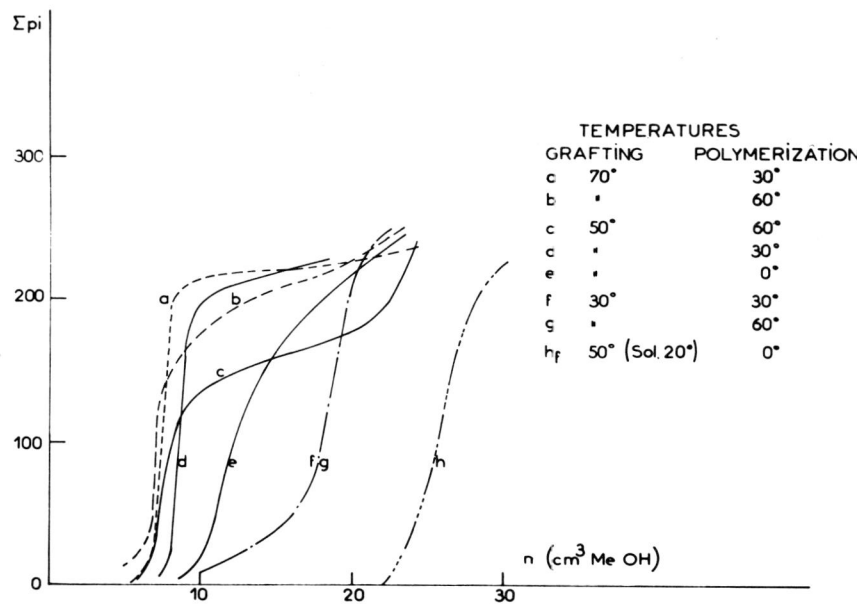

Figure 6. Effect of polymerization temperature on fractionation curve for PVC–BD-AN copolymers

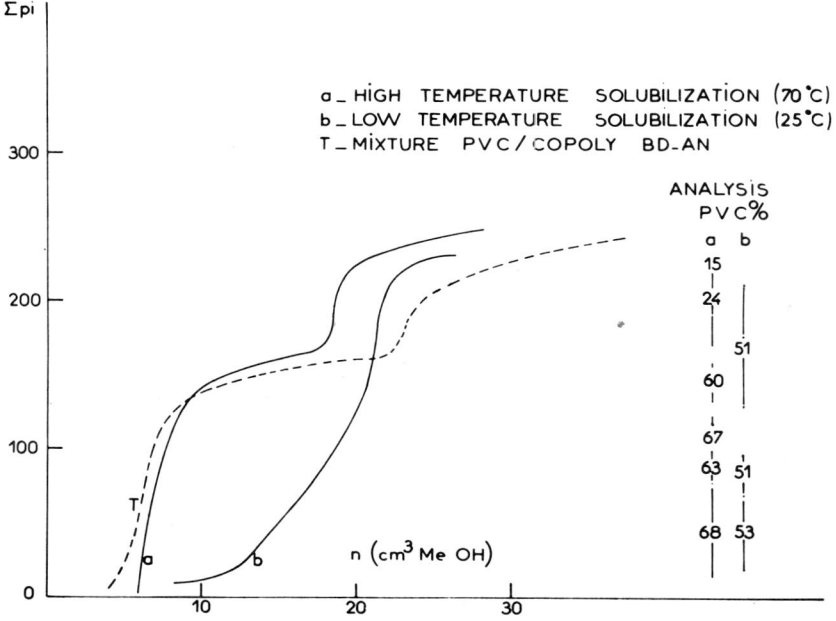

Figure 7. PVC–BD-AN copolymers prepared with addition of benzene

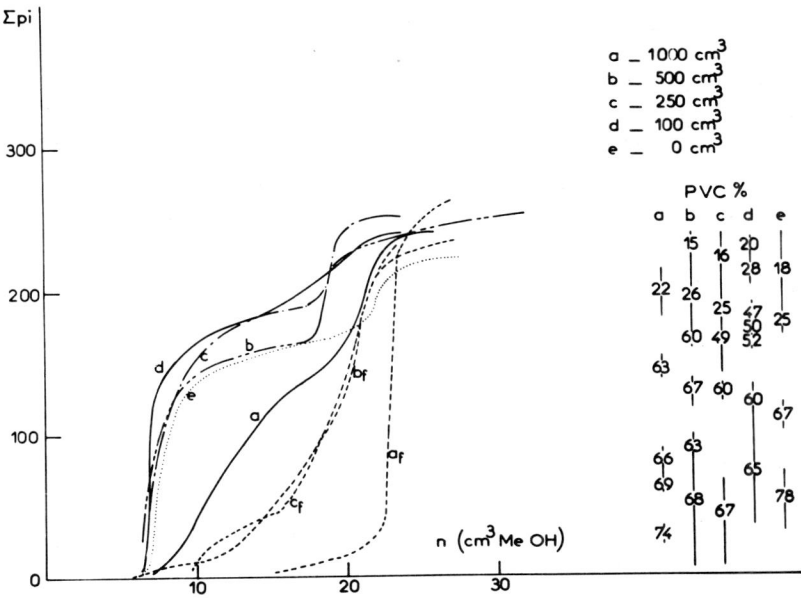

Figure 8. PVC–BD-AN copolymers prepared with addition of dichloro-ethane

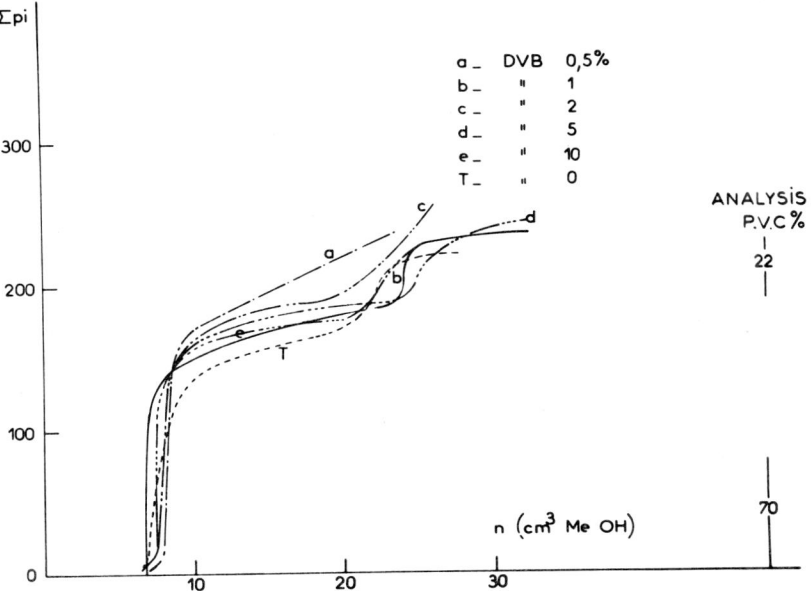

Figure 9. PVC–BD-AN copolymers prepared with addition of divinyl-benzene

Before grafting, butadiene and acrylonitrile were diluted with benzene (0.25 part of benzene for one part of monomer). Other conditions were constant. The precipitation curve shown in Figure 7 is not very different from that of the standard graft polymer (without benzene). The PVC percentage in the first group of fractions is ca. 60–70%. The slope of the curve has two maxima which correspond to the two groups of fractions. The polymer is soluble at 20°C; precipitation is continuous above $n = 10$–12 with a maximum value of the slope at $n = 21$–22.

Dichloroethane was added in place of benzene (Figure 8). Some polymerizations were done using different weights of dichloroethane ranging from 0.05 to 0.5 part for one part monomer (i.e., 100–1000 cc for 2000 cc monomer). The shapes of the precipitation curve depend strongly on the quantity of dichloroethane added. For 0.5 part of dichloroethane, precipitation is nearly regular from $n = 8$ to $n = 24$–25, but there is a slight inflection near $n = 18$. The PVC percentage of the fractions is also shown in Figure 8. There is not much difference between the samples. The values are ca. 63–78% for group A and 15–28% for group B. Above 0.125 (250 cc) dichloroethane graft copolymers are soluble at 20°C. The curve is comparable with that obtained using benzene except for high dichloroethane ratios (0.5 part). Then precipitation occurs abruptly at $n = 22$–24 in a narrow range of n values.

Divinylbenzene. Dichloroethane and benzene can modify both polymer morphology (by swelling action) and polymerization kinetics (by dilution and transfer action). Divinylbenzene (DVB) usually modifies the chemical structure of copolymers by forming bridges between macromolecular chains. On the fractionation curves shown in Figure 9, precipitation appears abruptly around $n = 7$, regardless of the DVB proportion. Precipitation is much more abrupt than for the reference sample (without DVB); the PVC percentage is then about 70% for that fraction. A second precipitation range appears usually between $n = 20$–25. The PVC proportion is about 20%.

Samples prepared with 1–5% DVB were dissolved at 20°C and fractionated (Figure 10). Precipitation is abrupt near $n = 8$–9. At $n = 10$, 90% of the polymer flocculates.

Mercaptan. In contrast to the action of DVB, mercaptan decreases the molecular weight and reduces or prevents crosslinking and grafting (6). Two polymerizations were performed using 1 and 5% TDM (Figure 11).

With 1% TDM, precipitation begins at $n = 7$ and then is regular; the PVC proportion, in fractions of n less than 20, is higher than in the reference sample (85 to 88%, compared with 67–78%). With 5% TDM, the curve covers a wide range of n values. The residue is still high (10%)

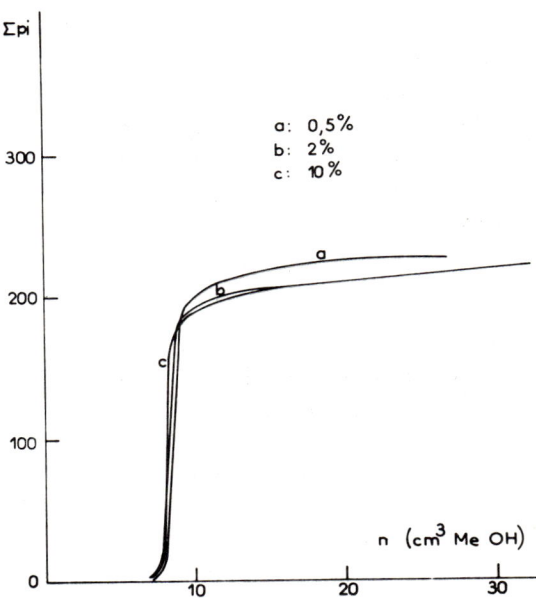

Figure 10. Fractionation curves at 20°C of samples prepared with 1-5% divinylbenzene

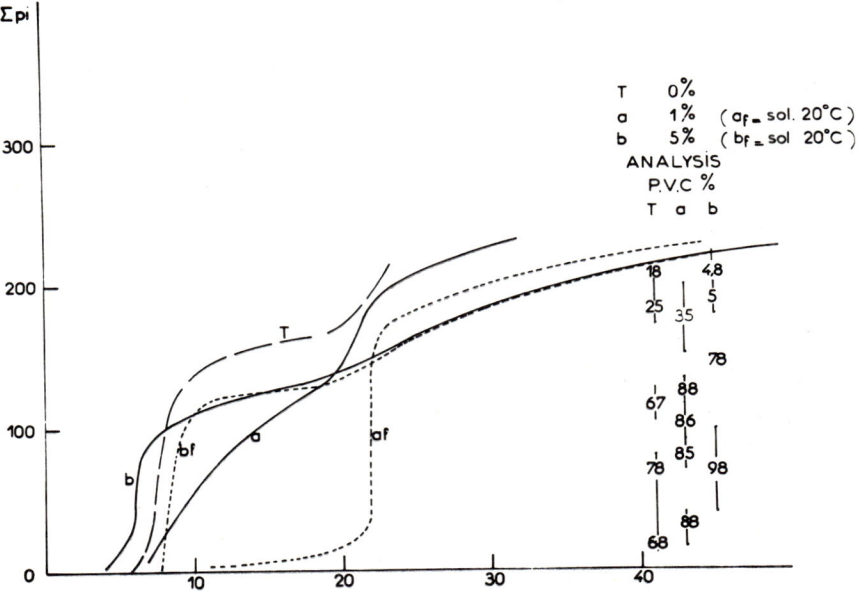

Figure 11. PVC-BD-AN samples prepared with addition of mercaptan

at $n = 50$. The first fractions are rich in PVC (90–98%); last fractions contain negligible quantities of PVC.

Both polymers are soluble at 20°C. For the sample with 1% mercaptan more than 50% of the polymer precipitates abruptly at $n = 22$. For 5% mercaptan the precipitation begins at $n = 6$; 50% of the sample is recovered at $n = 12$; beyond that, the precipitation curve joins that obtained with a solution prepared at 70°C.

Discussion

These experimental results show that the behavior of butadiene–acrylonitrile copolymers varies considerably with the conditions of solution and synthesis. First we must explain these phenomena, and then we deduce some conclusions as to macromolecular structure.

Solutions Prepared at 20°C. Some of the polymers studied are soluble at 20°C. This is true for polymers prepared using additives such as swelling agent, crosslinking agent, and chain transfer agent. This difference in solubility character may not be related to macromolecular structures but to morphology, accessibility of solvent inside the resin, or other incidental reasons.

The fractionation curves of solutions prepared at 20°C (ϕ solutions) are always different from the curves of solutions prepared at 70°C (γ solutions). Examination of a polymer synthesized using dichloroethane as swelling agent shows the following results:

(a) The observed phenomenon is not reversible; ϕ fractions dissolved at 70°C and γ fractions dissolved at 20°C give the same curves (Figure 12).

(b) ϕ to γ transformation is progressive and complete at 50°C.

(c) Addition of a PVC homopolymer solution modifies the fractionation curve. A ϕ solution is mixed with a solution prepared at 70°C of PVC homopolymer ($[\eta] \sim 0.8$), total concentration being kept constant (1%). The polymer ratio varies from 0/250 to 100/150. For a constant value of $n = 15$ (where normally nearly all PVC precipitates), the quantity of graft polymer precipitated by PVC increases with increasing PVC ratio (Table IV). The limit is equal to the weight precipitated with a γ solution. Figure 13 shows the influence on the precipitation curve of PVC added. It is clear that part of PVC graft polymer is drawn down by PVC homopolymer, but we cannot exclude the possibility of a dissolving action of the PVC owing to the graft copolymer (normally 90% of PVC homopolymer precipitates at $n = 10$) (9).

To explain this behavior we advance two hypotheses:

(1) Hot dissolution modifies the chemical structure of the graft copolymer. It is true that polymers containing butadiene are very thermosensitive, but the conditions of transformation from ϕ to γ are slight enough to exclude this hypothesis.

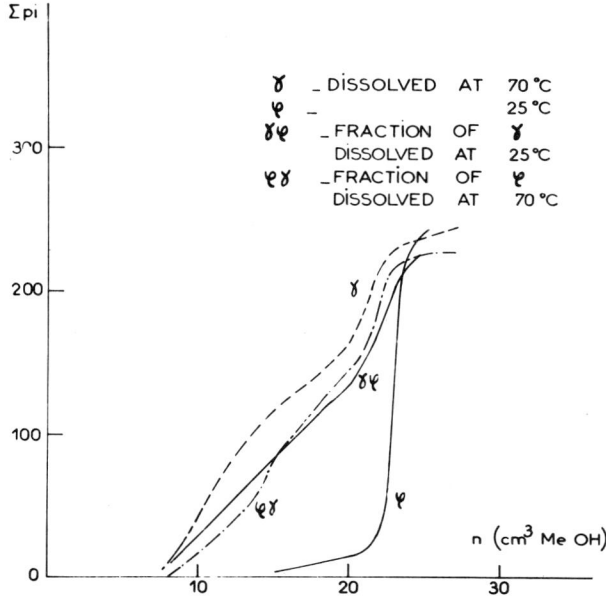

Figure 12. Effect of temperature of solubilization on precipitation curves

Table IV. Precipitation of Mixture of PVC and Graft Copolymer 498 at $n = 15$

498, wt	CPV, wt	Total weight precipitated	Weight of 498 coprecipitated	% Graft coprecipitated
250	0	13	—	—
200	50	129	79	39
175	75	166	91	52
150	100	206	106	71
100	150	226	79	79

(2) Hot dissolution destroys the macromolecular network. In an organic swelling solvent, PVC chains are partially solvated. When temperature rises, dissolution occurs by destruction of PVC network (7). Anagnostopoulos (1) showed that the apparent PVC dissolution temperature in dichloroethane is about 52°–57°C; it is exactly the temperature range above which ϕ polymer disappears. BD–AN chains are extended but not the PVC chains. Micelles are held in solution mainly by BD–AN grafted chains forming a peripheral shell. We have a stable colloidal gel which can be flocculated by adding PVC homopolymer. Above 52°–57°C, PVC and BD–AN chains extend, and the initial PVC network cannot be restored; this phenomenon is not reversible. [The dichloroethane/polymer ratio is probably too low in the polymerization at 50°C

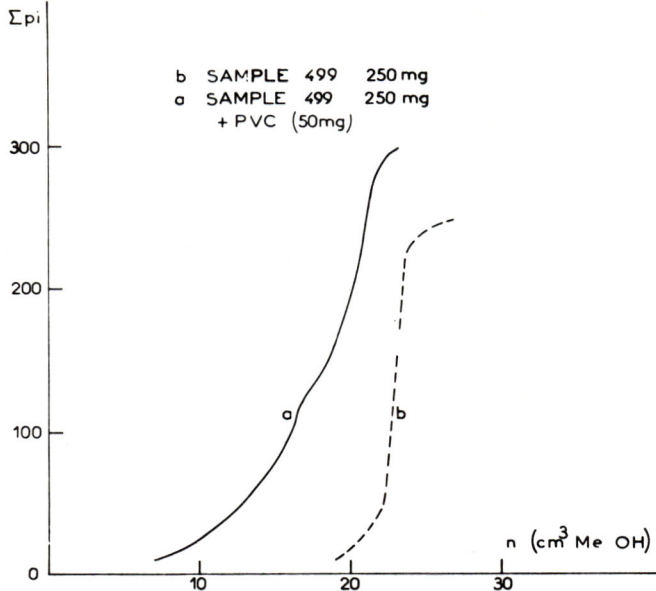

Figure 13. Precipitation after solubilization at low temperature for PVC/graft copolymer mixture

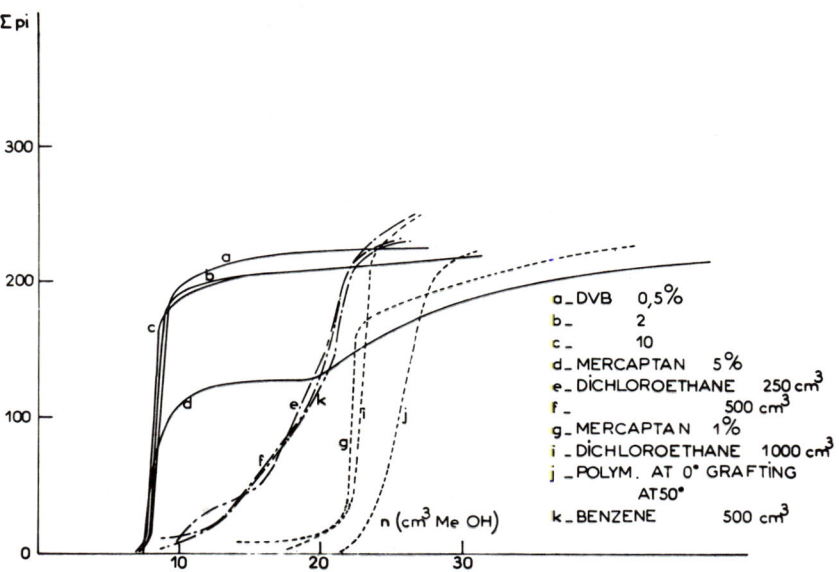

Figure 14. Fractionation of PVC–BD–AN copolymers after low temperature solubilization

to ensure destruction of PVC network. Anagnostopoulos' work is performed with excess solvent.]

The ϕ curves show different shapes according to preparation conditions. Generally, there are two types (Figure 14). In the first nearly all copolymer precipitates abruptly at a definite value of n ($= n'$); n' differs according to sample preparation. The polymers may be divided in two groups:

(1) $n' \sim 22$–24. Polymers prepared with high quantities of dichloroethane or with 1% TDM, PVC prepared at 0° and grafted at 50°C.

(2) $n' \sim 8$–10. Polymers prepared with divinylbenzene.

In the second curve shape precipitation is continuous between $n = 10$ and $n = 25$. The polymers are prepared with small quantities of dichloroethane or with benzene. The behavior of polymers prepared with 5% TDM is shown by both types.

COPOLYMERS PREPARED WITH CROSSLINKING MONOMER. Precipitation is abrupt near $n = 8$–9. Addition of DVB crosslinks butadiene–acrylonitrile chains giving a tridimensional network with linear PVC chains. Crosslinking reduces solubility. Then the precipitation curve must be displaced toward low n values until total insolubility occurs. Whatever the DVB ratio, the polymer remains soluble in dichloroethane. Polymerization is done in emulsion, and the diameter of particles is about 0.03–0.05 mμ. Grafting does not increase this size much. Thus, grafted particles are small and crosslinked; they look like microgels containing entangled PVC chain on which are grafted crosslinked butadiene–acrylonitrile chains.

COPOLYMERS PREPARED WITH TDM. *tert*-Dodecylmercaptan is a transfer agent; it reduces both number and length of grafted chains. When the TDM ratio is high, grafting may be nil. Precipitation curves of polymer prepared with 5% TDM are explained easily if the grafting ratio on PVC is small. Probably the very small quantity of grafted polymer is sufficient to displace the PVC homopolymer precipitation curve (probably by a solubilizing effect), but above $n = 15$, ϕ and γ curves are superimposed.

COPOLYMERS PREPARED WITH BENZENE. Copolymers prepared with benzene (0.25 part) or small quantities of dichloroethane (<0.25 part) show continuous precipitation above $n = 10$. For polymerization with benzene, vinyl chloride content of the fraction is always about 51–53.5% compared with a mean value of 52.4 for the sample. How can we explain this?

An analogy of the shape of the precipitation curves of these samples and of mixtures of graft copolymers PVC should lead us to put forth the hypothesis of the presence of PVC homopolymer which would be "solubilized" by the graft copolymer. However, this quantity would be at most 40% of the PVC which should be expressed by less PVC in the fraction corresponding to higher n value. In other respects, one cannot see why the amount of grafting would be smaller in the presence of benzene than in the presence of 1% mercaptan.

Graft copolymers prepared at 50°C without swelling agent are usually not soluble at low temperature. Sometimes partial solubility has been observed. When polymerization is performed with large quantities of dichloroethane or with small quantities of TDM, the copolymer is entirely soluble, and the precipitation range is about $n = 20$–24. Polymers prepared with small quantities of swelling agent are soluble, but the precipitation range is large. All polymerizations were performed to a high conversion ratio—*i.e.*, condition where butadiene chains crosslink. The number of crosslinks decreases by dilution of the monomer or by adding a transfer agent (which can be the added solvent). Consequently precipitation curves of polymers prepared with small quantities of swelling agent could be explained by the fact that they are less crosslinked than standard copolymers but more than copolymers prepared using mercaptan or dichloroethane. Precipitation occurs then according to the crosslinking ratio.

When using high dichloroethane ratios or 1% TDM, precipitation occurs when BD–AN chains collapse—*i.e.*, at high values of n owing to their more linear macromolecular structure. From the shape of the fractionation curve it is possible to conclude that all PVC chains are grafted in these samples? Examination of Figure 13 leads to the conclusion that such copolymer may contain only small quantities of PVC.

Solutions Prepared at 70°C. Generally, precipitation curves may be divided in two types:

(a) Precipitation is continuous in the range $n = 10$–20 (addition of dichloroethane (0.5 part), mercaptan (1%), grafting at low temperature).

(b) Precipitation is discontinuous and characterized by two groups of fractions (A and B) and two steps of precipitation of unequal importance: (addition of 5% mercaptan and DVB grafting at high temperature).

As for solubilization at 20°C we examine the addition of PVC homopolymer. We worked first on polymer prepared without additive and grafted at 50°C to which was added PVC homopolymer ($[\eta] \sim 0.8$). Solution concentration was always 1%, and the graft copolymer/PVC homopolymer ratio varied from 250/0 to 0/250 (Figures 15 and 16).

For all curves $\beta = f(n)$, where $\beta = M - (P_{\text{PVC}} + P_\text{G})$ with $M =$ weight of polymer precipitated at n, $P_{\text{PVC}} =$ weight of PVC homopolymer normally precipitated at n, and $P_\text{G} =$ weight of graft copolymer normally precipitated at n, there is a maximum for $n = 6.5/7$, which is also the precipitation step of PVC homopolymer. We verified that quantities of

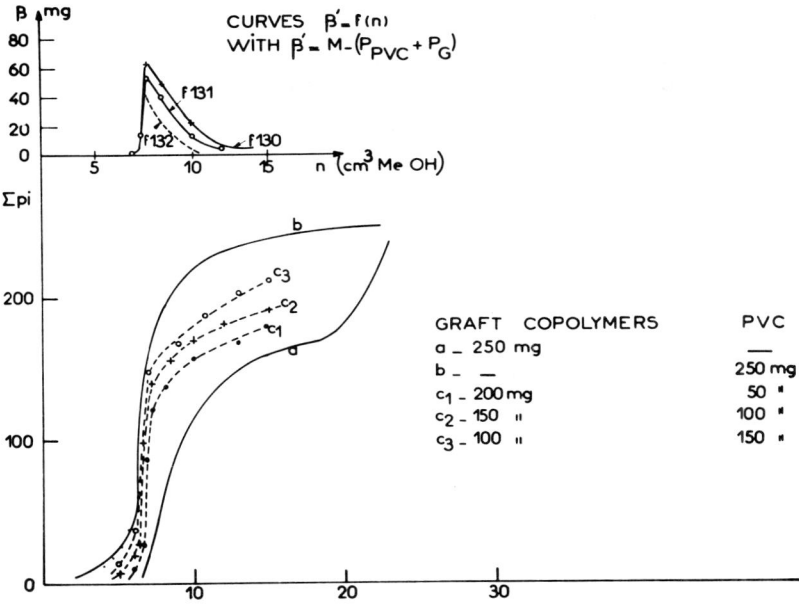

Figure 15. Top: fractionation of mixtures of PVC–graft copolymers. Bottom: fractionation curves for different compositions of PVC–graft copolymer mixtures

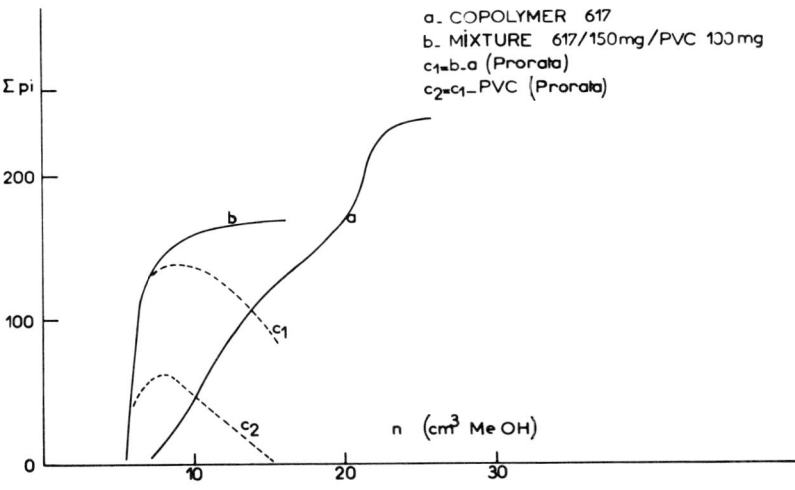

Figure 16. Fractionation of mixture of PVC and graft copolymer

PVC homopolymer as small as 4% of total weight are sufficient to ensure modification of the precipitation curve. Thus, addition of PVC gives an abrupt precipitation about $n = 6$ and coprecipitation of the graft copolymer with the PVC.

The same conclusion is found when the copolymer is prepared using large quantities of dichloroethane and the precipitation step occurs for higher n value. Consequently we must deduce, in such cases, that added PVC is different from PVC contained in the copolymer. One is justified in concluding that PVC is grafted by butadiene copolymer, (analysis verifies this) and that all PVC chains are grafted in the samples prepared with high dichloroethane ratios.

We now try to explain the influence of additives. Addition of small quantities of dichloroethane leads to curves with two inflections, one near $n = 7$ and another near $n = 16$–20. It is also the shape of copolymer prepared without additive.

The inflection point about $n = 7$ is different from the PVC homopolymer inflection point. A close examination of fractionation curves shows that this point is about that of PVC homopolymer (*i.e.*, around $n = 5$–6) only for mixtures of PVC homopolymer and graft copolymer and when grafting is performed using 5% TDM.

Dichloroethane can act in two ways: as swelling agent or chain transfer agent. Butadiene and acrylonitrile cannot swell PVC (which does not imply that only the external part of micelles are grafted—penetration may be capillary). Dichloroethane can do it partially. Then, with dichloroethane, monomer penetration is easier; grafting may involve more PVC macromolecules.

Dichloroethane can act as chain transfer agent; the grafting ratio must be reduced. Experimentally, the grafting ratio is roughly constant whatever the dichloroethane ratio, but dichloroethane can act also as diluent for the monomers, reducing reactions of crosslinking but not the grafting ratio.

Our explanation is as follows:

(1) The precipitation step at $n = 7$–8 arises from microgels (*11*).

(2) High dichloroethane ratios reduce microgel formation by dilution and/or chain transfer action.

Addition of increasing quantities of mercaptan at the grafting stage reduces the length and number of grafted chains (*6*); the grafting ratio then decreases. On the other hand, crosslinking reactions are reduced or avoided. When the mercaptan ratio increases very much, grafting no longer takes place; polymer is constituted by PVC homopolymer and low molecular weight butadiene–acrylonitrile copolymer. Experimental results obtained using 5% mercaptan agree with this hypothesis. There is pure PVC (verified by fractionation curves and chlorine analysis), and

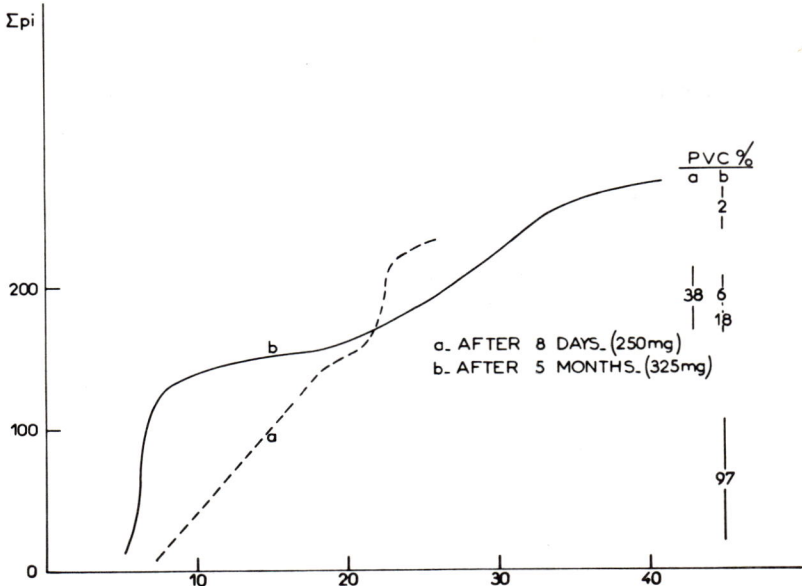

Figure 17. Degradation by aging (solution)

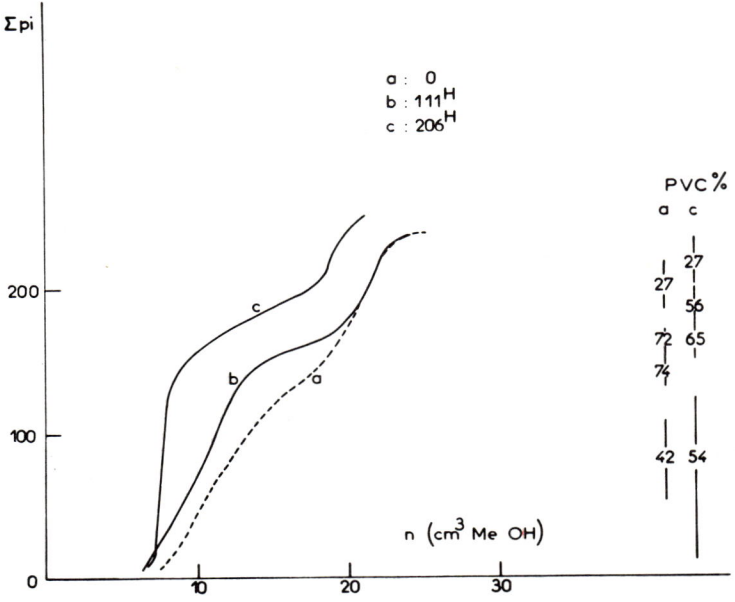

Figure 18. Thermal degradation

the precipitation range of the butadiene–acrylonitrile fraction is wide. We cannot certify if this fraction is only a mixture of non-grafted chains.

Addition of divinylbenzene instead of mercaptan gives a copolymer with two precipitation zones about $n = 5.5$–7 and $n = 20$–22. DVB crosslinks BD–AN chains and reduces their solubility. When the crosslinking ratio is very high, only the PVC backbone can extend. It seems impossible that PVC homopolymer may be present because: (a) the precipitation step (7–7.5) is slightly different from PVC homopolymer (5–6); (b) it is logical to think that properties are continuous. If polymers prepared without DVB but with dichloroethane or with chain-transfer agent are grafted, then polymer prepared with DVB must be also grafted. The fact that some polymer precipitates about $n = 20$–24 disagrees with the preceding explanation. We believe that this second fraction comes from the population of the smallest particles or the more grafted peripheral shell.

GRAFTING TEMPERATURE EFFECT. Temperature can influence the reaction rates in different ways: initiation, propagation, transfer, termination. For grafting reaction, the length and number of grafted chains depend on rate constant of these reactions. However for radiochemical grafting, the initiation rate is not temperature dependent (2, 3, 4).

For butadiene polymerization, the contribution of the second double bond increases with temperature and the conversion ratio giving crosslinked chains (12).

Experimental results show that at low grafting temperature (30°C) the solubility of grafted polymers is highly modified by butadiene–acrylonitrile chains. Precipitation curves look like those of polymers prepared using dichloroethane and therefore seem compatible with a less crosslinked BD–AN chain structure. At high grafting temperature (70°C), the precipitation curve resembles that obtained with DVB crosslinked copolymers.

These results are explained by a high thermal crosslinking at high temperature and a low thermal crosslinking at low temperature.

So, it seems reasonable to think that in polymers grafted at 30°C, all PVC chains are grafted. High temperatures favor grafting. Consequently, samples grafted at temperatures above 30°C must be more grafted. Therefore, the abrupt precipitation step near $n = 7$ can be attributed to highly grafted PVC and not to a PVC homopolymer.

Consequently, all experimental results can be explained. Polymers prepared without any additive are graft copolymers (containing perhaps separate BD–AN copolymer) whose grafted chains are crosslinked. Decrease of crosslinking ratio linked possibly with a decrease of grafting ratio leads to a more soluble polymer and conversely, increasing of crosslinking leads to a less soluble polymer.

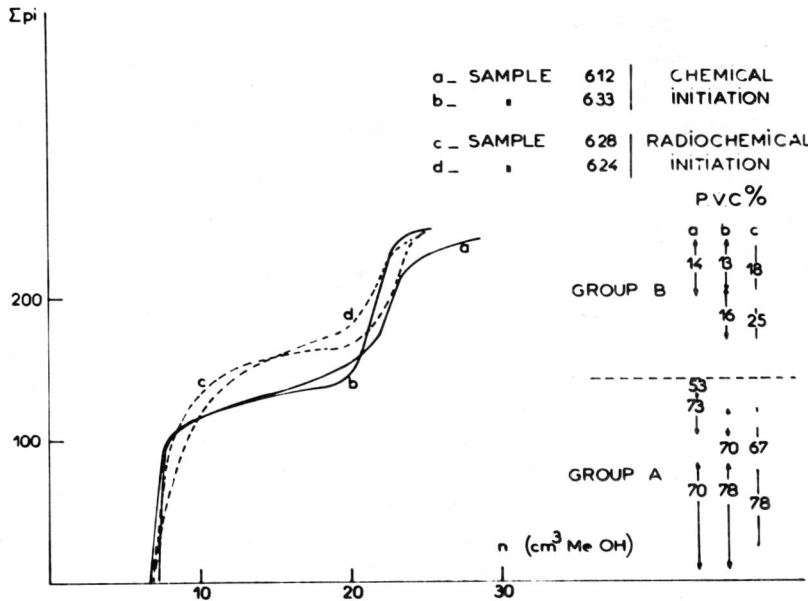

Figure 19. Fractional curves of polymer prepared chemically and radiochemically

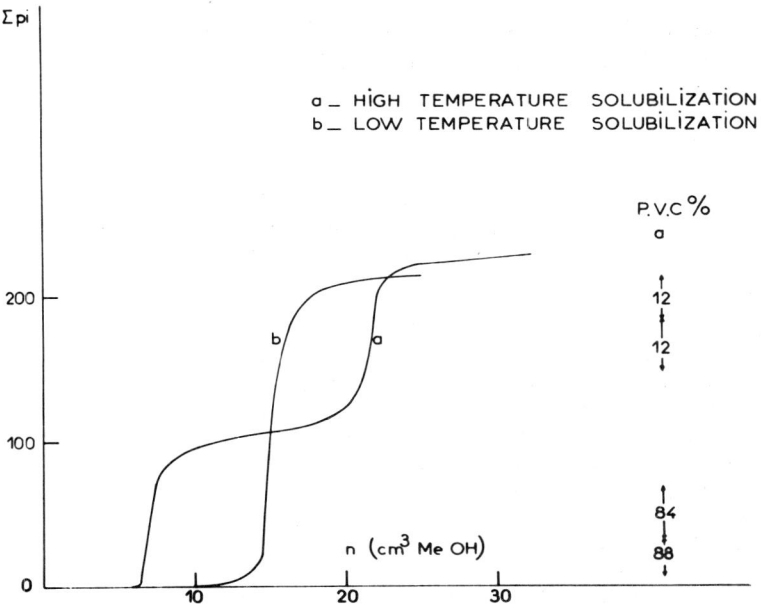

Figure 20. PVC–BD–AN copolymers prepared with persulfate mercaptan initiator

These results are corroborated by a study of the degradation of these graft copolymers. It is known (12) that butadiene-based copolymers are thermally degraded by a radical process (chemical or photochemical). If degradation is studied in dilute systems, polymer–polymer interactions are small; chains will break. Effectively the shapes of fractionation curves (Figure 17) exhibit a degrafting process. If degradation is studied in a condensed phase, where polymer–polymer interactions are great, crosslinking is also important (Figure 18) (note that structure of group A cannot be ascertained owing to degraded chains).

Chemical Grafting. By way of comparison, Figure 19 shows some fractionation curves of polymer prepared chemically. At identical grafting temperatures the precipitation curves are different. In general one could say that chemical polymers correspond to radiochemical polymers prepared at temperatures 10° to 20°C higher. The amount of PVC in fractions of group A is comparable, but the size of these fractions is slightly smaller for a redox system. As in the case of radiochemical polymers one again finds the particular behavior of solutions prepared at low temperature. There again the precipitation zones are displaced strongly towards low n values.

For a different chemical formula (PVC 5 parts–butadiene 4 parts–acrylonitrile 1 part) we have noted the curves obtained with a persulfate–mercaptan system for hot- and cold-prepared solutions (Figure 20). For the hot-prepared solution the amount of PVC of group A fractions is high but corresponds to values observed in radiochemical samples.

Conclusions

The preceding study constitutes an attempt to explain the structure of copolymers obtained by polymerizing a mixture of butadiene and acrylonitrile in presence of a poly(vinyl chloride) latex. Several points have not been elucidated, but some important conclusions can be drawn:

(1) The copolymers obtained behave very differently from that of a mixture of PVC and butadiene–acrylonitrile copolymer even if one takes into account the fact that a strict comparison (maintaining identical average molecular weights) is not practically possible. The butadiene–acrylonitrile chains seem to be linked to those of PVC; the hypothesis of grafting stated in this text should be retained.

(2) In copolymers prepared in the absence of large quantities of powerful transfer agents, the results obtained can be explained admitting that a large part, if not all, the PVC is grafted. Very precise conclusions are difficult since the presence of graft polymers is likely to modify slightly the solubility characteristics of the homopolymer. It seems that a large part of the butadiene–acrylonitrile copolymer is equally grafted, but no figures could be obtained.

(3) The structure of graft copolymers prepared in the absence of a transfer agent is notably a microgel type structure—*i.e.*, the polymer is crosslinked at the level of the particles in the emulsion. This structure is therefore more pronounced with increasing crosslinking (experiments in the presence of difunctional monomers or at elevated temperatures), and less pronounced with less crosslinking (experiments with transfer agents). The study of fractionation in solution demonstrate the influence of reticulation.

Literature Cited

(1) Anagnostopoulos, C. E., Cornu, A. V., Gamrath, H. R., *J. Appl. Polymer Sci.* **4,** 181 (1960).
(2) Burland, W. J., Hoffman, A. S., "Block and Graft Polymers," Reinhold, New York, 1960.
(3) Ceresa, R. J., "Block and Graft Polymers," Butterworth, London, 1962.
(4) Chapiro, A., "Radiation Chemistry of Polymeric Systems," Interscience, New York, 1962.
(5) Desreux, V. *et al.*, *Chem. Weekblad* **48,** 247 (1952).
(6) Hayes, R. A., *J. Polymer Sci.* **11,** 531 (1953).
(7) Jasse, B., *Bull. Soc. Chim. France* **1968,** 878.
(8) Kato, K., *Polymer* **1967,** 33; *Kolloid-Z.* **220** (1), 24 (1967).
(9) Merret, F. M., Wood, R. S., *Trans. Proc. Inst. Rubber* **32,** 27–42 (1956).
(10) Pechiney-Saint Gobain Co., U. S. Patent **3,327,022** (1965); U. S. Patent **3,330,886** (1965); French Patent **1,309,809** (1959).
(11) Shashoua, V. E., Van Holde, K. E., *J. Polymer Sci.* **33,** 395 (1958).
(12) Whitby, G. S., "Synthetic Rubber," Wiley, New York, 1954.

RECEIVED February 19, 1970.

20

Thermal Stabilization of Poly(vinyl chloride) through Graft Copolymerization of cis-1,4-Polybutadiene

NORMAN G. GAYLORD and AKIO TAKAHASHI

Gaylord Research Institute Inc., Newark, N. J. 07104

> *The thermal stability of poly(vinyl chloride) is improved greatly by the* in situ *polymerization of butadiene or by reaction with preformed* cis-1,4-*polybutadiene using a diethylaluminum chloride–cobalt compound catalyst system. The improved thermal stability at 3–10% add-on is manifested by greatly reduced discoloration when the modified poly(vinyl chloride) is compression molded at 200°C in air in the absence of a stabilizer, hydrogen chloride evolution at 180°C is retarded, and the temperature for the onset of HCl evolution and the peak decomposition temperature (DTA) increase, i.e. 260°–280°C and 290°–325°C, respectively, compared with 240°–260°C and 260°–280°C for the unmodified homopolymer, in the absence of stabilizer. The grafting reaction may be carried out on suspension, emulsion, or bulk polymerized poly(vinyl chloride) with little or no change in the glass transition temperature.*

The utilization of vinyl chloride polymers is directly related to their processibility. To obtain an adequately low melt viscosity for improved processibility without reducing the average molecular weight or the molecular weight distribution, it is necessary either to incorporate an external plasticizer or polymeric additive, utilize a comonomer in the polymerization and/or process at elevated temperatures. The use of an external plasticizer leads to a flexible thermoplastic while copolymerization or the use of additives leads to flexible or rigid polymers depending on the concentration of comonomer or additive. In either case, the second-order or glass transition temperature is generally decreased.

Increasing the processing temperature is an obvious method for decreasing the melt viscosity, making possible high speed fabrication, including injection molding, extrusion, and blow molding, of rigid poly(vinyl chloride) (PVC). However, at elevated temperatures the polymer has poor thermal stability as manifested by discoloration and a loss of properties. The degradation noted upon the exposure of PVC to elevated processing temperatures results from thermal and thermo-oxidative dehydrochlorination. Consequently, stabilizers are incorporated into the polymer to retard or delay the initiation or propagation of dehydrochlorination as well as to scavenge or react with the evolved hydrogen chloride. The stabilizers commonly used include metal compounds, such as lead, barium, cadmium, tin, calcium, and zinc compounds as well as epoxides and organic phosphorus compounds. The disadvantages of such added stabilizers include their potential toxicity, color, incompatibility, extractability, migration, and/or cost.

The role of structural and environmental factors in the thermal degradation of PVC has been examined in numerous reviews (2, 8, 11, 16, 19). Oxygen has a most deleterious effect in that it increases dehydrochlorination and promotes chain scission. Although the degraded polymer may subsequently undergo oxidative bleaching, the embrittlement which accompanies chain scission and crosslinking (the latter arising from the polyene sequences) results in deterioration of the mechanical properties.

In the presence of hydroquinone the rate of dehydrochlorination of poly(vinyl chloride) in oxygen is reduced considerably (20, 24). The presence of an oxygen scavenger in lieu of a radical scavenger or antioxidant would be expected to have an even greater effect.

This hypothesis has been confirmed by the greatly improved thermal stability of PVC as a result of the formation of a graft copolymer of cis-1,4-polybutadiene onto poly(vinyl chloride). The improved thermal stability is demonstrated by the almost total absence of discoloration on molding the graft copolymer into a film at 200°C in air, the reduced rate of dehydrochlorination on heating in an inert atmosphere at 180°C, and higher onset and peak temperatures for hydrogen chloride evolution as determined by differential thermal analysis.

Experimental

Polymerization of Butadiene in Presence of PVC. Et_2AlCl–Cobalt Compound Catalyst. A suspension of 20 grams of PVC and 18.7 mg (0.03 mmole) of cobalt(II) stearate in 300 ml of chlorobenzene in a three-necked flask equipped with a reflux condenser, Teflon-coated magnetic stirring bar, thermometer, and gas inlet and outlet was stirred at room temperature for 1 hour while nitrogen was bubbled through. The suspension was cooled to 5°–8°C, 2 grams of butadiene (passed through columns packed with anhydrous potassium hydroxide and molecular sieves and condensed in a small flask) were added, followed by 6 mmoles of Et_2AlCl. The reaction mixture was stirred under nitrogen for 30 minutes at 5°–8°C. Upon adding 300 ml of methanol containing 0.2% of 2,6-di-*tert*-butyl-*p*-cresol, the white, powdery product was filtered readily, washed with methanol containing antioxidant, and dried to constant weight at 40°–50°C *in vacuo* to yield 21.5 grams of product.

Extraction of the product with refluxing *n*-hexane for 24 hours gave 21.0 grams of hexane-insoluble material and 0.5 gram of soluble material. The add-on was therefore 5%.

AlEt$_3$-*tert*-Butyl Chloride–Cobalt Compound Catalyst. PVC, 20 grams, was suspended in 190 ml chlorobenzene, and the resultant slurry was cooled to 5°C. The addition of 2 grams of butadiene was followed by the successive addition of 0.01 mmole of cobalt(II) bis(salicylaldehyde imine), 2 mmoles of *tert*-butyl chloride, and 2 mmoles of triethylaluminum. The reaction mixture was stirred at 8°C for 60 minutes and poured into 1 liter of methanol. The vacuum-dried product weighed 21.9 grams, representing an add-on of 5.1%.

Reaction of *cis*-1,4-Polybutadiene and PVC. Et$_2$AlCl–Cobalt Compound Catalyst. Commercial *cis*-1,4-polybutadiene prepared with a Et$_2$AlCl–cobalt compound catalyst system was freed of antioxidant by solution in benzene and precipitation with methanol. The *cis*-1,4,polybutadiene had an intrinsic viscosity in benzene at 25°C of 2.4 and a greater than 96% cis-1,4 content.

Two grams of *cis*-1,4-polybutadiene were dissolved in 100 ml of chlorobenzene. When all the rubber was in solution, 20 grams of PVC were added, and the slurry was stirred for 30 minutes while nitrogen was bubbled through the heterogeneous mixture. After the mixture was cooled to 5°C, 5.4 mg (0.03 mmole) of cobaltous bis(salicylaldehyde imine) and 11.6 mmoles of Et$_2$AlCl were added. After 40 minutes a small amount of methanol was added to stop the reaction, and the mixture was precipitated into a large amount of methanol. The dried reaction product weighed 22.0 grams, and after extraction with refluxing hexane for 24 hours the dried residue represented 98.2% of the product.

Et$_2$AlCl Catalyst. Under the same conditions as above 20 grams of PVC and 2 grams of *cis*-1,4-polybutadiene, prepared with an alkylaluminum–titanium tetraiodide catalyst system (95% *cis*-1,4 content, intrinsic viscosity at 25°C in benzene 2.2) in 200 ml chlorobenzene were allowed to react in the presence of 2 mmoles of Et$_2$AlCl at 5°–10°C for 60 minutes. The reaction product was isolated by precipitation in methanol and dried to yield 22.0 grams of modified poly(vinyl chloride). Hexane extraction under reflux for 24 hours removed 8% of hexane-soluble material.

Tests for Thermal Stability. HCl Evolution. In a modification of ASTM Method D793-49 (1965), utilizing the apparatus described therein, 5 grams of polymer were heated at 180°C in a flask immersed in an oil bath, and preheated nitrogen was passed over the sample. Under these conditions the sample slowly evolved hydrogen chloride which was carried by the nitrogen and bubbled into 70 ml of distilled water. The pH of the solution was measured continuously with a pH meter and converted to millimoles of HCl by a calibration curve constructed by adding known quantities of HCl to 70 ml distilled water. Another method of measurement involved bubbling the evolved HCl into a dilute sodium hydroxide solution and back titrating with dilute hydrochloric acid. The calibration curve was generally used, although both tests were frequently used as a check on a given polymer sample.

The results were plotted as millimoles of hydrogen chloride evolved as a function of time and the shape of the curve, and the time required for HCl evolution representing 0.1 mole % (0.058 wt %) decomposition of the poly(vinyl chloride) was noted.

COMPRESSION MOLDING AT 200°C IN AIR. The polymer in powder form was preheated in the press at 200°C for 1 minute, followed by pressing at 200°C for 1 minute at 6000 psig.

DIFFERENTIAL THERMAL ANALYSIS. A sample of film pressed in air at 200°C was heated at the rate of 10°C/minute from 25°–500°C in a nitrogen atmosphere. The T_g and the temperatures for the onset of HCl evolution and the peak endotherm were determined from the DTA plot.

Grafting of cis-1,4-Polybutadiene onto PVC

Grafting by *in situ* Polymerization of Butadiene. The polymerization of butadiene to a high *cis*-1,4-polybutadiene with a catalyst system containing diethylaluminum chloride and a cobalt compound is now a well established technique (*1, 9, 15, 18, 22*). This catalyst system is particularly effective when the cobalt compound is soluble in the reaction medium.

The presence of water is necessary for preparing high *cis*-1,4-polybutadiene by the Et_2AlCl–Co compound catalyst system (*10*). Aluminum chloride, pyridine, or ethanol may be substituted for water (*1*). Organic halides are effective activators not only for the Et_2AlCl–Co compound catalyst (*10*) but also promote the formation of high molecular weight, high *cis*-1,4 polymer with a trialkylaluminum–Co compound catalyst system (*23*).

The cationic polymerization of isobutylene (*12*) and styrene (*13*) is initiated readily by Et_2AlCl in the presence of an alkyl halide, RCl. The interaction of the catalyst and cocatalyst is presumed to produce the carbonium ion R^+, which initiates polymerization, and the corresponding gegenion $Et_2AlCl_2^-$. Alkyl halides with low R–Cl bond dissociation energies—*e.g.* tertiary, substituted allylic, and benzylic halides—are among the most effective cocatalysts.

In view of the presence of similar labile chlorine atoms in PVC and the promoting effect of such halides in the preparation of *cis*-1,4-polybutadiene, the polymerization of butadiene was carried out in the presence of PVC using the Et_2AlCl–Co compound catalyst system. The reaction proceeded readily, and the product was identified as a graft copolymer of *cis*-1,4-polybutadiene on PVC (*7*).

Since the objective was the preparation of a modified PVC containing a relatively few appended chains of polybutadiene, the reaction was carried out heterogeneously by suspending the PVC in chlorobenzene. The suspension was cooled to 5°–10°C, butadiene, a cobalt compound, and Et_2AlCl were added, and the mixture was stirred at 5°–10°C for 30–60 minutes before the addition of methanol to terminate the reaction.

Et_2AlCl could be replaced by the sesquichloride or by a mixture of a trialkylaluminum and a reactive halide such as benzyl chloride or *tert*-butyl chloride. The effective cobalt compounds were those which are known to yield *cis*-1,4-polybutadiene—e.g. cobalt stearate, cobalt acetylacetonate, cobalt bis(salicylaldehyde imine), cobalt chloride–pyridine, etc. Et_2AlCl concentration could be varied within the range 0.3–5% by weight based on PVC, and the cobalt compound concentration was 0.002–0.01 mole per mole of Et_2AlCl.

The butadiene concentration was generally 3–10% by weight of PVC to limit the extent of grafting and to minimize changes in the basic polymer properties—most particularly, the second-order transition temperature. The preferred extent of grafting was less than 5%.

The reaction temperature could be varied between 0° and 40°C. However, a temperature of 5°–10°C gave a greater than 80% reaction within 0.5–1 hour. At higher temperatures the reaction was completed within 15 minutes, but careful temperature control was necessary to avoid degradation and gelation of the reaction product.

Halogenated aromatic hydrocarbons such as chlorobenzene were preferred as reaction media owing to their ability to act as solvents for butadiene and polybutadiene as well as the catalyst components and as swelling agents for PVC. However, halogenated aliphatic compounds as well as mixtures of chlorobenzene with other halogenated compounds could also be used. The amount of solvent used was generally between 5 and 20 times by volume based on the weight of PVC and was determined by the efficiency of agitation and the ability to maintain control of the reaction temperature.

Upon completion of the reaction the reaction mixture was treated with a diluent which is a nonsolvent for PVC. Suitable diluents included aliphatic or aromatic hydrocarbons such as hexane, heptane, or benzene or compounds containing an active hydrogen atom such as acetic acid or a lower alkanol such as methanol or ethanol. Methanol was the preferred diluent by virtue of its miscibility with the preferred reaction medium (chlorobenzene), its ability to react readily with and deactivate an aluminum alkyl or alkylaluminum halide, and its low boiling point and water solubility.

MECHANISM. Although it is widely known that the Et_2AlCl–cobalt compound system catalyzes the polymerization of butadiene to a high cis-1,4-polybutadiene, there has been no general agreement as to the mechanism of polymerization. Thus, propagation through carbonium ion (3) and carbanion (4), as well as concurrent carbonium ion/carbanion (10) mechanisms have been proposed. However, it is generally accepted that the cobalt compound forms a complex with the butadiene in which the monomer is bound in its cis configuration and presents the cisoid butadienylcobalt complex to the growing chain end. Therefore, the cobalt compound is not expected to have a function in the grafting reaction but to influence the polymerization of butadiene to a cis-1,4-polybutadiene.

The cocatalytic activity of alkyl halides in the cationic polymerization of styrene in the presence of stannic chloride (17), in the polymerization of butadiene with Et_2AlCl–cobalt compound (10) and R_3Al–cobalt compound (23) catalyst systems and in the cationic polymerization of isobutylene (12) and styrene (13) in the presence of Et_2AlCl is well documented. It is reasonable to propose that a reaction between Et_2AlCl and a labile chlorine atom on PVC results in the generation of a carbonium ion on the polymer backbone.

$$\text{PVC}\underset{\underset{\text{Cl}}{|}}{\rule{2cm}{0.4pt}} + Et_2AlCl \rightarrow \text{PVC}\underset{+}{\rule{2cm}{0.4pt}} + Et_2AlCl_2^- \quad (1)$$

The polymerization of butadiene (BD) on this site proceeds to yield a cis-1,4-polybutadiene through the addition of a cisoid monomer in the form of the cobalt complex.

$$\text{PVC}\underset{+}{\rule{2cm}{0.4pt}} + \text{BD (cobalt complex)} \rightarrow \text{PVC}\underset{\underset{(BD)_x(\text{cis})}{|}}{\rule{2cm}{0.4pt}} \quad (2)$$

An alternative mechanism involves the addition of the polymeric carbonium ion to a double bond in cis-1,4-polybutadiene, the latter formed in situ by the polymerization of butadiene by the Et_2AlCl–cobalt compound catalyst system.

$$
\begin{array}{c}
|\\
CH_2\\
|\\
-C^+\\
|\\
CH_2\\
|\\
HC-Cl\\
|
\end{array}
\;+\;
\begin{array}{c}
|\\
CH_2\\
|\\
CH\\
||\\
CH\\
|\\
CH_2\\
|
\end{array}
\;\rightarrow\;
\begin{array}{c}
|\\
CH_2\\
|\\
-C\\
|\\
CH_2\\
|\\
HC-Cl\\
|
\end{array}
\begin{array}{c}
|\\
CH_2\\
|\\
-CH\\
|\\
HC^+\\
|\\
CH_2\\
|
\end{array}
\quad (3)
$$

An analogous mechanism has been proposed in the "molecular jump" reaction in which the average molecular weight of cis-1,4-polybutadiene is increased and branching is promoted by adding an alkyl halide to a polymer containing a Friedel-Crafts catalyst such as R_2AlCl (5, 21).

This mechanism has been confirmed by the reaction between PVC and cis-1,4-polybutadiene in the presence of Et_2AlCl.

Grafting by Reaction with cis-1,4-Polybutadiene. Although the PVC–cis-1,4-polybutadiene reaction could be carried out homogeneously, a heterogeneous system was preferred to control the extent of reaction and simplify the procedure for isolating the reaction product. The reaction was consequently carried out under conditions such that the reaction medium was a solvent for cis-1,4-polybutadiene but only suspended or swelled the PVC.

Halogenated aromatic hydrocarbons such as chlorobenzene or mixtures with aliphatic or aromatic hydrocarbons or halogenated derivatives thereof were suitable reaction media. The volume of solvent was generally 5–20 times the weight of PVC.

The catalyst for the reaction between PVC and the high cis-1,4-polybutadiene was a dialkylaluminum halide—e.g., Et_2AlCl—or sesquihalide. The catalyst could be added from an external source or could be generated in situ—e.g., by the reaction of an aluminum alkyl with either titanium tetrachloride or a reactive organic halide such as benzyl chloride or tert-butyl chloride. The concentration of Et_2AlCl was varied from 0.2–5% by weight based on PVC, although the preferred concentration was between 0.5 and 1.5 wt %.

Although the graft copolymer was obtained readily by the reaction of cis-1,4-polybutadiene with PVC in the presence of Et_2AlCl alone, the addition of 0.001–0.1 mole of a cobalt compound per mole of the Et_2AlCl yielded a gel-free product with superior properties. The preferred cobalt compound concentration was between 0.002 and 0.01 mole per mole of aluminum compound. The effective cobalt compounds were those generally used in polymerizing butadiene to cis-1,4-polybutadiene using the Et_2AlCl–cobalt compound catalyst system.

The reaction was carried out using up to 10% by weight of polybutadiene based on PVC. However, to avoid changing the properties of PVC other than the thermal stability, the preferred extent of reaction was 3–6%.

The reaction temperature could be varied from below 0° to $+40°C$. However, it was preferred to carry out the reaction with the polybutadiene at 5°–10°C. The reaction time at the lower temperature was approximately 60–80% complete in 1 hour at 5°–10°C. Although the reaction was usually carried to completion, it was often desirable to terminate the reaction at this point by adding methanol. In this way a 5% charge of polybutadiene resulted in a 3–4% add-on, while a 10% charge resulted in a 6–8% reaction. The unreacted polybutadiene could be extraced with hexane or other suitable solvent. However, it was normally not necessary to remove the unreacted polybutadiene which was compatibilized by the PVC–polybutadiene reaction product.

The high *cis*-1,4-polybutadiene used to modify PVC could be prepared by any of the well known processes including polymerization of butadiene using catalyst systems based on aluminum alkyl–titanium tetraiodide, dialkylaluminum chloride–cobalt compound–water, aluminum alkyl–cobalt compound–organic halide, lithium metal, organolithium compounds, etc. Although PVC with improved thermal stability could be prepared by reaction with polybutadiene with a cis-1,4 content of at least 35%, the greatest improvement was obtained with a polybutadiene with a cis-1,4 content of at least 75%, preferably greater than 90%.

When commercially available polybutadienes were used, it was generally necessary to remove the antioxidants and other stabilizers which are usually incorporated to improve storage stability. This was accomplished by conventional methods—*e.g.*, by dissolving the polymer in benzene or chlorobenzene and precipitating in methanol or washing the polymer solution with aqueous alkali if the stabilizer was a phenolic compound, or with aqueous acid if the stabilizer was an amine derivative. If the stablilizer was removed by aqueous washing, it was necessary to remove residual water from the polymer solution by drying over suitable desiccants such as lithium aluminum hydride, calcium hydride, etc.

A convenient method for carrying out the reaction of a high *cis*-1,4-polybutadiene with PVC was to polymerize butadiene using a suitable catalyst system—*e.g.*, Et_2AlCl–cobalt stearate–*tert*-butyl chloride or Et_3Al–cobalt chelate–benzyl chloride, and then to add an appropriate quantity of the resultant polybutadiene solution to a suspension of PVC in chlorobenzene. Additional Et_2AlCl could then be added to the reaction mixture, although this was unnecessary if the initial concentration was adequate.

MECHANISM. Under the experimental conditions and concentrations previously described, the addition of Et_2AlCl to a chlorobenzene solution of high *cis*-1,4-polybutadiene, followed by precipitation with methanol, did not reduce the complete hexane or tetrahydrofuran solubility of the polybutadiene. However, when PVC was present during the addition of Et_2AlCl to the *cis*-1,4-polybutadiene, under the same conditions, only 0–40% of the polybutadiene was extractable by hexane, the amount depending upon the reaction time. When the hexane-insoluble residue was extracted with tetrahydrofuran, greater than 90% dissolved. The soluble material was identifiable as unmodified PVC while the hexane-insoluble, tetrahydrofuran-insoluble residue was shown by elemental analysis to contain hydrocarbon residues—*i.e.*, it was a reaction product of PVC and *cis*-1,4-polybutadiene. When the initially added *cis*-1,4-polybutadiene was of low molecular weight, the precipitated product from the reaction with PVC in the presence of Et_2AlCl yielded a hexane-insoluble residue, which was essentially completely soluble in tetrahydrofuran and contained hydrocarbon residues.

When a solution of cis-1,4-polybutadiene was prepared directly by polymerizing butadiene with a catalyst system which contained sufficient Et_2AlCl, either *per se* or as a result of *in situ* generation, and the solution was then added to a suspension of PVC in chlorobenzene, precipitation with methanol after a suitable reaction period, resulted in a reaction product which had little or no solubility in refluxing hexane. The soluble fraction, when obtainable, was identifiable as a high molecular weight polybutadiene with a cis-1,4 content of greater than 90%. When the initial polybutadiene solution was precipitated with methanol without reaction with PVC, the isolated polymer was soluble in hexane and was identifiable as the same high molecular weight polybutadiene with a greater than 90% cis-1,4 content.

When butadiene was polymerized in the presence of a PVC suspension in chlorobenzene using a catalyst system based on Et_2AlCl, the methanol-precipitated reaction product had little or no solubility in refluxing hexane. The soluble fraction, when obtainable, was identifiable as a greater than 90% cis-1,4-polybutadiene. In this case, the hexane-insoluble residue was essentially completely soluble in tetrahydrofuran, probably owing to the low molecular weight of the polybutadiene in the PVC–cis-1,4-polybutadiene reaction product

Although it was difficult to detect structural characteristics other than those of PVC in a polybutadiene–PVC reaction product containing less than 5% polybutadiene, products containing 5–10% polybutadiene were shown to contain cis-1,4 unsaturation by infrared spectroscopic analysis.

Thus, it appears that the products obtained by either polymerizing butadiene in the presence of PVC or by the reaction of high cis-1,4-polybutadiene with PVC, in either case utilizing a catalyst containing Et_2AlCl, had the same composition—*i.e.*, they were graft copolymers of PVC and high cis-1,4-polybutadiene mixed with unmodified PVC.

The mechanism proposed in Reaction 3—*i.e.*, the generation of a polymeric carbonium ion by the reaction of Et_2AlCl with PVC and the addition of the carbonium ion to a double bond in cis-1,4-polybutadiene —would appear to be applicable to the polymer–polymer grafting reaction. The monomer–polymer grafting reaction may involve polymerization of butadiene on the polymeric carbonium ion site or the reaction between polybutadiene generated *in situ* and the polymeric carbonium ion.

Thermal Stability of cis-1,4-Polybutadiene-Grafted PVC

Discoloration. When PVC, free of plasticizer or stabilizer, is compression molded in air at 200°C under adequate pressure (*e.g.*, 500–6000 psig), thermal degradation results in discoloration to a pink to brown colored film. When the modified PVC containing as little as 3% of grafted *cis*-1,4-polybutadiene was pressed under the same conditions, the resultant film was essentially colorless or no more than faintly discolored.

Since discoloration is related to dehydrochlorination and is generally considered to result from the development of sequences of conjugated unsaturation of sufficient length to absorb in the visible region of the spectrum—*i.e.*, a minimum of five to seven double bonds (8)—substitution on the PVC chain would prevent the development of long polyene sequences and reduce discoloration. However, dehydrochlorination resulting in numerous short polyene sequences is not necessarily materially reduced, and the polymer becomes embrittled as a result of chain scission and crosslinking, although only slightly discolored.

Consequently, reduced or retarded hydrogen chloride evolution is a more reliable measure of the thermal stability of PVC. When the unmodified polymer was pressed to a film in air at 200°C on a black iron or untreated steel plate, the film was deeply colored, and the metal surface beneath the polymer film was corroded and became covered with a film of rust. In contrast, under the same conditions, the pressed film from the graft copolymer was very slightly discolored, and the metal surface remained free of corrosion and rust even after two years exposure to the atmosphere.

Thus, the thermal stabilization of PVC which resulted from the heterogeneous grafting of as little as 3–5% *cis*-1,4-polybutadiene was more than a simple additive effect and indicates a synergistic interaction. This was demonstrated further by dissolving up to 10% *cis*-1,4-polybutadiene in a chlorobenzene suspension or solution of PVC and isolating the polymer blend by precipitation with methanol. Films pressed from the polymer blend were generally deeply colored and contained incompatible, probably gelled or crosslinked, areas.

To improve the thermal stability of PVC, it is common to add stabilizers, generally metalloorganic compounds, in concentrations of 1–5 parts per 100 parts of PVC. Although the *cis*-1,4-polybutadiene grafted PVC, in the absence of an added stabilizer, yielded essentially colorless or only faintly discolored films, completely colorless films were obtained when the conventional stabilizers were added in concentrations of 0.1–0.3 parts per 100 parts of modified PVC. Organotin stabilizers were not necessary and, in some cases, actually resulted in greater color development than when they were absent.

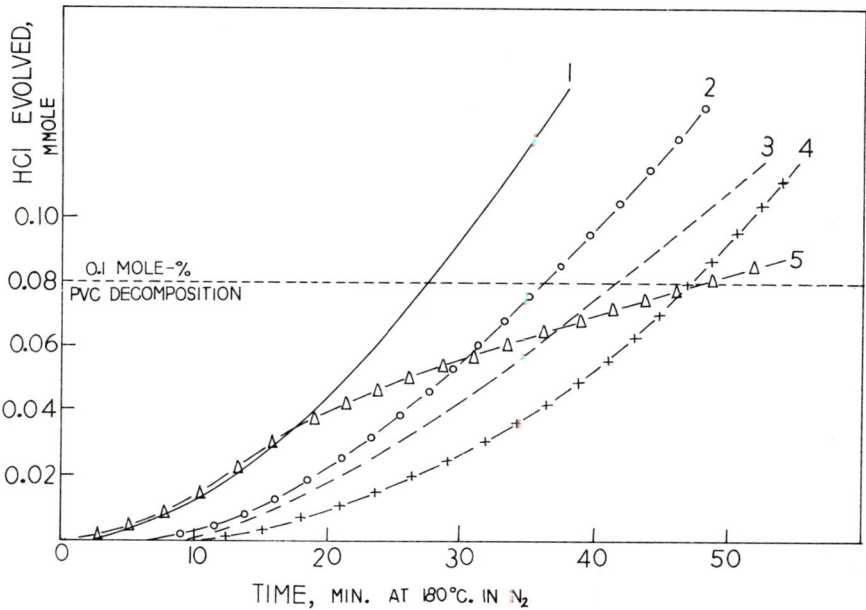

*Figure 1. Evolution of hydrogen chloride at 180°C (**nitrogen as carrier gas**) from suspension PVC (1), suspension PVC + stabilizer (2), cis-1,4-polybutadiene–PVC (suspension) graft copolymer from monomeric butadiene (Type M) (3), Type M graft copolymer + stabilizer (4), and graft copolymer from cis-1,4-polybutadiene (Type P) (5)*

	As Shown	Normal
Stabilizer, phr: Ferro 59-V-11 (Ca-Zn)	0.10	2–3
Ferro 5376 (organic)	0.05	1–1.5

A further indication of the enhanced thermal stabilization inherent in the grafted polymer was the reduction in the concentration of conventional stabilizer required to prevent discoloration of an unmodified PVC, when the latter was blended with the graft copolymer or the mixture of graft copolymer and unmodified PVC resulting from the heterogeneous grafting reaction.

Hydrogen Chloride Evolution. Two additional tests for thermal stability, as measured by hydrogen chloride evolution, demonstrated the improved stability of the grafted PVC.

The first was a modification of ASTM D793-49 (1965) which describes a procedure for determining the short-time stability at elevated temperatures of plastics containing chlorine. Utilizing the apparatus described therein, 5 grams of polymer in the form of powder, as isolated from the grafting reaction or as obtained commercially, was heated to 180°C, and the evolved hydrogen chloride was carried by preheated nitrogen into water or a dilute aqueous alkali solution. The results were plotted as millimoles of hydrogen chloride evolved as a function of time, and the shape of the curve and the time required for 0.1 mole % decomposition of the polymer were used as a measure of the thermal stability.

The curve obtained with PVC prepared by suspension polymerization generally indicated autocatalytic thermal dehydrochlorination, and the time required for 0.1 mole % decomposition was usually less than 35 minutes (Figure 1). When PVC was prepared by bulk polymerization, the dehydrochlorination plot was generally slightly more linear, showed little autocatalytic character, and the time for 0.1 mole % decomposition was approximately 40 minutes (Figure 2).

When the product of the *in situ* polymerization of monomeric butadiene to *cis*-1,4-polybutadiene in the presence of PVC, referred to as Type M PVC, as well as the product of the reaction of *cis*-1,4-polybutadiene with PVC, referred to as Type P PVC, were subjected to the hydrogen chloride evolution test at 180°C, irrespective of whether the base PVC was prepared by suspension or bulk polymerization, the curve was essentially linear and the time for 0.1 mole % decomposition was generally more than 45 minutes and often as much as 100 minutes (Figures 1 and 2).

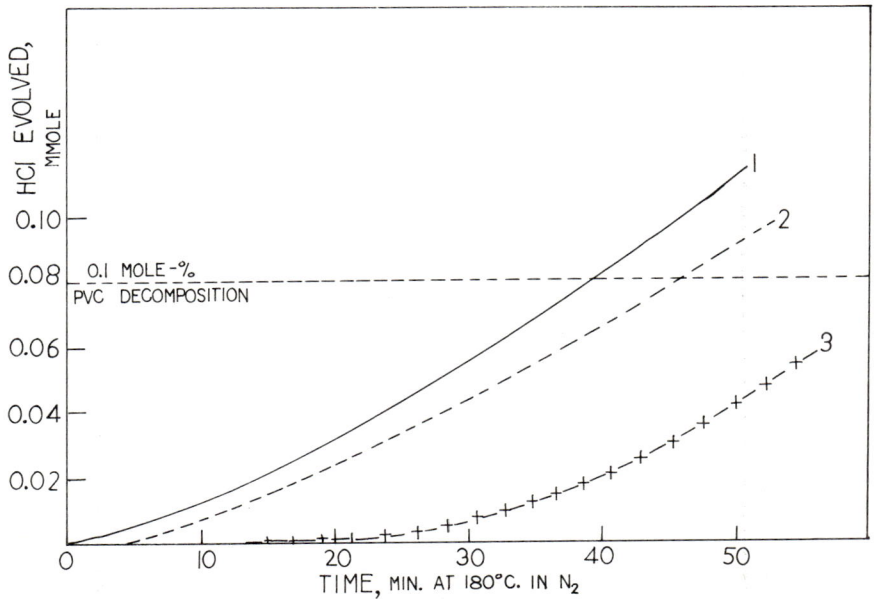

Figure 2. Evolution of hydrogen chloride at 180°C (nitrogen as carrier gas) from bulk PVC (1), cis-1,4-polybutadiene–PVC (bulk) graft copolymer from monomeric butadiene (Type M) (2), and Type M graft copolymer + stabilizer (3) (stabilizer composition as in Figure 1)

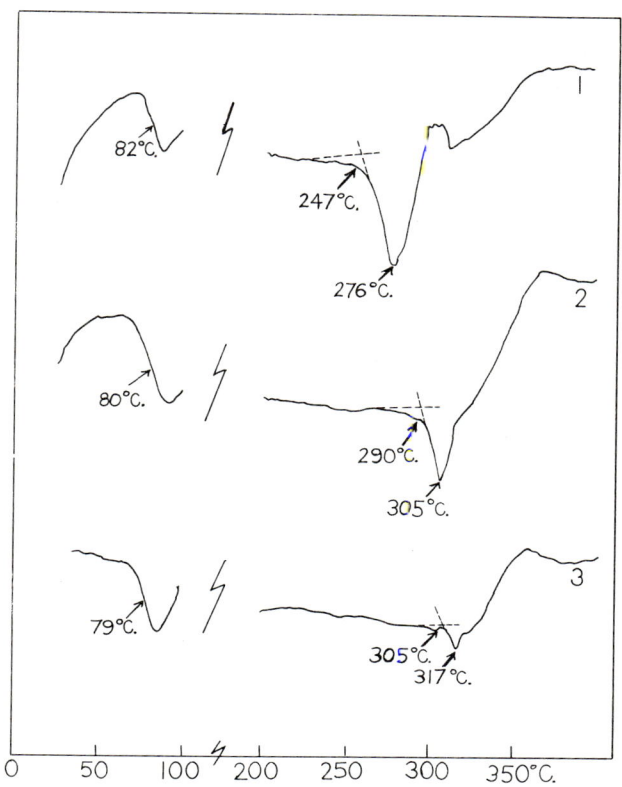

Figure 3. Differential thermal analysis (10°C/minute in nitrogen) of film (pressed at 200°C in air) from suspension PVC (1) and cis-1,4-polybutadiene–PVC (suspension) graft copolymer from monomeric butadiene (Type M) (2) and cis-1,4-polybutadiene (Type P) (3)

The addition of small amounts of various stabilizers to untreated or polybutadiene-grafted PVC decreased the rate of hydrogen chloride evolution. Thus, as shown in Figure 1, 0.15 phr which is less than 10% of the recommended amounts of a mixture of non-toxic calcium–zinc stabilizers, increased the time for 0.1 mole % dehydrochlorination of suspension polymerized PVC from 28 to 36 minutes and of Type M PVC prepared from the suspension polymer from 41 to 47 minutes. Type P PVC prepared from the same polymer required 49 minutes for the same extent of dehydrochlorination.

Bulk PVC underwent 0.1 mole % decomposition at 180°C in 39 minutes, Type M PVC prepared from the same polymer required 46 minutes, while the addition of stabilizer to the Type M PVC extended the time to more than 60 minutes (Figure 2).

Differential thermal analysis was used as a further measure of thermal stability. The polymer powder was pressed at 200°C in air to form a film. The film sample was heated at 10°C/minute from 25°–500°C in a nitrogen atmosphere. The T_g, the temperature at the onset of decomposition— i.e., dehydrochlorination—as well as the peak endothermic temperature of the polymer which had already been subjected to a temperature of 200°C during film formation, were obtained from the DTA plot (Figures 3 and 4).

The T_g of untreated PVC ranged from 80°–84°C. Type M and Type P PVC containing up to 5% cis-1,4-polybutadiene generally had a T_g from 78°–82°C, with the lower T_g corresponding to the 5% grafted level

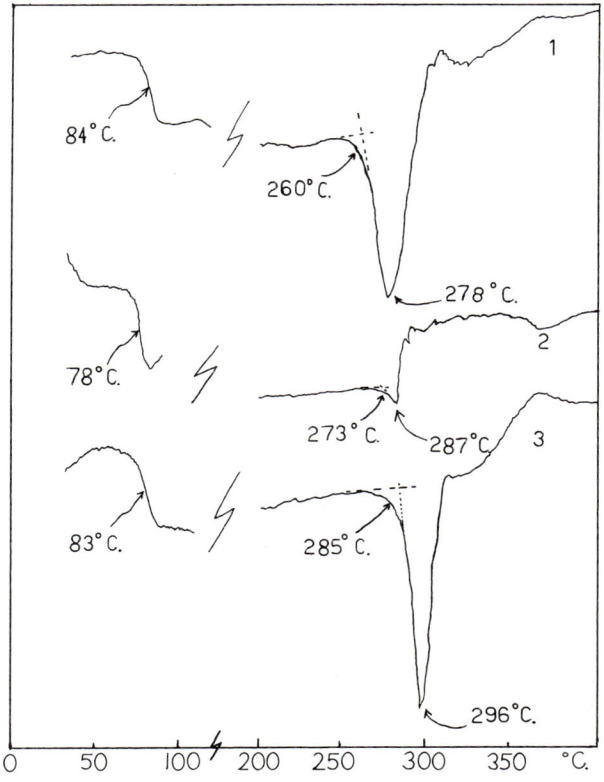

Figure 4. Differential thermal analysis (10°C/minute in nitrogen) of film (pressed at 200°C in air) from bulk PVC (1) and cis-1,4-polybutadiene–PVC (bulk) graft copolymer from monomeric butadiene (Type M) (2) and cis-1,4-polybutadiene (Type P) (3)

and the incorporation of relatively high molecular weight cis-1,4-polybutadiene.

Whereas the onset of decomposition, under the conditions described above, for untreated PVC was observed at approximately 240°–260°C and the peak decomposition temperature was approximately 260°–280°C, the grafted PVC prepared from the same untreated base polymer showed the onset of decomposition at approximately 260°–280°C and a peak decomposition temperature of 290°–325°C.

Although the addition of a stabilizer may retard the initial rate of dehydrochlorination sufficiently to yield a colorless or faintly discolored film at 200°C, the results of DTA and the hydrogen chloride evolution test are unchanged. Thus, the addition of 0.5 part cadmium stearate per 100 parts of unmodified bulk polymerized PVC resulted in a composition which yielded an essentially colorless film at 200°C. However, the hydrogen chloride evolution test at 180°C indicated 0.1 mole % decomposition after 40 minutes, unchanged from the value for the unstabilized polymer. Similarly, DTA showed an unchanged T_g of 81°C, onset of decomposition at 245°C, and a peak decomposition temperature of 280°C. This clearly points up the magnitude of the stabilization effect resulting from the grafting reaction.

MECHANISM OF STABILIZATION. The improved thermal stability of the cis-1,4-polybutadiene-grafted PVC may arise as a result of the treatment to which the polymer is subjected and/or to the structural changes generated thereby.

Suspension of PVC in a swelling solvent such as chlorobenzene may remove the lower molecular weight fraction which would be expected to contain the greatest number of reactive or thermally sensitive chain ends. The limited contribution of this fractionation effect was confirmed by swelling PVC in chlorobenzene, adding methanol, and isolating the polymer under the conditions used in the grafting reaction.

Treatment of PVC, suspended in chlorobenzene, with Et_2AlCl followed by the addition of methanol greatly improved the thermal stability (6). This will be the subject of a separate publication. However, the nature of the improvement was different—i.e., film pressed in air at 200°C was yellow not colorless, but processing stability as determined by the torque rheometer test in the Brabender Plasticorder at 195°C was far superior and was measured in hours rather than minutes. The removal of labile chlorine atoms by Et_2AlCl undoubtedly contributed to heat resistance. However, in the absence of a further reaction, such dehalogenation probably became dehydrohalogenation, contributing to color development.

Linear polymers from conjugated dienes readily undergo oxidation in air, in some cases even at room temperature (14). The allylic carbon is even more susceptible to hydrogen abstraction than the C–H in a saturated hydrocarbon, and a peroxy radical is formed.

$$-\underset{H}{\overset{H}{C}}=\underset{H}{\overset{H}{C}}-CH_2- \xrightarrow{O_2} -\underset{H}{\overset{H}{C}}=\underset{H}{\overset{H}{C}}-\underset{OO\cdot}{\overset{H}{C}}- \quad (4)$$

The latter initiates a chain reaction by attacking another allylic group and is converted to a hydroperoxide.

$$-\underset{H}{\overset{H}{C}}=\underset{H}{\overset{H}{C}}-\underset{OO\cdot}{\overset{H}{C}}- + -\underset{H}{\overset{H}{C}}=\underset{H}{\overset{H}{C}}-CH_2- \rightarrow -\underset{H}{\overset{H}{C}}=\underset{H}{\overset{H}{C}}-\underset{OOH}{\overset{H}{C}}- + -\underset{H}{\overset{H}{C}}=\underset{H}{\overset{H}{C}}-\overset{\cdot}{C}H- \quad (5)$$

The hydroperoxide decomposes to form an allyloxy radical and a hydroxyl radical, each of which continues the chain reaction.

$$-\underset{H}{\overset{H}{C}}=\underset{H}{\overset{H}{C}}-\underset{OOH}{\overset{H}{C}}- \rightarrow -\underset{H}{\overset{H}{C}}=\underset{H}{\overset{H}{C}}-\underset{O\cdot}{\overset{H}{C}}- + HO\cdot \quad (6)$$

$$-\underset{H}{\overset{H}{C}}=\underset{H}{\overset{H}{C}}-\underset{O\cdot}{\overset{H}{C}}- + -\underset{H}{\overset{H}{C}}=\underset{H}{\overset{H}{C}}-CH_2- \rightarrow -\underset{H}{\overset{H}{C}}=\underset{H}{\overset{H}{C}}-\underset{OH}{\overset{H}{C}}- + -\underset{H}{\overset{H}{C}}=\underset{H}{\overset{H}{C}}-\overset{\cdot}{\underset{H}{C}}- \quad (7)$$

$$HO\cdot + -\underset{H}{\overset{H}{C}}=\underset{H}{\overset{H}{C}}-CH_2- \rightarrow -\underset{H}{\overset{H}{C}}=\underset{H}{\overset{H}{C}}-\overset{\cdot}{C}H- + HOH \quad (8)$$

The peroxy and allyl radicals may couple to generate dialkyl peroxides which decompose to form additional allyloxy radicals.

$$\begin{array}{c} -\underset{H}{\overset{H}{C}}=\underset{H}{\overset{H}{C}}-\underset{OO\cdot}{\overset{H}{C}}- \\ \\ -\underset{H}{\overset{H}{C}}=\underset{H}{\overset{H}{C}}-\overset{\cdot}{\underset{H}{C}}- \end{array} \rightarrow \begin{array}{c} -\underset{H}{\overset{H}{C}}=\underset{H}{\overset{H}{C}}-\underset{O}{\overset{H}{C}}- \\ | \\ O \\ | \\ -\underset{H}{\overset{H}{C}}=\underset{H}{\overset{H}{C}}-\underset{H}{\overset{H}{C}}- \end{array} \rightarrow 2 -\underset{H}{\overset{H}{C}}=\underset{H}{\overset{H}{C}}-\underset{O\cdot}{\overset{H}{C}}- \quad (9)$$

The formation of cyclic peroxides has been proposed (*14*) in the oxidation of poly(2,3-dimethylbutadiene). The same reaction in *cis*-1,4-polybutadiene would also yield additional allyloxy radicals.

$$\text{[structures of cyclic peroxide intermediates and allyloxy radicals]}$$

(10)

cis-1,4-Polybutadiene in the grafted PVC may therefore act as an oxygen scavenger, protecting the PVC from oxidative attack. The sensitivity of the diene polymer to oxidation makes it necessary to incorporate an antioxidant of the type normally used to protect such polymers during storage and handling in the grafted PVC.

In addition to its protective role against oxygen, grafted *cis*-1,4-polybutadiene may also serve indirectly as a scavenger of hydrogen chloride. The autocatalytic nature of the dehydrochlorination reaction in PVC has been attributed to the accelerating effect of hydrogen chloride (*8, 25*). It has been shown (*25*) that acceleration does not occur in PVC containing hydroxyl groups either as an additive or bound to the chain.

The allyloxy radicals generated from hydroperoxide, dialkyl peroxide and cyclic peroxide intermediates, in accordance with Reactions 6, 9, and 10, respectively, are converted to allyl alcohol moieties, as shown in Reaction 7. The latter are particularly susceptible to reaction with hydrogen chloride.

$$-\overset{H}{\underset{}{C}}=\overset{H}{\underset{}{C}}-\overset{H}{\underset{OH}{C}}- + HCl \rightarrow -\overset{H}{\underset{}{C}}=\overset{H}{\underset{}{C}}-\overset{H}{\underset{Cl}{C}}- + H_2O \qquad (11)$$

The active species in the accelerating effect of hydrogen chloride is presumed (25) to be the chloride ion which acts as a strong base. In a protic environment—*e.g.*, in the presence of water or a hydroxyl-containing compound—the hydrogen chloride becomes a strong acid and the autocatalyzed decomposition is prevented.

It is possible that the Diels-Alder reaction between the unsaturation in the polybutadiene and the unsaturation generated in the PVC as a result of dehydrochlorination results in substitution on the backbone of the latter, preventing the formation of the long polyene sequences which are responsible for discoloration.

$$-CH_2-\underset{}{C}=\underset{H}{\underset{|}{C}}-\underset{Cl}{\underset{|}{C}}-CH_2-\underset{}{C}=\underset{H}{\underset{|}{C}}-CH_2- \rightarrow \tag{12}$$

$$-CH_2-C=C-C=C-C=C-CH_2-$$

$$\begin{array}{c} -CH_2-C=C-C=C- \\ -CH_2-\underset{Cl}{\underset{|}{C}}-C=C-CH_2-\underset{Cl}{\underset{|}{C}}- \end{array} \longrightarrow \begin{array}{c} -CH_2-C\diagup^{C=C}\diagdown C- \\ -CH_2-\underset{Cl}{\underset{|}{C}}-\underset{H\ H}{\underset{|\ |}{C}}-\underset{}{\underset{|}{C}}-CH_2-\underset{Cl}{\underset{|}{C}}- \end{array} \tag{13}$$

The high level of unsaturation in the *cis*-1,4-polybutadiene *per se* and after dehydrochlorination as well as the unsaturation generated in the PVC has its drawbacks in the potential susceptibility to crosslinking. This may be responsible for the unexpectedly small improvement in the processibility of the *cis*-1,4-polybutadiene–PVC graft copolymer, as indicated by the torque rheometer test.

Conclusions

The *in situ* polymerization of butadiene with a Et_2AlCl–cobalt compound catalyst system in a suspension of PVC in chlorobenzene yields a cis-1,4-polybutadiene–PVC graft copolymer. A similar copolymer is formed when cis-1,4-polybutadiene reacts with PVC in the presence of Et_2AlCl. A graft copolymer containing 3–10% cis-1,4-polybutadiene has greatly improved thermal stability as indicated by reduced discoloration on heating in air and higher temperatures for the onset of hydrogen chloride evolution and the peak decomposition temperature.

Literature Cited

(1) van Amerongen, G. J., ADVAN. CHEM. SER. **52**, 136 (1966).
(2) Chevassus, F., DeBroutelles, R., "The Stabilization of Poly(vinyl chloride)," E. Arnold, London, 1963.
(3) Childers, C. W., *J. Am. Chem. Soc.* **85**, 229 (1963).
(4) Cooper, W., Eaves, D. E., Vaughan, G., *Makromol. Chem.* **67**, 229 (1963).
(5) Engel, F., Schäfer, J., Kiepert, K. M., *Rubber Plastics Age* **45**, 1499 (1964).
(6) Gaylord, N. G., Takahashi, A., Belgian Patent **729,626** (March 10, 1969).
(7) Gaylord, N. G., Takahashi, A., Belgian Patent **729,627** (March 10, 1969).
(8) Geddes, W. C., *Rubber Chem. Technol.* **40**, 178 (1967).
(9) Gippin, M., *Ind. Eng. Chem., Prod. Res. Develop.* **1**, 32 (1962).
(10) *Ibid.*, **4**, 160 (1965).
(11) Gordon, G. Ya., "Stabilization of Synthetic High Polymers," Israel Program for Scientific Translations, Jerusalem, 1964.
(12) Kennedy, J. P., *Proc. Intern. Symp. Macromol. Chem., Tokyo-Kyoto, 1966*, Preprint 2.1.04, p. I-47.
(13) Kennedy, J. P., *J. Macromol. Sci. (Chem.)* **A1**, 632 (1968).
(14) Kössler, I., Stolka, M., Vodehnal, J., Gaylord, N. G., *J. Macromol. Sci. (Chem.)* **A1**, 1487 (1967).
(15) Longiave, C., Castelli, R., Croce, G. F., *Chim. Ind. (Milan)* **43**, 625 (1961).
(16) Madorsky, S. L., "Thermal Degradation of Organic Polymers," Interscience, New York, 1964.
(17) Mathieson, A. R., in Plesch, P. H., "The Chemistry of Cationic Polymerization," Chap. 6, p. 243, Pergamon, Oxford, England, 1963.
(18) Natta, G., Porri, L., Fiore, L., British Patent **849,589** (Sept. 28, 1960).
(19) Neiman, M. B., "Aging and Stabilization of Polymers," Consultants Bureau, New York, 1965.
(20) Rieche, A., Grimm, A., Mucke, H., *Kunststoffe* **52**, 265 (1962).
(21) Ring, W., Cantow, H.-J., *Makromol. Chem.* **89**, 138 (1965); *Rubber Chem. Technol.* **40**, 895 (1967).
(22) Takahashi, A., Kambara, S., *J. Polymer Sci. Pt. B*, **3**, 279 (1965).
(23) Takahashi, A., Takahashi, K., Hirose, T., Kambara, S., *J. Polymer Sci. Pt. B*, **5**, 415 (1967).
(24) Talamini, G., Cinque, G., Palma, G., *Mat. Plast.* **30**, 317 (1964).
(25) van der Ven, S., de Wit, W. F., *Angew. Makromol. Chem.* **8**, 143 (1969).

RECEIVED February 11, 1970.

21

Properties of Graft and Block Copolymers of Fibrous Cellulose

JETT C. ARTHUR, JR.

Southern Regional Research Laboratory, Southern Utilization Research and Development Division, Agricultural Research Service, U. S. Department of Agriculture, New Orleans, La. 70119

> *Graft and block copolymers of cotton cellulose, in fiber, yarn, and fabric forms, were prepared by free-radical initiated copolymerization reactions of vinyl monomers with cellulose. The properties of the fibrous cellulose–polyvinyl copolymers were evaluated by solubility, ESR, and infrared spectroscopy, light, electron, and scanning electron microscopy, fractional separation, thermal analysis, and physical properties, including textile properties. Generally, the textile properties of the fibrous copolymers were improved as compared with the properties of cotton products.*

Fibrous celluloses, both as natural fibers and regenerated cellulosic fibers, comprise more than two-thirds of the world's textile fibers which include apparel, household, and industrial products. Cotton cellulose, even in developed countries, is still the major textile fiber. In the United States cotton's percentage share of the textile fiber market has declined in recent years; however, about four billion pounds of cotton are used annually in America in textile products. The blending of cellulosic fibers with man-made fibrous polymers to obtain textile fabrics with desired properties is commonly practiced. Other approaches have been to alter the properties of cellulosic fibers through chemical modification, such as acetylation, physical modification, such as mercerization, and copolymerization with vinyl monomers (*3, 4, 9, 11, 17*).

The modification of the properties of fibrous cotton cellulose through free-radical initiated copolymerization reactions with vinyl monomers has been investigated at the Southern Laboratory for a number of years. Both graft and block copolymers are formed. Under some experimental conditions the molecular weight of the polyvinyl polymer, covalently

Table I. Solubility of Cellulose in Polyacrylonitrile Copolymers of Cellulose in Cupriethylenediamine (0.5M) at 25°C

Copolymer[a]	Fraction of Cellulose Soluble, %	Nitrogen, % Total	Nitrogen, % Insoluble Fraction
A	99	—	—
B	48	6.6	10.2
C	97	4.8	—
D	0	13.0	11.5
E	0	—	—

[a] A, purified cotton cellulose; B, cellulose (75%)–polyacrylonitrile (25%) copolymer (γ-radiation initiated); C, cyanoethylated cellulose (D.S. 0.7); D, cyanoethylated cellulose (D.S. 0.7) (62%)–polyacrylonitrile (38%) copolymer (γ-radiation initiated); E, cellulose (75%)–polyacrylonitrile (25%) copolymer (ceric ion initiated).

linked to cellulose, may be equal to or greater than the molecular weight of the cellulose molecule (12, 42). Previously, the basic mechanisms and principles involved in the free-radical reactions of cellulose with vinyl monomers were discussed (1, 2, 8, 10, 19, 20, 25, 40). In this chapter we summarize the properties of graft and block copolymers of fibrous cotton cellulose, including chemical structure, morphology, and physical properties. Some comparisons of the effects of free-radical initiation of copolymerization reactions of cellulose with vinyl monomers on the properties of cotton products are made.

Chemical Structure

Solubility. The effects of free-radical initiation of acrylonitrile copolymerization with cellulose on the solubility of the cellulose in the copolymer product in cupriethylenediamine are shown in Table I. The solubility of the cellulose in the product prepared by the radiation method was greater than the solubility of the cellulose in the product prepared by the ceric ion method (34). We have reported previously that the moles of cellulose per mole of grafted polyacrylonitrile in products prepared by the radiation method, using aqueous $ZnCl_2$, ranged from about 5 to 86 and in products prepared by the ceric ion method was about 0.4 (42). The higher molecular degree of substitution of polyacrylonitrile in the latter product probably accounts for the lower solubility of the cellulose in the product. Cyanoethylation of cellulose plus copolymerization of acrylonitrile with the modified cellulose, initiated by radiation, also gave a product in which the cellulose was not soluble in cupriethylenediamine. In this case probably most of the hydroxyl groups on the cellulose molecule, normally accessible for complexing with cupriethylenediamine and in effecting dissolution of the cellulose, reacted (15, 16, 34).

Solubility of cellulose in the products has also been used to indicate the presence of covalent bonds between cellulose and the polyvinyl copolymer. Thin sections of fibrous cellulosic copolymers, as prepared for electron photomicrography, are examined both before and after successive extractions with solvents for cellulose and for the polymer. The presence of undissolved cellulose and polymer in the extracted thin section is interpreted as evidence for the presence of covalent bonds and for grafting (52).

Instrumental Methods. Infrared spectroscopy has often been used to investigate the chemical structure of cellulose copolymers. Since free-radical initiated reactions of cellulose usually involve oxidative depolymerization of the cellulose molecule, increases in infrared absorption arising from the increase in concentration of C=O groups are recorded (5). Attempts to identify covalent links between cellulose and polyvinyl polymer by recording infrared spectral data have been qualitatively successful. Typical infrared data for cellulose–polyacrylonitrile copolymers are shown in Table II. Cellulose, crystalline lattice type I, was copolymerized with acrylonitrile. The products were extracted with N,N-dimethylformamide to remove homopolymer prior to their examination. The increase in the concentration of the characteristic C≡N group of polyacrylonitrile is readily detected. For both copolymers B and E decreases in O—H and C—H concentrations are recorded, which may indicate covalent bonds between cellulose and polyacrylonitrile (14). Previously, we reported that the mechanisms of free-radical initiation, as shown by ESR spectroscopy data, were probably (1) dehydrogenation, and to a lesser extent depolymerization, in the case of γ-irradiation and (2) complexing of ceric ion with OH groups on C_2 and C_3 of the cellulose molecule in the case of redox initiation (1, 5, 6). Polarized infrared spectral data of cellulose copolymer products gave similar results (56).

Table II. Infrared Spectral Data of Polyacrylonitrile Copolymers of Cellulose

	Optical Density/mg. Cellulose				
Copolymer[a]	2.8–3.4μ O–H stretching	3.4μ C–H stretching	4.45μ C–N stretching	7.25μ C–H deformation	8.6μ C–OH deformation
A	0.534	0.213	0.000	0.281	0.457
B	0.363	0.169	0.091	0.243	0.357
C	0.371	0.183	0.063	0.246	0.384
D	0.475	0.234	0.157	0.325	0.475
E	0.373	0.221	0.061	0.224	0.331

[a] Footnote same as for Table I.

Table III. Density of Polyacrylonitrile Copolymers of Cellulose

Copolymer[a]	Polyacrylonitrile, %	Breaking Strength of Fibrous Copolymer, $\times 10^{-3}$ gram	Density, grams/ml
A	0	4.5	1.533
B	14	4.1	1.462
C	19	4.0	1.434
D	25	4.4	1.421

[a] γ-Radiation initiated copolymerization. Monomer solution: 32 parts of acrylonitrile in 68 parts of aqueous 80% $ZnCl_2$. Dosage: A, 0; B, 2.1×10^{19} ev/gram; C, 3.1×10^{19} ev/gram; D, 4.2×10^{19} ev/gram.

Applications of ESR spectroscopy to investigations of free-radical initiated copolymerization reactions of cellulose with vinyl monomers have been reported (1, 2). NMR and infrared spectroscopy have been used to examine products obtained from degradation of cellulose copolymers to characterize the nature of the chemical bonds between cellulose and polyvinyl polymer (35, 42). Covalent bonds were indicated in most cases.

X-ray examination of cellulose, which had been γ-irradiated, showed no change in the degree of crystallinity of the cellulose (2, 5). Similarly, x-ray examinations of cellulose copolymers have not demonstrated conclusively that the formation of copolymers decreased the crystallinity of cellulose (54, 55, 57).

Fractional Separation. The densities of cellulose copolymers have been used to differentiate between copolymers and mixtures of cellulose and homopolymers. Typical variations in the densities of fibrous cellulose–polyacrylonitrile copolymers are shown in Table III. The whole graft copolymer, which contained the highest percentage of polyacrylonitrile, had the lowest density. The values are in fairly good agreement with those that would have been predicted from the known densities of cellulose and amorphous polyacrylonitrile. Increased concentration of amorphous polyacrylonitrile in and decreased density of the copolymer did not significantly affect the breaking strength of the fibrous copolymer (13).

The degradation of the cellulose fraction of the copolymer and subsequent recovery of the polyvinyl polymer have often been used to characterize the polymer. For example, cellulose may be acetylated and acid hydrolyzed to remove it from the copolymer. Then the recovered polymer can be dissolved, in solvent normally used for the polymer, and the molecular weight of the polymer determined viscometrically (12, 42). As reported previously for polymers, such as polyacrylonitrile, a functional group on the polymer may be altered during the fractionating. These changes have been determined by infrared spectroscopy. For free-radical initiated copolymerization reactions, the molecular weights of polyacrylonitrile copolymers may range from about 30,000 to more than 1,000,000 (42). For polystyrene copolymers molecular weights of about 300,000–500,000 have been reported (12). The initial molecular weight of cellulose during free-radical initiated copolymerization reaction may be as high as 700,000 (4, 5). During the reaction, some oxidative depolymerization of cellulose occurs, so that in many instances the copolymer has a higher molecular weight than the cellulose to which it is chemically bonded (1, 5).

When copolymers are prepared by the free-radical initiation of reactions of irradiated cellulose (on which the free-radical site is located) with binary mixtures of vinyl monomers, fractionation of cellulose and

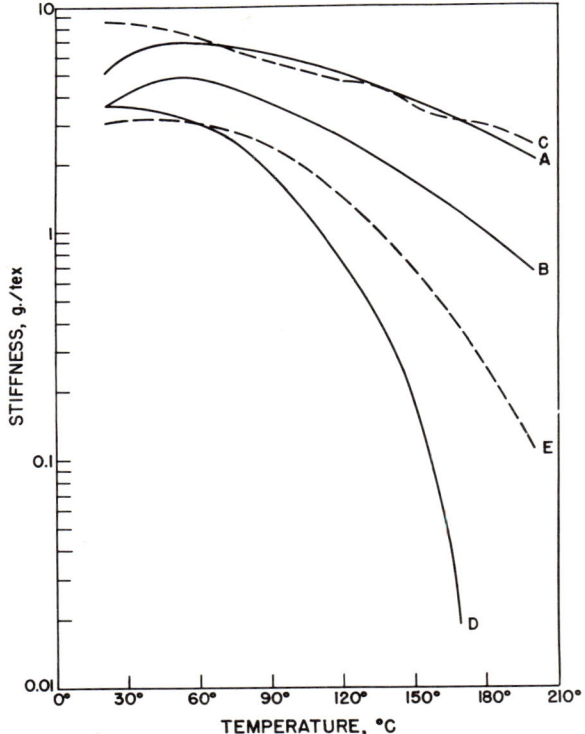

Figure 1. Effect of temperature on the stiffness of fibrous cellulose–polyvinyl copolymers

A: *purified cotton*
B: *fibrous cellulose–polyacrylonitrile*
C: *cyanoethylated cotton*
D: *fibrous cyanoethylated cellulose–polyacrylonitrile*
E: *fibrous cellulose–polystyrene*

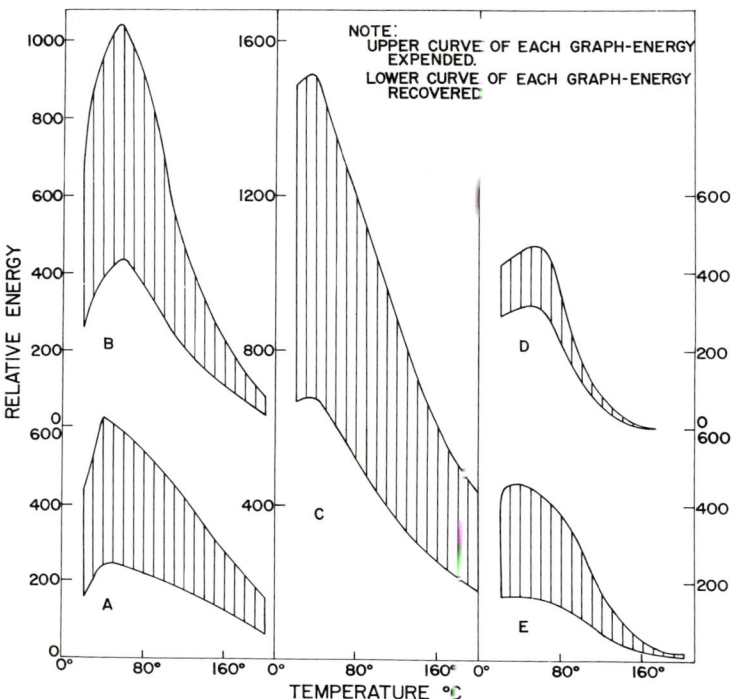

Figure 2. Effect of temperature on the relationship between energy expended to elongate fibrous cellulose–polyvinyl copolymers and energy recovered on relaxation of the copolymers. (See Figure 1 for identification of copolymers.)

polymer indicates that grafted block polymers may be bonded to cellulose. The composition of the grafted polymer depended on the composition of the binary mixture of monomers but was not necessarily the same as that composition. For a given binary mixture the composition of the grafted polymer was independent of the extent of graft copolymerization (37).

Thermal Analysis. Fibrous cellulose copolymers exhibit softening or second-order transition temperatures, as shown in Figure 1. In both purified cotton (Figure 1A) and cyanoethylated cotton (Figure 1C), the stiffnesses of the fibers decreased slightly with increasing temperature. The presence of a small amount of moisture in these fibers probably had a plasticizing effect which resulted in these relationships. The stiffnesses of fibrous cellulose–polyacrylonitrile (Figure 1B), cellulose–polystyrene (Figure 1E), and cyanoethylated cellulose–polyacrylonitrile (Figure 1D) decreased with increasing temperature. These changes indicate that the cellulose copolymers became soft and extensible with increasing temperature and passed through second-order transition temperatures. Cyanoethylated cellulose–polyacrylonitrile exhibited the lowest second-order transition temperature of the copolymers examined probably because of the high amorphous content of the copolymer. Fibrous cellulose–polystyrene copolymers showed a drop in resiliency index and stiffness with increasing temperature at about 100°C; cellulose–polyacrylonitrile and cyanoethylated cellulose–polyacrylonitrile, at about 80°–100°C (*14, 33*).

On relaxation the recovery of energy expended to elongate the fibrous copolymers to about 1.5% of their initial length at temperatures ranging from 21° to 200°C is shown in Figure 2. The lined sections in each graph of Figure 2 show the energy expended during elongation at a given temperature that was not recovered during relaxation at that temperature. Summation of the energy relationships over the entire temperature range show that the recovery of energy expended ranged from 38 to 43% for cellulose, cyanoethylated cellulose, cellulose–polyacrylonitrile, and cellulose–polystyrene. For cyanoethylated cellulose–polyacrylonitrile copolymer the recovery of energy was about 66%. These results indicate that it is possible to prepare cellulose–polyvinyl copolymers which have thermoplastic and/or thermoelastic properties (*24, 33*).

Elastomers, prepared by free-radical initiated copolymerization of ethyl acrylate with cellulose to several hundred percent extent of grafting of poly(ethyl acrylate) onto cellulose, exhibited rubber-like behavior and second-order transition temperatures. Cellulose–poly(ethyl acrylate) elastomers had transition temperatures below −35°C, about −20°C, and below 5°C when measured in ethyl acetate, dry air, and water, respectively (*43, 44*).

Morphology

As reported previously, the morphology of fibrous cellulose–polyvinyl copolymers, determined by electron microscopy, depends on the method of free-radical initiation of the copolymerization reaction, the experimental conditions during the reaction, and the type of vinyl monomer used. Variations in the shape of the fibrous copolymer cross section, in layering effects in the copolymer structure, and in location and distribution of the polyvinyl polymer within the fibrous structure were shown (*1, 2, 7, 29, 52*).

Recently, scanning electron microscopy has been used in our laboratory to investigate the effects of abrasion on the morphology of fibrous cellulose copolymers (*36*). For example, cotton cellulose was woven into fabric form (print cloth construction and weight). Cellulose copolymer fabrics were prepared by irradiating a sample of this fabric, followed by copolymerization of the irradiated fabric with a binary mixture of acrylo-

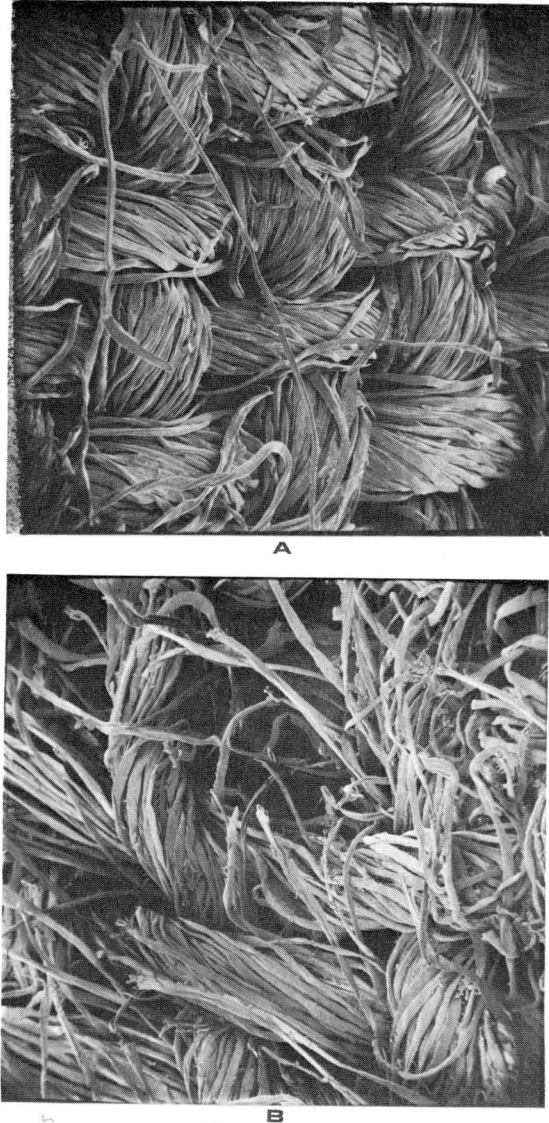

Figure 3. Scanning electron microphotograph of cellulose fabric sample before (A) and after (B) flex abrasion

nitrile and butyl methacrylate (37). Under a flex abrasion test, untreated fabric ruptured after 2000 cycles. Scanning electron microphotographs of the surfaces of samples of this fabric and the abraded fabric are shown

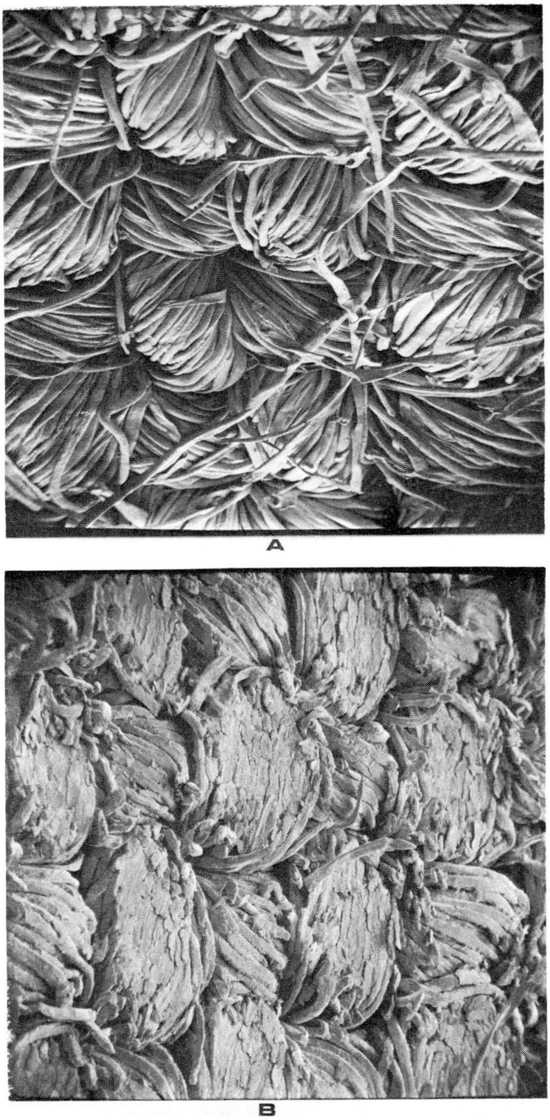

Figure 4. Scanning electron microphotograph of cellulose–polyacrylonitrile–poly(butyl methacrylate) copolymer fabric sample before (A) and after (B) flex abrasion

in Figure 3. As shown in Figure 3B, the fibers were ruptured and splintered by abrasion. Under the same flex abrasion test, the cellulose copolymer fabric was not ruptured even after 24,000 cycles. The surfaces of the copolymer fabric and the abraded copolymer fabric are shown in scanning electron microphotographs in Figure 4. The copolymer fibers (Figure 4A) are smoother than the untreated fibers (Figure 3A). The abraded copolymer fibers have been sheared but appear to be fused together (Figure 4B). They are not ruptured as in the case of the untreated fibers (Figure 3B). This probably accounts for the high flex abrasion resistance of the copolymer fabric as compared with that of the untreated fabric.

Physical Properties

Free-Radical Initiation. The effects of the method of free-radical initiation of the copolymerization reaction of acrylonitrile with fibrous cellulose on the properties of copolymer fabrics are shown in Table IV (*31*). At the same level of polymer add-on (25–27%), the molecular weight of

Table IV. Effect of Method of Initiation of Copolymerization Reaction on the Properties of Polyacrylonitrile–Cotton Copolymer Fabrics (Print Cloth)

Property	Copolymerization Reaction Conditions[a]					
	Control	A	B	C	D	E
Grafted polymer Add-on, %	—	25	27	25	26	26
Molecular weight of grafted polymer, $\times 10^{-5}$	—	5.9	5.1	0.33	0.84	0.84
Breaking strength, $\times 10^{-4}$ gram	1.9	2.3	2.3	1.8	1.9	1.8
Elongation at break, %	22	26	34	21	18	22
Tear resistance, gram	573	507	650	447	393	413
Flex abrasion, sample/control	1.0	1.7	2.4	1.1	1.4	1.0
Flat abrasion, sample/control	1.0	6.1	6.3	2.0	3.4	2.3
Shape of fibrous cross-section	kidney	kidney	round	kidney	kidney	kidney
Polymer distribution in fibrous cross-section	—	outer layers	uniform	uniform	uniform	uniform

[a] A: post-irradiation grafting; γ-radiation 1 megarad; cellulose immersed in 80% $ZnCl_2$. B: simultaneous irradiation grafting; γ-radiation 0.2 megarad; cellulose immersed in 32% acrylonitrile in 80% aqueous $ZnCl_2$. C: simultaneous irradiation grafting; γ-radiation 1 megarad; cellulose immersed in 30% acrylonitrile in N,N-dimethylformamide. D: ceric ion (0.005M) in HNO_3 (0.05M) grafting; 4% acrylonitrile. E: ferrous ion (0.1%) hydrogen peroxide (0.03%) grafting; 8% acrylonitrile.

the grafted polymer varied from 33,000 to 590,000 and depended on the method of free-radical initiation and experimental conditions. There is no obvious relationship between molecular weight of the grafted polymer and the properties of the copolymer fabrics.

Copolymerization reaction condition IVB gave copolymer fabrics with increased breaking strength, tear resistance, and flex and flat abrasion resistances as compared with the control fabric and the other copolymer fabrics. This improvement may be caused partly by effects of the experimental conditions on the morphology of the copolymer fabrics. As reported, the fibrous cross section was rounded, and the grafted polymer was distributed uniformly in the cross section. Reaction condition IVA gave copolymer fabrics in which the initial shape of the fibrous cross section was unchanged, and the grafted polymer was concentrated in the outer layers of the cross section. Copolymer fabrics prepared by this method had improved flex and flat abrasion resistances and breaking strength as compared with the control fabric. However, tear resistance decreased compared with the control fabric, and flex abrasion resistance was lower than that for copolymer fabric prepared by method IVB. Method IVB could also be developed into a continuous process, so that the fabric would first be immersed in a solution of vinyl monomer and then irradiated. For all of the methods indicated there would be a reduction in the molecular weight of the cellulose as a result of oxidative depolymerization.

Type of Vinyl Monomer. As reported previously, one of the important factors in determining the morphology of fibrous cellulose copolymers is the type of vinyl monomer used (27, 28, 30). We have also discussed the effects of type of monomer and chemical modification of cellulose on the thermal stress properties of copolymers. Similarly, the physical properties of the fibrous copolymers depend on these factors. For example, when the elastic recovery properties of fibrous cellulose copolymers, described in Table I, were examined, only copolymer D exhibited a change in these properties, as shown in Table V. At an elongation at break as high as 10%, changes in permanent set and delayed recovery were recorded. These changes were observed for copolymer D in both fiber and yarn forms (21, 24). In use as a textile, the elongation represents about the normal stress that would be applied to the fiber.

The effects of type of vinyl monomer used on the properties of copolymer yarns are shown in Table VI. Compared with the cotton control, cellulose–poly(vinyl acetate) copolymers exhibited the largest increase in elongation at break and breaking toughness and the largest decrease in stiffness of the fibrous copolymers examined. Cellulose–polystyrene and cellulose–poly(methyl methacrylate) copolymers exhibited the largest decrease in breaking toughness. Cellulose–polyacrylonitrile copolymer exhibited only a small decrease in breaking toughness. All of the fibrous copolymers examined had lower stiffness values than the cotton control.

Table V. Elastic Recovery Properties of Fibrous Cyanoethylated Cotton Cellulose (D.S. 0.7) (62%)–Polyacrylonitrile (38%) Copolymer

Elongation of Fibrous Copolymer, % of elongation at break	Recovery, % of Actual Elongation		
	Purified Cotton Yarn	Cyanoethylated Cotton Yarn	Cyanoethylated Cotton–Polyacrylonitrile Copolymer Yarn
Immediate Recovery			
5	30	34	34
10	23	29	24
25	9	10	9
50	6	8	5
75	6	7	4
Delayed Recovery			
5	49	49	63
10	58	55	71
25	66	69	74
50	46	46	48
75	29	30	29
Permanent Set			
5	21	17	3
10	19	16	5
25	25	21	17
50	48	46	47
75	65	65	67

Table VI. Properties of Cotton Cellulose–Polyvinyl Copolymer[a] Yarns

Property	Cotton Control	Copolymer (Add-on)					
		PS (44%)	PAN (26%)	PVA (47%)	PMMA (48%)	CN–Cotton DS 0.7	CN–Cotton PAN (38%)
Breaking strength, $\times 10^{-3}$ g	4.5	3.5	4.0	4.0	3.6	2.8	3.7
Elongation at break, %	13	13	23	108	24	12	28
Breaking toughness, g/tex	0.80	0.45	0.73	1.09	0.44	0.40	0.64
Average stiffness g/tex	144	74	43	5	35	76	21

[a] Copolymerization reaction initiated by γ-radiation. PS, polystyrene; PAN, polyacrylonitrile; PVA, poly(vinyl acetate); PMMA, polymethyl methacrylate); CN–cotton, cyanoethylated cotton; DS, degree of substitution.

The effects of type of vinyl monomer on the properties of copolymer fabrics are shown in Table VII. Two different fabric constructions, print cloth and twill, were used. After the copolymer fabrics were prepared, they were given the usual treatment with dimethyloldihydroxyethylene urea to crosslink the cellulose. The wash-wear appearance ratings, flat abrasion resistances, and, in most cases, wrinkle-recovery angles of copolymer twill fabrics were improved over fabric controls. For copolymer

Table VII. Properties of Textile Fabrics Containing Cellulose Graft Copolymers

Grafted Polymer[a]	Polymer Add-on, %	DMDHEU[b] Add-on, %	Wash-Wear Rating after 30 Cycles	Flat Abrasion (Fill), Cycles	Wrinkle Recovery Angle, $(W + F)°$ Conditioned
Twill Fabric					
—	—	7	4.0	290	263
PAN	8	9	4.5	520	252
PMMA	14	9	4.5	400	286
PBMA	6	7	4.6	470	278
PLMA	4	3	4.2	400	279
Print Cloth Fabric					
—	—	6	4.4	110	274
PAN	9	7	4.4	90	293
PMMA	13	6	3.6	90	303
PBMA	4	9	4.1	120	311
Blended Twill Fabric					
PS–C[c]	6	9	4.7	—	275
PS–C[d]	6	11	4.6	—	248
PMMA–C[e]	11	9	4.3	—	270

[a] PAN, polyacrylonitrile; PMMA, poly(methyl methacrylate); PBMA, poly(butyl methacrylate); PLMA, poly(lauryl methacrylate); PS, polystyrene; C, cotton.
[b] DMDHEU, dimethyloldihydroxyethylene urea.
[c] Blend: cotton fibers (80 parts) + fibrous cotton–PS (29% add-on) copolymer (20 parts).
[d] Blend: cotton fibers (60 parts) + fibrous cotton–PS (15% add-on) copolymer (40 parts).
[e] Blend: cotton fibers (80 parts) + fibrous cotton–PMMA (53% add-on) copolymer (20 parts).

print cloth fabrics, wrinkle-recovery angles higher than those for copolymer twill fabrics were obtained. Three blended twill fabrics were woven from blends of fibrous cellulose copolymers and cotton fibers. Wash-wear rating appearances for these blended fabrics as high as 4.7 (5.0 is the maximum rating) were obtained (*18, 38, 39*).

Soil Release. The modification of the physical properties of cellulosic textile products, particularly by crosslinking or treatment with polymeric finishes, usually decreases the ease of soil removal from the textiles. As shown in Figure 5, the crosslinking of cotton print cloth with dimethyloldihydroxyethylene urea (DMDHEU), which increases its wash-wear rating appearance and wrinkle-recovery angle, decreases the ease of removal of both oily soil and aqueous soil. If softeners are added, the removal of the soil is decreased further (*32*).

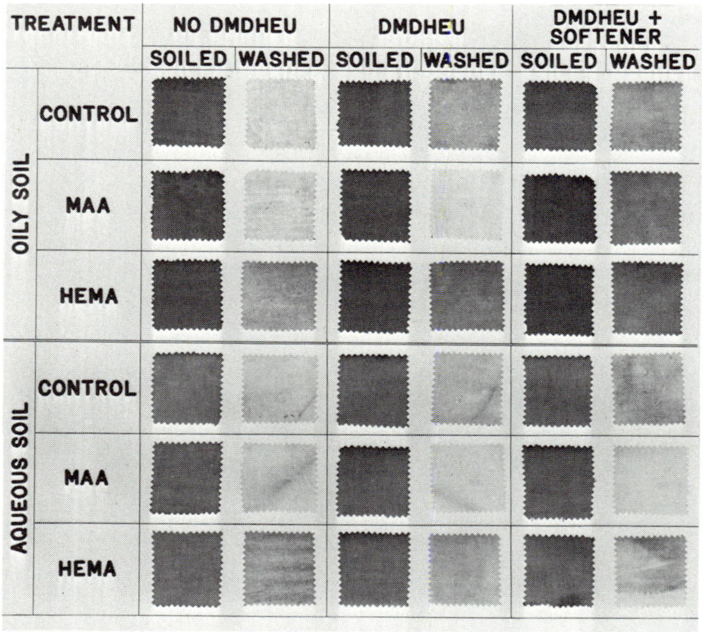

Figure 5. Soil release properties of cellulose and cellulose–polyvinyl copolymer fabrics before and after crosslinking

DMDHEU: dimethylol dihydroxyethylene urea
MAA: methacrylic acid
HEMA: hydroxyethyl methacrylate

The preparation of cellulose copolymers with vinyl monomers, which have functional groups that are hydrophilic in nature, has given textile products with improved soil release properties. For example, as shown in Figure 5, crosslinked cellulose–poly(methacrylic acid) copolymer fabrics exhibited both increased oily and aqueous soil release during washing compared with the crosslinked control fabrics. However, for oily soil, when softeners were added to the wash water, this improvement in soil release decreased. Crosslinked cellulose–poly(hydroxyethyl methacrylate) copolymer fabrics exhibited no improvement in oily soil release and only slight improvement in aqueous soil release compared

Table VIII. Rot Resistance of Cotton Cellulose–Polyvinyl Copolymer Yarns

Grafted Polymer[a]	Copolymerization Reaction Conditions[b]	Polymer Add-on, %	Strength Retained, %		
			Exposure in Soil Bed, weeks		
			1	4	8
Cotton control	—	—	0	0	0
PVA	15% VA in 70% aq. $ZnCl_2$	47	0	0	0
PMMA	15% MMA in 80% aq. $ZnCl_2$	54	6	0	0
PAN	15% AN in 70% aq. $ZnCl_2$	44	69	32	24
PAN	15% AN in 75% aq. $ZnCl_2$	57	67	28	16
PAN	15% AN in 80% aq. $ZnCl_2$	50	41	9	5
PAN	32% AN in 80% aq. $ZnCl_2$	14	98	37	21
PAN	32% AN in 80% aq. $ZnCl_2$	49	—	76	44
PAN	32% AN in 80% aq. $ZnCl_2$	69	93	85	87

[a] PVA, poly(vinyl acetate); PMMA, poly(methyl methacrylate); PAN, polyacrylonitrile.
[b] Copolymerization reaction initiated by γ-radiation.

Table IX. Weather Resistance of Cotton Cellulose–Polyvinyl Copolymer Yarns

Grafted Polymer[a]	Polymer Add-on, %	Strength Retained, %		
		Exposure in Weather, months		
		2	4	6
Cotton	—	69	53	45
PVA	56	78	59	46
PAN	58	67	53	40
PMMA	61	68	59	51

[a] Copolymerization reaction initiated by γ-radiation. PVA, poly(vinyl acetate); PAN, polyacrylonitrile; PMMA, poly(methyl methacrylate).

with crosslinked control fabrics. This area of investigation is being explored actively (*32*).

Rot and Weather Resistance. Cellulosic products are often used where degradation of the cellulose by microorganisms and by enzymes which they elaborate is an important factor. The rot resistance of sand bags is an example. Other uses of cellulosic products involve degradation initiated by light, microorganisms, and/or airborne pollutants, such as is observed in weathering. Some typical rot and weather resistances of cellulose copolymers are shown in Tables VIII and IX. Cellulose–polyacrylonitrile copolymers show increased rot resistance compared with cotton. Copolymerizing vinyl acetate or methyl methacrylate with cellulose did not significantly affect the rot resistance of the products (30).

Table X. Effects of Degree of Polymerization of the Irradiated Cellulose and Extent of Grafting on the Elastic Recovery Properties of Fibrous Cellulose–Poly(ethyl acrylate) at 25°C

Extent of Grafting, % add-on	Fraction of Extension Recovered on Relaxation, %				
	Extension, %				
	10	50	100	300	500
Cellulose DP 120					
—	51	—	—	—	—
290	55	—	—	—	—
500	60	55	—	—	—
Cellulose DP 100					
—	52	—	—	—	—
290	63	57	—	—	—
500	71	62	66	—	—
Cellulose DP 75					
—	50	—	—	—	—
150	57	—	—	—	—
300	80	75	80	89	89
510	81	76	80	89	89
720	84	81	88	95	95
900	87	82	87	94	95
Cellulose DP 40					
—	49	—	—	—	—
300	100	97	94	90	88
550	100	98	95	90	89
990	100	98	97	93	90

The weather resistances of the copolymers were about the same as that of cotton. The differences in mechanisms of the initiation of degradation of cellulose by rotting or by weathering probably accounts for these results (22, 23, 26, 41, 48, 49, 50, 51, 53). Apparently, under the conditions of weathering used, degradation of cellulose by microorganisms was not an important factor (30).

Elastomers. Ethyl acrylate has been copolymerized onto cellulose fibers, preirradiated by high energy radiation, to prepare elastomers (43,

44, 45, 46, 47). The elastomers exhibited rubber-like behavior. From an examination of electron photomicrographs of cross sections of the elastomers, the fibrillar structure of the cellulose fibers apparently formed a network, and poly(ethyl acrylate) was distributed uniformly among the fibrils. The rigid crystalline regions of the cellulose fibers apparently stabilized the amorphous, grafted poly(ethyl acrylate) to determine the mechanical properties of the elastomers (*43, 44*). For example, typical elastic recovery properties for these elastomers are shown in Table X.

Conclusions

Fibrous cellulose copolymers, prepared by free-radical initiated copolymerization reactions of vinyl monomers with cellulose, retain some of the properties of cellulose and have some new properties which increase the usefulness of cellulosic fibers as textile products. The morphology of the fibers is selectively and permanently changed; the rupturing of fibers during abrasion is much less in fibrous copolymers than in untreated fibers, thereby increasing the abrasion resistance of the products. The fibrous copolymers have second-order transition temperatures and, in some cases, are thermoelastic. At a high extent of grafting of polymer onto cellulose, elastomers are formed which have rubber-like elastic recovery properties. The surface properties of copolymers, by selecting vinyl monomers with hydrophilic types of functional groups, can be changed to increase the soil release properties of the products, as compared with untreated products. The rot resistances of copolymers, dependent on the type of vinyl monomer used, are higher than those of untreated celluloses. Crosslinked copolymer cellulosic fabrics have higher wash-wear appearance ratings than crosslinked cellulosic fabrics, which would influence consumer acceptance.

Literature Cited

(1) Arthur, J. C., Jr., ADVAN. CHEM. SER. **91**, 574 (1969).
(2) Arthur, J. C., Jr., *Advan. Macromol. Chem.* **II**, 1 (1970).
(3) Arthur, J. C., Jr., *Cotton Trade J., Intern. Ed.* **1964-65**, 41.
(4) Arthur, J. C., Jr., "Encyclopedia of Polymer Science and Technology," H. F. Mark, N. G. Gaylord, N. M. Bikales, Eds., Vol. 4, p. 244, Interscience, New York, 1966.
(5) Arthur, J. C., Jr., "Energetics and Mechanisms in Radiation Biology," G. O. Phillips, Ed., p. 153, Academic, London, 1968.
(6) Arthur, J. C., Jr., "Energy Transfer in Radiation Processes," G. O. Phillips, Ed., p. 29, Elsevier, Amsterdam, 1966.
(7) Arthur, J. C., Jr., *J. Macromol. Sci., Chem.* **A4**, 1052 (1970).
(8) Arthur, J. C., Jr., Baugh, P. J., Hinojosa, O., *J. Appl. Polymer Sci.* **10**, 1591 (1966).
(9) Arthur, J. C., Jr., Blouin, F. A., *Am. Dyestuff Reptr.* **51**, 1024 (1962).

(10) Arthur, J. C., Jr., Blouin, F. A., *J. Appl. Polymer Sci.* **8**, 2813 (1964).
(11) Arthur, J. C., Jr., Blouin, F. A., *U. S. At. Energy Doc.* **TID-7643**, 319 (1962).
(12) Arthur, J. C., Jr., Daigle, D. J., *Textile Res. J.* **34**, 653 (1964).
(13) Arthur, J. C., Jr., Demint, R. J., *Textile Res. J.* **30**, 505 (1960).
(14) *Ibid.*, **31**, 988 (1961).
(15) Arthur, J. C., Jr., Demint, R. J., U. S. Patent **3,109,798** (Nov. 5, 1963).
(16) *Ibid.*, **3,157,460** (Nov. 17, 1964).
(17) Arthur, J. C., Jr., Demint, R. J., McSherry, W. F., Jurgens, J. F., *Textile Res. J.* **29**, 759 (1959).
(18) Arthur, J. C., Jr., Harris, J. A., Mares, T., *Textile Indus.* **132**, 77 (1968).
(19) Arthur, J. C., Jr., Hinojosa, O., Bains, M. S., *J. Appl. Polymer Sci.* **12**, 1411 (1968).
(20) Arthur, J. C., Jr., Hinojosa, O., Tripp, V. W., *J. Appl. Polymer Sci.* **13**, 1497 (1969).
(21) Arthur, J. C., Jr., Grant, J. N., *Textile Res. J.* **36**, 934 (1966).
(22) Arthur, J. C., Jr., Mares, T., U. S. Patent **3,443,879** (May 13, 1969).
(23) Arthur, J. C., Jr., Mares, T., George, M., *Textile Res. J.* **35**, 1116 (1965).
(24) Arthur, J. C., Jr., Markezich, A. R., McSherry, W. F., *Textile Res. J.* **33**, 896 (1963).
(25) Baugh, P. J., Hinojosa, O., Arthur, J. C., Jr., *J. Appl. Polymer Sci.* **11**, 1139 (1967).
(26) Baugh, P. J., Hinojosa, O., Mares, T., Hoffman, M. J., Arthur, J. C., Jr., *Textile Res. J.* **37**, 942 (1967).
(27) Blouin, F. A., Arthur, J. C., Jr., *Polymer Preprints* **6**, 359 (1965).
(28) Blouin, F. A., Arthur, J. C., Jr., *Textile Res. J.* **33**, 727 (1963).
(29) Blouin, F. A., Cannizzaro, A. M., Arthur, J. C., Jr., Rollins, M. L., *Textile Res. J.* **38**, 811 (1968).
(30) Blouin, F. A., Morris, N. J., Arthur, J. C., Jr., *Textile Res. J.* **36**, 309 (1966).
(31) *Ibid.*, **38**, 710 (1968).
(32) Byrne, G. A., Arthur, J. C., Jr., unpublished data.
(33) Demint, R. J., Arthur, J. C., Jr., Markezich, A. R., McSherry, W. F., *Textile Res. J.* **32**, 918 (1962).
(34) Demint, R. J., Arthur, J. C., Jr., McSherry, W. F., *Textile Res. J.* **31**, 821 (1961).
(35) Gagnaire, D., Vincendon, M., *Bull. Soc. Chim. France* **1966**, 204.
(36) Goynes, W. R., *Proc. Engis Stereoscan Colloquim* **1969**, 149.
(37) Harris, J. A., Arthur, J. C., Jr., *J. Appl. Polymer Sci.*, in press.
(38) Harris, J. A., Arthur, J. C., Jr., *Textile Ind.* **134** (1970).
(39) Harris, J. A., Mares, T., Arthur, J. C., Jr., Hoffman, M. J., *Textile Ind.* **133**, 117 (1969).
(40) Mares, T., Arthur, J. C., Jr., *J. Polymer Sci.* **B7**, 419 (1969).
(41) Mares, T., Arthur, J. C., Jr., *Textile Res. J.* **39**, 303 (1969).
(42) Morris, N. J., Blouin, F. A., Arthur, J. C., Jr., *J. Appl. Polymer Sci.* **12**, 373 (1968).
(43) Nakamura, Y., Arthur, J. C., Jr., Negishi, M., Doi, K., Kageyama, E., Kudo, K., *J. Appl. Polymer Sci.* **14**, 929 (1970).
(44) Nakamura, Y., Arthur, J. C., Jr., Negishi, M., Doi, K., Kageyama, E., Kudo, K., *Polymer Preprints* **11**, 281 (1970).
(45) Nakamura, Y., Hinojosa, O., Arthur, J. C., Jr., *J. Appl. Polymer Sci.* **13**, 2633 (1969).
(46) *Ibid.*, **14**, 789 (1970).
(47) Nakamura, Y., Hinojosa, O., Arthur, J. C., Jr., *Polymer Preprints* **10**, 788 (1969).
(48) Phillips, G. O., Arthur, J. C., Jr., *Textile Res. J.* **34**, 497 (1964).
(49) *Ibid.*, p. 572.

(50) Phillips, G. O., Hinojosa, O., Arthur, J. C., Jr., Mares, T., *Textile Res. J.* **36,** 822 (1966).
(51) Reine, A. H., Arthur, J. C., Jr., *Textile Res. J.* **40,** 90 (1970).
(52) Rollins, M. L., Cannizzaro, A. M., Blouin, F. A., Arthur, J. C., Jr., *J. Appl. Polymer Sci.* **12,** 71 (1968).
(53) Sarkar, I. M., Arthur, J. C., Jr., George, M., *Textile Res. J.* **38,** 1145 (1968).
(54) Shirakashi, K., Ishikawa, K., Miyasaka, K., *Sen-i Gakkaishi* **19,** 177 (1963).
(55) *Ibid.*, p. 182.
(56) Sydykov, T. S., Shablygin, M. V., Livshits, R. M., Rogovin, Z. A., *Vysokomol. Soedin.* **8,** 2035 (1966).
(57) Toda, T., *J. Polymer Sci.* **58,** 411 (1962).

RECEIVED April 17, 1970.

22

Substrate Particle Size in ABS Graft Polymers

CHARLES F. PARSONS and EDMOND L. SUCK, JR.

Development Division, Chemicals and Plastics Group, Borg-Warner Corp., Washington, W. Va. 26181

> *Homogeneous particle size latices were prepared as substrates for ABS graft copolymers. Phase composition studies determined that the number of graft chains per unit area of substrate surface was essentially constant over the range of particle sizes and substrate concentration in the graft reaction. The average distance between graft chains was 51 ± 9 A. Notched Izod impact strength increased with increasing particle size when measured at a substrate level of 20 and 30 wt %.*

Dispersion of rubbery polymers in glassy resins produces materials which combine the useful properties of high impact resistance and high modulus. The early ideas of Merz, Claver, and Baer (4) have been regarded by most workers as the first attempt to explain the impact resistance of rubber-rigid resin composites. Bragaw has proposed that rubber particles in rigid polymers greatly increase the energy absorbing volume by causing cracks and/or crazes to branch dynamically at rubber sites (1). He also suggests that an optimum balance of impact strength and stiffness derives from regular particle size rubber particles. For ABS, this size is predicted to be greater than 5000 A. Although the experimental evidence presented by Bragaw is consistent with both predictions, ABS resins have been prepared using substrate particles with a mean diameter considerably less than 5000 A and impact strengths greater than predicted by theory. These ABS resins were prepared by grafting styrene–acrylonitrile (SAN) to a diene rubber latex. This paper reports results of a study of impact reinforcement by four grafted rubber substrates which had a narrow distribution of latex particle size.

A discussion of structure-to-properties relationships of graft ABS polymers by Frazer (3) emphasized the influence of substrate structure. At that time it was recognized that the average particle size of the substrate latex influenced the number of grafted chains and that relationships could be obtained between substrate particle size and graft structure with the mechanical properties of the graft polymer. Identical impact strengths were obtained for ABS graft polymers prepared from substrates of different average particle size when different conditions were imposed to alter grafting favorably. By making certain changes in grafting conditions (e.g., type of initiator) the average substrate particle size will not only influence impact resistance but also tensile strength, modulus, orientation, photoxidation resistance, creep, clarity, heat distortion, and flexural modulus.

As pointed out by Frazer, many workers have noted the interaction among particle size, grafting, and impact strength, but these findings have never been collated adequately. Impact strength of graft ABS polymers can be shown to increase with an increase in substrate level, or, depending upon the level of grafting, impact strength may increase with increased substrate particle population at constant substrate level. In other words, impact strength increases with increasing interfacial area. Dinges and Schuster (2) stated that the probability of a thrust of energy or a fracture crack hitting a substrate particle and the dependent notch impact strength should increase with the decreasing particle size of the grafted particles. Their initial experimental data showed exactly the opposite effect: notched Izod impact strength decreased with decreasing particle size in their particular ABS systems. Reasoning that their model must be incomplete, they investigated grafting variations and demonstrated an interaction of particle size and graft structure on impact strength which permitted them to conclude that their original model and mechanism was correct. However, to have the same quantity of substrate in the path of the fracture crack, the total level of the small particle size substrate had to be increased much higher than the large particle size substrate. In our opinion, the increase of rubber content of the small particle size substrate is misleading.

For our work we measured Izod impact strength of ABS resin blends at two levels (20 and 30%) of substrate and four substrate particle sizes. Graft structure was maintained so that the molecular weights of the graft and free copolymer were reasonably constant, and the area per grafted chain on the surface of the substrate particle was relatively equal regardless of substrate particle size. Consequently, at this particular graft level, impact strength should depend on substrate particle size and rubber level.

Experimental

Diene substrate latices were prepared by seeding techniques following the scheme shown in Figure 1. Persulfate anion was the initiator, and an alkyl aryl sulfonate was the emulsifier. Reaction time varied between 13 and 33 hours. Since all the substrates were prepared by similar techniques, the volumetric swell index and gel gravimetric data (toluene) varied little with particle size. Small differences in swell and gel characteristics should not be a variable in this study because of the polarity of the grafting monomer mixture and rapidity of the graft reactions.

Figures 2, 3, and 4 are electron photomicrographs of three of the substrate latices. Because of softening of the particles, it was difficult to obtain more representative pictures. The effects of the double seed can be observed in Figure 2.

Graft ABS polymers were prepared by the reaction of an approximate azeotropic mole ratio of styrene–acrylonitrile monomer mixture in the presence of different weights of diene substrate latex. Initiator, chain transfer agent, and soap levels were constant. Grafting reactions were

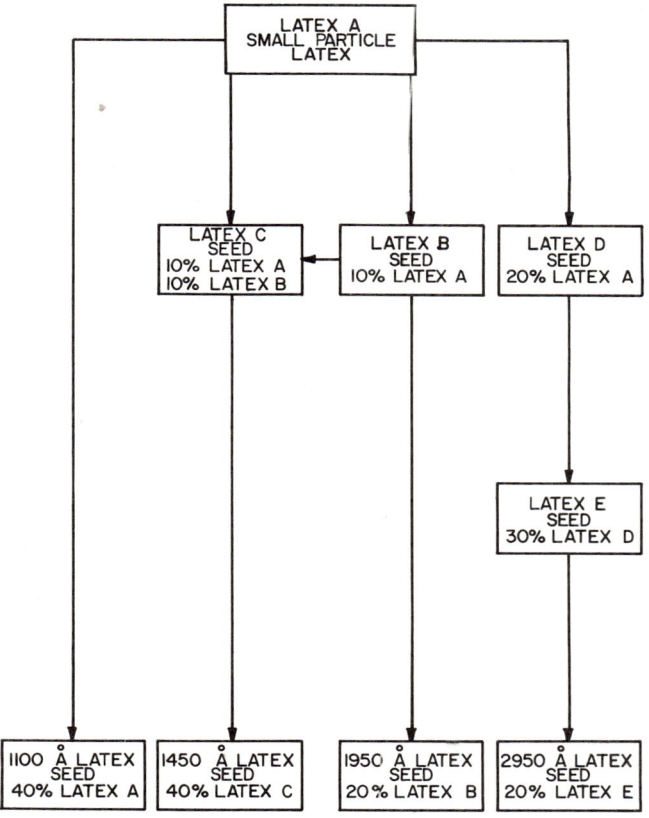

Figure 1. Preparation of substrate latices

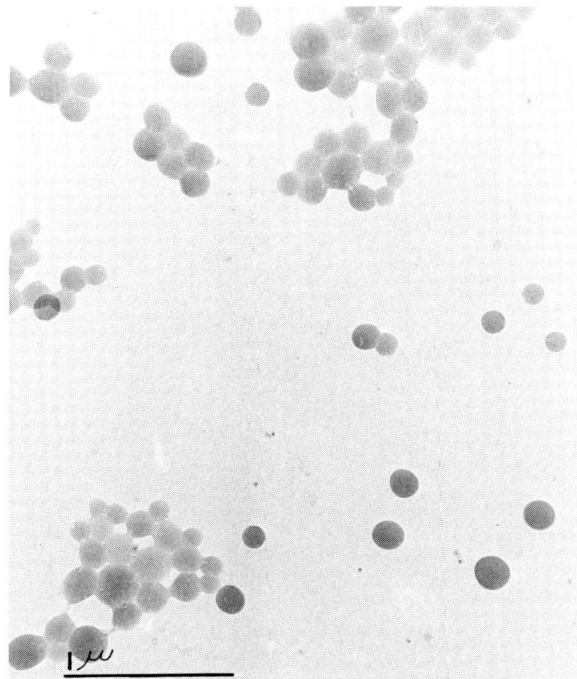

Figure 2. 1450 A substrate

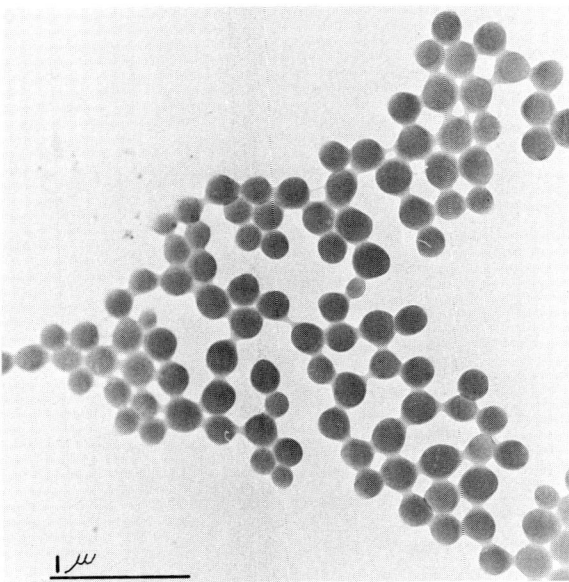

Figure 3. 1950 A substrate

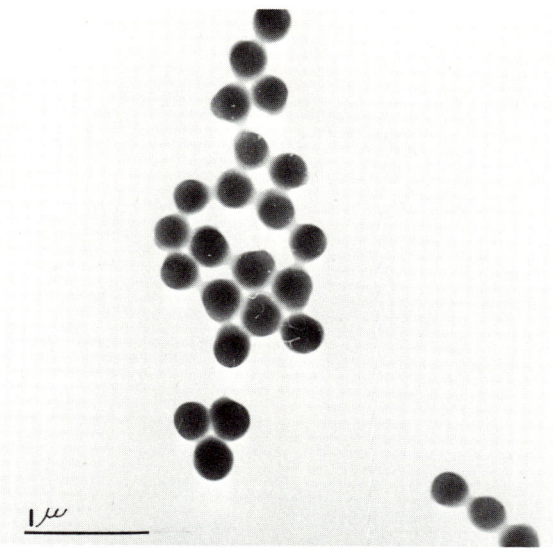

Figure 4.   2950 A substrate

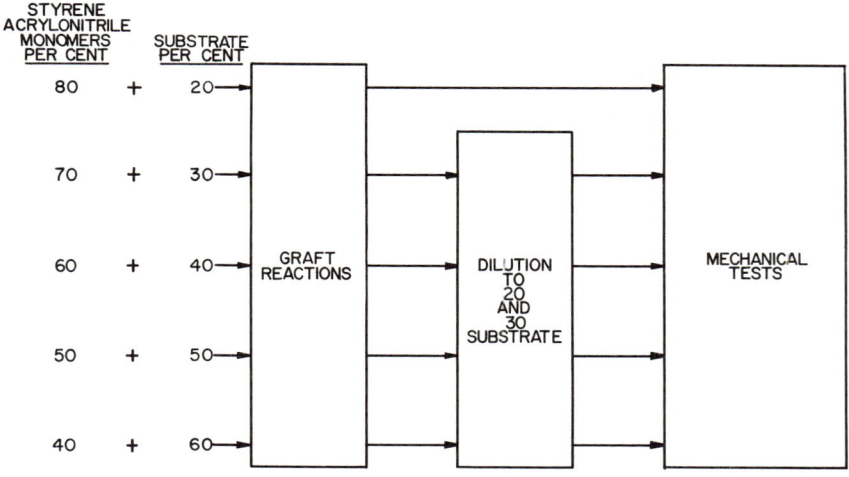

Figure 5.   Preparation of graft polymers

carried out by standard techniques of polymerization in the presence of substrate latex. Monomers, substrate latex, redox initiator system, etc. were charged to the reaction vessel. After a sufficient reaction time at 60°C, a portion of the graft latex was removed and treated with an antioxidant system. The latex was coagulated with calcium chloride solution. Resin powders were dried in an air-circulating oven at 50° to 60°C for approximately 24 hours. Samples for ABS phase composition studies were dried an additional 16 hours at reduced pressure. Procedures described by Moore et al. (5) were followed for phase separation and structure analysis.

The graft polymers were diluted with an SAN copolymer to 20 and 30 wt % substrate. The dry blends were mixed on a two-roll mill for a maximum of 5 minutes (165°–175°C) without added lubricants. Compression-molded samples from the milled slabs were evaluated for impact and tensile strength. Figure 5 is a schematization of this procedure.

Additional electron photomicrographic efforts have indicated no aggregation of substrate particles during grafting. Grafting apparently permits the substrate particles to retain their dimensions during milling and molding.

Discussion

Phase separation of the resins from the graft reactions gave material for determining molecular weight and the weight percent of the graft copolymer. The SAN number average molecular weights contained in Table I were obtained from viscosity measurements on the separated

Table I. Number Average Molecular Weight of Styrene–Acrylonitrile Copolymer, $\times\ 10^{-5}$

Substrate in Graft Reaction, Wt. %	Particle Size, A			
	1100	1450	1950	2950
20	—	—	1.00	1.20
30	—	1.10	0.93	2.00
40	—	1.20	1.20	1.15
50	1.10	1.10	0.76	1.47
60	—	1.15	1.10	1.20

Table II. Weight Percent of Grafted Styrene–Acrylonitrile Copolymer

Substrate in Graft Reaction, Wt. %	Particle Size, A			
	1100	1450	1950	2950
20	—	—	5.6	1.9
30	—	11.2	8.2	6.0
40	—	13.9	5.7	5.0
50	22.7	13.7	12.7	10.6
60	21.8	16.3	14.3	11.9

solubilized copolymer and previously prepared correlation curves. The equality of the molecular weight of the SAN in the graft phase and in the free solubilized copolymer has been shown by Dinges and Schuster and assumed in our study. The weight percent of SAN copolymer grafted to the diene substrate during each graft reaction is given in Table II.

Usually the ratio of the weight of grafted copolymer to the weight of diene substrate is calculated for each graft reaction. Variations of grafting will change this ratio and permit the tailoring of polymer systems. Variation of this ratio in these experiments was less than ±0.05 for each particle size and decreased as particle size increased. It was also evident (Table II) that the amount of copolymer grafted decreased as the particle size increased at any one particular substrate level. This is a further indication that grafting is proportional to available substrate surface area. To prove that uniform grafting had occurred, the relative substrate surface area per grafted copolymer chain was calculated for each graft reaction. The moles of grafted copolymer, Avogadro's number, the particles' diameter, and the weight and density of the diene substrate are the basis for the calculations. The density is important since actual surface area per grafted copolymer chain will depend on the number of particles per gram of rubber which depends on the estimation of substrate particle density. Accuracy of measuring the particle size and monomer swelling of the particle during the graft reaction may affect the density. We used 0.9 gram/cc for these calculations. Table III shows the range of areas

Table III. Average Substrate Surface Area Per Graft Chain, $A^2 \times 10^{-3}$

Substrate in Graft Reaction, Wt. %	Particle Size, A			
	1100	1450	1950	2950
20	—	—	2.02	4.74
30	—	2.25	1.93	3.74
40	—	2.63	4.77	3.44
50	2.41	3.05	1.70	2.59
60	—	3.23	2.62	2.26

Table IV. Average Distance Between Graft Chains, A

Substrate in Graft Reaction, Wt. %	Particle Size, A			
	1100	1450	1950	2950
20	—	—	45	69
30	—	47	44	61
40	—	51	69	59
50	49	55	41	51
60	—	57	51	48

per chain as determined from the molecular weight and weight of graft copolymer given in Tables I and II. The square root of these areas or the average distance between chains for each graft reaction is given in Table IV. Of the examples presented, three sets of experiments definitely are outside the regular order of data. Two are attributed to the low weight of grafted copolymer (2950 A particle size at 20 wt % substrate in the graft reaction and 1950 A particle size at 40 wt % substrate in the graft reaction), the third graft polymer had a much lower molecular weight graft chain than any of the other graft reactions (1950 A particle size at 50 wt % substrate in the graft reaction). The first two had a mean distance of 69 A between graft chains while the remainder average 51 ± 9 A. This value approaches the limits of experimental accuracy at the present state of technique and equipment. Within limits of experimental measurements of impact and tensile strength at two substrate levels (Tables V and VI), this variation in interchain distance is too small to affect significantly impact or tensile strength when the number average molecular weight of the grafted chain is approximately 100,000. The present study was not undertaken to determine any preferred molecular weight, distance per graft chain, optimum impact strength, or combi-

Table V. Notched Izod Impact Strength (ft-lbs/inch) of Polyblends at 20 and 30% Substrate Levels

Substrate in Graft Reaction, Wt. %	Particle Size, A			
	1100	1450	1950	2950
	20% Substrate Level			
20	—	—	3.3	4.1
30	—	3.0	3.2	4.6
40	—	2.4	2.8	4.2
50	1.8	2.4	2.2	4.0
60	1.4	2.4	2.8	5.4
	30% Substrate Level			
30	—	6.8	6.2	7.5
40	—	6.5	6.5	7.5
50	4.5	6.0	6.7	7.0
60	5.3	6.3	6.1	7.5

Table VI. Tensile Strengths (psig) of Polyblends

Substrate in Graft Reaction, Wt. %	Particle Size, A			
	1100	1450	1950	2950
	20% Substrate Level			
20	—	—	6200	6400
30	—	6075	6225	6300
40	—	6300	6350	6225
50	6800	6325	6500	6325
60	6600	6400	6325	5975
	30% Substrate Level			
30	—	4375	4500	4525
40	—	4450	4600	4575
50	5000	4400	4325	4600
60	4725	4325	4550	4400

nations of specific properties. Both smaller and larger areas per grafted chain and different molecular weights of grafted copolymer can be achieved. Additional effort is required to determine if grafting can be controlled within even narrower limits over a similar range of particle sizes and a wider range of molecular weights.

It is our belief that Table III or IV should be the preferred method to describe graft structures when molecular weight and particle size data are available.

Impact Strength. Figure 6 shows the dependence of the average Izod impact strength on substrate particle size at 20 and 30 wt % substrate of the polyblends. At the 20 wt % rubber substrate level the extra-

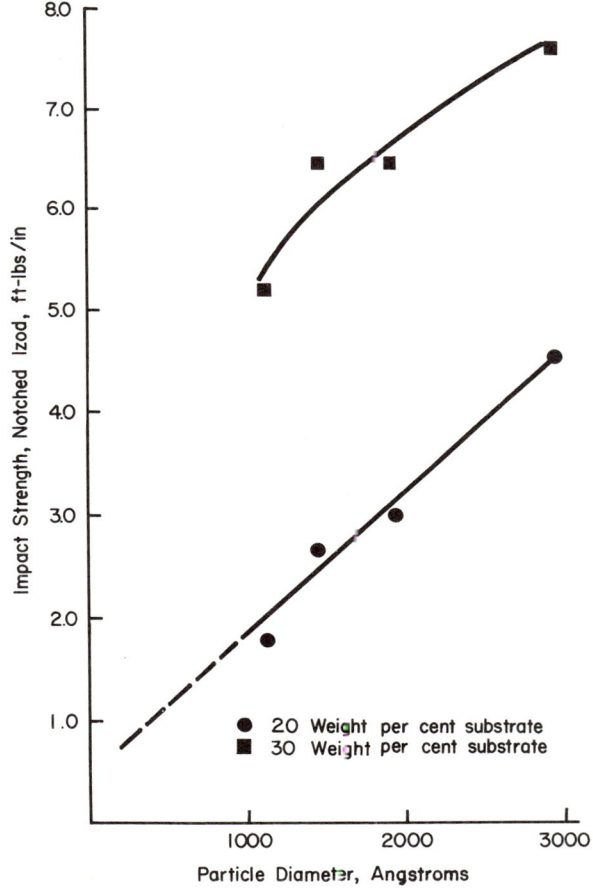

Figure 6. Notched Izod impact strength vs. particle diameter

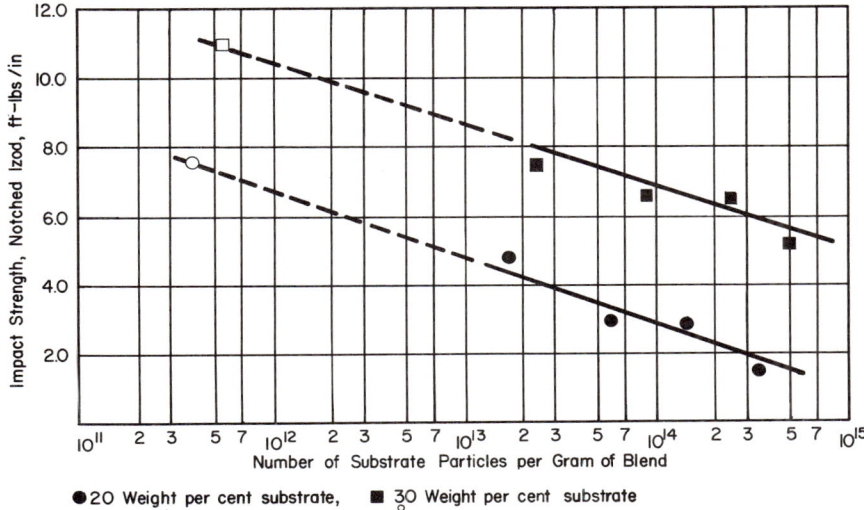

● 20 Weight per cent substrate, ■ 30 Weight per cent substrate
○ and □ Calculated number of 5000 Å particles.

Figure 7. Notched Izod impact strength vs. number of substrate particles per gram of blended resin

polated intercept of 0.4 ft-lb/inch is the equivalent of the impact strength of unmodified styrene–acrylonitrile copolymer. At 30 wt % of rubber substrate the increase of notched Izod impact strength with increasing substrate particle size is less linear at these individual particle sizes. Extrapolation of the 30% impact values to the copolymer impact value requires a substantial increase in impact strength when substrate particles are less than 1000 A in diameter. The shape of the impact–particle size curve in this region is subject to conjecture and may depend heavily on graft structure.

For a prediction of the impact strength at a particle diameter of 5000 A (Bragaw's hypothesis) at this level of grafting and molecular weight, impact strength was plotted *vs.* the number of substrate particles in the polyblend. Figure 7 is a semilogarithmic plot of this data. Notched Izod impact strengths of 9.5 and 11.0 ft-lb/inch for 20 and 30 wt % substrate are predicted. The level of grafting (area or molecular weight per grafted chain) could change the location or slope of these lines. As for lower rubber substrate levels, both Figures 6 and 7 suggest that large particle size substrates are preferred.

Tensile Strengths. Table VI contains the tensile strengths of the polyblends corresponding to the impact values of Table V. Fluctuations are small since the weight ratio of grafted copolymer to substrate is reasonably constant. Tensile strengths of the 1100 A substrate polyblends were higher than any of the other three particle sizes. These observations are consistent with the inverse relationship of impact and tensile strength.

Conclusions

Impact strength of ABS graft resins will increase with increasing particle size of the substrate latex when the substrate surface area per grafted copolymer chain is maintained at equivalent values. At 20 wt % substrate levels the increase in impact strength appears linear with increasing particle size, but at 30 wt % substrate, nonlinearity is indicated when the substrate particle is less than 1100 A. The inverse relationship of tensile and impact strength is preserved.

Acknowledgments

We are pleased to recognize the technical assistance of D. O. Conley. Helpful discussions were held with W. J. Frazer, V. E. Malpass, P. J. Fenelon, P. L. Wineman, and E. Baer. We thank E. Lanterman and M. Draginis of the Roy C. Ingersoll Research Center, Borg-Warner Corp. for the electron photomicrographs.

Literature Cited

(1) Bragaw, C. E., ADVAN. CHEM. SER. 99, 86 (1971).
(2) Dinges, E., Schuster, H., *Makromol. Chem.* 101, 200 (1967).
(3) Frazer, W. J., *Chem. Ind.* 1399 (1966).
(4) Merz, E. H., Claver, G. C., Baer, M., *J. Polymer Sci.* 22, 325 (1956).
(5) Moore, L. D., Moyer, W. W., Frazer, W J., *Appl. Polymer Symp.* 7, 67 (1968).

RECEIVED June 17, 1970.

23

Kinetics of Aggregation and Dimensions of Supramolecular Structure in Noncrystalline Block Copolymers

M. HOFFMANN, G. KAMPF, H. KROMER, and G. PAMPUS

Farbenfabriken Bayer AG, 509 Leverkusen, Germany

Styrene–butadiene copolymers were investigated for relationships between molecular structure, phase separation, and microheterogeneous structure. Phase separation starts at slightly greater concentrations than in mixtures of homopolymers. During solvent evaporation, aggregates shrink to unfavorable dimensions. By annealing above their T_g, they become more uniform in size and shape, achieve sharper boundaries, and rearrange to well ordered arrays. In solutions containing <50% copolymer, phase separation is usually complete within seconds, but undiluted copolymers must be annealed several hours at T_g of the aggregates to obtain an equilibrium supramolecular structure. Size distribution, shape, and long range order of the styrene domains depend on molecular structure and uniformity. The diameter of the domains depends on domain shape, coil sizes of the sequences, and molecular architecture. The shapes and sizes of the domains and their short range order are calculated from the molecular structure.

Quantitative relationships between molecular structure and the properties of block copolymers can be gained only by considering the supramolecular structure. A highly regular molecular structure is necessary for forming a precise long range supramolecular order. Therefore we used anionically polymerized block copolymers of butadiene and styrene as model substances for noncrystalline block polymers in general. Such polymers can be prepared with uniform structure (33).

Methods for proving and characterizing the aggregation of similar sequences are treated first. The critical concentrations of phase separa-

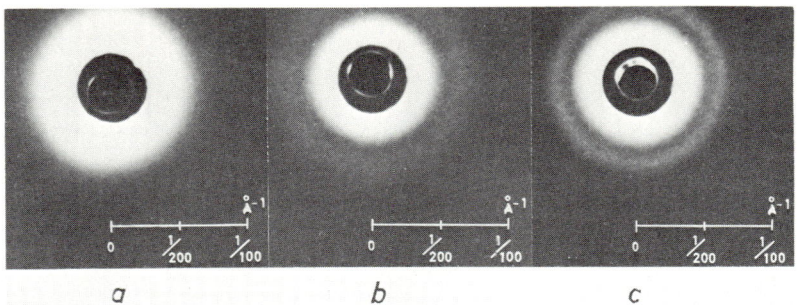

Figure 1. Small angle x-ray scattering diagrams
(a) P 45, from CH_2Cl_2 + CH_3COCH_3, quickly evaporated by injection into a vacuum
(b) P 45, solution evaporated at room temperature, polymer annealed for 1 hour at 100°C
(c) P 45, solution evaporated at 120°C, polymer annealed for 1 hour at 120°C
The polymer contains 41% styrene blocks.

tion in block polymers are compared with those in polymer mixtures. Further experiments concern the velocity of aggregation and reaggregation. Depending on the kinetics of aggregation, quite different degrees of perfection of the microheterogeneous structure are obtained. Finally, quantitative and general relationships between form and size of the aggregates and the molecular structure are derived by considering the change in free energy caused by close packing of the aggregates. The results agree with experiments.

Proof and Characterization of Aggregation of Like Blocks

By aggregate we mean a domain in which the concentration of one kind of block is considerably greater than average. We do not, however, mean statistical variations of concentration and anomalously high concentrations in isolated coils. Suitable methods for proving the existence of such aggregates have been reviewed (20).

X-ray small-angle scattering (29) is a method generally applicable to the study of microheterogeneous structure and has been used to investigate block copolymers (25, 66). Figure 1 shows some typical scattering diagrams. Compared with electron microscopy and dynamical mechanical measurements, this method has the advantage that it can also be used for polymers with soft aggregates, liquids, and solutions. It yields mean values (29) of the dimensions and distances of many aggregates and thus saves the counting and the statistical evaluation customary with electron micrographs. In principle it is possible to determine the particle size distribution and the lack of phase boundary sharpness by this method. The quantitative evaluation, however, requires assumptions regarding the form of the aggregates and is based on the comparison of experimental scattering diagrams with calculated scattering curves of models. The scattering curves of many simple geometric models (sphere, cylinder, ellipsoid, etc.) have been calculated (*40*). In most cases an unambiguous determination of the shape of the particles and of the average size will be possible only if further experimental results, primarily the volume fraction ϕ of the particles, are combined with the x-ray diffraction data.

Electron microscopy (*59, 60*) is not as generally applicable because sample preparation is sometimes difficult. For the butadiene–styrene copolymers used here, contrasting of the butadiene phase with OsO_4 is particularly suitable (*41*). Thin films, several tenths of a micron thick, are cast from dilute solutions and placed in the vapor of an aqueous OsO_4 solution. A similar process is followed for thin sections of macroscopic samples. If the morphological structure of these films in planes normal to the film surface is of interest, such films must be embedded before cutting in a material of similar hardness which does not swell or dissolve the sample to be investigated.

Figure 2 shows that the different methods of preparation give the same results. The butadiene phase is dark, the styrene phase pale.

To what extent are the ordered structures observable in the electron microscopic photographs of the polymer films cast from solution influenced by the substrate? Any significant influence can be excluded for the following reasons:

(a) The samples cast on glass plates and evaporated under solvent atmosphere or nitrogen show, without annealing, a poorly developed supramolecular structure. A far-reaching order can be observed after annealing (for example at 100°C), irrespective of whether the films had been annealed on the glass plates or without any substrate. The same structures were also found when the polymer solutions were cast on gelatin film and the floated-off film was subsequently annealed.

(b) The same mean long periods were measured on the annealed films described under (a) by small-angle electron diffraction (*43*) as with small-angle diffraction of x-rays on macroscopic samples.

(c) Thin sections were prepared from macroscopic samples by a special deep-freeze microtome operated at −186°C which showed the same ordered structures as stated under (a).

In transmission electron micrographs all the particles situated at various depths in the specimen appear to be projected into the image plane because the depth of focus is considerably greater than the thickness of the specimen. This may alter the interpretation. It is possible, however, by short treatment with OsO_4 to limit the contrasting to the layers near the surface (Figure 7c), or very thin films can be prepared so that all easily visible aggregates lie in one plane. The mean aggregate

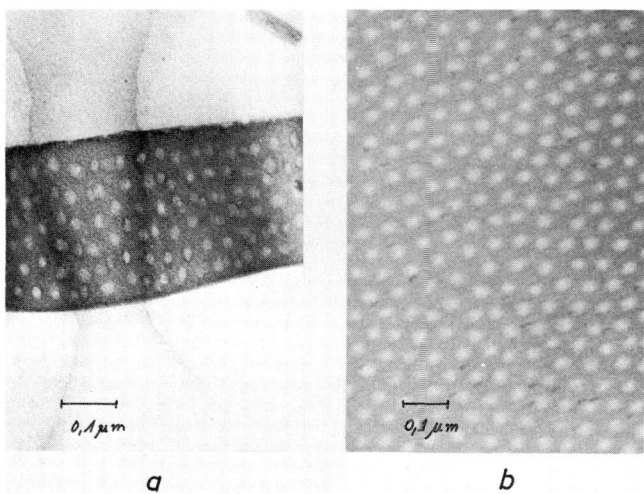

Figure 2. Electron micrographs of a mixture (1:1) of P 42 and P 5, stained with OsO_4 in the vapor above a 0.1% solution of OsO_4 for 6 hours at 25°C

(a) From benzene solution (0.5%), evaporated at room temperature, polymer annealed 1 hour at 120°C
(b) Thin section of the same polymer after the same annealing and contrasting procedure
Polymer P 42 contains 15% styrene blocks, while P 5 is a polybutadiene.

diameters D and the distances A can then be read from the electron micrographs, and it is possible to calculate from these values the volume fraction ϕ of the aggregates, provided the shape and the geometrical arrangement of the particles are known. We assume that the centers of the spheres are arranged in the same way as in a cubic or hexagonal close-packed structure, without the spheres touching one another. Analogously, the cylinders form a lattice of the symmetry of the densest (hexagonal) cylinder packing. Lamellar aggregates are assumed to be plane parallel and infinitely extended.

Spheres: $$\varphi_K = 0.74 \cdot \left(\frac{D_K}{A_K}\right)^3 \qquad (1)$$

Cylindrical rods: $$\varphi_Z = 0.91 \cdot \left(\frac{D_Z}{A_Z}\right)^2 \qquad (2)$$

Layers: $$\varphi_S = D_S/A_S \qquad (3)$$

Figure 2a shows aggregates situated at various depths, so that the styrene content cannot be evaluated according to Equation 1. Figure 2b supplies the following values: $D_K = 310$ A; $A_K = 570$ A; $\phi_K = 0.12$. The volume fraction of the styrene blocks, determined by chemical analysis, is 9%, in satisfactory agreement with ϕ_K. Figure 6a gives $\phi_Z = 0.17$ for the cylinders, which corresponds to the experimental ϕ_Z. (The cylinders lie parallel to the surface in thin films so that their diameters can be measured readily.) However, contrasting can increase the volume of the butadiene phase by up to 15% (42), and the actual shapes need not correspond precisely to the models. The x-ray data (Table IV) are more reliable in this respect and show that the observed ϕ_{agg} is always nearly equal to the value determined by chemical analysis. More precise information on the shape of the aggregates and the lack of sharpness of the phase boundaries was obtained by photometric evaluation of the electron micrographs (43).

Like x-ray scattering, light scattering (5, 28, 62) or turbidity are also generally suitable for proving the existence of aggregation. If, however, the system to be investigated consists of closely and regularly packed aggregates (e.g., undiluted block copolymers or concentrated solutions of them), the light scattered by the particles is attenuated strongly by destructive interference (33, 58, 61). Light scattering is a fast and sensitive method which may be used for specimens containing very much diluent. We have found with this method that the diameters of the aggregates do not change substantially when block polymer P 42 is blended with very much polybutadiene P 5 (33). On the other hand, the aggregation of P 42 in benzene solutions (0–50%) could not be recognized at 25°C by the scattering intensity of the polarized light, whereas the concentration dependence of depolarization showed a discontinuity at the critical concentration.

If the softening temperatures of the kinds of blocks differ by more than 40°C, aggregation can be recognized by the appearance of two ranges of softening in the temperature dependence curves of the relaxation modulus (1, 10, 11, 12, 13, 45, 48, 55, 71), the relaxation velocity (4, 45, 49, 67, 68), or the melt viscosity (33). If the variation of viscosity with concentration is measured at a temperature between the two softening ranges, an anomalous concentration dependence is obtained as a consequence of the aggregation starting at a certain concentration, provided the matrix of the dispersion is formed by the softer polymer and its share of solvent (33) (Figure 3).

When there is no aggregation, the concentration dependence of the viscosity does not vary much with styrene content and block length. It corresponds somewhat to that of polybutadiene P 5.

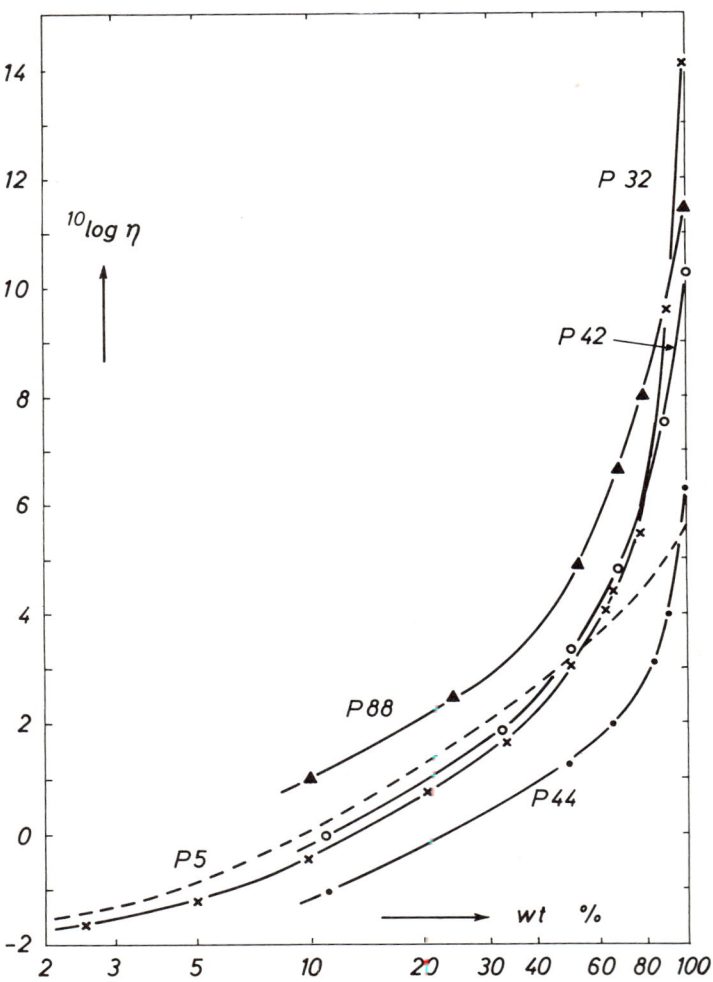

Figure 3. Concentration dependence of viscosity for benzene solutions of some block copolymers and one polybutadiene (P 5) at 25°C. Viscometers with long capillaries were used and viscosities were extrapolated to zero shear gradient (34).

P 5, 0% styrene blocks, M = 67000
P 88, 7.5% styrene blocks, M = 159000
P 42, 15% styrene blocks, M = 83000
P 44, 19% styrene blocks, M = 23500
P 32, 26% styrene blocks, M = 70000

Causes of Aggregation of Like Blocks and Rules of Phase Separation

Aggregation is a consequence of the usual incompatibility of the two kinds of blocks. Incompatibility of different polymers has been the subject of thermodynamic considerations (24, 36). The theory of phase separation in polymer mixtures which may contain solvent is confirmed somewhat by experiments. Therefore, we have experimentally examined the theory (23, 37, 38, 46, 52) developed for block polymers and have tried to clarify the problem as to what extent experiments concerning phase separation in polymer mixtures allow valid conclusions on aggregation in block polymers. We mixed polystyrenes of molecular weights corresponding to those of the styrene blocks (33) with polybutadiene P 87 ($M = 39000$) so that the styrene content was 17 or 29%, respectively. At 25°C, solutions of these mixtures in benzene are clear, up to weight fractions g_{25} of the polymer mixture (Table I), and they are turbid at higher concentrations; g_{25} depends on the quantity and the molecular weight of the polystyrene.

Table I. Phase Separation Studies

Polystyrene, MW	Polystyrene, wt %	g_{25}	$(dg/dT)_{25} \times 10^2$
2000	17	1.00	—
5500	17	0.390	0.155
	29	0.343	
12000	17	0.236	0.100
	29	0.195	
32000	17	0.139	0.063
	29	0.131	

In Figure 4, g_{25} values are plotted against the molecular weight of the polystyrene and compared with the critical concentrations at which aggregation occurs in solutions of block copolymers according to Figure 3 or according to small-angle scattering of x-rays and depolarization of the light scattering.

In contrast to some theoretical predictions (23, 46, 52) aggregation or phase separation in block copolymers occurs at a slightly higher total concentration than in polymer mixtures. Covalent bonding of the two kinds of blocks thus slightly increases the mutual solubility. Polystyrene blocks with $M = 2000$ dissolve in polybutadiene ($M = 75000$) up to concentrations of about 20% (33). Therefore, special mechanical properties are, in general, only to be expected in sequence copolymers above a certain block length (in most cases $M_s > 10^3$).

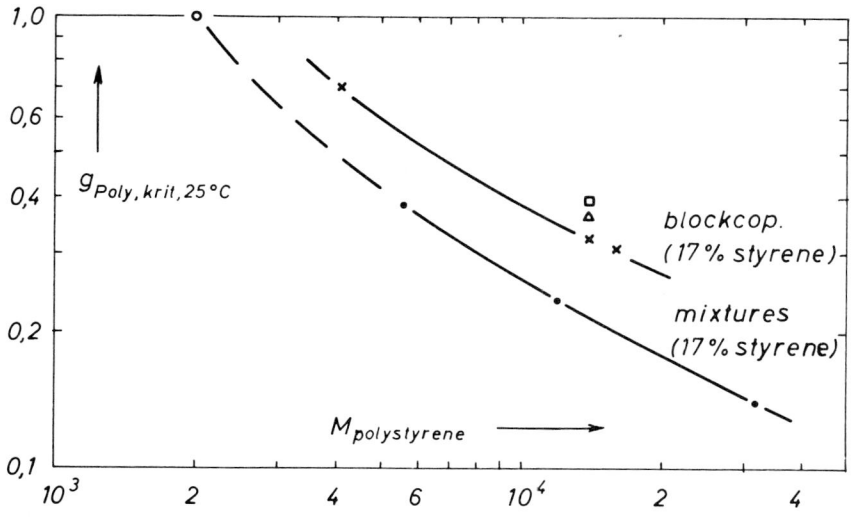

Figure 4. Critical concentrations, g_{25}, of phase separation for block copolymers and polymer mixtures in benzene solutions at 25°C as functions of the polystyrene molecular weight

Detection methods:
- □ x-ray small-angle scattering
- △ depolarization of scattered light
- × viscosity
- ● macroscopic phase separation

Further details of the phase separation of polymer mixtures were obtained by analyzing the high styrene phase sedimented at a weight concentration g_B of benzene (Table II).

In accordance with general experience with immiscible polymers, the styrene phase contains very little polybutadiene. In the same manner, only very little polystyrene can then be dissolved in the polybutadiene phase. The aggregates of the block polymers are certainly also free

Table II. Analysis of Styrene Phase

Polystyrene MW	Polystyrene in Polymer wt %	g_B, %	Polybutadiene in Swollen Precipitate, %	Benzene in Swollen Precipitate wt %	Precipitate vol %
5500	29	45		43	22
12000	29	62	0.13	59	25
32000	29	78	0.17	68	20

of the other kind of blocks. Thus, for example, styrene aggregates cannot be stained with OsO_4, and calculations of ϕ_{agg} (*see* Figure 2) yield approximately the analytical quantity of styrene blocks—*i.e.*, between $\phi = \phi_S$ and $\phi = 1.05 \phi_S$. Finally, the aggregates in the presence of benzene, as well as the precipitate of Table II, will probably contain benzene.

If solvents are used which do not possess a high dissolving power for both kinds of blocks (high second virial coefficients of osmotic pressure), phase separation occurs at considerably lower concentrations, and the solvent content of the aggregates is lower than that of the matrix.

Kinetics of Aggregation

The speed of aggregation depends on the solvent. If it is a precipitant for one sort of block and if this block comprises a sufficiently large part of the molecule, aggregates of this kind of block are formed even in dilute solutions. P 42 in *n*-hexane at 60°C, for instance, forms genuine solutions with a viscosity of 0.56 dl/gram and a light scattering value which yields the correct molecular weight ($M_w = 82000$). If such solutions ($c = 0.3$ gram/liter) are cooled quickly, a strong turbidity occurs within tenths of a second and remains constant afterwards, giving a molecular weight of 2.55×10^6. The osmometric molecular weight is 160,000 and the intrinsic viscosity 0.53 dl/gram. The aggregates thus form very quickly in dilute solutions and have a broad size distribution. Light scattering shows that their sphere diameter at $c = 0.3$ gram/liter is 720 A, which corresponds well to double the thickness of a single molecular coil in hexane ($h_B = 280$ A, h_S 85 A).

When the concentrations are smaller, the capability of forming aggregates decreases; for example at $c = 0.1$ gram/liter to $M_w = 1.7 \times 10^6$ and at $c = 0.03$ gram/liter to $M_w = 0.5 \times 10^6$. These degrees of aggregation remain constant for several days and seem to be equilibrium values. In better solvents—*e.g.*, in cyclohexane at 25°C—there is no aggregation, but only a decrease in the interaction between the polymer and the solvent, osmotically measured as $d(p_{osm}/c)/dc$, with increasing styrene content. The interaction decreases, as expected, to a greater extent than corresponds to the styrene content and has dropped, for 30% styrene, to half the original value. It is possible in such or better solvents to produce moderately concentrated solutions, without any aggregation (*see* Figure 3). In such cases only the critical concentration (*see* Figure 4) and the rate of evaporation or cooling determine the aggregation. Molecular mobility does not play a major role for we found in experiments with polymer mixtures that turbidity occurred on cooling even at high concentrations (at least 50%) within seconds. Such speeds of aggregation are understandable. For example, molecules with $M \sim 10^4$, which means $D \cdot \eta_1 = 10^{-8}$ (35), still diffuse over distances of 300 A even in a medium of $\eta_1 \approx 10^3$ poise in about 0.5 second. The first stage of aggregation will be reached in most cases very quickly, at speeds approximately proportional to the solvent viscosity.

Reaggregation, however, occurs much slower, primarily in undiluted polymers and at high polymer concentrations, because the styrene content of the aggregates elevates the transition temperature and thus reduces the chain mobility. The reaggregation slows down as it progresses.

The two-block polymer P 45 (about 40% styrene) was annealed, starting from a very poor superstructure (see Figure 1a), until the outer ring, shown in Figure 1b and which is caused by the particle form factor, just became visually recognizable. The annealing time necessary for obtaining this stage of aggregation depends strongly on temperature:

T, °C	t, min.
110	5
100	25
90	180
80	1800
70	36000

The temperature dependence of the velocity corresponds to the temperature dependence of the viscosity of polystyrene near the glass transition temperature.

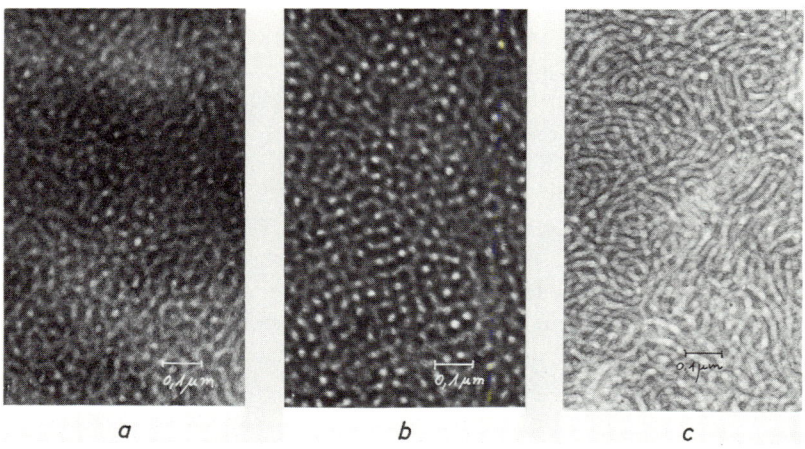

a b c

Figure 5. Electron micrographs of films of block copolymers which were cast from aliphatic hydrocarbons ($C_5 + C_6$) at 25°C and contrasted at 25°C for 1 hour with OsO_4 vapor

(a) P 41 (17% styrene blocks)
(b) P 32 (26% styrene blocks)
(c) P 32, specimen prepared by slow evaporation of a benzene solution

The following equation holds approximately:

$$\ln t(\min) = -15.7 + \frac{1950}{T\ (°C) + 5} \quad (4)$$

The value of the glass transition temperature thus determined is about 50°C lower than the value observed in macroscopic specimens of styrene homopolymer of molecular weight equal to that of the styrene blocks.

Electron micrographs of the same polymer show more or less pronounced supramolecular structures depending on the preparation, the nature of the solvent, and the rate of evaporation. Fast evaporation from dilute solutions at temperatures far below the softening temperature of the aggregates results in a supramolecular structure we shall call the starting structure. Irregular aggregates with diffuse boundaries and broad distributions of the shapes, sizes, and distances are characteristic features of the starting structure (Figure 5). If these samples are annealed at temperatures around 100°C, a supramolecular structure is obtained after some hours which is much more regular than the starting structure. It no longer changes on continued annealing, and we call it therefore the end structure. Figure 6 shows some examples.

During annealing and reaggregation, structures can be observed which are still faulty (Figure 7). These faults, which also include diffuse phase boundaries, are richer in free energy than their surroundings and facilitate reaggregation. According to Figure 7b it is estimated that in annealed samples the diffuse interface of the aggregates is not thicker than about 10–20 A. It is remarkable that the degree of perfection of the end structure of an annealed block copolymer depends on the quality of the starting structure and thus, for example, on the solvent (3, 17, 18, 30, 53, 70). If solutions of methylene chloride and acetone are sprayed into a vacuum, even intensive annealing is far from resulting in an end structure which is as good as that obtained when the polymer solution (xylene) is evaporated slowly at elevated temperatures (120°C). Thus, after a certain time of annealing, reaggregation comes to a standstill, independent of the degree of order that has been reached. Obviously, the displacement of butadiene chains out of the styrene aggregates takes place more quickly than the building up of long range order. Because of the gradually increasing inhibition of chain mobility, the well ordered equilibrium structure cannot always be reached. However, the mean dimensions of the supramolecular structure after tempering do not depend on the nature of the solvent or the evaporation technique and thus are typical for each polymer. This is contrary to some assumptions in the literature (37, 38). Literature data (2, 3, 4, 21, 30) on the kinetics of reaggregation are in satisfactory agreement with our experiments. The regeneration of the mechanical properties of strongly deformed SBS

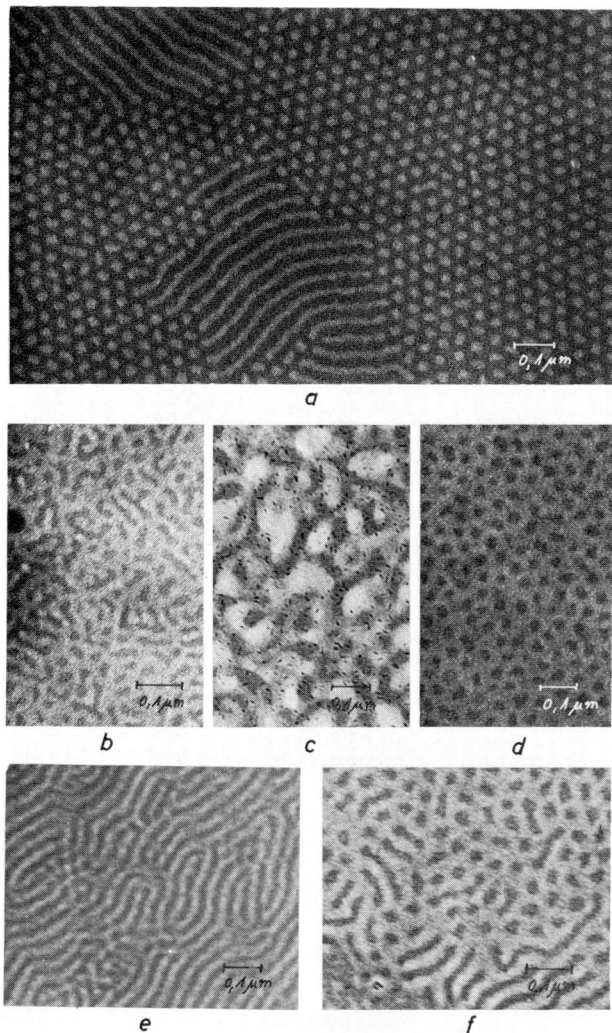

Figure 6. Electron micrographs of films of some block copolymers cast from various solvents and contrasted with OsO_4 vapor after annealing

(a) P 41, 17% styrene blocks, BS, 1 hour 100°C, hydrocarbon
(b) P 41, 17% styrene blocks, BS, 1 hour 110°C, methyl ethyl ketone
(c) P 32, 26% styrene blocks, BS, 1 hour 120°C, benzene
(d) P 46, 30% styrene blocks, BS, 1 hour 100°C, benzene
(e) P 53, 22% styrene blocks, S/BS, 6 hours 100°C, benzene
(f) P 54, 32% styrene blocks, S/BS, 1 hour 100°C, benzene

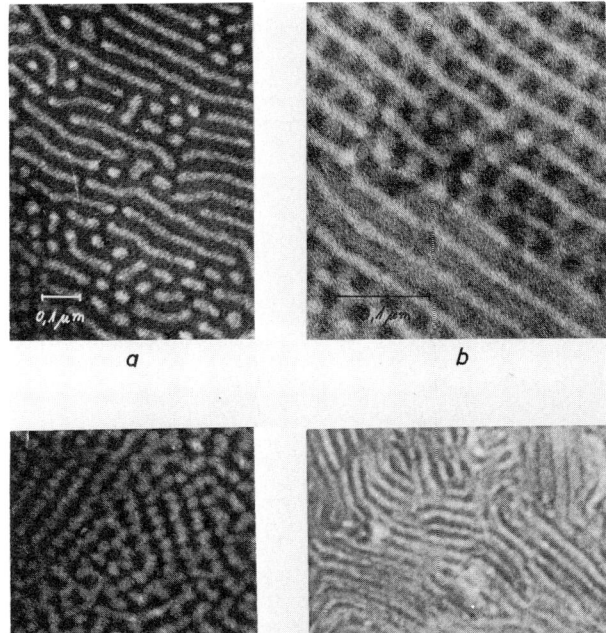

Figure 7. Electron micrographs of intermediate stages of supramolecular order

(a) P 42, 15% styrene blocks, BS, 1 hour 100°C, hydrocarbon, short rods and thicker spheres
(b) P 41, 17% styrene blocks, BS, 1 hour 120°C, hydrocarbon, different ways of packing cylinders
(c) P 42, 15% styrene blocks, BS, 1 hour 100°C, hydrocarbon, spheres arranged along lines
(d) P 32, 26% styrene blocks, BS, 25°C, benzene, branched thread-like aggregates which tend to form layers

polymers at room temperature will probably, however, also depend on other mechanisms (recombination of radicals).

Aggregate Shapes and Long-Range Order of the Aggregates

According to our observations and to literature data (4, 19, 22, 24, 47, 49, 50, 54, 56, 63, 64, 65), the following aggregate shapes occur in the end structures: spheres, ellipsoids, long, substantially cylindrical rods in various regular arrangements, as well as layers, and the corresponding shapes of the matrix. The arrangement of the aggregates is a statistical one only in mixtures of block polymers with butadienes. In most cases, however, we find long range order. With spheres and frequently also with rods, arrangements of the centers are found like in the corresponding closest packings. The aggregates do not touch one another because the butadiene chains belonging to every styrene aggregate form a shell around the aggregate and enforce a minimum distance. Such packings have a surprisingly precise long range order going far beyond the length of extended molecules (*see* Figure 6a). The aggregates are obviously so uniform in their forms and sizes that they build up a lattice of far-reaching order. Such regular aggregates can be formed only by polymers of very uniform molecular structure. The long range order thus is disturbed greatly if polymers are mixed, even similar polymers (Figure 8).

If many aggregate sizes are measured in Figures 8b and 8c, the resulting size distribution exhibits two maxima, corresponding to the values of the components of the mixture. Together with the irregular distribution of the styrene domains in Figure 8b, this proves that in such mixtures a separation into the components occurs, primarily on annealing. It is remarkable that this demixing in the system shown in Figure 8c is no

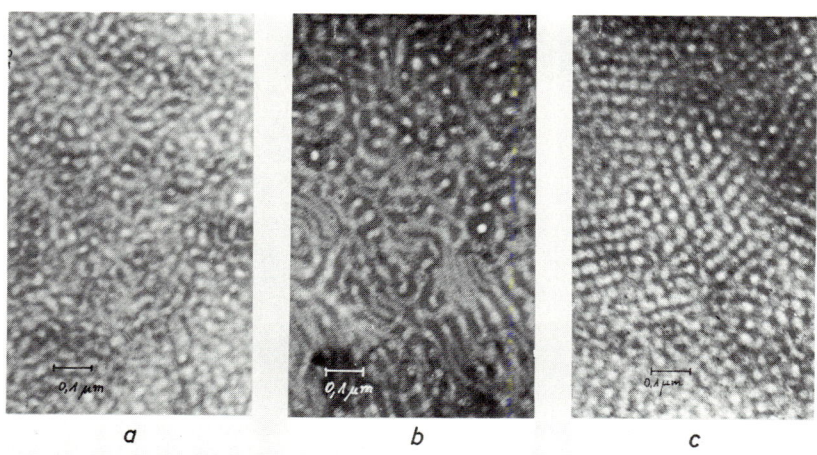

Figure 8. Electron micrographs of mixtures (1:1) of two-block copolymers cast from benzene

(a) P 42 (15% styrene blocks, M = 83000) + P 32 (26% styrene blocks, M = 70000), not annealed
(b) P 42 + P 32, annealed for 1 hour at 100°C
(c) P 41 (17% styrene blocks, M = 48000) + P 42, annealed for 1 hour at 100°C

longer caused by a different chemical composition of the components but only by their different molecular weights.

A further form of the arrangement of aggregates is considerably important from the standpoint of mechanical properties. Thread-shaped aggregates, especially the pearl strings occurring on incomplete annealing, often touch one another if the styrene content is sufficiently high, and thus form a hard network inside a soft matrix (*33*). When deformations are strong, this network, with its very long relaxation time, is presumably broken. The polymer, which in cases of little deformations is elastic and has a high modulus of elasticity, can therefore be deformed plastically at higher strains and remains soft after such deformations. Anomalous behavior of the butadiene–styrene block copolymers may thus be understood (*33*). A substantial reduction of the glass transition temperature of the aggregates by the stress applied (*6, 7, 31, 39, 51, 68*) need not be assumed in order to explain the stress–strain behavior and the non-Newtonian viscosity (*33*) of the two-block polymers.

Size and Shape of the Aggregates as a Function of Molecular Structure

Table III presents data on the molecular structure, and Table IV gives mean values from several measurements of the supramolecular structure of these polymers. Surprisingly, every polymer appeared in several end structures. The two-block polymer P 41, for example, shows in 50 preparations (films) spotty aggregations just as often as streaky ones. The energy difference between the two shapes does not seem to be great. It seems that, in general, more than one characteristic supramolecular structure may exist in a given block copolymer. Moreover the diameters D (and necessarily the distances A) of the two kinds of aggregates are distinctly and reproducibly different. Both shapes are found to have narrow size distributions—e.g., in our two-block polymers mostly mean deviations of $\pm 13\%$ of the mean value. The two aggregate thicknesses thus depend upon molecular weight in a different way. Interpreting the aggregate sizes will not be as easy as has hitherto been tried (*4, 6, 7, 9, 21, 22, 31, 32, 39, 47, 49, 50, 51, 54, 63, 64, 68*).

In our search for quantitative relationships between the aggregate dimensions and the molecular structure, we first discuss the conformation of the sequences in the aggregates. If the sequences were densely folded or pressed together to compact nodules (*69*) without substantial mutual penetration, the aggregate thicknesses should—since no butadiene sequences are allowed to be present in the aggregate—not be much bigger than double the edge length:

$$d = \sqrt[3]{\frac{M_s}{N_L \cdot \rho_s}}$$

Table III. Molecular

Preparation	Type	Styrene, %	Styrene Block, %	η_{Tol}
87	B	0	0	0.74
5	B	0	0	1.08
88	BS	10	7.5	1.79
93	BS	10	6.5	1.82
41	BS	20	17.5	0.68
42	BS	20	15.1	1.06
94	BS	20	17.5	1.16
39	BS	20	17.7	0.80
95	BS	30	25.5	0.46
30	BS	30	24.7	0.63
32	BS	30	26.0	0.80
96	BS	30	(25.5)	1.03
31	BS	30	24.0	1.48
46	BS	37	30	0.58
45	BS	46	41	0.83
97	BS	50	41	0.90
98	BS	80	68	0.54
99	BS	80	68	0.50
40	B/S[b]	20	17.8	0.94
38	B/S	20	22.6	0.65
44	S/B[c]	20	19	0.42
69 u	S/B[c]	20	19	0.64
52	BS/B	28	(22)	0.79
53	S/BS	27	22.2	0.86
55	S/BS	40	(32)	0.57
54	S/BS	40	(32)	0.72
90	BS/BS	50	42	0.61
91	BS/BS	20	18	0.76
92	S/BS/BS/BS	20	18	0.76
68	SB$_4$ Star	20	18	1.51

[a] Integral distribution of molecular weights.
[b] The stroke indicates a new charging of monomer.

of cubic nodules. In P 41 there is $2d = 63$ A, while experimentally for spheres $D_K = 212$ A (x-ray) and for long cylinders $D_Z = 176$ A is found. Similar differences between $2d$ and D are found with all preparations. Consequently, compressed coils do not exist.

The extended chain is another conformation which has been discussed often and proved to exist—e.g., in linear polyethylene crystallized at high pressures. It has also been assumed for amorphous polymers in the form of strands of parallel molecules [kink model (57)]. The maximum length L (x-ray) of a sequence of vinyl monomers of basic molecular weight M_o is $\frac{M}{M_o} \cdot 2.52$ A. Straight strands of molecules can be

Structure

	$M_S{}^a$			M_B	M
10%	50%	90%			
—	—	—	—	39000	39000
—	—	—	—	67000	67000
			12000	147000	159000
			12000	163000	175000
5000	10000	15000	10000	38000	48000
			15000	69000	83000
			17000	78000	95000
	13000		11000	47000	58000
	13000		12000	22000	32000
	14000		15000	35000	49000
14000	21000	26000	20000	50000	70000
			25000	72000	97000
21000	36000	46000	36000	121000	157000
8500	10500	13500	13000	35000	48000
23000	41000	47000	36000	54000	90000
			32000	48000	80000
			56000	26000	82000
			52000	24000	76000
	13000		13000	59000	72000
	16000		15000	32000	44000
			4400	19000	23500
			7200	33000	40000
	17000		16000	24000	63000
	14000		11000	52000	71000
5200	8200	9500	9000	32000	50000
6200	9700	11500	10000	46000	66000
8000	13000	17000	14000	20000	68000
—	3000	11000	5000	22000	54000
3500	5700	9000			49000
5000	10000	14000	11000	194000^d	238000

[c] Polymerized with addition of ether.
[d] Star with four branches of $M = 48500$.

arranged only in layers without any packing difficulties. It is obvious that spherical or cylindrical aggregates cannot be built up from completely extended styrene sequences because the aggregates do not contain butadiene and must have the same density everywhere. Moreover, the experimental D_K and D_Z are far smaller than $2L$ and even smaller than L. It must therefore be assumed that such molecular strands, if they occur at all, are kinked or curved.

Before drawing further conclusions regarding the conformation, we have to take into account that in such aggregates the conformation requirements of both kinds of blocks must be fulfilled. The relations between aggregate size and block length may be expected to be especially simple if two-block polymers of the same composition but of different

Table IV. Supramolecular

Preparation	$h_{o,S}$	$h_{o,B}$	φ_s	Aggregate Composition
88	79	490	0.073	St
93	80	530	0.063	St
41	71	220	0.168	St
42	90	310	0.146	St
94	100	345	0.168	St
39	79	250	0.170	St
95	79	161	0.244	St
30	90	211	0.237	St
32	106	260	0.250	St
96	120	320	0.244	St
31	150	435	0.229	St
46	83	200	0.284	St
45	150	270	0.385	St
97	140	255	0.40	St
98	195	160	0.70	Bu
99	187	150	0.70	Bu
40	83	283	0.171	St
38	90	200	0.216	St
44	45	140	0.181	St
69 u	60	190	0.181	St
52	93	170	(0.21)	St
53	75	265	0.212	St
55	69	200	0.30	St
				Bu
54	71	250	0.30	St
				Bu
90	87	153	0.395	St
91		162	0.172	St
92	~65	~140	0.172	St
68	75	380	0.172	St

[a] h_o : end-to-end distance in athermic solutions. φ_s : volume fraction of styrene

Structure[a]

Shape	Diameter (A), x-ray	elec. micros.	$A(A)$, x-ray	$\dfrac{D}{h_o}$ x-ray
K	224		414	2.83
K	217		391	2.70
K	214	267	336	3.00
Z		176		
K	244	324	361	2.71
Z		210		
K	260		369	2.50
K	215		361	2.72
Z	157		288	1.98
Z	200		336	2.22
K		221		
Z	206	195	370	1.94
S		166		
Z	252		440	2.10
Z	316		565	2.11
Z	178	203		2.15
S		231		
S	200		490	1.33
S	174		445	1.24
Z	198	209	336	1.24
Z	195		328	1.30
K	255		370	3.07
K	224		361	2.49
K		132	211	
K	172	174	276	2.86
Z	141	133	256	1.50
K		196		
Z	167	163	313	2.22
K	214		261	3.10
Z	187	160	261	2.71
S		219		
Z	138		261	0.69
S	160		227	0.80
K	274		320	3.71
Z	230	227	320	3.25
S		225		
S	202		278	0.81
K		290		
Z		186	243	
Z		145	288	
K		201		
Z		120		
K	250	300	414	3.32
Z		222		

blocks. The shapes are spheres ($= K$), long cylinders (Z) or layers (S).

molecular weight are considered. The quotient of the block weights M_s/M_B is then approximately constant so that the shape of the aggregates should be the same; therefore, the size of the aggregates should be linked in a particularly simple way—e.g., through a scale factor—with the size of the molecules. A group of such polymers consists of the samples P 95 to P 31 in Table IV, which contain 30% styrene. The thickness of the thread-shaped aggregates of these polymers is not proportional to M or M_s but rather to $\sqrt{M_s}$.

Molecular strands would therefore have to be considerably kinked or curved if they could occur at all in atactic polymers with voluminous side groups because the ratio D_Z/L_S is not constant but decreases strongly with rising molecular weight of the blocks. The dependence of the experimental particle diameter on the molecular weight corresponds fairly well, according to Table IV, to the molecular weight dependence of the statistical coil size. For our present problem, we will assume statistically coiled sequences penetrating each other at random. We do not exclude the possibility that the chains may be running parallel over short distances $<<h$, which are perhaps related to the length of stereospecific sequences.

We further assume that the blocks in the aggregates as well as in the continuous phase would have the lowest possible free energy if their coil dimensions would equal those of the corresponding homopolymers in athermic solutions. Conformation restrictions (8, 15, 16, 72), which are caused by a block end being connected with a second block that does not mix with the first one, are thus neglected. (At least one end of each block is situated in the interface.) For blocks of polystyrene and polybutadiene the following equations are valid (19, 24, 33, 56, 65) (with the end-to-end distance $h = \sqrt{h_i^2}$):

$$h_{o,S} = 0.345 \cdot M_s^{0.58} \; ; \; h_{o,B} = 0.50 \cdot M_B^{0.58} \qquad (5)$$

If the molecules consist of s styrene sequences (molecular weight M_S) and b butadiene sequences, M_S/M_B and $h_{o,S}/h_{o,B}$ can be expressed by the volume fraction ϕ_s of the styrene sequences. The densities ρ(polystyrene 1.075; polybutadiene 0.96) are required for this purpose.

$$\frac{\varphi_S}{1 - \varphi_S} = \frac{s \cdot M_S}{\rho_S} : \frac{b \cdot M_B}{\rho_B} = \frac{s \cdot M_S \cdot \rho_B}{b \cdot M_B \cdot \rho_S} \qquad (6)$$

$$\frac{h_{o,B}}{h_{o,S}} = 1.45 \left(\frac{M_B}{M_S}\right)^{0.58} = 1.35 \left[\frac{s(1 - \varphi_S)}{b \cdot \varphi_S}\right]^{0.58} \qquad (7)$$

The formation of micellae of blockpolymers consisting of a nucleus (containing the styrene sequences) and a shell may now be considered from a thermodynamic point of view by the following theories (23, 37, 38, 46, 52) which assume a statistical coiling of the sequences. The corresponding homopolymers of the same degree of polymerization, completely separated into macroscopic phases with negligible surface energy, are taken as a reference state for the Gibbs free energy G. A microheterogeneous dispersion of the styrene phase in the butadiene phase has an internal surface of $\phi_s \cdot \frac{3}{D_K}$. The end-to-end distances h may differ from those (h_o) in the reference state:

$$h_S = p \cdot h_{o,S} \ ; \ h_B = q \cdot h_{o,B} \qquad (8)$$

Then the change in free energy by dispersing is approximately (for $p \geqslant 1$ and $q \geqslant 1$)

$$\Delta G_K = \varphi_S \cdot \frac{6}{D_K} \cdot \Delta W + \frac{3 \cdot \rho \cdot RT}{2 \cdot M} \left\{ s(p^2 - 1) + b(q^2 - 1) \right\} + \Delta G_m \qquad (9)$$

ΔW is the interfacial energy per unit area, and ΔG_m takes into account further restrictions of conformation (23, 46, 52); $6/D_K$ must be replaced by $4/D_Z$ for long cylindrical aggregates and by $2/D_S$ for layers. The second term on the right side of Equation 9 increases with decreasing molecular weight. If it is accepted that the blocks tend to form statistical coils with end-to-end distance h_o in the absence of conformation restrictions, it follows that the free energy per block increases if the space accessible to such a block is limited in one dimension, for instance by two plane and parallel interfaces having a distance smaller than h_o. On the other hand, at distances $\gg h_o$ of the opposite interfaces the energy per chain will also increase because the blocks are fixed to at least one interface and thus forced to form less probable conformations, elongated in the direction of the normal to the interfaces, in order to fill space between the interfaces completely with matter.

For these reasons and contrary to some opinions expressed recently (23, 37, 38, 46, 52), a minimum of free energy exists for each set of blocks at some definite distance of the interfaces which is approximately equal to the coil size. If the aggregates are not layers, there is a broad distribution of distances between the interfaces in different directions so that the average of ($q^2 - 1$) must be used in Equation 9. The distance between opposite interfaces may be somewhat larger than $p \cdot h_{o,S}$ or $q \cdot h_{o,B}$ because of the overlapping of coils (*see below*). Considering the ways in which coils with Gaussian distributions of segments can overlap, we estimate the optimum distance of interfaces in the following way. If two coils approach each other until the concentration of segments is approximately constant everywhere between their centers of gravity, the distance of their centers of gravity is about $0.55 \ h_o$. The concentration of segments decreases in outward direction and, at about 0.35 h_o, is only one-third of the maximum value between the centers of gravity. Thus 0.55 h_o + 2 · 0.35 h_o = 1.25 h_o may be considered as the "diameter of the aggregate." This derivation neglects the fact that h_o is increased if one end of the coil is fixed within an interface (8, 15, 16, 72). On the other hand, such a dilatation of a coil may be diminished by the necessity to fill the space between interfaces. Our experiments show (*see below*) that 1.2 h_o is a reasonable value for the actual end-to-end distance between plane interfaces. In the case of cylinders one has to discuss the distribution of segments over the cross section. The segments of six coils arranged at the corners of a regular hexagon with suitable dimensions will overlap in such a way that the concentration of segments is nearly constant inside the hexagon and a small distance to the outside. Such an arrangement is similar to the circular cross section of a cylinder and has a "diameter" of about 1.5 h. An arrangement of coils similar to a sphere may be built up using 3.6 = 18 coils and has a "diameter" of about 1.8 h. For these reasons, the following relations are expected to hold as a first approximation:

$$D_K = 1.8\ p_K h_o\ ;\ D_Z = 1.5\ p_Z h_o\ ;\ D_S = 1.2\ p_S h_o \qquad (10)$$

Analogously, the preferred average distance between opposite interfaces across the matrix should be about 1.2 $q h_{o,B}$.

The experimental results of Table IV show for the case of layered aggregation that the value of p_s is approximately 1.07 and q_s about 0.96. Neither ΔW nor the limitation (8, 15, 16, 72) of the conformation space seem to influence substantially the values of p and q in this case. Our experiments do not confirm the assumption $D = 2h_o$ but give the approximate value $D_s \approx 1.2\ h_o$ for layers.

The two energy terms related to the styrene blocks and the butadiene blocks, respectively, have minima at different values of D ($A - D$ is proportional to D for constant ϕ_S). The sum of both terms shows an energy minimum at such a value of D that both p and q differ from 1. A minimum of free energy is obtained without considering the influence of the interfacial energy ΔW.

We have therefore tried to explain the existence of an equilibrium diameter without taking into account ΔW. As the effective values of the distances between interfaces cannot easily be determined for the matrix (except layered structures), we developed the following approach to overcome this difficulty (for symbols used *see* text referring to Equations 5–9).

Let us first discuss the structure of an isolated aggregate or micella: It consists of an inner domain, the "nucleus" (*e.g.*, styrene blocks) and a surrounding "shell" (*e.g.*, butadiene blocks). Such a shell of thickness

$1.2 \cdot q \cdot h_{o,B}$ and volume V_{shell} may contain a maximum number of N_{shell} blocks of one kind and the nucleus (volume V_{nucl}) up to N_{nucl} blocks of the other kind. For isolated spherical (K) domains we find:

$$\left(\frac{N_{\text{shell}}}{N_{\text{nucl}}}\right)_K = \frac{V_{\text{shell}} \cdot \rho_B \cdot M_S}{V_{\text{nucl}} \cdot \rho_S \cdot M_B} = \frac{(D_K + 2 \cdot 1.2 \cdot q_K \cdot h_{o,B})^3 - D_K^3}{D_K^3} \cdot$$

$$\frac{\rho_B \cdot M_S}{\rho_S \cdot M_B} = \left[\left(1 + \frac{2 \cdot 1.2 \cdot q_K \cdot h_{o,B}}{1.8 \cdot p_K \cdot h_{o,S}}\right)^3 - 1\right] \cdot \frac{\rho_B \cdot M_S}{\rho_S \cdot M_B} \qquad (11)$$

With Equations 5, 6, and 7 one obtains

$$\left(\frac{N_{\text{shell}}}{N_{\text{nucl}}}\right)_K = \left\{\left[1 + 1.8 \cdot \frac{q_K}{p_K}\left(\frac{s(1 - \varphi_s)}{b \cdot \varphi_s}\right)^{0.58}\right]^3 - 1\right\} \cdot \frac{b \cdot \varphi_s}{s(1 - \varphi_s)} \qquad (12)$$

The analogous derivation for cylindrical (Z) aggregates gives:

$$\left(\frac{N_{\text{shell}}}{N_{\text{nucl}}}\right)_Z = \left\{\left[1 + 2.15 \cdot \frac{q_Z}{p_Z}\left(\frac{s(1 - \varphi_s)}{b \cdot \varphi_s}\right)^{0.58}\right]^2 - 1\right\} \cdot \frac{b \cdot \varphi_s}{s(1 - \varphi_s)} \qquad (12a)$$

For $p = 1$ and $q = 1$ the ratio $N_{\text{shell}}/N_{\text{nucleus}}$ is substantially greater than b/s (ratio of numbers of B blocks to S blocks in the molecule). This means that the butadiene chains attached to the surface of a styrene nucleus fill the space determined by their coil sizes to only a very small degree. The butadiene shells of adjacent aggregates therefore penetrate each other to such an extent that the shells and thus the whole space between the styrene nuclei are completely filled with matter. (In undiluted block copolymers, isolated aggregates with non-overlapping shells would have anomalous conformations and so an increased free energy.)

The overlapping can formally be taken into account by introducing overlap factors, F_K and F_Z, which are a measure of the penetration of the shells and influence the values p and q of Equation 9:

$$\left(\frac{N_{\text{shell}}}{N_{\text{nucl}}}\right)_K = F_K \cdot \frac{b}{s} \; ; \; \left(\frac{N_{\text{shell}}}{N_{\text{nucl}}}\right)_Z = F_Z \cdot \frac{b}{s} \qquad (13)$$

If F_K and F_Z are regarded as constant values, q/p depends only on the volume fraction ϕ_s but not on the molecular weight (neglecting ΔW as a first approximation). This is confirmed by experimental data (Table IV). Moreover, Table IV shows that F_K and F_Z, calculated by Equation 12 from experimental p-values ($1.2 \, q = 1$), are nearly constant for different ϕ_s. F_K appears to be about 9 and F_Z about 4.

The experimental values of F_K and F_Z can be interpreted by simple geometric considerations. A cross section of a hexagonal cylinder packing with the volume fraction $\phi_s = 0.25$ shows, for example, that the shells of three neighboring cylinders overlap entirely in the space between them. The shells of the three next-but-one cylinders also protrude partly into this field and thus increase the mean overlap factor F_Z from 3 to approximately 4. F_Z does not depend greatly upon $h_{o,B}/D$ and ϕ_s, so that the value $F_Z = 4$ may be expected to be a suitable first approximation for all hexagonal cylinder packings. If nearly Gaussian-shaped segment density distributions are assumed around the centers of the butadiene coils, such overlapping of the shells gives a nearly constant over-all concentration of matter everywhere between the aggregate nuclei.

For a closest packing of spherically symmetric particles consisting of nucleus and shell, the value of the overlap factor, $F_K = 9$, can be shown to be plausible in an analogous manner. Deviations from these values of F_K and F_Z should increase the free energy in Equation 9.

With the aid of the conditions

$$F_K = 9$$
$$F_Z = 4 \qquad (14)$$
$$1.2 \cdot q = 1$$

and Equations 12 and 13 the quantities p_K and p_Z and thus, according to Equation 10, the diameters may be calculated if the volume fraction ϕ_s of the aggregates and the structural type of the molecule (s styrene blocks and b butadiene blocks) are known.

The data in Table V show good agreement between experimental and calculated p values and confirm the validity of the assumptions used —i.e., $\Delta W \approx 0$ and $1.2 \cdot q \approx 1$. It should be remembered that the average experimental values of F_K and F_Z may reflect to some extent an influence of $\Delta W \neq 0$ or $1.2 \cdot q \neq 1$ on p. Nevertheless, the introduction of F already leads to a minimum of ΔG for $p \neq 1$.

Table V also includes values for q_{min} which is a measure of the deformation of the butadiene sequences along the line connecting the centers of aggregate nuclei.

$$A_K - D_K = 1.2\, q_{min,K} \cdot h_{o,B} \;;\; A_Z - D_Z = 1.2\, q_{min,Z} \cdot h_{o,B} \qquad (15)$$

Thus q_{min} as well as the average q are proportional to p for a constant value of ϕ_s. If in the case of spherical aggregates $1.2 \cdot q_{min}$, which is always smaller than 1, becomes smaller than 0.5, the two-block polymers tend to avoid this strong deformation of the butadiene sequences by forming cylindrical aggregates. Both particle shapes are observed in cases when q_{min} is calculated to be approximately 0.5 to 0.7 both for spheres and cylinders.

Equation 9 gives the excess free energy and may be simplified by expressing q and D in terms of p. Neglecting ΔG_m we find a value of 12

Table V. Comparison of Experimental and Calculated Data[a]

Type	φ_{agg}	Aggregate Composition	Shape	$\dfrac{D}{h_o}$ exp.	F_{exp}	$\dfrac{D}{h_o}$ calc.	$1.2\, q_{min}$ calc.
BS	0.07	St	K	2.8	11.3	3.07	0.61
		St	Z			1.90	0.82
	0.17	St	K	2.7	8.7	2.65	0.50
		St	Z			1.92	0.74
	0.21	St	K	2.5	9.5	2.56	0.46
		St	Z			1.94	0.72
	0.25	St	K			2.51	0.43
		St	Z	2.0	3.9	1.96	0.70
	0.39	St	K			2.38	0.32
		St	Z	1.9	4.5	2.07	0.62
		St	S	1.2		1.2	1.2
	0.29	Bu	K			1.35	0.40
		Bu	Z	1.2		1.09	0.68
BSB	0.21	St	Z	1.5	3.2	1.30	0.72
SBS	0.21	St	Z	2.2	6.3	2.90	0.72
	0.30	St	Z	2.1	6.9	2.98	0.67
(SB)₄	0.17	St	K	3.3	13.4	2.9	0.63

[a] Abbreviations same as in Table IV. D is the diameter of the aggregates formed by the indicated sequences.

erg/cm² for the interfacial energy ΔW in the case of P 41. This value is certainly determined mainly by the influence of F on p. We propose to neglect ΔW and to use Equations 12 to 14 only.

The experimental values of p are greater than 1, so that a considerable deformation of the styrene coils must be expected. On the other hand, the most probable form of a statistical coil does not seem to be spherically symmetric (44). Rather, the coil shape resembles an ellipsoid, the maximum length being $1.25\,h$ and the short axes about 0.45 and 0.71 h. Arranging such coils in an analogous manner as described above, we estimate the diameters to be $1.5\,h$ (layer), $2.0\,h$ (cylinder), and $2.4\,h$ (sphere). Moreover, Equations 12, 13, and 14 do not depend on the variables $h_{o,B}$ and $h_{o,S}$ themselves, but only on the ratio $h_{o,B}/h_{o,S}$. It can therefore not yet be concluded from the agreement between calculation and experiment that the assumption $h_o = h_{ath}$ is correct. It is possible that in undiluted polymers and their concentrated solutions the coil diameters h are greater by a factor independent of the molecular weight and of the chemical nature than the values measured in athermic diluted solutions. A parallel arrangement of the chains for lengths $<<h$ would, after all, be able to increase the statistical segment by, for example, the factor 1.7 and thus h by the factor 1.3. The influence of a limitation of

the conformation space (end of chain in interface) on the p-value (8, 15, 16, 72) should be considered as well. Above all one should evaluate q more exactly to find its influence on p.

With these considerations, the dimensions and shapes of the aggregates of block polymers have been traced back satisfactorily to the molecular structure. The derivations consider both kinds of sequences. Precise agreement between experiment and theory is not expected because we considered simple, idealized particle shapes, made simplifying assumptions concerning the chain conformation, neglected the interface energy and introduced errors into the calculations by inaccurate values of molecular weights, particle diameters, etc For example, the p_K values determined from electron-microscopic photographs (*see* Figure 8) of P 41 and P 42 are distinctly greater than the x-ray values, presumably because they are flattened ellipsoids instead of spheres.

Compared with two-block polymers the conditions for an energy minimum become even more complicated for the aggregation of three-block polymers. Thus, for instance, both ends of the medium sequence must lie in the interface, so that the choice of possible conformations is further restricted (for example $p > 1$ only if both ends of the medium sequence lie on the same side of the aggregate nucleus). Nevertheless, the aggregate sizes of three-block polymers can be calculated satisfactorily with the simple theory. For polymers with more than three sequences, aggregation is rendered so difficult that even in cases of uniform molecular structure no regular end structures are found (in four-block polymers there are no longer two, but four kinds of sequences). The same difficulties are encountered in mixtures of block copolymers. In such cases, however, the difficulties are, as a rule, avoided by demixing.

Mixing the block copolymers with homopolymers will partly remove the restrictions governing the conformations of sequences if the h-value of the added homopolymer is smaller than that of the sequence. However, the homopolymers soluble in the matrix increase the distances of diffusion necessary for aggregation and reaggregation and disturb the formation of long range order and of big aggregates.

Solvents dissolving both kinds of sequences do not increase ϕ_{agg} but modify the h_o values slightly. When such solutions are being concentrated, the aggregates formed first contain about 60% of solvent. On further evaporation, they shrink by giving off solvent. If their diameters decrease below the value $p \cdot h_o$, reaggregation becomes necessary. For example, thread-shaped aggregates with uniform diameters may be built up at lower concentrations and may be converted later to the pearl strings often observed in starting structures.

The diameters, shapes and distances of the aggregates have thus been derived from the molecular structure, and the process of aggregation has been elucidated. Our results may be generalized easily and applied to other block copolymers.

Acknowledgment

This paper is dedicated to H. Holzrichter on his 60th birthday.

Literature Cited

(1) Angelo, R., Ikeda, R. M., Wallach, M. L., *Polymer* **6**, 141 (1965).
(2) Bailey, J. T., Bishop, E. T., Hendricks, W. R., Holden, G., Legge, N. R., *Rubber Age* **98**, 69 (Oct. 1966).
(3) Beecher, J. F., Marker, L., Bradford, R. D., Aggarwal, S., *Am. Chem. Soc., Div. Polymer Chem., Preprints* **8**, 1532 (1967).
(4) Beecher, J. F., Marker, L., Bradford, R. D., Aggarwal, S., *J. Polymer Sci., Pt. C* **26**, 117 (1969).
(5) Bohm, L., *Rubber Chem. Technol.* **41** (2), 495 (1968).
(6) Bonart, R., *J. Macromol. Sci. Phys.* **2** (1), 115 (1968).
(7) Bradford, E., Vanzo, E., *J. Polymer Sci. Pt. A1*, **6**, 1661 (1968).
(8) Chandrasekhar, S., *Rev. Mod. Phys.* **15**, 5 (1953).
(9) Childers, C. W., Kraus, G., *Rubber Chem. Technol.* **40**, 1183 (1967).
(10) Cooper, S. L., Tobolsky, A. V., *Textile Res. J.* **36** (9), 800 (1966).
(11) Cooper, S. L., Tobolsky, A. V., *J. Appl. Polymer Sci.* **10**, 1837 (1966).
(12) *Ibid.*, **11**, 1361 (1967).
(13) Corish, P. J., *Rubber Chem. Technol.* **40**, 324 (1967).
(14) Debye, P., Bueche, F., *J. Chem. Phys.* **20**, 1337 (1952).
(15) DiMarzio, E. A., *J. Chem. Phys.* **42**, 2101 (1965).
(16) DiMarzio, E. A., *Am. Chem. Soc., Div. Polymer Chem., Preprints* **9**, 256 (1968).
(17) Dony, A., Rossi, J., Gallot, G., *Compt. Rend.* **267**, 1392 (1968).
(18) Dony, A., Gallot, B., *Compt. Rend.* **268**, 1218 (1969).
(19) Elias, H. G., Etter, O., *Makromol. Chem.* **66**, 56 (1963).
(20) Estes, G. M., Cooper, S. L., Tobolsky, A. V., *Rev. Macromol. Chem.*, to be published.
(21) Fischer, E., Henderson, J. F., *J. Polymer Sci., Pt. C* **26**, 149 (1969).
(22) Fischer, E., *J. Macromol. Sci., Pt. A* **2**, 1285 (1968).
(23) Fedors, R. F., *J. Polymer Sci., Pt. C* **26**, 189 (1969).
(24) Flory, P. J., "Principles of Polymer Chemistry," p. 554, Cornell University Press, Ithaca, N. Y., 1953.
(25) Franta, E., Skoulios, A., Rempp, P., Benoit, H., *Makromol. Chem.* **87**, 271 (1965).
(26) Gallot, B., Mayer, R., Sadron, C., *Compt. Rend.* **C263**, 42 (1966).
(27) Gallot, B., Mayer, R., Sadron, C., *Rubber Chem. Technol.* **40**, 932 (1967).
(28) Gessner, B. D., *J. Appl. Polymer Sci.* **11**, 2499 (1967).
(29) Guinier, A., Fournet, G., "Small-Angle Scattering of X-Rays," Wiley, New York, 1955.
(30) Henderson, J. F., Grundy, K. H., Fischer, E., *J. Polymer Sci., Pt. C* **16**, 3121 (1968).
(31) Hendus, H., Illers, K. H., Ropte, E., *Kolloid Z.* **216, 217**, 110 (1967).
(32) Holden, G., Bishop, E. T., Legge, N. R., *J. Polymer Sci., Pt.* **26**, 37 (1969).
(33) Hoffmann, M., Pampus, G., Marwede, G., *Kautschuk. Gummi* **22**, 691 (1969).
(34) Hoffmann, M., Rother, K., *Makromol. Chem.* **80**, 95 (1964).

(35) Hoffmann, M., in Tobolsky, A. V.: "Mechanische Eigenschaften und Struktur von Polymeren," p. 81, Berliner Union, Stuttgart, 1967.
(36) Huggins, M. L., *J. Phys. Chem.* **46**, 151 (1942).
(37) Inoue, T., Soen, T., Hashimoto, T., Kawai, H., *J. Polymer Sci., Pt. A2*, **7**, 1283 (1969).
(38) Inoue, T., Soen, T., Hashimoto, T., Kawai, H., *Am. Chem. Soc., Div. Polymer Chem., Preprints*, **10**, 538 (1969).
(39) Inoue, T., Soen, T., Kawai, H., Fakatsu, M., Kurata, M., *J. Polymer Sci., Pt. B*, **6** (1), 75 (1968).
(40) "International Tables for X-Ray Crystallography," Vol. 3, Kynoch Press, Birmingham, 1962.
(41) Kato, K., *Polymer* **9**, 419, 225 (1968).
(42) Kaempf, G., *Intern. Kongr. Elektronenmikroskopie*, 7th, Grenoble, 1970.
(43) Kaempf, G., *Ber. Bunsenges. Phys. Chem.* **74** (1970).
(44) Kuhn, H., Moning, F., Kuhn, W., *Helv. Chim. Acta* **36**, 731 (1953).
(45) Kraus, G., Childers, C. W., Gruver, J. T., *J. Appl. Polymer Sci.* **11**, 1581 (1967).
(46) Krause, S., *J. Polymer Sci., Pt. A2*, **7**, 249 (1969).
(47) Lewis, P., Price, C., *Nature* **223**, 494 (1969).
(48) Marei, A. I., Sidorovich, E. A., *Polymer Mechanics* **1** (5), 55 (1965).
(49) Matsuo, M., Ueno, T., Horino, H., Chujyo, S., Asai, H., *Polymer* **9**, 425 (1968).
(50) Matsuo, M., Sagae, S., Asai, H., *Polymer* **10**, 79 (1969).
(51) Matsuo, M., *Japan. Plastics* **2** (3), 6 (1968).
(52) Meier, D. J., *J. Polymer Sci., Pt. C* **26**, 81 (1969).
(53) Merrill, S. H., *J. Polymer Sci.* **55**, 343 (1961).
(54) Morton, M., McGrath, J. E., Juliano, P. C., *J. Polymer Sci., Pt. C* **26**, 99 (1969).
(55) Nielsen, L. E., *J. Am. Chem. Soc.* **75**, 1435 (1953).
(56) Orofino, T. A., Michey, J. W., Jr., *J. Chem. Phys.* **38**, 2512 (1963).
(57) Pechhold, W., *Kolloid Z.* **228**, 1 (1968).
(58) Peterlin, A., "Die Physik der Hochpolymeren," H. A. Stuart, Ed., Vol. 2, p. 350, Springer-Verlag, Berlin, 1953.
(59) Picht, J., Heydenreich, J., "Einfuehrung in die Elektronenmikroskopie," VEB-Verlag Technik, Berlin, 1966.
(60) Reimer, L., "Elektronenmikroskopische Untersuchungs- und Praeparationsmethoden," Springer-Verlag, Berlin, 1959.
(61) Rhodes, M. B., Stein, R. S., *J. Polymer Sci. Pt. A2*, **7**, 1539 (1969).
(62) Rosen, S. L., *Polymer Eng. Sci.* **7**, 115 (1967).
(63) Sadron, C., *Chim. Pure Appl.* **4**, 347 (1962).
(64) Sadron, C., *Chim. Ind. Genie. Chim.* **96** (1), 507 (1966).
(65) Schulz, G. V., Baumann, H., *Makromol. Chem.* **60**, 120 (1963).
(66) Skoulios, A. E., Tsouladze, G., Franta, E., *J. Polymer Sci., Pt. C*, **4**, 507 (1964).
(67) Smith, T. L., Dickie, R. A., *J. Polymer Sci., Pt. C* **26**, 163 (1969).
(68) Vanzo, E., *J. Polymer Sci., Pt. A*, **1** (4), 1727 (1966).
(69) Vollmert, B., Stutz, H., *Angew. Makromol. Chem.* **1968**, 182.
(70) Wilkes, G. L., Stein, R. S., *Intern. Union Pure Appl. Chem., Meetg., Toronto*, 1968.
(71) Williams, B. L., Weissbein, L., Singh, A., *Rubber Age* **100** (7), 57 (1968).
(72) Zachmann, H. G., Spellucci, P., *Kolloid Z.* **213**, 39 (1966).

RECEIVED February 16, 1970.

24

Dispersion of Solid Particles in Organic Media

G. E. MOLAU and E. H. RICHARDSON

The Dow Chemical Co., Polymer Science, Physical Research Laboratory, Midland, Mich. 48640

> *Block copolymers are dispersing agents which stabilize dispersions of solid particles in organic media. Dispersions of titanium dioxide in toluene stabilized by partially carboxylated styrene–butadiene block copolymers are studied as model systems. The block copolymer is modified by adding thioglycolic acid to some of the butadiene units. Through these carboxylated sites, the butadiene ends of the block copolymer molecules are adsorbed selectively at the substrate surface, while the styrene ends are dissolved in the dispersion medium. We believe that a "polymeric double layer" is formed, and that the unusually high stability of the dispersions against settling is a direct consequence of the interaction of the double layers of different particles.*

Dispersions of solid particles in organic media are of considerable technological importance, particularly in oil-based paints and printing inks, but their stabilization has not been studied as extensively as the stabilization of dispersions in aqueous systems. Romo (*16*) studied the stability of dispersions of titanium dioxide in pure solvents and in solutions containing organic resins, and Crowl and Malati (*2*) stabilized dispersions of titanium dioxide and iron oxide in benzene with polyesters of adipic acid and neopentyl glycol. McGown and Parfitt (*9, 10*) dispersed titanium dioxide in solutions of oil-soluble surfactants in xylene and treated the stabilization mechanism on the basis of the Derjaguin-Landau-Verwey-Overbeek theory. The influence of water on dispersions of titanium dioxide in organic solvents has been investigated by McGown and Parfitt (*9*) and by Zettlemoyer, Micale, and Lui (*19*).

We have shown (*12, 14*) that block and graft copolymers (BG copolymers have emulsifying properties in "polymeric oil-in-oil emulsions."

These emulsions are liquid–liquid systems comprising immiscible polymer solutions in nonpolar solvents and BG copolymer emulsifiers. The emulsifying power of BG copolymers has been attributed (12) to coalescence barriers formed by accumulation of the BG copolymers in the emulsion interface. This interface apparently has the structure of a double layer consisting of the different subchains of the BG copolymers which are solvated by the organic solvent. The chemically different sequences in BG copolymers are separated in different layers in the interface because polymer chains of different chemical structures are usually incompatible (3, 4), particularly in nonpolar solvents.

It appeared attractive to extend the work on emulsification of liquid–liquid systems by BG copolymers to solid–liquid systems. As a first approach a model system was studied which comprises titanium dioxide dispersed in toluene with modified styrene–butadiene block copolymers as dispersants. These studies are reported here.

The work was planned on the basis of a model of a dispersed solid particle onto which one type of sequences of a BG copolymer is adsorbed selectively while the other type sequence is dissolved in the dispersion medium. A sketch of this model is shown in Figure 1. The model is the result of applying the same arguments which had been advanced (12) in discussing the mechanism of stabilization of polymeric oil-in-oil emulsions by BG copolymers to the problem of stabilization of dispersions of solid particles in organic media. Previously, essentially the same arguments had led to the demonstration of micelle formation of styrene–butadiene block copolymers in organic media under certain conditions (15).

In the model a solid particle is coated with a "polymeric double layer" formed by a number of BG copolymer molecules consisting of monomer units A and monomer units B. The sequences of A units are

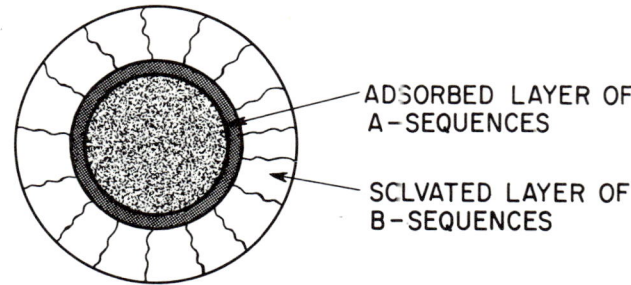

Figure 1. Model of a solid particle in an organic medium. The particle is coated with a BG copolymer consisting of A sequences (adsorbed) and B sequences (dissolved).

absorbed onto the surface of the solid substrate, while the sequences of B units are dissolved completely in the surrounding organic medium. No electrical charges are involved in the formation of coalescence barriers according to this model. The stabilizing principles are based entirely on the interactions of the polymer chains with each other and with the surrounding organic medium. The principles underlying this model have been discussed in detail (12), and the model itself and experimental data showing the dispersant activity of block copolymers in dispersions of solid particles in organic media have been presented (13). Recently, a patent (1) appeared which also describes the use of block or graft copolymers for coating pigment particles and dispersing the coated particles in organic liquids.

Experimental

The components of the model system titanium dioxide, toluene, and styrene–butadiene block copolymers were selected for the following reasons: titanium dioxide is available in various particle sizes because it is a widely used pigment; toluene is a good solvent for the styrene–butadiene block copolymers and has a low viscosity, so that the stabilizing effect of the block copolymers can be observed without overlapping stabilization effects resulting from a high viscosity of the dispersion medium; styrene–butadiene block copolymers were chosen because samples were available which had structures designed to give good performance as emulsifiers in polymeric oil-in-oil emulsions.

Materials. Titanium dioxide was pigment-grade material (du Pont Ti-Pure R-900 brand) with spherical particles in the range of 0.2–0.5 microns. In most experiments, the material was used without further treatment. In one experiment (specified in the text), the material was dried for 18 hours at 145°C.

The block copolymers were AB type styrene–butadiene block copolymers prepared by anionic polymerization. Unless specified differently in the text, a block copolymer of number average molecular weight 110,000 containing 70 wt % styrene and 30 wt % butadiene (S30B) was used. Samples of different composition used in one series of experiments had approximately the same molecular weight.

The block copolymers were modified by introducing carboxylic groups into the butadiene sequences to provide sites for selective adsorption of one end onto the TiO_2 surface. The modification was accomplished by adding thioglycolic acid, making use of the well-known addition of mercaptans to carbon–carbon double bonds.

$$\sim CH_2-CH=CH-CH_2\sim \; + \; HS-CH_2-COOH \rightarrow$$

$$\sim CH_2-\underset{H}{CH}-\underset{\underset{CH_2COOH}{S}}{CH}-CH_2\sim$$

Addition of Thioglycolic Acid (TGA) to Styrene–Butadiene Block Copolymers. The addition of thioglycolic acid to homopolymers of butadiene has been described by Marvel and co-workers (8). Starting with conditions described in their paper, various procedures were tested for achieving various degrees of carboxylation of the block copolymers. In a typical recipe, 20 grams of block copolymer (S30B) were dissolved in 280 grams of 1,4-dioxane, and 0.6 gram of thioglycolic acid (Evans Chemetics, Inc.) was added. The solution was stirred for 24 hours in a nitrogen atmosphere. The carboxylated block copolymer was recovered by precipitation with either methanol or n-hexane followed by washing with nonsolvent. The degree of carboxylation, expressed as mole % conversion of butadiene units to adduct units, was determined from the oxygen content determined by neutron activation analysis. To assure that the oxygen analysis was not falsified by methanol entrapped in the polymer, the last reprecipitation of the samples was always carried out with n-hexane. The degree of carboxylation (DC) obtained with 0.6 gram thioglycolic acid was 1.1% conversion of butadiene units; for other degrees of carboxylation, the amount of thioglycolic acid in the same basic recipe was varied.

The carboxylated block copolymers were redissolved in the solvent in which they were to be used (usually in toluene) immediately after reprecipitation because they crosslink when they are stored in the solid state.

Adsorption of Carboxylated Block Copolymer in o-Dichlorobenzene. In a three-necked flask equipped with stirrer, reflux condenser, and N_2 inlet, 200 grams of a 1% solution of carboxylated styrene–butadiene block copolymer in o-dichlorobenzene and 2 grams TiO_2 were stirred at various temperatures for 3 hours. (Typical temperatures in this treatment were 110°, 150°, and 178°C.) After the heat treatment the TiO_2 was isolated from the block copolymer solution by centrifuging for 1–2 hours at 2,000–2,200 rpm. After decanting the supernatant solution, the solid particles were washed with solvent and dried *in vacuo* for about 16 hours at 50°C.

Adsorption of Carboxylated Block Copolymer in Toluene. The later adsorption experiments were carried out in a Waring Blendor which had been wrapped with heating tape. Solutions could be conveniently heated to and maintained at 100°C with this arrangement. 50 Grams of TiO_2 and 500 grams of a 1% solution of the carboxylated block copolymer were agitated for half an hour. When heat treatment was applied, the block copolymer solution was heated to the desired temperature before adding TiO_2. Typical temperatures were 100°C and room temperature. The resulting dispersion was used directly for the settling test without isolation and washing of the coated substrate by centrifugation. In all recipes blank runs were made to test the effect of the heat treatment on pigment dispersibility in the absence of block copolymer.

Determination of Adsorbed Block Copolymer. The titanium dioxide–block copolymer composite was isolated from the dispersion medium by centrifugation. After decanting the supernatant, the centrifugate was redispersed twice in toluene and collected again by centrifugation to ascertain removal of block copolymer not adsorbed. The final centrifugate was washed with methanol, dried *in vacuo*, and submitted for carbon–

hydrogen analysis. The amount of adsorbed block copolymer, which was the only portion of the polymer–substrate composite containing carbon, was calculated from the carbon content of the total composite.

Settling Test for Measuring Dispersion Stability. A quantity of 20 ml of a freshly prepared dispersion was placed in a graduate cylinder equipped with a stopper and allowed to stand motionless at room temperature. As the particles settled slowly to the bottom of the graduate, fractionation occurred because the fraction of larger particles settled at a faster rate than the fraction of smaller particles. Two zones became discernible. The boundary between the zones could be seen more easily by shining an oblique beam of light into the dispersion. The upper zone consisted of a very stable dispersion of fine particles, while the lower zone consisted of a more concentrated dispersion containing coarse as well as fine particles. The finer particles remained dispersed throughout the test and for long periods after the coarser particles had settled out. In control experiments, in which titanium dioxide without adsorbed block copolymer was dispersed under otherwise the same conditions, all particles including the fine particles settled out within a few seconds, and the supernatant was entirely clear.

The rate of settling was measured by recording the motion of the pseudo-boundary between the two zones as a function of time. Plots of the volume fraction of the lower zone as a function of settling time were essentially linear. When most of the coarser particles had settled out, the volume fraction of the lower zone became constant—*i.e.*, the slopes of all of the straight lines changed abruptly, and the oblique lines combined into one horizontal line. The times at the breaking points in the lines were recorded as "total settling time" (TST), usually expressed in days, and were taken as a measure of dispersion stability.

Since the finer particles in the dispersion remained dispersed for long times, and the settling rate of only the coarser particles was measured by the settling test, the recorded total settling times represent the minimum of attainable stability of a given dispersion because the test measures the worst performers in the system.

Results

In the presentation and discussion of the data, the compositions of the block copolymer samples are indicated by symbols such as S30B · TGA (block copolymer of 70 wt % styrene and 30 wt % butadiene modified by adding thioglycolic acid (TGA) onto part of the butadiene units). The degree of carboxylation (DC) of the butadiene chains is expressed as the percentage of butadiene units modified by a thioglycolic acid unit. When the dispersion is prepared, s grams of substrate (TiO_2) are added to a solution of p grams of block copolymer. The substrate/polymer ratio is denoted by the symbol S/P. The amount of block copolymer actually adsorbed onto the substrate, which is determined by carbon analysis of the block copolymer/TiO_2 composite, is denoted by the symbol L ("Langmuir"), where L is the number of mg of block copolymer adsorbed onto 1 gram of TiO_2. The "total settling time," which has been defined in the experimental section is denoted by TST.

As a first approach to the adsorption of carboxylated styrene–butadiene block copolymers onto titanium dioxide particles, experimental procedures and conditions were used which had been developed by Schechter (18) in his studies on the adsorption of fatty acids onto titanium dioxide and on the effect of the fatty acid coating on the dispersibility of the particles in organic solvents. Schechter had treated titanium dioxide with solutions of fatty acids in o-dichlorobenzene at 150°–200°C for varying periods of time. After this heat treatment, he isolated the coated titanium dioxide particles by filtration, washing with ether, and drying in vacuo. Redispersion of the coated particles in n-heptane and measuring the rate of settling as described in the experimental section gave total settling times of 20–60 minutes, depending on the chain length of the fatty acid.

Following the guidelines established by Schechter's work, we dispersed titanium dioxide particles in 1% solutions of carboxylated styrene–butadiene block copolymers and stirred the dispersions at elevated temperatures in a nitrogen atmosphere. Typical data are shown in Table I. The dispersions (primary dispersions) in o-dichlorobenzene were quite stable. The titanium dioxide particles were isolated from these primary dispersions by centrifugation and were washed with toluene and finally with methanol. After drying in vacuo, samples of the block copolymer–titanium dioxide composites were submitted for carbon analysis. The

Table I. Adsorption of S30B · TGA onto TiO_2 in o-Dichlorobenzene[a]

Exp. No.	T, °C	S/P	DC, %	L, mg/gram
RA6	110	0.8	11.6	26
RA9	150	0.8	11.6	40
RA10	178	0.8	11.6	34
RD42	150	1.0	11.6	12
RD41	150	1.0	12.6	12
RD43	150	1.0	15.9	15
RD44	150	1.0	17.4	14
RD46	150	1.0	25.8	29
RD48	150	1.0	25.8	35
RD47[b]	150	1.0	1.4	17
RD36[c]	150	1.0	13.6	26

[a] Stirred for 3 hours under nitrogen in a round-bottomed flask.
[b] Stirred for 30 minutes at room temperature in a Waring Blendor, then stirred for 3 hours in a round-bottomed flask.
[c] Carboxylated polybutadiene, PB · TGA.

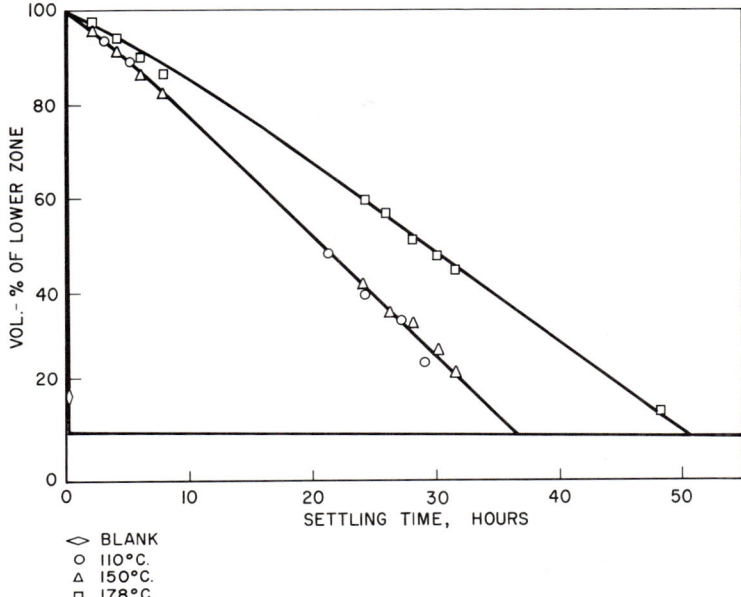

Figure 2. Secondary dispersions of TiO_2 in toluene. The particles were coated with S30B · TGA in o-dichlorobenzene.

amount of adsorbed block copolymer was calculated from the result of the carbon analyses in combination with the compositions of the adsorbed block copolymers.

The dried block copolymer-coated titanium dioxide particles were dispersed in toluene (secondary dispersions) and subjected to the settling test described in the experimental section. In the first few experiments, dispersions with 9.1% solids were prepared; later, under standardized conditions, the solids content of the dispersions was 10 wt %. Adsorption and settling data obtained under these conditions are given in Table I. The settling data of the first three samples of this table are plotted in Figure 2.

Control experiments were made to ascertain that the increase in dispersion stability in the presence of carboxylated block copolymer was indeed caused by adsorbed block copolymer and was not a side effect of the treatment—e.g., the result of a breakdown of the titanium dioxide particles from severe agitation or the result of factors associated with the exposure of the samples to elevated temperatures. Titanium dioxide was heated and agitated in o-dichlorobenzene under the same conditions as the samples listed in Table I but without block copolymer. After the agitation stopped, all particles, including the fraction of very small particles, settled out within less than one minute. When titanium dioxide was agitated for half an hour in a Waring Blendor, either at room temperature or at 100°C, no increase in dispersion stability was observed; all particles settled out immediately after the agitation stopped, and settling was complete within less than one minute. This experiment shows that severe agitation alone does not increase dispersion stability, even if it caused a decrease in particle size. When titanium dioxide was agitated for half an hour in a 1% solution of uncarboxylated block copolymer in toluene at 100°C in a Waring Blendor, again no increase in dispersion stability was observed; all particles settled out completely within less than one minute after agitation has stopped. This experiment shows that the carboxylation of one end of the block copolymer molecules is necessary to achieve dispersion stability. Since the block copolymer is entirely inactive without carboxylation, we can conclude that with carboxylated block copolymers in stable dispersions the block copolymer is adsorbed onto the titanium dioxide at its carboxylated end.

The first series of experiments were carried out in a round-bottomed flask equipped with a simple stirrer, reflux condenser, and nitrogen inlet. Later experimentation with other forms of agitation showed that better pick-up of block copolymer was obtained with more severe forms of agitation—e.g., in a Waring Blendor (—e.g., see experiment RD47 in Table I). This observation can be rationalized by assuming that severe agitation breaks up aggregates of TiO_2 particles, thus exposing more surface area to the block copolymer solution. After the initial experiments, the round-bottomed flask was abandoned as a vessel for the adsorption reaction. In all subsequent experiments, the adsorption was carried out in a Waring Blendor, usually by agitating the dispersions for 30 minutes

Table II. Adsorption of S30B · TGA in Toluene at 100°C[a]

Exp. No.	S/P	DC	D, mg/gram	TST days	Curve[b] No.
RD55	20	0.3	—	2	1b
RD61	10	0.3	1	16	1c
RD57	20	0.8	—	9	2b
RD52	10	0.8	1	22	2c
RD68	50	1.3	5	7	3a
RD67	20	1.3	8	17	3b
RD62	10	1.3	13	28	3c
RD65	50	9.9	—	<1	4a
RD66	20	9.9	28	3	4b
RD64	10	9.9	34	8	4c

[a] Stirred in a Waring Blendor for 30 minutes.
[b] The curve no. identifies the curves in Figure 3.

at 100°C. Data resulting from experiments carried out under these conditions are presented in Tables II and III. The settling data obtained with the samples of Table II are plotted in Figure 3. The data in both tables and the settling curves in Figure 3 illustrate the effect of the substrate to polymer ratio (S/P) and of the degree of carboxylation (DC) on the amount of adsorbed block copolymer (L) and on the dispersion stability as reflected in the total settling times.

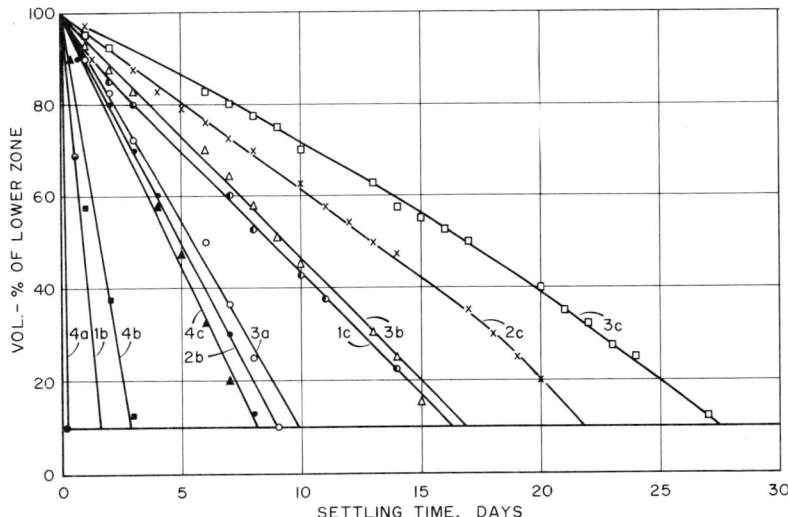

Figure 3. Primary dispersions of TiO_2 in toluene. The particles were coated with S30B · TGA at 100°C in toluene. The curves are identified in Table II.

Both the S/P ratio and the degree of carboxylation control the block copolymer pick-up, L, but a larger pick-up does not necessarily lead to a more stable dispersion. The data in Table II indicate that the total settling times increase with increasing pick-up at each degree of carboxylation, when this increase in pick-up results from a reduction of the S/P ratio. High pick-up can also be obtained at a given S/P ratio by increasing the degree of carboxylation, but even at high pick-ups obtained in this manner the total settling times are usually low at degrees of carboxylation above 10–12%. In some samples, good dispersion stability was observed at degrees of carboxylation as low as 0.1% and as high as 11.2% (see Table IV). In present experiments we prefer degrees of carboxylation of about 1–2% because block copolymers so modified give good dispersion stability under most combinations of variables.

Table III. Adsorption of S30B · TGA in Toluene at 100°C[a, d]

Exp. No.	DC, %	L, mg/gram	TST, days
RD72	0.9	8	14
RD70	1.1	8	23
RD75	0.6	9	15
RD71	1.5	9	21
RD61	0.3	11	16
RD52	0.8	11	22
RD76	0.6	12	23
RD99	0.6	12	24
RD69	1.0	12	25
RD62	1.3	13	28
RD51[b]	0.8	12	13
RD100[c]	0.6	12	26

[a] Stirred in a Waring Blendor for 30 minutes.
[b] Stirred in a Waring Blendor for 30 minutes followed by stirring in a round-bottomed flask for 3 hours.
[c] The TiO_2 was dried at 145°C for 18 hours.
[d] S/P = 10 for all experiments in this table.

Table III lists some experiments carried out under the same conditions with block copolymers of low degrees of carboxylation. Comparing the settling times with both the pick-ups and the degrees of carboxylation, it appears that the settling time increases with increasing pick-up at a given degree of carboxylation, and at constant pick-up the settling time increases with increasing degree of carboxylation. On the other hand, variations in degree of carboxylation are not always paralleled by variations in pick-up; it seems that any degree of carboxylation in the range of 0.3–1.5% could produce any pick-up in the range of 8–13 mg/gram. It is not clear whether the apparent lack of correlation between degree of carboxylation and pick-up in this small range reflects some intangible differences in the adsorption experiments or whether it simply reflects the uncertainty of the analytical data. The degrees of carboxylation were determined by neutron activation analysis of oxygen and are probably quite accurate. The pick-ups have been calculated from carbon analyses of the titanium dioxide–block copolymer composites, and we are not certain whether or not the experimental errors in the carbon analysis are large enough to invalidate any conclusions drawn from the trends in the data in Table III. If we consider all variations of the variables in Table III to be within the range of experimental error and refrain from interpreting trends in the data, we can calculate average values of the variables for the entire table: DC = 0.8%; L = 11 mg/gram; TST

= 21 days. As experiment RD100 indicates, drying the titanium dioxide before use had no effect on either pick-up or settling time.

The improvement in dispersion stability reflected by the high total settling times of 20–30 days achieved with the experiments listed in Tables II and III resulted not only from the optimization of variables such as DC, S/P, and L but also from a change in procedure. In the earlier experiments, the observation was made that the primary dispersions obtained during the heat treatment in o-dichlorobenzene were much more stable than the secondary dispersions in toluene prepared by isolating the block copolymer-coated TiO_2 particles, washing and drying them, and redispersing them in pure toluene. Therefore, the isolation of the block copolymer-coated particles was abandoned, and heat treatment directly in toluene was tried, even though the lower boiling point of toluene necessitated a reduction in temperature to 100°C. At the same time, the reaction vessel was changed: a Waring Blendor heated with heating tape was used instead of a stirred flask. The settling tests were carried out directly with the primary dispersions in all further experiments.

After all other conditions had been optimized, the effect of temperature during the adsorption step was studied again, particularly because the temperature reduction from 150° to 100°C had no deleterious effect on dispersion stability. Adsorption of the block copolymer onto the titanium dioxide at room temperature would be easy to carry out in practical applications and might be worth even a sacrifice in dispersion stability. Figure 4 shows settling data of dispersions prepared in a Waring

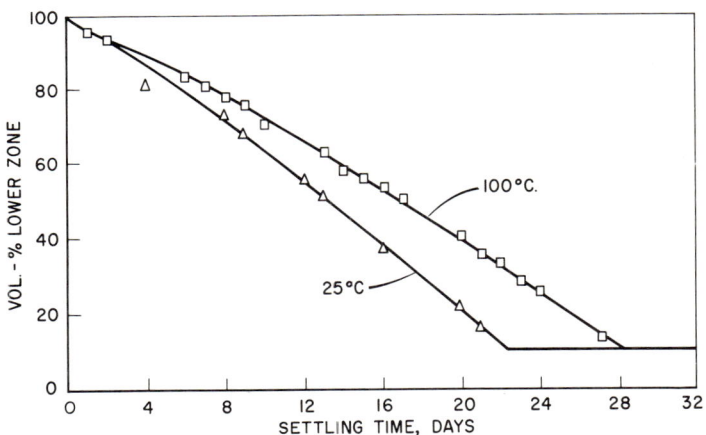

Figure 4. Primary dispersions of TiO_2 with S30B · TGA in toluene prepared at 100°C and at room temperature, respectively

Table IV. Adsorption of SX%B · TGA in Toluene at Room Temperature[a]

No.	% B	DC, %	L, mg/gram	TST, days
RD114	10	11.2	16	16
RD116	20	6.3	21	18
RD103	30	0.6	—	23
RE11	30	1.1	27	34
RE24	60	0.1	12	32
RE32	30	31.8	38	0
RE74	30	31.8	35	0

[a] Stirred in a Waring Blendor for 30 minutes. S/P = 10 for all experiments in this table.

Blendor at 100°C and at room temperature, respectively. The sample prepared at room temperature (TST = 22 days) settled somewhat faster than the sample prepared at 100°C (TST = 28 days), but the loss in total settling time was considered tolerable, and the heat treatment was abandoned entirely.

A number of adsorption experiments carried out at room temperature in a Waring Blendor are listed in Table IV. The table combines data obtained with block copolymers of different compositions. The carboxylation of the block copolymers in the first five adsorption experiments in Table IV was carried out under the same reaction conditions. With the block copolymers containing 30 wt % butadiene, these reaction conditions give a degree of carboxylation of 1–2%, but with the block copolymers containing 10 and 20% butadiene the same reaction conditions resulted in higher degrees of carboxylation. The reason for this difference in behavior of the block copolymers with lower butadiene content is apparently their higher solubility in the reaction medium. In the addition of thioglycolic acid to double bonds in a polymer under a given set of reaction conditions, the attained conversion depends largely on the solubility of both the original polymer and the carboxylated polymer in the reaction medium. The last two block copolymer samples in Table IV were carboxylated under reaction conditions designed to give highly carboxylated block copolymers.

The dispersions prepared in the first five experiments listed in Table IV were very stable. The settling data of these samples are plotted in Figure 5. The amount of adsorbed block copolymer in the treatment at room temperature is in about the same range as the pick-up previously obtained with the heat treatment methods, and the total settling times of the best dispersions in Table IV are even somewhat higher than the values previously obtained. At very high degrees of carboxylation (experiments RE32 and RE74 in Table IV), the dispersions are, again, unstable as in the experiments with higher carboxylated block copolymers carried out at elevated temperatures. It is remarkable that experiment RD114 (Table IV) gave a dispersion of quite acceptable stability, even though the block copolymer is carboxylated to an extent (DC = 11.2%) that would have been too high under previous conditions. The data are not sufficient to decide whether block copolymers at lower butadiene content can, or even must, be carboxylated to a higher degree to be adsorbed sufficiently. If the variation in the degrees of carboxylation were

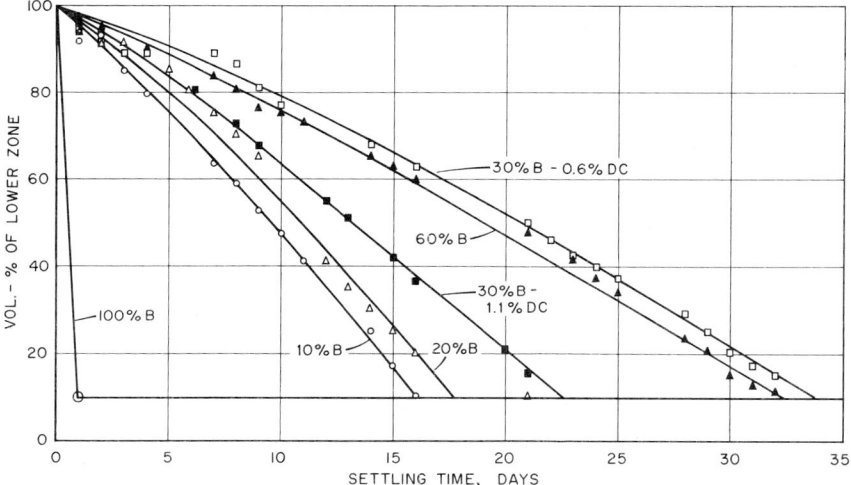

Figure 5. Primary dispersions of TiO_2 in toluene prepared at room temperature with SX%B · TGA block copolymers of various compositions (see Table IV)

disregarded in Table IV, the pick-up data and the total settling times at varying block copolymer composition considered alone would suggest that the best dispersion stability is obtained with block copolymers of intermediate compositions—*i.e.*, between 30 and 60% butadiene. Intuitively, this conclusion is quite appealing. However, the somewhat higher degrees of carboxylation obtained with the block copolymers of lower butadiene content present a difficulty in determining the effect of block copolymer composition on dispersion stability from the data in Table IV because the dispersion stability decreases as the degree of carboxylation increases to values above 10%. A study of the thioglycolic acid addition reaction with styrene–butadiene block copolymers of varying composition would be necessary to prepare a series of block copolymer samples of different composition, but all carboxylated to the same degree, to arrive at data showing the effect of block copolymer composition on dispersion stability more conclusively.

The two extremes on the styrene–butadiene block copolymer composition scale are homopolymers of butadiene or styrene, respectively. To test the usefulness of homopolymers as dispersants, polybutadiene (PB) was carboxylated by adding thioglycolic acid, and polystyrene (PS) having carboxylic groups was prepared by copolymerizing small amounts of acrylic acid (AA) into the styrene chain. Adsorption experiments with these carboxylated homopolymers are listed in Table V. In the first

Table V. Adsorption of Carboxylated Homopolymers

No.	Polymer	T, °C	S/P	DC, %	L,mg/gram
RD36	PB TGA[a]	150[c]	1	13.6	26
RD117	PB TGA[a]	25[d]	10	2.0	39
RE30	PS 1% AA[b]	25[d]	10	1.4	—
RE21	PS 2% AA[b]	25[d]	10	2.9	—

[a] Polybutadiene, $M_n = 71,000$.
[b] Random copolymers of styrene and 1 wt % or 2 wt % acrylic acid polymerized in mass without catalyst at 85°–145°C.
[c] Stirred for 3 hours under nitrogen in a round-bottomed flask (o-dichlorobenzene).
[d] Stirred for 30 minutes in a Waring Blendor (toluene).

experiment (RD36), carboxylated PB was adsorbed onto titanium dioxide by heat treatment in o-dichlorobenzene at 150°C—i.e., under the experimental conditions in Table I. In the second experiment (RD117), the optimum conditions established later (see Table IV) were used. While the amounts of adsorbed carboxylated homopolymer are comparable with the amounts of block copolymer adsorbed under equivalent conditions, the dispersions show no stability at all under the conditions of the o-dichlorobenzene treatment and extremely poor stability (TST much less than one day) under the conditions of the direct dispersions prepared by the treatment at room temperature in toluene. In the equivalent experiments with carboxylated polystyrenes in toluene at room temperature, dispersions of the same poor stability were obtained (TST much less than one day).

Discussion

This work has demonstrated that carboxylated styrene–butadiene block copolymers are excellent dispersants for titanium dioxide particles in toluene. Combining our results with carboxylated block copolymers and homopolymers and Schechter's (18) results with fatty acids, we can arrange the dispersants in the order of increasing effectiveness as follows:

Dispersant	Total Settling Time
Fatty acids	1 hour
Homopolymers	1 day
Block copolymers	1 month

Titanium dioxide particles without any surface treatment settle out completely within a few seconds.

Although the considerable increase in total settling time of the dispersions in going from homopolymer to block copolymer dispersants cannot be taken as a proof that the working model sketched in Figure 1 describes the dispersing action of block and graft copolymers correctly in all details, the magnitude of this increase does suggest that block copolymers stabilize dispersions of solid particles by a different mechanism than adsorbed homopolymers or low molecular weight compounds such as fatty acids.

The improved dispersibility of titanium dioxide particles coated with fatty acids is probably largely a consequence of better wetting of the particles by the organic medium. In addition to the wettability factor, "entropic repulsion" will also contribute to some extent to the stabilizing effect of fatty acids. The idea of "entropic repulsion" as a stabilizing principle has been introduced by Mackor (6) and has been treated for low molecular weight hydrocarbons using a simple model based on rod-shaped molecules (6, 7). Since "entropic repulsion" results from a loss of configurational entropy of adsorbed chains between colliding particles, the effect can be expected to be much larger with polymers than with low molecular weight compounds. A statistical treatment of the effect for adsorbed polymers has been presented recently by Meier (11). A similar idea for the stabilization of dispersions by adsorbed polymer molecules has been introduced by Heller and Pugh (5). In this concept, "sterical hindrance" caused by the presence of adsorbed macromolecules retards or prevents the flocculation of dispersed particles because the particles cannot approach each other to distances small enough for van der Waals forces to cause flocculation.

The effects of improved wettability, entropic repulsion, and sterical hindrance undoubtedly play a role in stabilizing dispersed solid particles by block or graft copolymers. However, since the dispersions of titanium dioxide in toluene stabilized by carboxylated styrene–butadiene block copolymers are so much more stable than dispersions stabilized by carboxylated homopolymers under otherwise identical conditions, we must assume that an additional factor comes into play when block copolymers are used. The model in Figure 1 is an attempt to explain this additional factor on the basis of the known incompatibility of chemically different polymer sequences. When two particles collide with sufficient force to cause interpenetration of the double layers of the two particles, interactions unfavorable to flocculation will result when the A layer of one particle penetrates the B layer of another particle and vice-versa because A sequences and B sequences are incompatible in nonpolar polymers.

Although we believe that the interaction of the "polymeric double layers," as shown in Figure 1, is the most important effect in the stabilizing action of block and graft copolymers, we cannot exclude the possibility that ionic charges also play a role. This is because the styrene–butadiene block copolymers are adsorbed onto the titanium dioxide surface *via* carboxylic groups. Ionic forces may have either a stabilizing or a flocculating influence. In our systems, the dispersion stability decreases as the degree of carboxylation increases. When the degree of carboxylation is too high, the styrene–butadiene block copolymers act as flocculants rather than as dispersants. A simple explanation for this effect is conceivable considering that polymer chains adsorb onto solid surfaces as loops protruding from the substrate into the medium (*17*). If too many carboxylic groups are present in the butadiene portions of the block copolymer molecules, not all of them will be attached to the titanium dioxide surface. Some will be located at the outer portions of the chain folds which extend into the medium. These far-out carboxylic groups can adsorb onto other titanium dioxide particles, thus causing flocculation by bridging.

The data in Tables I, II, and IV show the influence of the degree of carboxylation on the amount of adsorbed block copolymer. Under all conditions more block copolymer per gram of titanium dioxide is adsorbed when more carboxylic groups are attached to the butadiene chains, but a high pick-up of block copolymer is not necessarily paralleled by a high dispersion stability. At low degrees of carboxylation, both the block copolymer pick-up and the total settling time increase with increasing degree of carboxylation. At high degrees of carboxylation, the pick-up of block copolymer is high, but the dispersions are quite unstable—at 20–30% carboxylation settling occurs in less than one day. The optimal degree of carboxylation is somewhere between 1 and 10%, probably at 1–2%. All dispersions made at 1–2% carboxylation were very stable, the most stable of all samples studied. At 10% carboxylation the dispersions were usually of poor stability. Above 10–12% carboxylation stability was lost almost entirely.

The deleterious effect of too high carboxylation seems to be less pronounced with block copolymers of low butadiene content (*see e.g.,* Table IV, exp. RD114). A simple explanation would be that highly carboxylated short chains can lie flat on the substrate surface rather than having chainfolds extending far into the dispersion medium; the chance of bridging would be much reduced in this case. The good dispersion stability of sample RD114 (Table VI) even at the somewhat high degree of carboxylation of 11.2% can also be rationalized if one considers that the styrene chains are longer and thus should extend farther into the dispersion medium when the butadiene chains are shorter (at constant over-all molecular weight of the block copolymer).

The S/P (substrate-polymer) ratio has been varied between values of 0.8 and 50 (see Tables I, II, and V). Under all conditions both pick-up and total settling time vary according to the S/P ratio. The optimal value is at about 10 grams of TiO_2 per gram of polymer. Using less TiO_2 neither decreases nor increases the dispersion stability, but using more TiO_2 leads to less stable dispersions. Both this effect of the S/P ratio and the loss in dispersion stability in going from primary to secondary dispersions suggest that the adsorption of the carboxylated block copolymer is not irreversible; an adsorption equilibrium seems to exist, which, of course, depends on the S/P ratio. In contrast to the adsorption of carboxylated styrene–butadiene block copolymers the adsorption of fatty acids onto titanium dioxide under the conditions used by Schechter (18) seems to be essentially irreversible. Schechter found that his fatty acids were not removed even by water or dilute acid, but they were removed by alkali.

The initial experiments at higher temperatures (Table I and Figure 2) seemed to indicate that a temperature of 150°C gives both highest pick-up and highest dispersion stability, and Figure 4 suggests that heat treatment, even at 100°C, is better than treatment at room temperature. However, a survey of all data later obtained at room temperature leads to the conclusion that temperature is the least important variable in this work. In many experiments, even better pick-ups were obtained at room temperature than at elevated temperatures. It seems now that the use of elevated temperatures in the adsorption treatment is not necessary for preparing stable dispersions under the present conditions, under which only anhydrous solvents were used. The effect of water on the dispersions has not been studied. In the light of Schechter's data heat treatment might be beneficial when the dispersions are exposed to water because it might reduce the sensitivity of the dispersions to desorption of the block copolymer by water.

In future work we intend to study the effect of the composition of the block copolymers on dispersion stability more extensively. Preliminary results with graft copolymers have indicated that graft copolymers act as dispersants in essentially the same way as block copolymers when one type of sequence is carboxylated and the other type sequences are dissolved in the dispersion medium.

Acknowledgment

The authors are grateful to W. Reifschneider and R. A. Hickner for their suggestions on the thioglycolic acid addition reaction, to D. L. Schechter for his permission to use his data on fatty acid adsorption, and O. U. Anders for determining the degrees of carboxylation by neutron activation analysis.

Literature Cited

(1) Cox, G. H., Osmond, D. W. J., Skinner, M. W., Young, C. H., U. S. Patent **3,393,162** (July 16, 1968).
(2) Crowl, V. T., Malati, M. A., *Discussions Faraday Soc.* **42**, 301 (1966).
(3) Dobry, A., Boyer-Kawenoki, F., *J. Polymer Sci.* **2**, 90 (1947).
(4) Flory, P. J., "Principles of Polymer Chemistry," Cornell University Press, Ithaca, N. Y., 1953.
(5) Heller, W., Pugh, T. L., *J. Polymer Sci.* **47**, 203 (1960).
(6) Mackor, E. L., *J. Colloid Sci.* **6**, 492 (1951).
(7) Mackor, E. L., van der Waals, J. H., *J. Colloid Sci.* **7**, 535 (1952).
(8) Marvel, C. S., Clarke, K. G., Inskip, H. K., Taft, W. K., Labbe, B. G., *Ind. Eng. Chem.* **45**, 2090 (1953).
(9) McGown, D. N. L., Parfitt, G. D., *Kolloid-Z.* **220**, 56 (1967).
(10) McGown, D. N. L., Parfitt, G. D., *Discussions Faraday Soc.* **42**, 225 (1966); *Kolloid-Z.* **219**, 48 (1967).
(11) Meier, D. J., *J. Phys. Chem.* **71**, 1861 (1967).
(12) Molau, G. E., *J. Polymer Sci.* **A3**, 1267, 4235 (1965).
(13) Molau, G. E., Gordon Research Conference on Chemistry at Interfaces, Kimball Union Academy, Meriden, N. H., July 21-26, 1968.
(14) Molau, G. E., Keskkula, H., Canadian Patent **754,636** (March 14, 1967).
(15) Molau, G. E., Wittbrodt, W. M., *Macromolecules* **1**, 260 (1968).
(16) Romo, L. A., *J. Phys. Chem.* **67**, 386 (1963).
(17) Rowland, F., Bulas, R., Rothstein, E., Eirich, F. R., *Ind. Eng. Chem.* **57**, 47 (1965).
(18) Schechter, D. L., The Dow Chemical Co., private communication.
(19) Zettlemoyer, A. C., Micale, F. J., Lui, Y. K., *Ber. Bunsenges.* **71**, 286 (1967).

RECEIVED March 4, 1970.

25

Time–Temperature Superposition in Block Copolymers

C. K. LIM, R. E. COHEN, and N. W. TSCHOEGL

Division of Chemistry and Chemical Engineering, California Institute of Technology, Pasadena, Calif. 91109

> *The mechanical properties of Shell Kraton 102 were determined in tensile creep and stress relaxation. Below 15°C the temperature dependence is described by a WLF equation. Here the polystyrene domains act as inert filler. Above 15°C the temperature dependence reflects added contributions from the polystyrene domains. The shift factors, after the WLF contribution, obeyed Arrhenius equations ($\Delta H_a = 35$ and 39 kcal/mole). From plots of the creep data shifted according to the WLF equation, the added compliance could be obtained and its temperature dependence determined independently. It obeyed an Arrhenius equation ($\Delta H_a = 37$ kcal/mole). Plots of the compliances derived from the relaxation measurements after conversion to creep data gave the same activation energy. Thus, the compliances are additive in determining the mechanical behavior.*

Complete mechanical characterization of polymeric materials at a given temperature generally requires data extending over 10 to 20 decades of time or frequency. Because such time or frequency spans are not accessible experimentally, the measurements are generally conducted over a relatively restricted range but at widely differing temperatures, and the response curves are shifted empirically into coincidence along the logarithmic time or frequency axis. This procedure is based on the principle of the equivalence of temperature and time effects. The principle is strictly valid only when all relaxation mechanisms are affected equally by a change in temperature—i.e., when the ratio of any relaxation time at temperature T to the same relaxation time at another temperature T_r, is the same for the entire distribution of relaxation times. This requirement is expressed by Equation 1

$$\tau_i(T) = a_{T_r}(T)\tau_i(T_r) \tag{1}$$

in which $\tau_i(T)$ is the ith relaxation time at temperature T, T_r is a reference temprature, and $a_{T_r}(T)$, commonly abbreviated to a_T, is the ratio.

Materials to which time–temperature superposition is applicable, are sometimes termed thermorheologically simple materials. The amount of shift, log a_T, which is required to bring measurements at a given temperature, T, into superposition with measurements at a reference temperature, T_r, is described, usually within a range of temperatures $T_g < T < (T_g + 100)$, by the WLF equation (6):

$$\log a_T = -\frac{c_1(T - T_r)}{c_2 + T - T_r} \tag{2}$$

In this equation c_1 and c_2 are primarily empirical constants characteristic of the material and dependent on the chosen reference temperature. They have, however, been given some theoretical interpretation (6). Below the glass transition temperature, T_g, the temperature dependence of the mechanical properties is often described by the Arrhenius equation

$$\log a_T = \frac{\Delta H_a}{2.303\ R}\left(\frac{1}{T} - \frac{1}{T_r}\right) \tag{3}$$

where R is the gas constant, and ΔH_a is the (apparent) activation energy (6).

Many amorphous homopolymers and random copolymers show thermorheologically simple behavior within the usual experimental accuracy. Plazek (23, 24), however, found that the steady-state viscosity and steady-state compliance of polystyrene cannot be described by the same WLF equation. The effect of temperature on entanglement couplings can also result in thermorheologically complex behavior. This has been shown on certain polymethacrylate polymers and their solutions (22, 23, 26, 31). The time–temperature superposition of thermorheologically simple materials is clearly not applicable to polymers with multiple transitions. The classical study in this area is that by Ferry and co-workers (5, 8) on polymethacrylates with relatively long side chains. In these the complex compliance is the sum of two contributions with different sets of relaxation mechanisms: the compliance of the chain backbone and that of the side chains, respectively.

Time–temperature superposition in materials with multiple transitions can be studied advantageously in block copolymers. Although exceptions have been noted (25), random copolymerization of monomers whose homopolymers would have different glass temperatures generally leads to substances forming a single thermodynamic phase with a single intermediate transition temperature. This can usually be predicted from the composition and the T_g's of the constituent homopolymers (*10, 17, 18*). By contrast, block copolymers, composed of relatively long segments of homopolymers with different T_g's, may show multiple glass transitions if the blocks are sufficiently dissimilar to form distinct thermodynamic phases. A similar effect can be achieved by blending two dissimilar homopolymers (*16*). However, in such blends the domains formed by the constituent present in the smaller amount are quite large and differ widely in size (*21, 32*). In block copolymers the constituents are linked into a single molecule, and the size of the domain is governed primarily by the molecular weight of the constituent block of the disperse phase. The domains are thus orders of magnitude smaller and quite regular in size (*2, 12, 14, 21*).

Phase separation and the formation of domains in blends and block copolymers results from the thermodynamic incompatibility of the constituents. The mixing of two phases is governed by the thermodynamic relation

$$\Delta G_m = \Delta H_m - T \Delta S_m \tag{4}$$

where ΔG_m, ΔH_m, and ΔS_m are changes in the free enthalpy (Gibbs free energy), the enthalpy, and the entropy of mixing, and T is the absolute temperature. For a long chain molecule ΔS_m is necessarily small. The change in the enthalpy of mixing of hydrocarbons is generally positive in the absence of specific strongly interacting groups. Thus, ΔG_m, the change in the free enthalpy, is usually also positive, and mixing cannot occur. The fact that the segments forming the two phases are portions of the same molecule has no effect as long as all constituent blocks have sufficiently large molecular weights.

Since the relaxation mechanisms characteristic of the constituent blocks will be associated with separate distributions of relaxation times, the simple time–temperature (or frequency–temperature) superposition applicable to most amorphous homopolymers and random copolymers cannot apply to block copolymers, even if each block separately shows thermorheologically simple behavior. Block copolymers, in contrast to the polymethacrylates studied by Ferry and co-workers, are not single-phase systems. They form, however, felicitous models for studying materials with multiple transitions because their molecular architecture can be shaped with considerable freedom. We report here on a study of time–temperature superposition in a commercially available triblock copolymer rubber determined in tensile relaxation and creep.

Materials and Experimental Methods

The material was Shell Kraton 102, a polystyrene/1,4-polybutadiene/polystyrene block copolymer in which the polystyrene end blocks form glassy domains acting as multiple crosslink points. Because it is not covalently crosslinked, such a triblock elastomer is soluble in organic solvents. Kraton 102 contains about 0.25% antioxidant (Ionol) but is otherwise said to consist of the block copolymer only (15). No work appears to have been reported on Kraton 102 although some related materials (Shell Kraton 101, Shell T226, and Shell T125) have been investigated (4, 30). Kraton 101 has a somewhat higher molecular weight than Kraton 102 (28). Shell T226 and T125 contain a plasticizer, a dye, and possibly other additives (30).

The number average molecular weight of Kraton 102 as determined in a Melabs membrane osmometer in toluene was 56,500. The styrene content was determined as 32 wt % by NMR spectroscopy (27) on a Varian A-60 spectrometer. A separate determination by the method of Kolthoff, Lee, and Carr (19), in which the polybutadiene segment is cleaved off by oxidation and the remaining polystyrene is weighed, gave 31% by weight. From these figures the number average molecular weight of the polystyrene end blocks is estimated as 9500 and that of the polybutadiene center block as 37,500.

A gel permeation chromatogram obtained by Heller (13) indicated that Kraton 101 contained 1% polystyrene, 22% polystyrene/1,4-polybutadiene, and 77% polystyrene/1,4-polybutadiene/polystyrene, apparently because the material is manufactured by coupling of the diblock. The diblock would be present if the coupling reaction is not 100% efficient. Kraton 101 would thus be essentially a blend of ABBA and AB block copolymers. Assuming that Kraton 102 is manufactured in the same way, the molecular weight of the ABBA block copolymer results as about 72,000 from the above data, the polystyrene blocks having a molecular weight of about 12,000, and the polybutadiene center block one of about 48,000.

The polybutadiene microstructure was examined by NMR spectroscopy (27). It was estimated that the polybutadiene segment contained about 7% 1,2-addition and 93% 1,4-addition. The relative amounts of cis and trans structures in the 1,4-adduct could not be resolved by this method.

Sheets of Kraton 102 were prepared by casting from solution in benzene, cyclohexane, and tetrahydrofuran. The lower (1,4-polybutadiene) transition temperature was $-88°C$ for all samples by differential thermal analysis (DTA) on a duPont thermal analyzer at a heating rate of $5°C$/minute. The upper transition temperature of the triblock could not be determined reliably. Torsion pendulum measurements at around 0.1 Hz on a sample cast from toluene solution onto glass fibers showed a peak in a plot of the damping coefficient against temperature at about 80°C. DTA in the duPont thermal analyzer as well as differential scanning calorimetry on a Perkin-Elmer scanning calorimeter (9) revealed no transition but showed what appeared to be an endothermic peak centered at 83°C. Such a peak could result from a stress release which is sometimes observed to accompany a glass transition (9). This value agrees well with 84°C found by Canter (4) using Perkin-Elmer and Leeds and Northrup differential thermal calorimeters for the upper transition of the related block copolymer in Shell T125. The difficulty in locating the glass transition of the polystyrene blocks in Kraton 102 appears to be caused by the short chain length and relatively low percentage of polystyrene. DTA revealed no transition between $-88°$ and $+83°$C.

The sample sheets were prepared by pouring a 10% solution of Kraton 102 in the appropriate solvent into a carefully leveled glass tray with a piece of plate glass cemented to the bottom. The entire tray was placed on a composite cork-rubber cushion to help damp out vibrations. The tray was covered with a piece of cardboard with numerous pinholes to allow the solvent to evaporate. After several days the sheet was removed and kept at room temperature for two weeks before strips were cut. To check for residual solvent, several strips were preweighed and then exposed to vacuum at room temperature for 48 hours. No change in weight was observed.

The mechanical properties of Kraton 102 were measured in tensile relaxation and creep. For the relaxation measurements on Kraton 102 cast from benzene solution twenty 21.0-cm long, 1.27-cm wide, and 0.2-cm thick strip specimens with a gage length of 10.0 cm were cut from the sheet in a milling machine using a knife-edged disc-shaped cutter, specially modified from a slitting saw. Measurements were made in an Instron tester at a fixed strain of 4% and at 16 different temperatures between $-70°$ and 70°C. The strips were extended at a rate of 0.85 cm/sec. Each experiment covered a time range of about three decades from 1 to 1000 seconds.

The relaxation measurements on sheets cast from cyclohexane solution were made in the same way at 15 temperatures from $-70°$ to 80°C. Two sheets were cast from tetrahydrofuran solution. Relaxation measurements were made at 4% strain at 10 temperatures between $-70°$ and 70°C, and at 8% strain at 9 temperatures between $-70°$ and 70°C.

Creep tests in uniaxial tension were made at 11 temperatures between $-70°$ and 75°C on strips which were cut 1.27-cm wide, 0.12-cm thick, and 20.0-cm long from a sheet cast from benzene solution. The distance between the grips was 19.0 cm. Bench marks were placed 15.0 cm apart. Displacements were read from the bench marks using a power driven cathetometer of 0.0002-cm sensitivity. The creep tests covered a minimum range of about four decades of time beginning at 10 seconds. The measurements were repeated at 15°, 30°, 40°, and 50°C on a second sheet, also cast from benzene solution. Measurements derived from these two sheets are distinguished by referring to them as Sheet I and Sheet II, respectively. The loads were kept small enough so that the final strains, ϵ_x, were generally between 2 and 8%.

A nitrogen blanket was used whenever measurements were made at temperatures above 50°C to prevent oxidation of the samples. Specimen dimensions were corrected for temperature expansion using a linear expansion coefficient of 0.0002 (°C)$^{-1}$. This is the coefficient for an SBR of about the same composition.

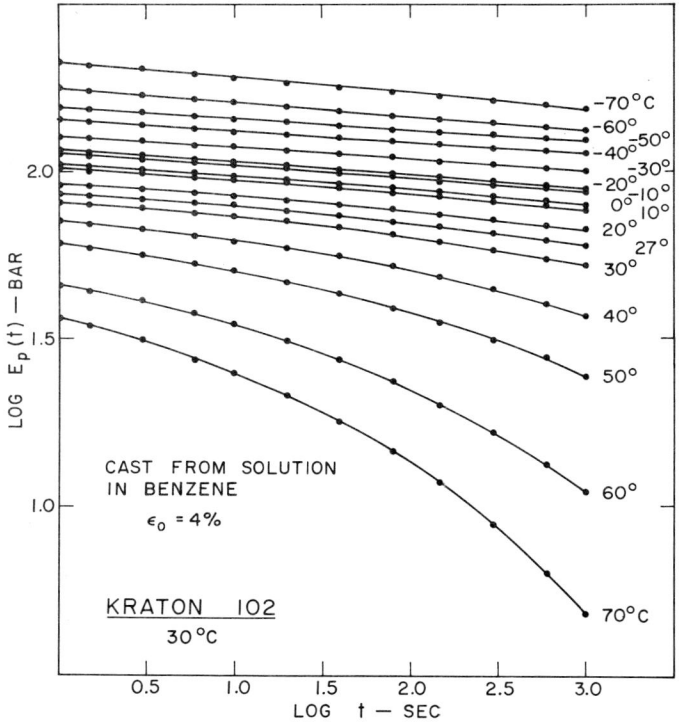

Figure 1. Reduced tensile relaxation modulus, $E_p(t)$, of Kraton 102 cast from benzene solution at different temperatures as a function of time, t

Results

Figure 1 shows a plot of the reduced tensile relaxation modulus, $E_p(t)$, against the time, t, in logarithmic coordinates for Kraton 102 cast from benzene. Similar plots were prepared for the results obtained on specimens cast from cyclohexane and tetrahydrofuran solution, respectively. The moduli were reduced to a common temperature, 30°C, according to

$$E_p(t) = (303/T)E(t) \tag{5}$$

neglecting the small correction for density changes (6). The segments were then shifted into superposition. The corresponding master curves are shown in Figure 2. $E_p(t)$ is given in bars. One bar equals 10^6 dynes/cm², or 14.5 psig.

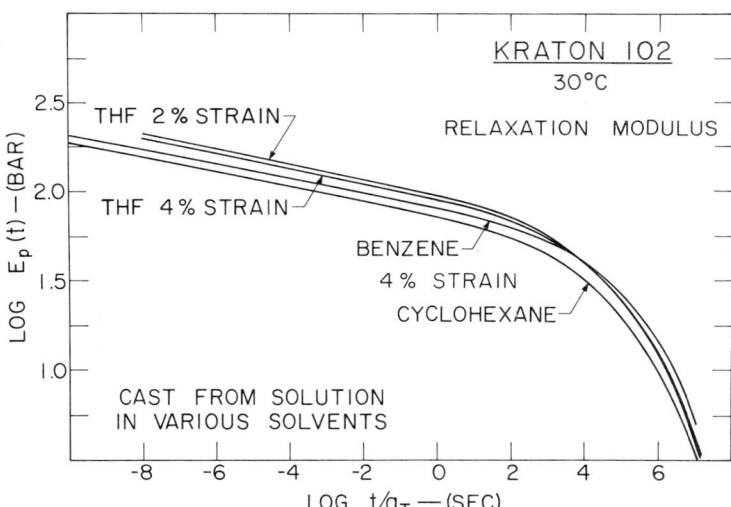

Figure 2. Master curves of the tensile relaxation modulus, $E_p(t)$, of Kraton 102 cast from benzene, cyclohexane, and tetrahydrofuran solutions, as a function of time, t, at 30°C

The creep compliances were reduced to 30°C using Equation 6.

$$D_p(t) = (T/303)D(t) \tag{6}$$

Figure 3 shows the plot of the reduced tensile creep compliance, $D_p(t)$, against t in logarithmic coordinates for the creep tests on Sheet I. A similar plot was made for the data obtained from Sheet II, and, in addition, for the relaxation data shown in Figure 1 after conversion to creep data using the relation (7):

$$D_p(t) = \frac{\sin(m\pi)}{m\pi E_p(t)} \tag{7}$$

in which m is the slope of log $E_p(t)$ vs. log t. This procedure gave rea-

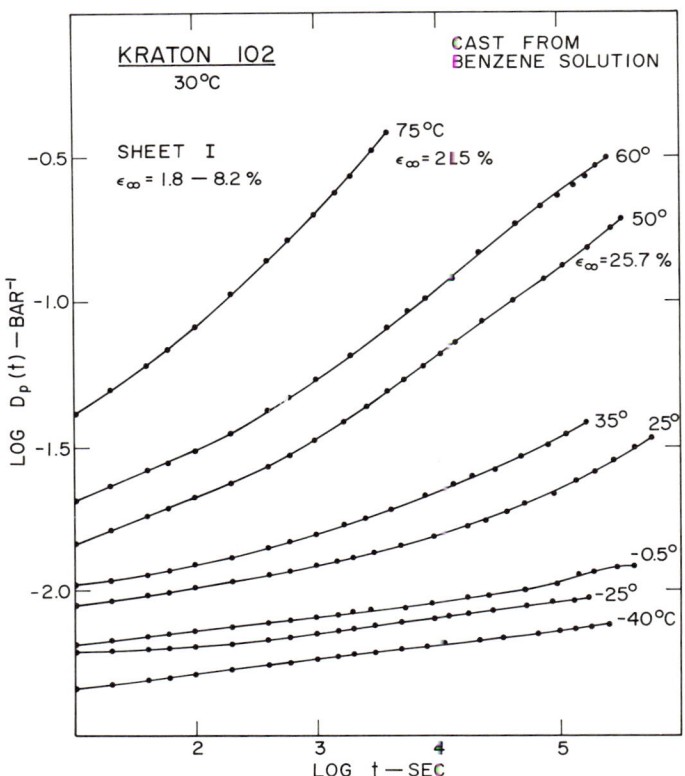

Figure 3. Reduced tensile creep compliance, $D_p(t)$, of Kraton 102 cast from benzene solution, at different temperatures as a function of time, t (Sheet I)

sonable results because the slopes were always smaller (7) than 0.8. The master curves are shown in Figure 4. $D_p(t)$ is given in reciprocal bars. One reciprocal bar equals 10^{-6} cm^2/dyne, or 0.069 sq in/lb.

The differences in the master curves for log $E_p(t)$ vs. log t obtained from Kraton 102 specimens cast from benzene, cyclohexane, and tetrahydrofuran solutions may be caused by differences in the composition or the morphology of the phases. Beecher et al. (2) have emphasized the role of solvent in determining the phase structure of cast block copolymer films. At 25°C the solubility parameters for the polymers (3) and solvents (1) are:

Material	Solubility Parameter
Cyclohexane	8.2
Polybutadiene	8.4
Tetrahydrofuran	9.1
Polystyrene	9.1
Benzene	9.2

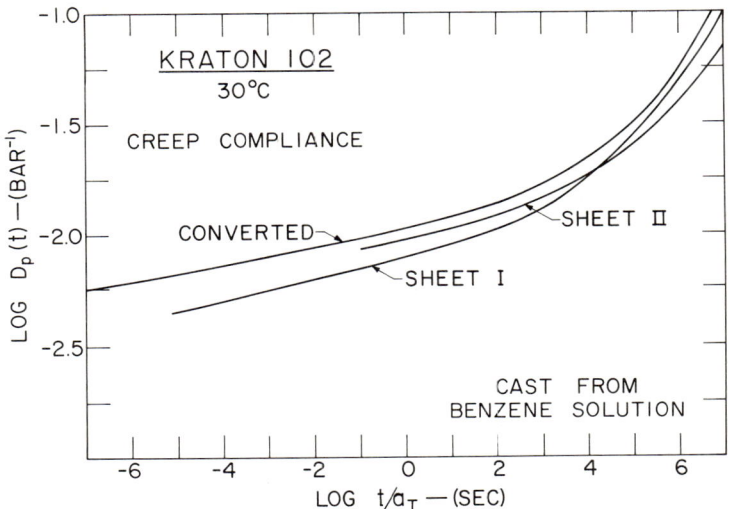

Figure 4. Master curves of the tensile creep compliance, $D_p(t)$, of Kraton 102 cast from benzene solution, as functions of time, t, at 30°C

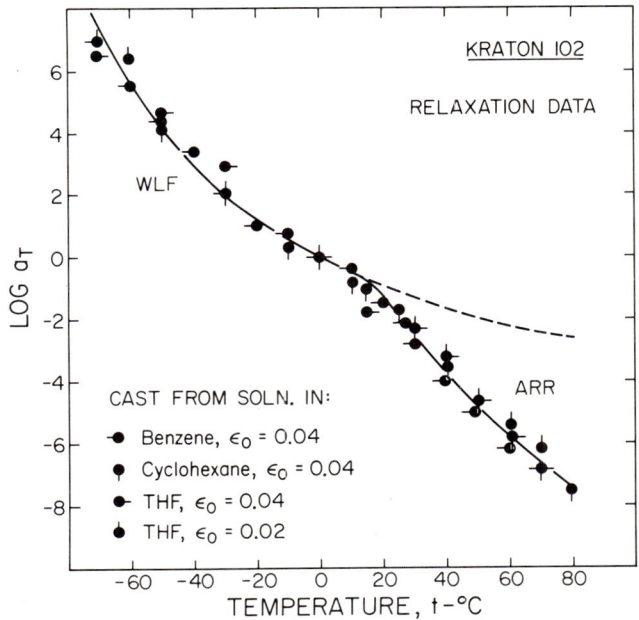

Figure 5. Temperature dependence of the relaxation modulus of Kraton 102 cast from benzene, cyclohexane, and tetrahydrofuran solutions

Figure 2, however, reveals no systematic change in properties with respect to the solubility parameters.

The master curves obtained from specimens cast from tetrahydrofuran solution at 2 and 4% strain, respectively, are slightly different. These differences, however, are probably within the experimental error. An idea of the reproducibility can be obtained from Figure 4, which shows the master curves of the creep compliances obtained on specimens cut from two sheets of Kraton 102 cast from benzene solution. Although the method of preparation appeared to be identical, there are noticeable differences between the two curves. Even larger differences exist between these curves and the master curve obtained from the relaxation data after conversion to creep. Again, there were no apparent differences in the method of preparation of the sheets from which the specimens for the relaxation and creep tests were cut.

These differences in the mechanical behavior are not reflected, within the experimental error, in the temperature dependence of the mechanical properties. As shown by the examples of Figures 1 and 3, the relaxation modulus and creep compliance data showed very little scatter and could be shifted smoothly into superposition along the logarithmic time axis. The amounts of shift, log a_T, required to effect superposition are plotted against the temperature, T, in Figure 5 for the relaxation data, and in

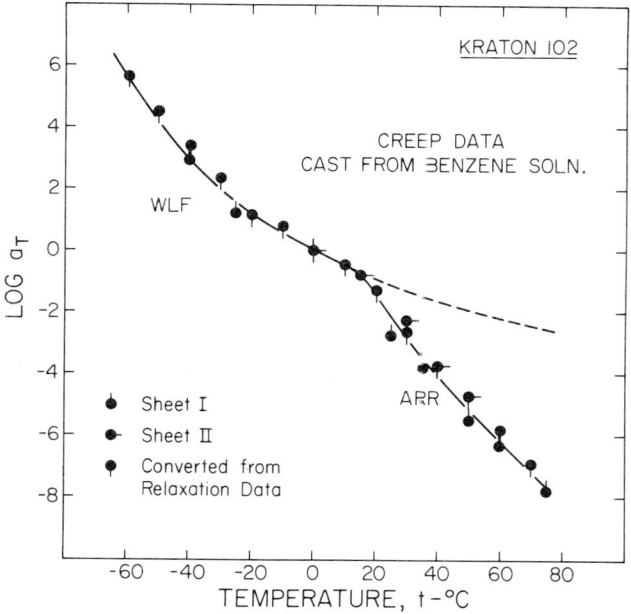

Figure 6. Temperature dependence of the creep compliance of Kraton 102 cast from benzene solution

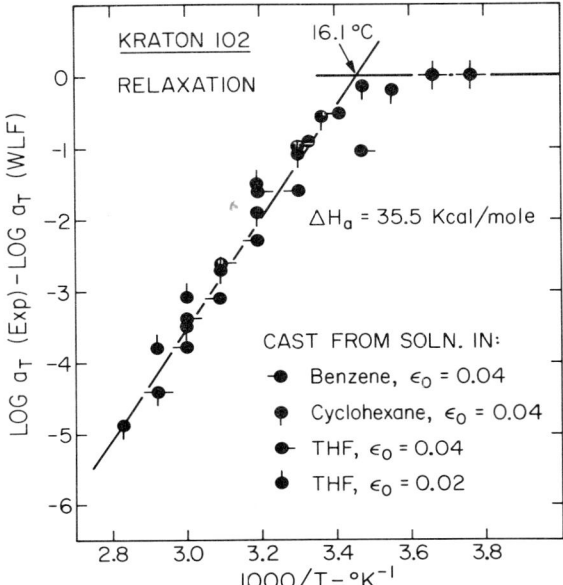

Figure 7. Difference between the experimentally determined shift factors, log $a_T(exp)$, and those predicted by the WLF equation, log $a_T(WLF)$, plotted against the reciprocal of absolute temperature. Relaxation data.

Figure 6 for the creep data, using 0°C as the reference temperature. The shift factors obtained from creep data on Sheets I and II are plotted on the same curve, within experimental error, although the two sets of data did not superpose well with each other. The same is true for the data at 2 and 4% strain on specimens cast from tetrahydrofuran solution. Quite generally, data could only be shifted with respect to each other when they were derived from the same sheet.

Below a characteristic temperature, T_o, of about 15° to 16°C, the shift factors appear to follow the WLF equation, Equation 2, with $c_1 = 7.1$, $c_2 = 135.9°C$, and $T_r = 0°C$. The coefficients were determined in the usual way (6). The temperature dependence of both the relaxation moduli and the creep compliances could be described with the same WLF equation within the experimental scatter. It appears that below T_o the triblock copolymer behaves essentially as a filled rubber, the polystyrene domains acting only as inert filler. However, the WLF equation which describes the temperature dependence of the mechanical properties in this region is not identical with that of pure 1,4-polybutadiene, for which Maekawa, Mancke, and Ferry (20) find $c_1 = 4.20$, $c_2 = 161.5°C$, while Kraus and Gruver (11) give $c_1 = 4.87$, $c_2 = 184.8°C$, when referred to 0°C as the reference temperature. The experiments of Maekawa, Mancke, and Ferry and of Kraus and Gruver were made on the same material, containing 40% cis and 53% trans 1,4-addition and the same (7%) 1,2-addition as Kraton 102.

At temperatures above T_o, the experimentally determined shift factors, log a_T, deviated noticeably from those predicted by the WLF equation, indicating the appearance of a new mechanism affecting the temperature dependence of the mechanical properties. If the deviations from WLF behavior are to be attributed to the polystyrene domains, the temperature dependence of this contribution should follow Arrhenius behavior because the polystyrene is in a glassy state up to about 80°C. To determine the nature of the temperature dependence of the deviations from WLF behavior, the differences between the observed shift factors and those calculated from the WLF equation are plotted against $1/T$ in Figures 7 and 8. The resulting straight line plots show that the temperature dependence of the new mechanism is, indeed, of the Arrhenius type within the experimental error. From the slopes, the activation energies

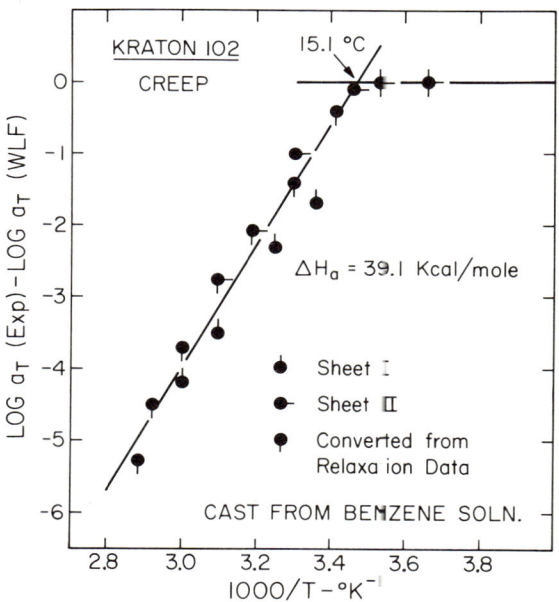

Figure 8. Difference between the experimentally determined shift factors, log $a_T(exp)$, and those predicted by the WLF equation, log $a_T(WLF)$, plotted against the reciprocal of absolute temperature. Creep data.

are obtained as 35.5 and 39.1 kcal/mole, respectively. The lines intersect the abscissa at 15.1° and 16.1°C, respectively. These values may be regarded as the temperatures, T_o, at which the contribution of the new mechanism to the temperature dependence of the mechanical properties of the block copolymer becomes noticeable.

The over-all temperature dependence can therefore be described by an equation of the WLF form up to T_o, and above T_o by an equation of the Arrhenius form superposed on the WLF equation. The solid lines in Figures 7 and 8 represent the equation

$$\log a_T = -\frac{c_1(T - T_r)}{c_2 + T - T_r} h(T) + \frac{\Delta H_a}{2.303\ R}\left(\frac{1}{T} - \frac{1}{T_o}\right) h(T - T_o) \quad (8)$$

where $h(T)$ is the unit step function (34), and the other quantities have the significance mentioned earlier. Because the material became too soft, the measurements could not be extended above the glass temperature of the polystyrene domains—i.e., above 80°C. Above this temperature the contribution of the polystyrene domains should turn from Arrhenius to WLF behavior. Equation 8 could be modified to account for this.

Discussion

Few data have appeared in the literature on time–temperature superposition in block copolymers. Beecher et al. (2) examined a polystyrene/cis-polyisoprene/polystyrene triblock of molecular weight 130,000 containing 22% styrene by weight. They showed a curve, without experimental points, of log a_T vs. T for this block copolymer which is in essential agreement with our curves in Figures 5 and 6. Smith and Dickie (30) investigated Kraton 101 in constant rate of extension experiments and found that their experimentally determined shift factors could be represented by a straight line in the region −30° to 60°C. At −40°C the datum point lay above the line. These experiments were made at large deformations, and the authors argue that the observed temperature dependence reflects disruption of the polystyrene domains.

Shen and Kaelble (29) found the same linear dependence in the region −60° and 60°C but state that below −50°C and above 80°C the temperature dependence of Kraton 101 could be described by the WLF equation with $c_1 = 16.14$, $c_2 = 56$, and $T_r = -97°C$ below −50°C, and $T_r = 60°C$ above 80°C. They ascribe the temperature dependence below −50°C to the pure polybutadiene phase and that above 80°C to the pure polystyrene phase. They then assume that at temperatures between −50° and 80°C the molecular mechanisms for stress relaxation are being contributed by an interfacial phase visualized as a series of spherical shells enclosing each of the pure polystyrene domains and characterized by a fairly sharp concentration gradient between shells of mixed polybutadiene and polystyrene segments. In our data there appears to be no evidence of an intermediate region of temperature dependence arising from an interfacial phase. As mentioned earlier, DTA measurements on Kraton 102 showed no evidence of a transition between −88° and 83°C.

Kraton 101 has a higher molecular weight than Kraton 102 (28). Contributions from an interfacial layer to the temperature dependence of a block copolymer might show up only if the molecular weights are large enough for the formation of the interlayer. If the existence of such a layer in Kraton 101 is corroborated by future experiments, one would have to conclude that Kraton 102 (or a block copolymer with similar molecular weights) is better suited for a study of the fundamentals of time–temperature superposition in block copolymers because the complications arising from the possible existence of a third interfacial phase are absent.

In an investigation of the birefringence and stress relaxation of Kraton 101 cast from solution in toluene and in methyl ethyl ketone, Wilkes and Stein (33) considered the relaxation modulus to be a weighted average of the moduli of the pure polybutadiene and polystyrene phases. Ferry and co-workers, in their investigations of time–temperature superposition in polymethacrylates with relatively long side chains, found the com-

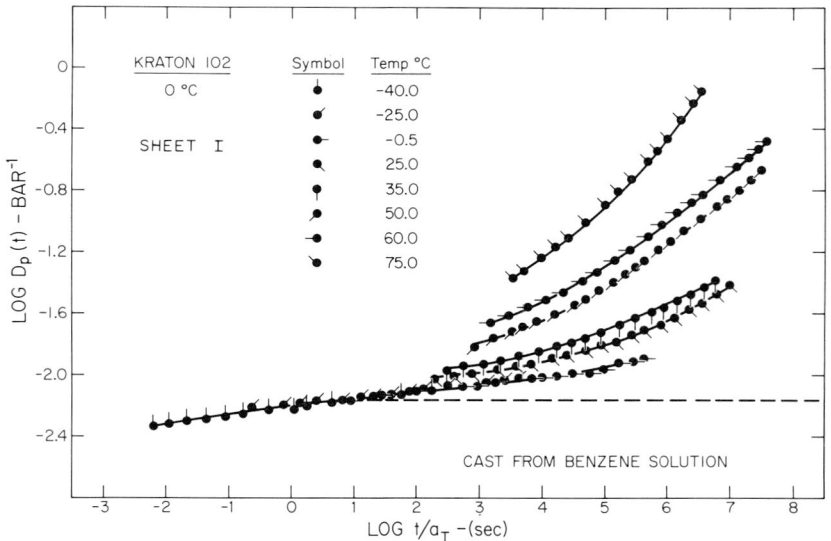

Figure 9. Reduced tensile creep compliance of Kraton 102 cast from benzene solution (Sheet I), as a function of t/a_T at 0°C (log a_T predicted by WLF equation)

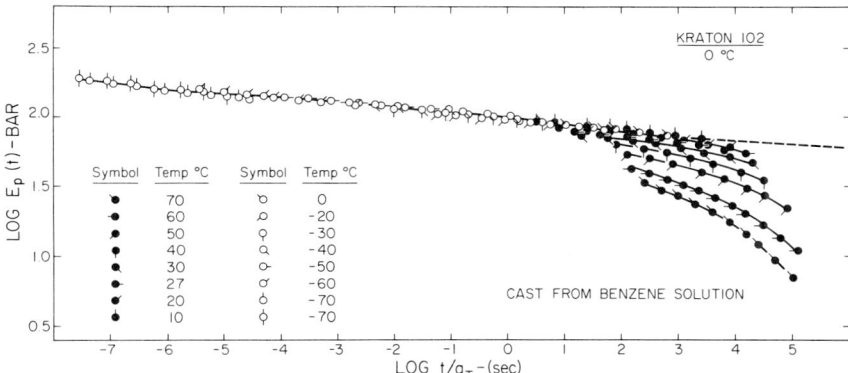

Figure 10. Reduced tensile relaxation modulus of Kraton 102 cast from benzene solution as a function of t/a_T at $0°C$ (log a_T predicted by WLF equation)

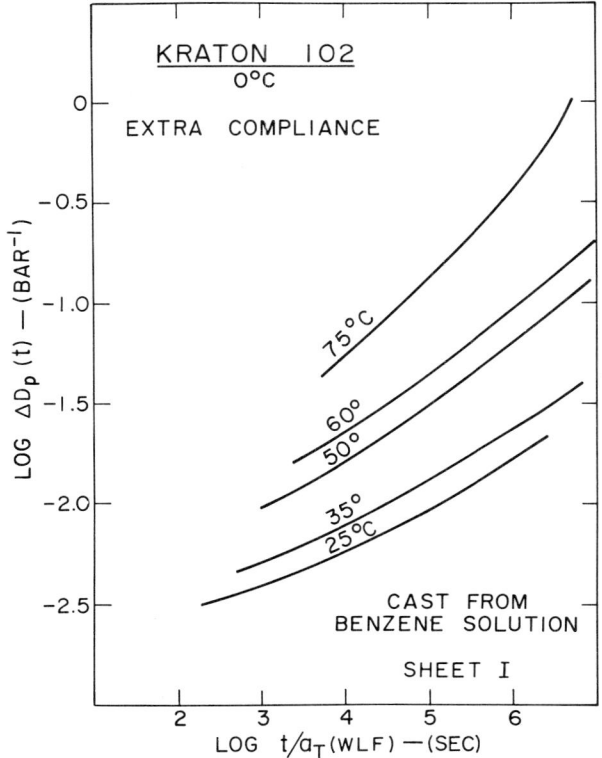

Figure 11. Extra compliance of Kraton 102 cast from benzene solution (Sheet I) at different temperatures as a function of reduced time, t/a_T

pliances to be additive. To determine which of the two possibilities applies to our data, we plotted log $D_p(t)$ or log $E_p(t)$, reduced to a reference temperature of 0°C, against t/a_T using the shift factors predicted by the WLF equation, log $a_T(\text{WLF})$, with the constants $c_1 = 7.1$ and $c_2 = 135.9°C$ determined from the experimental values plotted in Figures 5 and 6. This resulted in superposition into a master curve for the data below T_o and in a disconnected segment for each temperature above T_o. Figures 9 and 10 show the plots for the creep compliance and relaxation modulus data obtained from specimens cast from benzene solution. After suitable extrapolation of the base curves, values of the extra compliance, $\Delta D_p(t)$ or extra modulus, $\Delta E_p(t)$, appearing above T_o at different times were calculated for each temperature from the difference between the arithmetic value of the base curve and that of the appropriate segment. These values are shown as functions of log $t/a_T(\text{WLF})$ in Figures 11 and 12. The extra compliance is not a sensitive function of the position of the base line. We tried several base lines with only minor variation in the calculated apparent activation energy. The base line was drawn

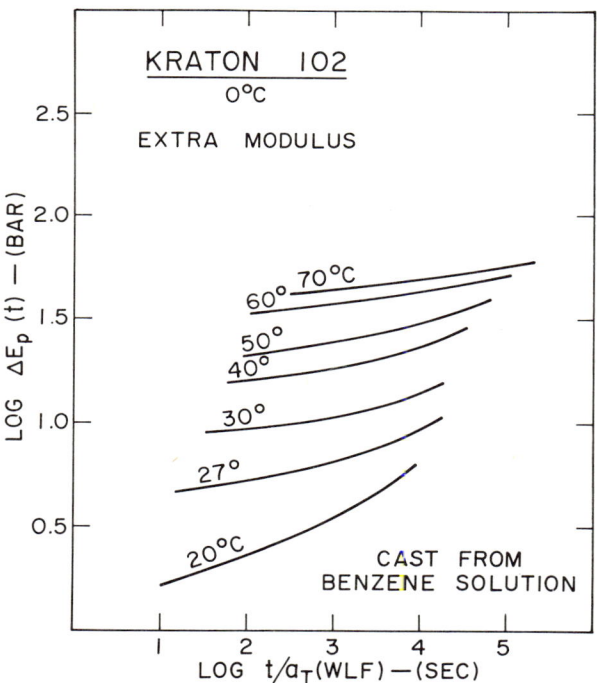

Figure 12. *Extra moduli of Kraton 102 cast from benzene solution at different temperatures as a function of reduced time,* t/a_T

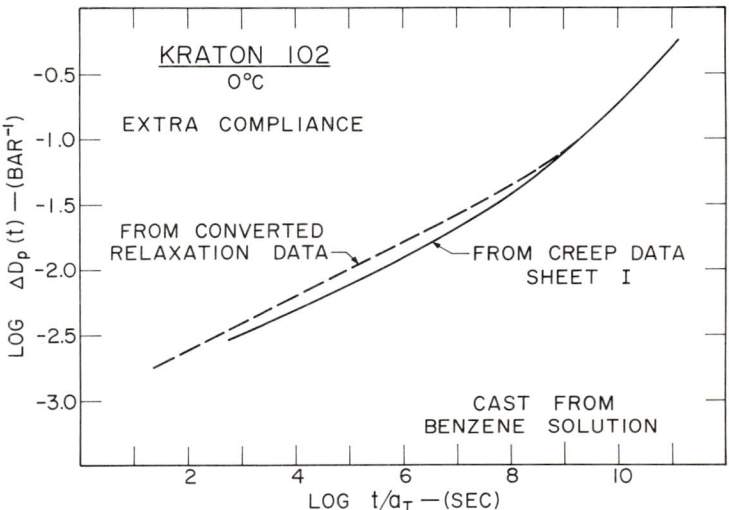

Figure 13. Master curves of the extra tensile creep compliance, $\Delta D_p(t)$, of Kraton 102, cast from benzene solution at 0°C

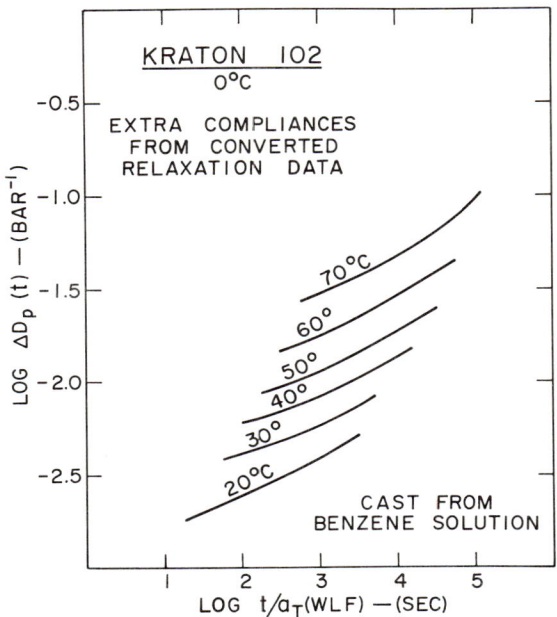

Figure 14. Extra compliance of Kraton 102 cast from benzene solution after conversion of relaxation to creep data at different temperatures as a function of reduced time, t/a_T

level on the assumption that if only crosslinked polybutadiene and inert filler were present in the sample, the base line, over the time range over which it extends, would represent the equilibrium compliance, D_e, of the filled polybutadiene, and that any departure from it must represent the compliance of the polystyrene domains.

It is clear from Figure 12 that the segments of log $\Delta E_p(t)$ vs. log t/a_T(WLF) can not be shifted into superposition with each other. The segments of log $\Delta D_p(t)$ could be superposed with ease. The corresponding master curve is shown in Figure 13, referred to the reference temperature of 0°C. This decomposition procedure clearly demonstrates that it is the compliances of the two mechanisms and not their moduli which are additive. The point is brought home even more forcefully by applying the same procedure to the relaxation data after conversion to creep. The resulting plot of log $\Delta D_p(t)$ vs. log t/a_T(WLF) is shown in Figure 14, and the shifted master curve is also plotted in Figure 13. Figure 15 shows an Arrhenius plot of the shift factors obtained by shifting

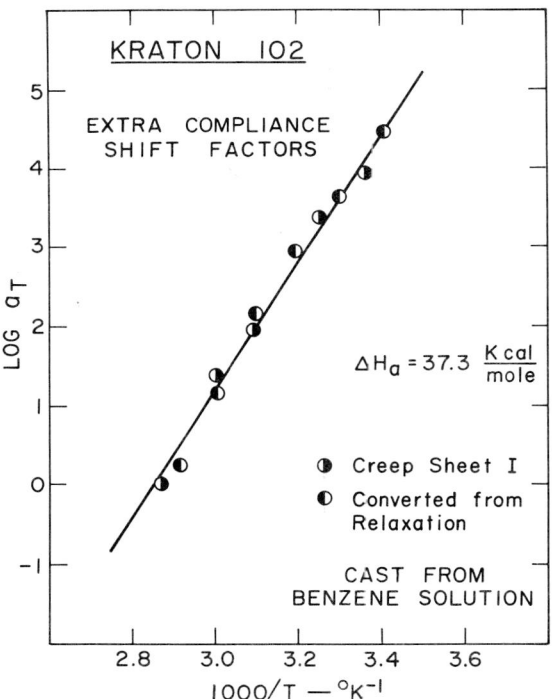

Figure 15. Shift factors, log a_T, for extra compliance of Kraton 102 cast from benzene solution, plotted against the reciprocal of absolute temperature

the extra compliances from the creep measurements and those from the relaxation measurements on the specimens cast from benzene solution after conversion to creep. They all lie on a straight line from which the apparent activation energy is obtained as 37.3 kcal/mole, in excellent agreement with the values calculated from the plots in Figures 7 and 8. These values are also in fairly good agreement with 40 kcal/mole calculated from the data by Smith and Dickie (30) for Kraton 101, and the values of 45 and 46 kcal/mole obtained by Wilkes and Stein (33) from birefringence and stress relaxation measurements.

Because of the uncertainties involved in the decomposition, this procedure would not appear to be a practical way to determine the ΔH_a value needed for Equation 8. It does, however, demonstrate three important points: (1) it is the compliances of the mechanisms that are additive; (2) T_o and ΔH_a can be obtained from plots such as those shown in Figures 7 and 8 of shift data determined in either relaxation or creep experiments without decomposition of compliance master curves; (3) Equation 8 describes time–temperature superposition in Kraton 102 adequately within the experimental accuracy.

Several questions were left unanswered in this investigation. One is the possible role of an interfacial phase which appears to be absent in Kraton 102. Another is the nature of the characteristic temperature, T_o, above which the contribution of the added compliance begins to be felt. Although it appears that this compliance arises from the polystyrene domains, it is not clear why it should appear precisely at 15°–16°C in Kraton 102.

As shown in Figures 5 and 7 the nature of the solvent does not appear to have any effect on T_o within experimental error. However, the solvent can have a profound influence on the morphology of cast block copolymer specimens. Thus, instead of the continuous polybutadiene phase normally observed, a continuous polystyrene phase appears to exist in Kraton 101 films cast from solution in MEK/THF mixtures (2). Methyl ethyl ketone has a solubility parameter of 9.3, only slightly higher than that of the solvents used in our work. It is clear from the data presented here that our films must have had continuous polybutadiene phases.

Dynamic experiments underway in our laboratory require a higher value of T_o, and it is possible that T_o depends on the magnitude of the strain. The strains used in the present investigation were of the order of 4%. Another possibility, suggested by R. G. Mancke, is that T_o might depend on the time scale of the measurements and would thus necessarily be different in dynamic and transient experiments. Work at present underway is expected to help clarify this issue.

Acknowledgment

The authors thank J. D. Ferry, R. F. Landel, and R. G. Mancke for helpful discussions, and L. T. Carmichael for invaluable assistance with the experimental measurements.

This work was supported by the Air Force Rocket Propulsion Laboratory, Edwards, Calif., and the Materials Laboratory, Wright-Patterson Air Force Base, Ohio, Air Force Systems Command, United States Air Force.

Literature Cited

(1) Bailey, J. T., *J. Elastoplastics* **1**, 2 (1969).
(2) Beecher, J. F., Marker, L., Bradford, R. D., Aggarwal, S. L., *J. Polymer Sci., Pt. C* **26**, 117 (1969).
(3) Burrell, H., Immergut, B., "Polymer Handbook," p. iv-341 ff, Interscience, New York, 1966.
(4) Canter, N. H., *J. Polymer Sci., Pt. A-2* **6**, 155 (1968).
(5) Child, Jr., W. C., Ferry, J. D., *J. Colloid Sci.* **12**, 389 (1957).
(6) Ferry, J. D., "Viscoelastic Properties of Polymers," 2nd ed., Chap. 11, Wiley, New York, 1970.
(7) *Ibid.*, p. 98.
(8) Ferry, J. D., Child, Jr., W. C., Zand, R., Stern, D. M., Williams, M. L., Landel, R. F., *J. Colloid Sci.* **12**, 53 (1957).
(9) Fyans, R. L., Perkin-Elmer Corp., Instrument Division, private communication (1969).
(10) Gordon, M., Taylor, J. S., *J. Appl. Chem.* **2**, 493 (1952).
(11) Gruver, T. J., Kraus, G., *J. Polymer Sci.* **2**, 797 (1964).
(12) Heinze, D., *Makromol. Chem.* **101**, 166 (1967).
(13) Heller, J., private communication (1969).
(14) Hendus, H., Illers, K.-H., Ropte, E., *Kolloid-Z., Z.F. Polymere* **216**, 110 (1967).
(15) Holden, G., private communication (1968).
(16) Horino, T., Ogawa, Y., Soen, T., Kawai, M., *J. Appl. Polymer Sci.* **9**, 2261 (1965).
(17) Illers, K.-H., *Kolloid-Z.* **190**, 16 (1963).
(18) Kanig, G., *Kolloid-Z.* **190**, 1 (1963).
(19) Kolthoff, I. M., Lee, T. S., Carr, C. W., *J. Polymer Sci.* **1**, 429 (1946).
(20) Maekawa, E., Mancke, R. G., Ferry, J. D., *J. Phys. Chem.* **69**, 2811 (1965).
(21) Matsuo, M., *Japan Plastics* 6 (July 1968).
(22) Newlin, T. E., Lovell, S. E., Saunders, P. R., Ferry, J. D., *J. Colloid Sci.* **17**, 10 (1962).
(23) Plazek, D. J., *J. Phys. Chem.* **69**, 3480 (1965).
(24) Plazek, D. J., *J. Polymer Sci., Pt. A-2* **6**, 621 (1968).
(25) Reding, F. P., Faucher, J. A., Whitman, R. D., *J. Polymer Sci.* **57**, 483 (1962).
(26) Saunders, P. R., Stern, D. M., Kurath, S. F., Sakoonkim, C., Ferry, J. D., *J. Colloid Sci.* **14**, 222 (1959).
(27) Senn, Jr., W. L., *Anal. Chim. Acta* **29**, 505 (1963).
(28) Shell Chemical Co., Technical Bulletin, Synthetic Rubber Division, New York (Oct. 1967).

(29) Shen, M., Kaelble, D. H., *J. Polymer Sci.*, in press.
(30) Smith, T. L., Dickie, R. A., *J. Polymer Sci., Pt. C* **26**, 163 (1969).
(31) Stern, D. M., Berge, J. W., Kurath, S. F., Sakoonkim, C., Ferry, J. D., *J. Colloid Sci.* **17**, 409 (1962).
(32) Turley, S. G., *J. Polymer Sci., Pt. C* **1**, 101 (1963).
(33) Wilkes, G. L., Stein, R. S., *J. Polymer Sci., Pt. A-2* **7**, 1525 (1969).
(34) Yagii, K., Tschoegl, N. W., *Trans. Soc. Rheol.* **14**, 1 (1970).

RECEIVED September 8, 1970.

26

Multicomponent Systems from Copolymers of Maleic Anhydride and Vinyl Monomers

RAYMOND B. SEYMOUR, HING SHYA TSANG,[1] ELVIS E. JONES,[2] PATRICK D. KINCAID, and ASHWIN K. PATEL

University of Houston, Houston, Tex. 77004

> *Macroradicals have been produced previously by the degradation or cleavage of macromolecules using energy sources, such as, heat, light, ultrasonics, electrical discharge, radiation, tension, compression and mastication of polymers. Macroradicals that are present during the propagation of free radical chains may be preserved if isolated from other free radicals or scavengers. The preparation of these macroradicals by the copolymerization of maleic anhydride and vinyl monomers in poor solvents and the production of block copolymers from these macroradicals are discussed. Data on the characterization of these block copolymers are also included.*

Macroradicals are thermodynamically stable and may be preserved indefinitely if isolated from other free radicals, telogens, or vinyl monomers. For example, trapped free radicals produced by the irradiation of cellulose were detected in significant concentration after four years (9). Macroradicals are also present during the free radical chain propagation of vinyl monomers, but because of their kinetic instability in the presence of monomers, telogens, or other free radicals they are seldom isolated. However, relatively stable macroradicals have been observed in the micelles present in the emulsion polymerization of vinyl monomers (29). Relatively stable macroradicals may also be produced in the absence of oxygen or scavengers by radiolysis (8), photolysis (24), high voltage spark discharge (7), heating (3), and a wide variety of mechanical techniques in which the polymer chains are stressed.

[1] Present address: Aerospace Division, Bendix Corp., Ann Arbor, Mich.
[2] Present address: Dow Chemical Co., Freeport, Tex.

Thus, macroradicals have been obtained by stretching fibers (20), deforming plastics by compression (37), ball mill grinding (11), freezing and grinding of polymer solutions (10), ultrasonic irradiation (1), mastication (19), dispersion in a microblender (25), and other mechanical techniques (36). Many reviews on the formation of macroradicals by degradative processes have also been published (5, 12, 13, 16, 33).

Relatively stable macroradicals are precipitated when they are insoluble in their monomers. Thus, poly(vinyl chloride) has been obtained by a process in which the solid polymer was removed continuously as it precipitated from the monomer (23). These precipitated macroradicals have been described as "popcorn" (21) or trapped free radicals (22). Macroradicals obtained by the polymerization of acrylonitrile which have been widely studied (4) have been used to prepare block copolymers (35).

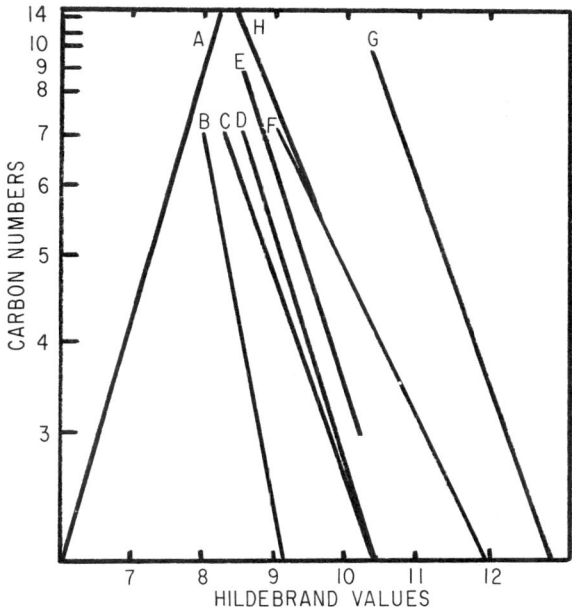

Figure 1. Relationship of Hildebrand solubility parameter values to molecular weight for different homologs

A: 1-alkenes
B: 1-chloroalkenes
C: methyl esters
D: formates and acetates
E: methyl ketones
F: 1-cyanoalkenes
G: n-alkanols
H: alkylbenzenes

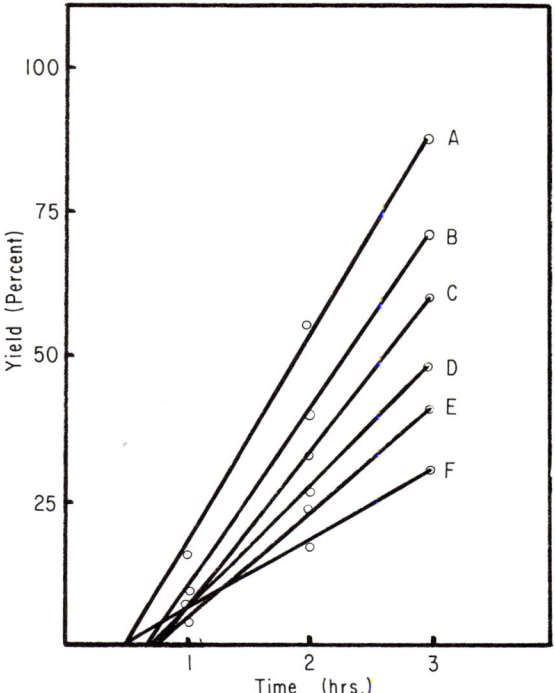

Figure 2. Rate of copolymerization of maleic anhydride and styrene in different solvents

A: benzene
B: xylene
C: cumene
D: methyl isobutyl ketone
E: p-cymene
F: acetone

Preparation of Macroradicals by Heterogeneous Solution Polymerization

Relatively stable macroradicals have also been obtained by the polymerization of vinyl chloride (15) or by the copolymerization of this monomer with vinyl acetate (32) in poor solvents—i.e., by heterogeneous solution polymerization. Appropriate solvents for this type polymerization can be selected on the basis of Hildebrand's solubility parameters (δ) (14). Data for these solubility parameters have been tabulated (6) and can be estimated from Figure 1.

Thus, it is possible to provide homogeneous solution polymerization systems providing the solubility parameters of the monomer, polymer, and solvent are known. Styrene and maleic anhydride were copolymerized by both techniques in the experiments described in this report.

Information on both homogeneous (17) and heterogeneous copolymerization of maleic anhydride and vinyl monomers is available (30).

The latter technique is more rapid than the former (27, 31). In 1930 Wagner-Jauregg showed that alternating copolymers are obtained when maleic anhydride is copolymerized with vinyl monomers (34). This is true for copolymerization in good solvents, but when the molar ratio of styrene to maleic anhydride is greater than 1, styrene may add to the alternating copolymer in poor solvents to produce block copolymers.

As shown in Figure 2, the rate of the heterogeneous copolymerization of styrene and maleic anhydride in benzene ($\delta = 9.2$) is faster than the homogeneous copolymerization of these monomers in acetone ($\delta = 9.9$). However, this rate decreases as the solubility parameter values of the solvents decrease in heterogeneous systems. Thus, the rate of copolymerization decreases progressively in xylene ($\delta = 8.8$), cumene ($\delta = 8.5$), methyl isobutyl ketone ($\delta = 8.4$), and p-cymene ($\delta = 8.2$). All of these rates were faster than those observed in homogeneous systems. The solubility parameter of the alternating styrene–maleic anhydride copolymer was $\delta = 11.0$.

The slow rate of copolymerization in acetone was related to the ease of termination of macroradicals by coupling. This coupling was hindered by the coiling of the macroradical chains in benzene, but propagation continued to take place since the monomers were able to pene-

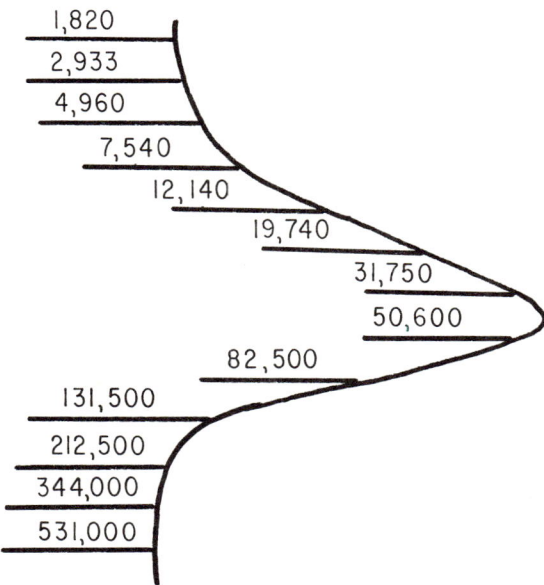

Figure 3. GPC molecular weight data for poly-(styrene–co–maleic anhydride) copolymerized in acetone

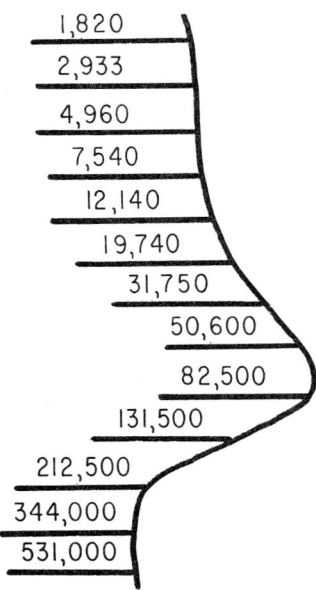

Figure 4. Molecular weight data for poly(styrene–co-maleic anhydride) copolymerized in benzene

trate the coils. Lower molecular weight macroradicals precipitated in the poorer solvents, and the propagation was progressively hindered as the coils became tighter in these poorer solvents.

The shift from homogeneous to heterogeneous copolymerization was also noted in acetone–benzene mixtures. The rate was slow, and the system was homogeneous when a benzene (50)–acetone (50) mixed solvent ($\delta = 9.6$) was used, but rapid heterogeneous copolymerization was observed in a benzene (80)–acetone (20) mixed solvent ($\delta = 9.3$).

As shown by the gel permeation chromatograms in Figures 3 and 4, the copolymer obtained from homogeneous copolymerization in acetone had a lower average molecular weight and a narrower molecular weight distribution than that obtained by heterogeneous copolymerization in benzene under similar conditions. Thus, the weight average molecular weight of the copolymer obtained from acetone was about 40,000 and that from benzene was about 70,000.

Some of the difference in average molecular was attributable to the continued propagation of the precipitated macroradicals. In addition, some coupling of the macroradicals probably occurred when they were dissolved in tetrahydrofuran to obtain the gel permeation chromatographic data. Thus, as shown in Figure 5, a lower average molecular weight was observed when a solution of hydroquinone in tetrahydrofuran was used as the solvent.

Comparable results were observed for the copolymerization of maleic anhydride and methyl methacrylate ($\delta = 10.8$ for copolymer), methyl acrylate ($\delta = 10.7$ for copolymer), and butyl methacrylate ($\delta = 10.7$ for copolymer). However, the copolymers of maleic anhydride and stearyl methacrylate ($\delta = 10.3$) and maleic anhydride and isobutyl methacrylate ($\delta = 10.4$) have lower solubility parameter values, and hence, a slow homogeneous copolymerization was observed when these monomers were copolymerized with maleic anhydride in benzene.

Attempts to change the copolymerization of styrene and maleic anhydride in benzene from a heterogeneous to a homogeneous process by using high concentrations of initiator or by adding weak chain transfer agents, such as carbon tetrachloride, were unsuccessful. However, homo-

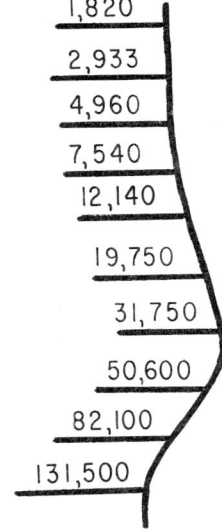

Figure 5. Molecular weight data for poly(styrene-co-maleic anhydride) copolymerized in benzene. Gel permeation chromatographic data obtained from THF solution containing hydroquinone

geneous copolymerization was noted when 30 wt % nitrobenzene was added to the monomer mixture in benzene.

Preparation of Block Copolymers from Macroradicals

The only product obtained by the copolymerization of styrene and maleic anhydride in acetone was the alternating copolymer even in the presence of more than equimolar quantities of either styrene or maleic anhydride. However, as shown by the data in Table I, larger quantities were obtained than could be accounted for by the formation of the alternating copolymer when excess styrene was used for the copolymerization in benzene solutions. In addition to the precipitates, there was also a trace of benzene-soluble product, which was shown to be polystyrene by infrared spectrometric (28) and pyrolytic gas chromatographic techniques (26).

Table I. Yields of Copolymers of Styrene and Maleic Anhydride Obtained by Heterogeneous Copolymerization in Benzene after 72 Hours at 50°C

Molar Ratio of Styrene to Maleic Anhydride	Percent Yield in Excess of that Accounted for by the Alternating Copolymer
1:1	0
4:1	20
17:3	23
9:1	23
19:1	19

Unlike the alternating copolymer, these high yield benzene-insoluble products were not completely soluble in acetone, but they were soluble in a mixture of acetone–benzene (2:1). However, unlike polystyrene, they were precipitated essentially completely when excess benzene was added to the acetone–benzene solutions. In contrast when excess benzene was added to a mixture of polystyrene and the alternating copolymer in an acetone–benzene solution, the copolymer precipitated and the polystyrene remained in the benzene-rich solvent. The polystyrene was recovered from this solution by adding excess methanol.

Typical chromatograms were observed when polystyrene was pyrolyzed in air and the pyrolytic products were analyzed by gas chromatography. A characteristic peak which was observed on the chromatograms obtained by the pyrolysis of maleic anhydride and the alternating styrene maleic anhydride copolymer but not with polystyrene was used as a reference peak. As shown in Table II, the ratio of the area under

Table II. Relationship of the Ratio of Areas under a Styrene Peak to the Reference Peak in the Gas Chromatograms of the Pyrolytic Products Obtained from Copolymers of Styrene and Maleic Anhydride in Benzene

Molar Ratio of Styrene to Maleic Anhydride	Ratio of Areas under the Styrene Peak to the Reference Peak
1:1	6.9
4:1	23.8
17:3	27.0
9:1	33.3
19:1	48.5

a styrene peak to this reference peak increased as the ratio of styrene to maleic anhydride in the feed increased.

It has been reported that pyrolysis gas chromatographic techniques could be used to differentiate between block and random copolymers (18). However, it was not possible to distinguish between the block copolymers and mixtures of polystyrene and the alternating copolymers of styrene and maleic anhydride by the PGC technique used in this investigation. However, differences were noted in the DTA thermograms of the alternating copolymer, the block copolymer, and the mixture of polystyrene and the alternating copolymer.

It has also been reported that an aromatic carbon–hydrogen out of plane deformation band at 759 cm^{-1} was sensitive to sequence distribution in styrene–maleic anhydride copolymers (2). A shoulder was noted at this frequency in the infrared spectra of the block copolymer, but it was not possible to demonstrate differences in the spectra of the alternating and block copolymers with the instrumentation available.

No product was obtained when attempts were made to copolymerize styrene and maleic anhydride in benzene at 50°C in the absence of bisazoisobutyronitrile. Likewise, no free radicals were detectable when these solutions were examined using EPR techniques. Negative results were also noted in solutions of the alternating copolymer prepared in acetone. However, the presence of free radicals was noted when the alternating copolymer produced by heterogeneous solution polymerization in benzene was examined. This peak was observed with freshly prepared and aged copolymer samples that had been stored in an inert environment. However, no peak was observed in product that had been washed with methanol.

The macroradicals obtained by the copolymerization of equimolar quantities of maleic anhydride and styrene were also used as initiators to form higher molecular weight copolymers and to prepare block copolymers. These macroradicals were effective as initiators after being stored for 180 hours at −20°C in an oxygen-free atmosphere. However, they were readily deactivated when a good solvent such as acetone was added.

As shown by the gel permeation chromatograph in Figure 6, the average molecular weight of poly(styrene–co-maleic anhydride) obtained by adding the macroradical to a benzene solution of the monomers was over 250,000. No copolymer was obtained under comparable conditions in the absence of the macroradicals. Attempts to use these macroradicals to produce copolymers in an acetone solution were unsuccessful.

Macroradicals obtained by the copolymerization of equimolar quantities of styrene and maleic anhydride in benzene or in cumene were also used as initiators to produce block copolymers with methyl methacrylate, ethyl methacrylate, and methyl acrylate. The yields of these block copolymers were less than those obtained with styrene, but as much as 38% of methyl methacrylate present in the benzene solution added to the macroradical to produce a block copolymer. The amount of ethyl methacrylate and methyl acrylate that was abstracted from the solution to form block copolymers was 35 and 20%.

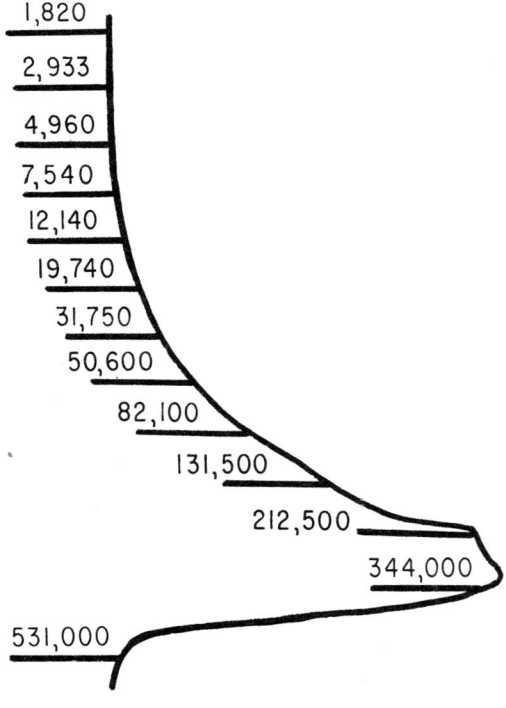

Figure 6. Molecular weight data for poly-(styrene–co-maleic anhydride) obtained in benzene using macroradicals as the initiator

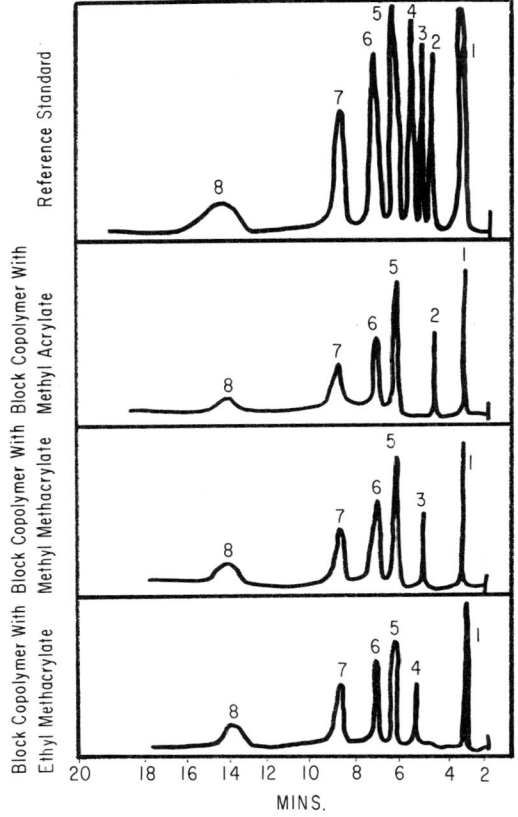

Figure 7. Pyrograms of block copolymers

1: *acetone*
2: *methyl acrylate*
3: *methyl methacrylate*
4: *ethyl methacrylate*
5: *benzene*
6: *toluene*
7: *ethylbenzene*
8: *styrene*

The formation of block copolymers from styrene–maleic anhydride and acrylic monomers was also indicated by pyrolytic gas chromatography and infrared spectroscopy. A comparison of the pyrograms of the block copolymers in Figure 7 shows peaks comparable with those obtained when mixtures of the acrylate polymers and poly(styrene–co–maleic anhydride) were pyrolyzed. A characteristic infrared spectrum was observed for the product obtained when macroradicals were added to a solution of methyl methacrylate in benzene. The characteristic bands for methyl methacrylate (MM) are noted on this spectogram in **Figure 8**.

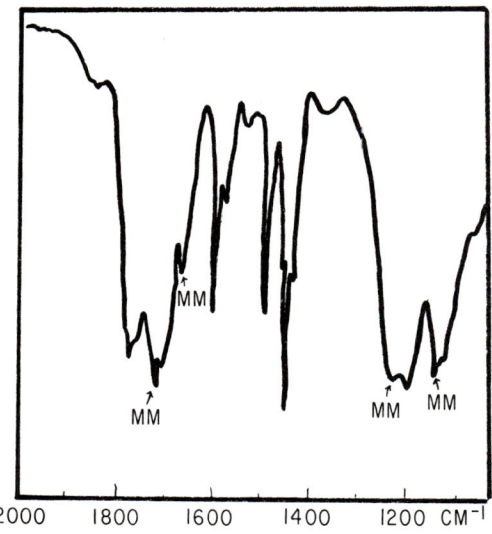

Figure 8. Infrared spectrum of block copolymer with methyl methacrylate

Experimental

Maleic anhydride was crystallized from sodium dried, thiophene-free benzene. The bisisobutyronitrile was crystallized from a mixed solvent containing equal volumes of benzene and toluene. The liquid monomers were purified by distillation under reduced pressure. Freshly distilled solvents were used.

The copolymers were prepared by the following method. Sufficient solvent was added to 0.5 gram of equimolar quantities of maleic anhydride and vinyl monomer and 0.0125 gram bisazoisobutyronitrile to make a total volume of 5 ml. The copolymerizations were conducted in agitated sealed glass tubes in the absence of oxygen. The product was recovered from homogeneous systems by adding methanol in a high speed blender. The precipitate was washed with methanol, filtered, and dried. The precipitate was removed from heterogeneous systems, washed well with fresh solvent after filtration and dried.

The gel permeation chromatographs were obtained by eluting a 2% solution of copolymer in a 10^4 A column manufactured by Water Associates. The calibration curve used to determine the molecular weights was obtained by plotting elution volumes against projected extended chain lengths of polystyrene.

Pyrolysis gas chromatographic investigations were made using a Wilkins model A100C aerograph equipped with a Servo-Ritter II Texas instrument recorder using helium as the carrier gas. The 10 ft × 1/4 inch diameter column was packed with acid washed chromosorb W (Johns Manville) with 20% SE-20 (General Electric Co.).

A rhenium tungsten code 13-002 pyrolyzing coil obtained from Gow-Mac Instrument Co. was used. Solutions of the samples were placed on the coil, and the solvent was allowed to evaporate before the residual film was pyrolyzed for about 10 seconds.

The infrared spectograms were obtained on a Beckman IR-10 instrument. Solutions of the styrene–maleic anhydride alternating copolymer and styrene block copolymer were used. A KBr pellet was used for the spectogram of the methyl methacrylate block copolymer.

Conclusions

Macroradicals obtained by the heterogeneous copolymerization of styrene and maleic anhydride in poor solvents such as benzene were used to initiate further polymerization of selected monomers. This technique was used to produce higher molecular weight alternating copolymers of styrene and maleic anhydride and block copolymers. Evidence for the block copolymers was based on molecular weight increase, solubility, differential thermal analysis, pyrolytic gas chromatography, and infrared spectroscopy.

Literature Cited

(1) Allen, P. E. M., Downer, J. M. et al., *Nature* **177**, 910 (1956).
(2) Ang, T. L., Harwood, H. J., *Polymer Preprints, ACS Div. of Polymer Chem.* **5** (1), 306 (1964).
(3) Arthur, J. C., Hinojosa, O., *Textile Res. J.* **36**, 385 (1966).
(4) Bamford, C. H., Jenkins, A. D. et al., *J. Polymer Sci.* **34**, 181 (1959).
(5) Battaerd, H. A. J., Tregear, G. W., "Graft Copolymers," Interscience, New York, 1967.
(6) Burrell, H., "Polymer Handbook," J. Brandrup, E. H. Immergut, Eds., Chapter 4, Interscience, New York, 1965.
(7) Ceresa, R. J., "Block and Graft Copolymers," p. 97, Butterworth, London, 1962.
(8) Chapiro, A., Mankowski, Z., *European Polymer J.* **2**, 163 (1966).
(9) Dilli, S., Ernst, I. T., Garnett, J. L., *J. Appl. Polymer Sci.* **11**, 836 (1967).
(10) Dubinskaya, A. M., Butyagin, P. Yu, *Vysokomol. Soedn. Ser. B* **9**, 525 (1967).
(11) Eckert, R. E., Maykrantz, T. R., Salloun, R. J., *J. Polymer Sci.* **6 B**, 213 (1968).
(12) Gould, R. F., Ed., ADVAN. CHEM. SER. **66** (1967).
(13) Grassie, N., "Chemistry of High Polymer Degradation Processes," Academic, New York, 1956.
(14) Hildebrand, J. H., Scott, R., "The Solubility of Nonelectrolytes," Reinhold, New York, 1949.
(15) Imperial Chemical Industries, Ltd., British Patent **366,897** (1929).
(16) Jelinek, H. H. G., "Degradation of Vinyl Polymers," Academic, New York, 1955.
(17) Johnson, J. H., Schaefgen, J. R., "Macromolecular Synthesis," Vol. 1, C. C. Overberger, Ed., Wiley, New York, 1963.
(18) Jones, C. E. R., Reynolds, C. E. J., *British Polymer J.* **1**, 197 (1969).
(19) Kraus, G., Rollmann, K. W., *J. Appl. Polymer Sci.* **8**, 2585 (1964).
(20) Matthies, P., Schlag, J., Schwartz, E., *Angew. Chem.* **77** (7), 323 (1965).

(21) Miller, G. H., Chock, E. P., *J. Polymer Sci.* **3A**, 3353 (1965).
(22) Morawetz, H., "Formation and Trapping of Free Radicals," A. M. Bass, J. P. Broida, Eds., Chapter 12, Academic New York, 1960.
(23) Pechiney-St. Gobain, Belgian Patent **647,821** (1964).
(24) Ranby, B., Carstensen, P., ADVAN. CHEM. SER. **66**, 256 (1967).
(25) Radtsig, V. A., Butyagin, P. Yu., *Polymer Sci.* **9**, 2883 (1967).
(26) Seymour, R. B., Anderson, A., *Thermochim. Acta* **1**, 137 (1970).
(27) Seymour, R. B., Tatum, S. D., Boriack, C. J., Tsang, H. S., *Texas J. Sci.* **21**, 13 (1969).
(28) Seymour, R. B., Tsang, H. S., Warren, D., *Polymer Eng. Sci.* **7** (1), 55 (1967).
(29) Smith, W. V., Ewart, R. H., *J. Chem. Phys.* **16**, 592 (1948).
(30) Trautvetter, W., German Patent **968,130** (Jan. 16, 1958).
(31) Tsuchida, E., Ohtani, V., Nakadai, H., Shirohara, I., *Chem. Soc. Japan J.* **70**, 573–79 (1967).
(32) Union Carbide Corp., U. S. Patent **2,075,429** (1937).
(33) Walling, C., "Free Radicals in Solution," Wiley, New York, 1957.
(34) Wagner-Jauregg, *Ber.* **63**, 3213 (1930).
(35) Yugi, M., Yayoi, O., *J. Polymer Sci.* **7A-1**, 2547 (1969).
(36) Zakrevskii, V. A., Baptizmanskii, Tomashevskii, E. E., *Soviet Phys., Solid State* **10**, 1341 (1968).
(37) Zhurkov, S. N., Zakrevskii, V. A. et al., *Radiospektrosk Tverd Tela Dokl. Vser. Soveschch. Krasnoyarsh USSR 1964*, 424 (c.a. **69**) 36628 (1968).

RECEIVED March 10, 1970.

Block and Random Copolymers by Oxidative Coupling of Phenols

GLENN D. COOPER, JAMES G. BENNETT, JR., and ARTHUR KATCHMAN

Plastics Department, General Electric Co., Selkirk, N. Y. 12158

Oxidation of mixtures of 2,6-disubstituted phenols leads to linear poly(arylene oxides). Random copolymers are obtained by oxidizing mixtures of phenols. Block copolymers can be obtained only when redistribution of the first polymer by the second monomer is slower than polymerization of the second monomer. Oxidation of a mixture of 2,6-dimethylphenol (DMP) and 2,6-diphenylphenol (DPP) yields a random copolymer. Oxidation of DPP in the presence of preformed blocks of polymer from DMP produces either a random copolymer or a mixture of DMP homopolymer and extensively randomized copolymer. Oxidation of DMP in the presence of polymer from DPP yields the block copolymer. Polymer structure is determined by a combination of differential scanning calorimetry, selective precipitation from methylene chloride, and NMR spectroscopy.

2,6-Disubstituted phenols react with oxygen in the presence of amine complexes of copper to yield linear poly(arylene oxides); the molding resin marketed under the trade name PPO is produced in this way by the oxidative polymerization of 2,6-dimethylphenol (14):

$$n \;\text{Ar-OH} + \tfrac{n}{2} O_2 \longrightarrow H{-}[\text{Ar-O}]_n{-}H + n\, H_2O \tag{1}$$

I

The poly(arylene oxides) represent a new class of polymers, with many useful properties. For this reason, as well as because of certain unique features of the polymerization mechanism, the oxidative coupling reaction has been studied intensively since it was first reported in 1959, and a number of reviews of the subject have appeared (*3, 7, 10, 12*). The reaction is a free radical chain process, with aryloxy radicals as intermediates. It has the characteristics of a polycondensation process—*i.e.*, polymer molecules couple with other polymer molecules as well as with monomer units (*8*). Polymerization is accompanied by a rapid redistribution between polymeric phenols (or monomer) of varying degrees of polymerization (*5*). Polymers have been prepared from a large number of phenols, but the reaction is most successful with phenols having small electron-releasing substituents in both the 2 and 6 positions. When bulky groups are present, as in 2,6-di-*tert*-butylphenol, carbon–carbon coupling predominates, and the major product is the diphenoquinone (*13*):

$$2 \; \text{[phenol]} + O_2 \longrightarrow \text{[diphenoquinone]} \qquad (2)$$

With phenols having open ortho positions (phenol, *o*-cresol) ortho coupling and other side reactions lead to branched rather than linear polymers (*15*). These reactions may be minimized by suitable choice of catalyst but cannot be eliminated entirely.

Copolymers of 2,6-dimethylphenol with 2-methyl-6-*tert*-butylphenol, 2,6-diisopropylphenol, 2-methyl-6-phenylphenol, and 2,6-diphenylphenol have been reported (*16*); but only the 2,6-dimethylphenol (DMP)–2,6-diphenylphenol (DPP) pair, which is described in this report, has been examined in detail. This system is particularly attractive because high molecular weight homopolymers can be obtained under suitable conditions from both monomers, facilitating the analysis of the copolymers. Both random and block copolymers have been obtained by varying the polymerization conditions (*1*).

Nature of Redistribution Reaction

The structure of copolymers produced by oxidative coupling is determined largely by the rate and other characteristics of the redistribution reaction, as is true of polyesters and other types of polymers which are subject to redistribution under the conditions of polymerization. The redistribution of polymeric phenols has been studied extensively, particularly with regard to the significance of this process in the mechanism of dimethylphenol polymerization. The reaction is believed to involve the formation and dissociation of an unstable quinone ketal, as shown below for two dimer radicals:

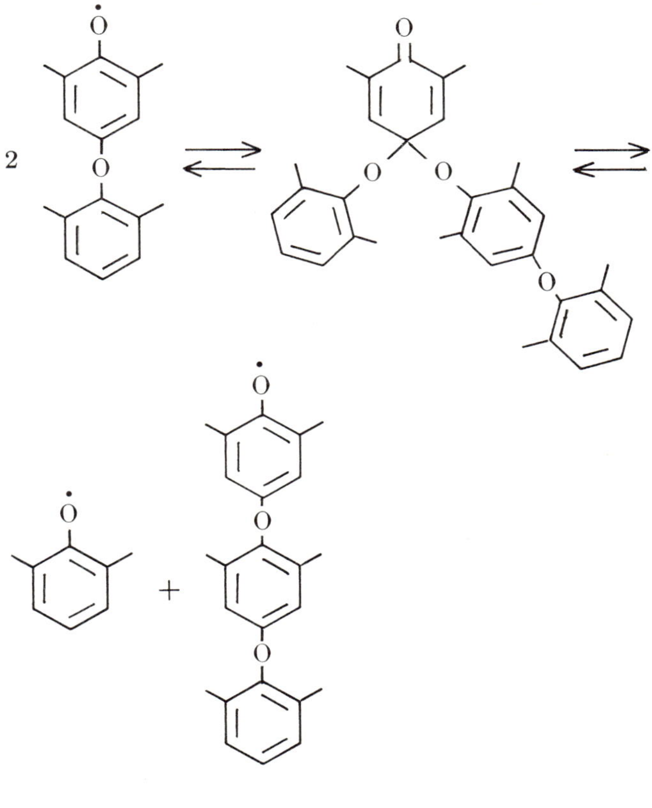

$$\text{(3)}$$

II

Reaction 3 provides a means for transferring an aryloxy unit from one radical to another. Hydrogen transfer reactions between phenols and aryloxy radicals occur so rapidly that the over-all reaction is a chain process which requires only initiation to proceed to the equilibrium mixture of oligomeric phenols, whose composition is determined by the over-all degree of polymerization of the system.

$$2 \text{ ArO(ArO)}_n\text{ArOH} + 1/2 \text{ O}_2 \rightarrow 2 \text{ ArO(ArO)}_n\text{ArÖ} + \text{H}_2\text{O} \quad (4)$$

$$2 \text{ ArO(ArÖ)}_n\text{ArO} \rightleftarrows \text{ArO(ArO)}_{n+1}\text{ArÖ} + \text{ArO(ArO)}_{n-1}\text{ArÖ} \quad (5)$$

$$\text{ArÖ} + \text{Ar'OH} \rightleftarrows \text{ArOH Ar'Ö} \quad (6)$$

$$X_m + X_n \rightleftarrows X_{m+1} + X_{n-1} \quad , \text{etc.} \quad (7)$$

Termination occurs when the reaction of two aryloxy radicals produces a single polymeric phenol, as when a monomer unit is added (Reaction 8).

III

Thus, the redistribution reaction does not change the degree of polymerization, does not consume oxygen other than that required for the initiation step, and can be observed independently of polymerization under suitable conditions (6); redistribution of high polymer with a monomeric phenol has been developed as a synthetic method for preparing substituted aryl ethers (18).

$$\text{Ar'OH} + \text{Ar(OAr)}_n\text{OH} \rightarrow \text{Ar'(OAr)}_m\text{OH} \quad m = 1, 2, \ldots \quad (9)$$

Just as redistribution can occur without polymerization, polymerization apparently can take place with little redistribution at sufficiently low temperatures, presumably by an intramolecular rearrangement of the ketal (19). In general, high temperatures and low amine concentrations favor redistribution. High temperature increases the extent of dissociation of the ketal, while base may be required for the coupling reaction, possibly by promoting the enolization of dienone (III) shown in Equation 8.

Although redistribution and coupling can be observed separately, oxidative polymerization under ordinary conditions involves both reactions and redistribution of oligomers to form monomer followed by removal of the monomer by coupling is an important mechanism of polymer growth. Redistribution in dimethylphenol polymerizations is extremely rapid. Addition of monomer to a polymerizing solution causes an immediate drop in the solution viscosity almost to the level of the solvent, as redistribution of polymer with monomer converts the polymer already formed to a mixture of low oligomers.

Copolymers from DMP and DPP

Oxidation of Mixtures of Monomers. The method most likely to yield random copolymers of DMP and DPP is the simultaneous oxidation of a mixture of the two phenols, although this procedure may present problems because of the great difference in reactivity of the two phenols. The production of high molecular weight homopolymer from DPP is reported to require both a very active catalyst, such as tetramethylbutanediamine–cuprous bromide, and high temperature, conditions which favor carbon–carbon coupling and diphenoquinone formation (Reaction 2) from DMP (11). With the less active pyridine–cuprous chloride catalyst at 25°C the rate of reaction of DMP, as measured by the rate of oxygen

Table I. Relative Rates of Oxygen Absorption (Py–CuCl Catalyst 25°C)

Phenol	Rate
DMP	1.00
DPP	0.03
DMP and DPP (1:1)	0.34

absorption, is more than 30 times that of DPP. It might be expected that under these conditions polymer initially would be produced almost exclusively from the more reactive monomer, with the DPP reacting in a second, much slower step to produce either block copolymer or a mixture of homopolymers, but the behavior actually observed in oxidation of mixtures of the two phenols does not confirm this prediction. An equimolar mixture of the two phenols absorbs oxygen at a rate intermediate to the rates of the pure monomers (Table I) with no break in the oxygen absorption curve which would indicate successive oxidation of the two phenols; the oxygen absorption curve is that which would be expected if the mixture contained a single phenol of intermediate activity.

The rate of disappearance of each of the two phenols during the oxidation is shown in Figure 1. If the two reacted independently, DMP would disappear completely before the absorption of 50% of the oxygen required to convert both phenols to high polymer, while the concentration of DPP would be reduced only slightly at this point. Although DMP was removed somewhat more rapidly than DPP in the early stages of oxidation, the difference is much less than would be expected on the basis of the reactivity of the two phenols in homopolymerization. Diphenylphenol was reduced by 16% during the first 25% of reaction, while dimethylphenol was still detectable after 75% of the total oxygen had been absorbed.

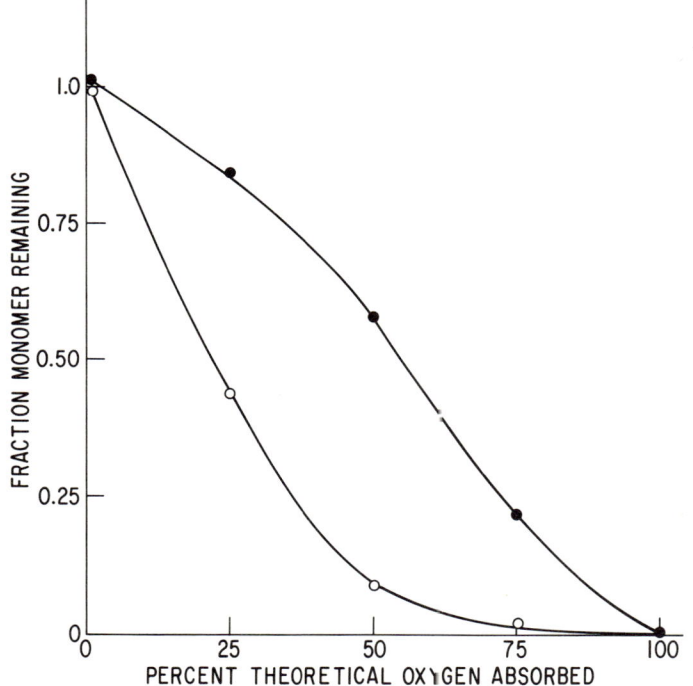

Figure 1. Disappearance of monomers on oxidation of equimolar mixture of dimethylphenol and diphenylphenol (pyridine–cuprous chloride catalyst)

A major factor in the interaction of the two phenols during oxidation, making the dimethylphenol appear less reactive and diphenylphenol more reactive than expected, must be the monomer–polymer redistribution reaction. Redistribution of diphenylphenol with the low oligomers of dimethylphenol removes diphenylphenol while returning dimethylphenol to the system.

$$\text{Ph-C}_6\text{H}_3(\text{OH})\text{-Ph} + \text{Me-C}_6\text{H}_2(\text{OH})(\text{Me})\text{-O-C}_6\text{H}_2(\text{Me})(\text{Me}) \longrightarrow \text{Me-C}_6\text{H}_3(\text{OH})\text{-Me} + \text{Me-C}_6\text{H}_2(\text{OH})(\text{Me})\text{-O-C}_6\text{H}_2(\text{Ph})(\text{Ph}) \quad (10)$$

These redistribution reactions of polymer molecules with other polymer molecules as well as with monomer, continue throughout the polymerization and should result in randomization of the polymer. Inasmuch as dimethylphenol is among the most reactive and diphenylphenol the least reactive of the phenols which have been oxidized successfully to linear high polymers, it appears likely that oxidation of any mixture of phenols will yield random copolymers.

Sequential Oxidation of DMP and DPP. The usual approach to formation of block copolymers is by the sequential polymerization of two or more monomers or by linking together preformed homopolymer blocks. In view of the importance of the redistribution process in the oxidative coupling of phenols there can be no assurance that successive polymerization of two phenols will yield block copolymers under any conditions. It is certain, however, that block copolymers can be formed only if the conditions are such that polymerization of the second monomer is much faster than redistribution of the added monomer with the polymer previously formed from the first. The extent of redistribution is followed conveniently by noting the effect of added monomer on solution viscosity, as indicated by the efflux time from a calibrated pipet.

When dimethylphenol is oxidized using a cuprous bromide–tetramethylbutanediamine catalyst and diphenylphenol is added to the polymerizing solution, the viscosity decreases rapidly, reaching a minimum within two minutes, and then increases very slowly as the mixture of monomers and low oligomers is polymerized. When this procedure is reversed, with dimethylphenol added to the growing diphenylphenol homopolymer, solution viscosity does not drop; instead the growth continues but at a faster rate (Figure 2). It appears that addition of the less reactive monomer (DPP) to the polymer from DMP should result in

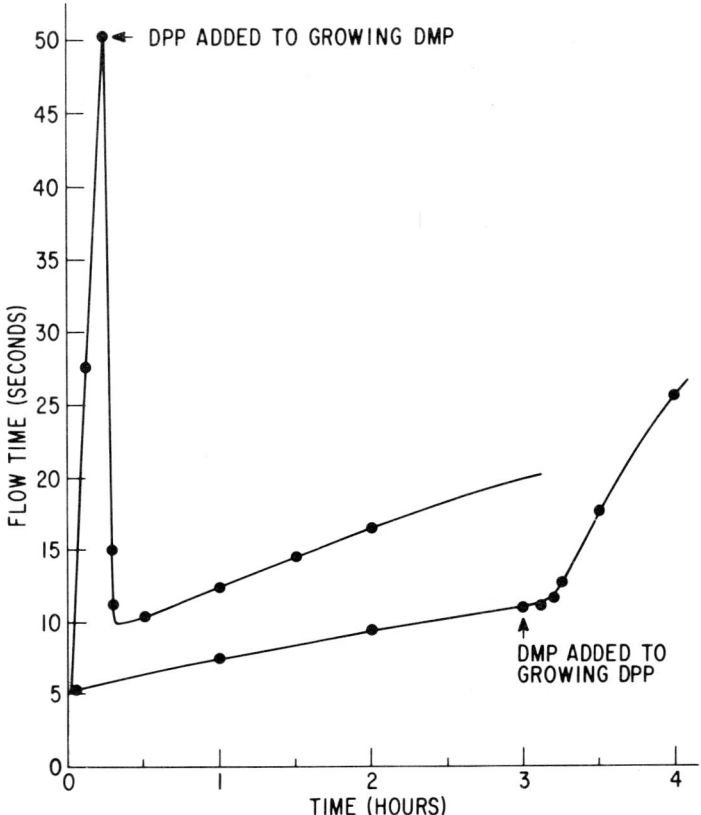

Figure 2. Effect of adding monomeric phenol on viscosity of polymerization mixture

random or extensively randomized copolymer, while the reverse procedure has at least a possibility of leading to block copolymers.

Polymerization Procedures. The polymerization reactions described below are typical of the procedures used to prepare copolymers. In other reactions the catalyst, catalyst concentration, reaction time, temperature, etc. were varied, but they all followed the general method of these examples.

PROCEDURE 1. POLYMERIZATION OF MONOMERS. A mixture of 0.282 gram of cuprous bromide, 0.284 gram of N,N,N',N'-tetramethyl-1,3-butanediamine, and 5.0 grams of anhydrous magnesium sulfate was stirred at 25°C in 140 ml of benzene, with a vigorous stream of oxygen introduced near the bottom of the flask. After five minutes 9.9 grams (0.04 mole) of 2,6-diphenylphenol and 4.9 grams (0.04 mole) of 2,6-dimethylphenol were added, and oxidation was continued for five hours. The mixture was filtered, and the polymer was isolated in 92% yield by precipitation in methanol; the composition, determined by integration of methyl and aromatic protons in the NMR spectrum, was 49 mole % DMP and 51% DPP. The intrinsic viscosity, measured in chloroform at 30°C, was 0.44 dl/gram.

PROCEDURE 2. ADDITION OF DMP TO POLYMERIZING DPP. This was carried out in the same way as in Procedure 1 but with only the DPP added initially. After three hours the DMP was added, and polymerization was continued for 50 minutes. The polymer was recovered in 92.5% yield. It had an intrinsic viscosity of 0.56 dl/gram and a composition of 46% DMP and 54% DPP.

PROCEDURE 3. ADDITION OF DPP TO POLYMERIZING DMP. The method of Procedure 1 was followed, with only the DMP initially present. After 30 minutes the reaction mixture was a thick gel, with much of the polymer out of solution. The DPP was added at this point, and oxidation was continued for four hours, during which the viscosity decreased gradually and the polymer returned to solution. Precipitation in methanol yielded 13.5 grams (92.5%) of polymer having an intrinsic viscosity of 0.56 dl/gram and a composition of 49 mole % DMP and 51% DPP.

PROCEDURE 4. POLYMERIZATION OF DMP WITH PREFORMED BLOCK OF DPP HOMOPOLYMER. Oxygen was introduced in a vigorously stirred mixture of 74.0 grams of DPP homopolymer (DP $\sim$ 60), 2.1 grams of cuprous bromide, 22.3 ml of diethylamine, and 50 grams of anhydrous magnesium sulfate in 1200 ml of benzene. A solution of 110 grams of DMP in 200 ml of benzene was added over a period of 10 minutes, and polymerization was continued for one hour. Dilute acetic acid was added to terminate the reaction, the organic layer was separated, and the polymer was isolated by precipitation in methanol, yielding 175 grams (95.5%) of a copolymer having an intrinsic viscosity of 0.50 dl/gram.

Determination of Copolymer Structure

Differential Scanning Calorimetry. Some structural information is provided by the thermal behavior of the polymer. The homopolymer of DPP crystallizes when heated above the glass transition temperature. A crystallization exotherm at the appropriate temperature therefore indicates the presence of DPP blocks, either as the homopolymer or in a block copolymer.

DSC traces of the polymers prepared in Procedures 1 and 2 are shown in Figure 3, along with that of a mixture of the two homopolymers. The polymer prepared by Procedure 2 shows the typical DPP crystallization exotherm, while that prepared by Procedure 1 does not crystallize and is therefore presumed not to contain DPP segments long enough to permit crystallization.

NMR Spectroscopy. NMR measurements are potentially capable of providing detailed structural information, providing there is sufficient separation between peaks and providing that each can be assigned to a specific structural arrangement.

The problem is most simply attacked through the dimethyl-substituted rings of the copolymer. If interactions more distant than a single segment are neglected, there are four possible environments for the methyl protons. A methyl-substituted unit may be situated between two other methyl-substituted units (MMM) or between two phenyl-substituted rings (PMP). Alternatively, it may be between a methyl and a phenyl-substituted unit. In this case there are two possible arrangements: the adjacent phenyl-substituted ring may be located toward the head of the chain (MMP) or towards the tail (PMM). The magnetic environment of the methyl protons differs at least slightly in each of these arrangements, so the methyl region of the NMR spectrum may show as many as four different peaks. An examination of the NMR spectra of "mixed dimers" (IV-VI) indicated that the chemical shifts should differ enough to permit resolution, and the method has been used successfully to determine the structure of the oxidation products of the mixed dimers (4):

IV V VI

The actual position of the methyl resonance in polymers is expected to differ slightly from the position calculated from the dimers. With the approximate position located in this way, however, it is not difficult to identify the peaks arising from each arrangement from a study of polymer spectra. The MMM sequence is observed at ε = 2.07–2.09 ppm in the spectra of a number of samples of PPO and of DMP–DPP copolymers containing a preponderance of DMP units. The PMP sequence can be located by examining the spectrum of a copolymer made by oxidizing a mixture containing 10 mole % DMP and 90 mole % DPP. On purely statistical grounds the PMP arrangement provides by far the most probable environment for a methyl-substituted unit, and the only peak observed in the methyl region occurs at 1.92 ppm, close to the calculated

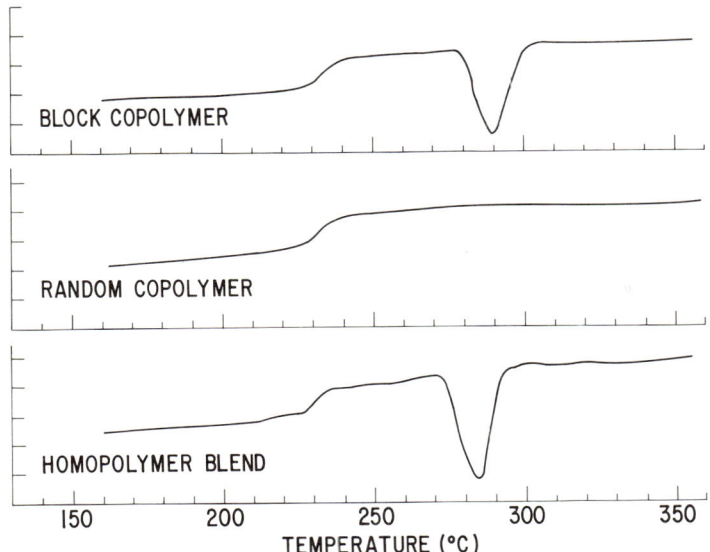

Figure 3. Differential scanning calorimeter traces of 1:1 DMP–DPP copolymers

Top: block copolymer prepared by Procedure 2
Center: random copolymer prepared by Procedure 1
Bottom: a blend of homopolymers

position of 1.94 ppm. Similar considerations applied to polymers containing a high proportion of DMP units allow the P$\underline{\text{MM}}$ and M$\underline{\text{MP}}$ frequencies to be located. The latter is subject to the greatest uncertainty, as it is separated by only 0.05 ppm from the M$\underline{\text{MM}}$ peak, usually the largest signal present.

The calculated chemical shifts for the various arrangements and the range over which each is actually observed in polymer spectra are given in Table II. Calculated shifts are based on the phenolic mixed dimers; slightly different values are obtained if the methyl ethers of the phenols were taken as models.

The aromatic protons of the polymer backbone may also be useful for determining structure, although the detailed analysis applied to the methyl groups is not feasible because of the complexity of this region of the spectrum. The effect of the neighboring rings can produce eight peaks for backbone protons (four from DMP and another four from DPP rings), some of which may be obscured by the protons of the pendant phenyl rings of the DPP units.

The spectra of a blend of the two homopolymers and of polymers prepared by Procedures 1 and 2 are shown in Figure 4. The polymer

Table II. Chemical Shifts of Methyl Protons in DMP–DPP Copolymers

Structure	Designation	Chemical Shift, ppm	
		Calcd.	Observed
(CH₃/CH₃, CH₃/CH₃, CH₃/CH₃)	MMM	2.08	2.07–2.09
(C₆H₅/C₆H₅, CH₃/CH₃, C₆H₅/C₆H₅)	PMP	1.94	1.92–1.95
(CH₃/CH₃, CH₃/CH₃, C₆H₅/C₆H₅)	MMP	2.17	2.13–2.15
(C₆H₅/C₆H₅, CH₃/CH₃, CH₃/CH₃)	PMM	1.85	1.84–1.87

prepared by simultaneous oxidation of both monomers has the spectrum expected of a random copolymer: four peaks arising from the methyl protons and at least six of the eight possible aromatic backbone peaks. In contrast, the spectrum of the polymer made by adding DMP to a growing DPP polymer is almost identical with that of the homopolymer blend. Peaks at $\partial = 7.25$ and 7.46 ppm correspond to the backbone protons of DPP and DMP homopolymers. The methyl protons are almost exclusively of the MMM type ($\partial = 2.08$ ppm); a small peak at $\delta = 1.86$ ppm possibly corresponds to a junction between blocks.

Solubility in Methylene Chloride. The methods described above can show the presence of blocks of DMP and blocks of DPP units, but they do not distinguish between block copolymers and blends of homopolymers. Gel permeation chromatograms of the copolymers are sharp and symmetrical, indicating that they are indeed copolymers rather than blends, but this alone is not conclusive as blends of the homopolymers do not produce binodal or badly skewed curves under the conditions used unless the two polymers differ considerably in molecular weight. A partial answer to this question is provided by the solubility behavior in methylene chloride. Dimethylphenol homopolymer dissolves readily in methylene chloride but precipitates quantitatively on standing for a short

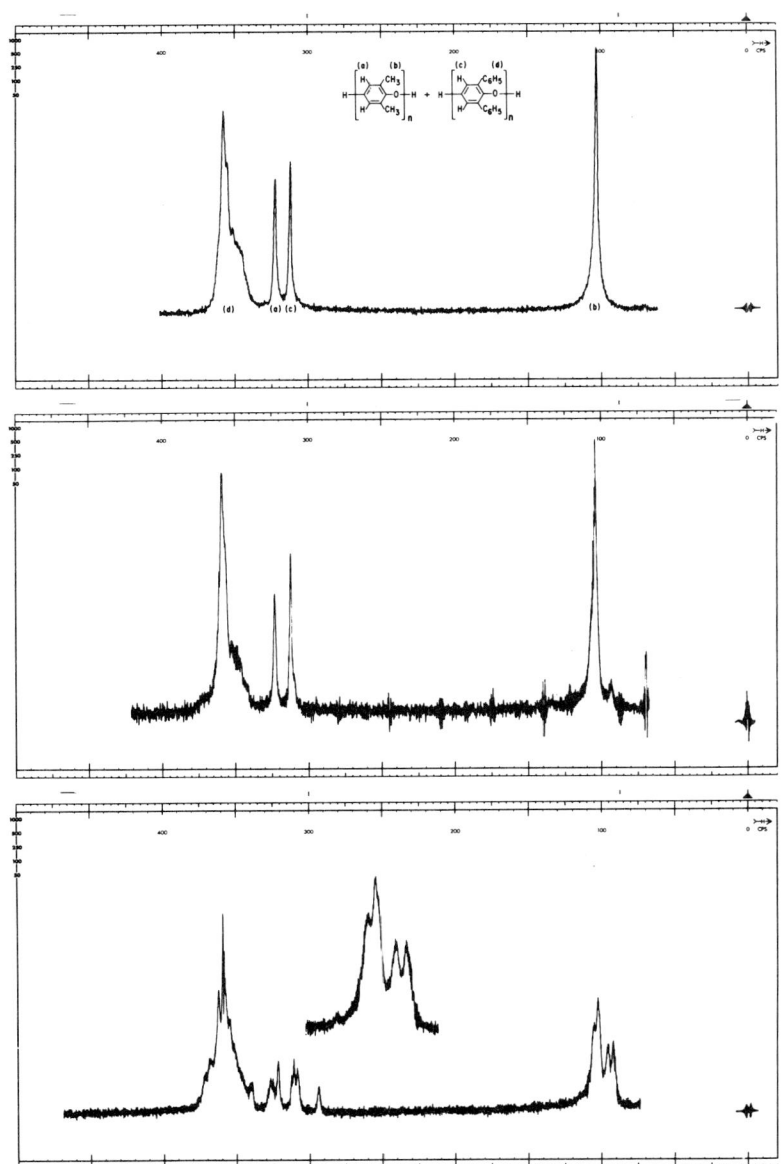

Figure 4. NMR spectra (100 megacycle) of 1:1 DMP–DPP copolymers
Top: homopolymer blend
Center: block copolymer prepared by Procedure 2
Bottom: random copolymer prepared by Procedure 1

time in the form of a polymer: CH_2Cl_2 complex (9). The homopolymer of diphenylphenol is soluble and does not precipitate. Thus, a blend of the two homopolymers can be separated easily.

Twenty percent solutions in methylene chloride of the two copolymers prepared by Procedures 1 and 2 were stable indefinitely, showing that no significant amount of dimethylphenol homopolymer was present and that the DMP blocks must be in the form of a block copolymer. Separate experiments using blends of DMP homopolymers with random copolymers or with DPP homopolymer showed that DMP homopolymer, even of very low molecular weight ($DP \sim 15$), could be detected easily if present to the extent of 5% of the total polymer.

Effect of Polymerization Method on Polymer Structure. The type of catalyst and reaction conditions used seem to make little difference in polymer structure when both monomers are oxidized together or when the less reactive DPP is polymerized before DMP is added. With three catalyst systems of widely differing activity (diethylamine–cuprous bromide, pyridine–cuprous chloride, and tetramethylbutanediamine–cuprous bromide) the product in the first case was always the random and in the second the block copolymer. When the more reactive monomer is oxidized first, however, the structure of the product may be affected by the nature of the catalyst. One of the problems in successive oxidation of the two monomers is that a catalyst system which causes polymerization of DPP at a reasonable rate is too active for controlled polymerization of DMP. The tetramethylbutanediamine–cuprous bromide catalyst used in Procedure 3 above caused extremely rapid polymerization of DMP, so that the polymer was of extremely high molecular weight and had partially precipitated from solution before the second monomer was added. The product of this polymerization appeared to have largely a block structure—the MMM peak was by far the strongest signal in the methyl region of the NMR spectrum, and the aromatic region was also consistent with a block copolymer, although with some randomization. Extraction of this material with methylene chloride showed, however, that 75% of the DMP portion was present as the homopolymer, and the product was therefore a blend of DMP homopolymer with a randomized DPP-rich copolymer.

The formation of DMP homopolymer in this case is readily explained. Under normal polymerization conditions not all of the polymer has the perfectly regular structure of I; in particular, some of the molecules do not contain the terminal hydroxyl group, which is necessary for redistribution, and are essentially inert. The fraction of imperfect polymer, which is probably formed in a termination reaction, varies with the polymerization conditions, but in general it tends to increase with increasing molecular weight. In the example described under Procedure 3, most of

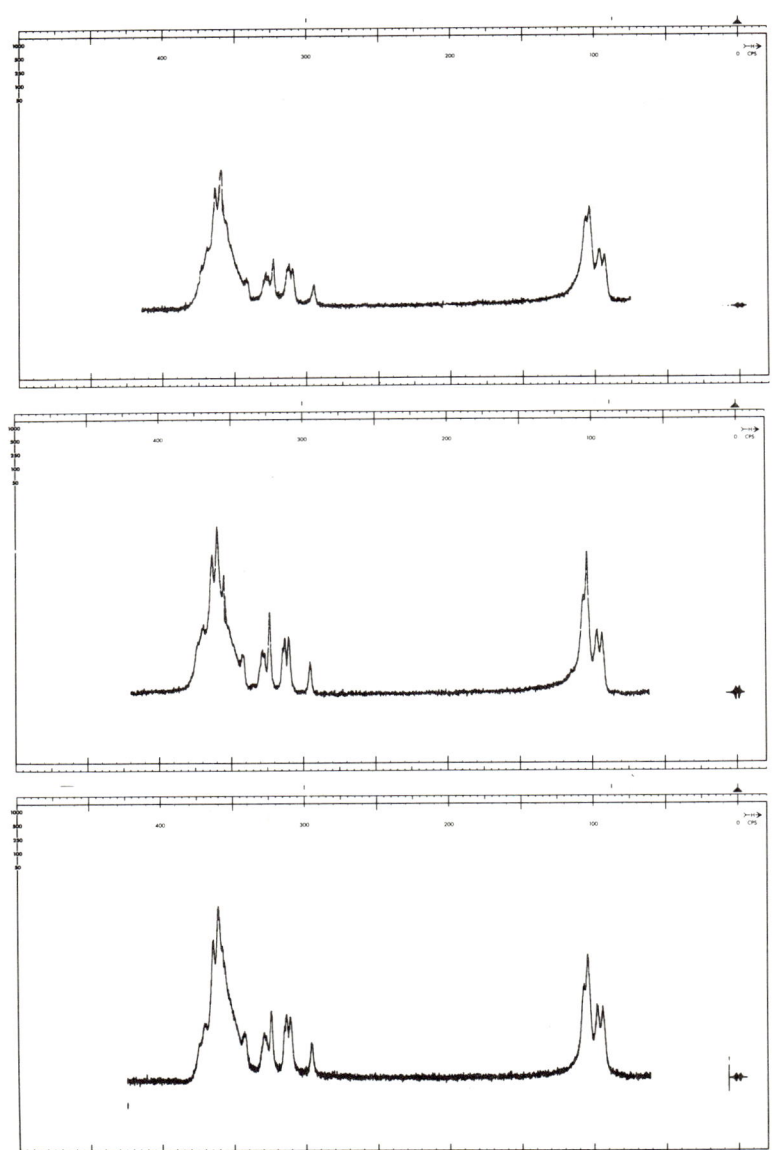

Figure 5. NMR spectra of random 1:1 DMP–DPP copolymers

Top: DPP added to growing DMP polymer, (DEA–CuBr catalyst)
Center: oxidation of mixture of monomers (TMBDA–CuBr catalyst)
Bottom: DPP oxidized with preformed DMP homopolymer (TMBDA–CuBr catalyst)

DMP homopolymer at the time the second monomer was added consisted of "dead" molecules, incapable of redistribution or of further normal polymerization. When the same procedure was followed, but with the less active diethylamine–cuprous bromide catalyst, only random copolymer was obtained, identical to that obtained by oxidation of the two monomers together. The same result was observed when DMP was oxidized with the diethylamine–cuprous bromide catalyst and tetramethylbutanediamine–cuprous bromide was added along with DPP to increase the polymerization rate (Figure 5).

A convenient method for avoiding the problems caused by the large difference in reactivity of the two monomers is by using preformed blocks—*i.e.*, by preparing and isolating the homopolymer under conditions most suitable for the polymerization of the particular monomer and then oxidizing a mixture of the polymer with the second monomer. When this procedure was followed, oxidation of DMP polymer with DPP always yielded random copolymer, regardless of the type of catalyst used, while oxidation of DPP polymer with DMP yielded only block copolymers.

Table III. Effect of Polymerization Procedure

DMP:DPP = 1:1

Procedure	Catalyst	Product
DMP and DPP	TMBDA–CuBr	Random copolymer
DMP and DPP	py–CuCl	Random copolymer
DMP and DPP	DEA–CuBr	Random copolymer
DPP, then DMP	TMBDA–CuBr	Block copolymer
DPP, then DMP	DEA–CuBr	Block copolymer
DMP, then DPP	TMBDA–CuBr	DMP homopolymer and randomized copolymer
DMP, then DPP	DEA–CuBr	Random copolymer
DPP block and DMP	TMBDA–CuBr	Block copolymer
DPP block and DMP	DEA–CuBr	Block copolymer
DMP block and DPP	DEA–CuBr	Random copolymer
DMP block and DPP	TMBDA–CuBr, 60°	Random, low MW
DMP block and DPP block	TMBDA–CuBr	Block copolymer

Block copolymers were also produced by oxidizing mixtures of the two homopolymers. A summary of the effect of polymerization conditions on the structures of polymers prepared using equimolar amounts of the two monomers is presented in Table III. The preformed blocks used in these examples were a DMP homopolymer prepared with a diethylamine–cuprous bromide catalyst and a DPP polymer prepared with tetramethylbutanediamine–cuprous bromide at 60°C. Each had an average degree of polymerization of approximately 50 units.

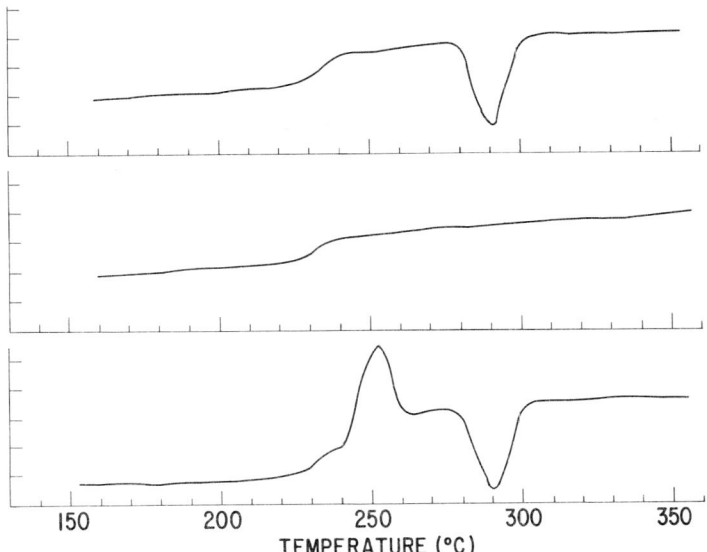

Figure 6. Effect of polymer treatment on thermal behavior of 1:1 DMP–DPP block copolymer

Top: polymer as precipitated (both blocks amorphous)
Center: after heating to 300°C (DMP amorphous, DPP crystalline)
Bottom: after treatment with 2:1 methanol–toluene (DMP crystalline, DPP amorphous)

Properties of Copolymers

Thermal Properties of 1:1 Copolymers. Random copolymers made from equimolar amounts of the two phenols have a glass transition temperature of $T_g = 226°C$. They are amorphous and will not crystallize either thermally or by solvent treatment (Figure 6).

Block copolymers of this composition are completely amorphous when isolated in the usual manner, by adding the polymer solution to a large volume of methanol or other antisolvent. They show a single $T_g = 226°C$. The T_g's of the two homopolymers are too close (225° for DMP (17) and 230°C for DPP (11)) to permit the observation of separate transitions for the DMP and DPP portions of the blocks. The DPP portion of the block crystallizes when heated to approximately 290°C, as does the DPP homopolymer. Melting of the crystalline DPP, which occurs at 480°C in the homopolymer, could not be observed in the copolymer because of the onset of decomposition at approximately 450°C.

DMP homopolymer, unlike that from DPP, does not crystallize thermally, and no evidence of such crystallinity has been observed in thermally crystallized block copolymers; the x-ray pattern of thermally crystallized copolymers is the same as that of the DPP homopolymer. The DMP homopolymer, however, does crystallize readily on treatment with suitable solvents, such as warm α-pinene or a mixture of toluene and methanol. Stirring the amorphous block copolymer in a 3:1 methanol–toluene solution causes the DMP portion to crystallize, while the DPP remains amorphous. When this copolymer is heated, crystallinity is reversed. The crystalline DMP melts at approximately 250°C; continued heating causes the DPP to crystallize at 290°C. Thus, by suitable treatment, the block copolymer can be obtained completely amorphous, with amorphous DPP and crystalline DMP, or with crystalline DPP and amorphous DMP. It has not yet been possible to obtain both portions of the block simultaneously crystalline.

Copolymer Composition. Some properties of a series of random and block copolymers covering a wide range of compositions are listed in Table IV. The random copolymers were prepared by oxidizing a mixture of the two monomers as described in Procedure 1; block copolymers were prepared by oxidizing DMP in the presence of a separately prepared and isolated DPP homopolymer. This is by far the most convenient method for preparing block copolymers and allows the initial block length of the DPP segments to be controlled. A typical polymerization is described in Procedure 4.

All of the copolymers both random and block, showed a single T_g between 220° and 230°C.

Table IV. Copolymers of Dimethylphenol and Diphenylphenol

Coloymer	DMP/PPD molar ratio	DPP block length, initial	Yield, %	Copolymer I.V., dl/g	T_g, °C	T_c, °C	Q_c cal/g of DPP
Block	86/14	60	97.4	0.45	224	317	4.5
Block	75/25	60	95.5	0.50	217	297	4.9
Block	50/50	120	94.7	0.40	225	289	5.0
Block	50/50	120	99.0	0.46	224	290	4.0
Block	50/50	60	97.2	0.38	223	297	4.3
Block	50/50	60	95.4	0.59	227	290	4.5
Block	50/50	60	96.7	0.39	227	297	5.2
	25/75	60	95.1	0.26	225	283	3.7
Block	25/75	300	88.6	0.74	230	283	—
Random	90/10	—	97.0	0.54	222	—	—
Random	75/25	—	89.5	0.60	226	—	—
	50/50	—	96.5	0.60	225	—	—
Random	10/90	—	88.5	0.45	244	289	2.1

The polymers listed in Table IV were isolated by addition to methanol and were completely amorphous. The crystallization temperature (the temperature at which the peak of the crystallization exotherm was observed when the polymer was heated at a rate of 40°C per minute) increased with decreasing DPP content, from 283° at 75 mole % DPP to 317°C at 15 mole %, and was not affected by changes in the initial DPP block length over the range of approximately 50 to 300 units. The fraction of DPP crystallizing appeared to be almost independent of both composition and block length; the heat of crystallization was 4–5 cal/gram DPP in all cases. Random copolymers containing 75% or less of DPP units showed no evidence of crystallization, but at high DPP levels, on the order of 90%, blocks of DPP units (or DPP homopolymer) long enough to allow some crystallization were produced.

All of the random copolymers and all block copolymers containing no more than 50% DMP units are completely stable in methylene chloride, demonstrating the absence of DMP homopolymer, but block copolymers containing 75% of DMP units precipitate from methylene chloride. The precipitate is not the homopolymer, however. It contains approximately the same ratio of DMP and DPP units as the original copolymer, and the composition does not change appreciably on reprecipitation. Apparently block copolymers having a high DMP content form insoluble complexes with methylene chloride.

Acknowledgment

The authors are greatly indebted to R. A. Kluge for assistance in interpreting differential scanning calorimetry and NMR results.

Literature Cited

(1) Bennett, J. G., Jr., Cooper, G. D., *Macromolecules* **3**, 101 (1970).
(2) Butte, W. A., Price, C. C., Hughes, R. E., *J. Polymer Sci.* **61**, f28 (1962).
(3) Cooper, G. D., *Ann. N. Y. Acad. Sci.* **159**, 1 (1969).
(4) Cooper, G. D., Bennett, J. G., Jr., *J. Org. Chem.*, in press.
(5) Cooper, G. D., Blanchard, H. S., Endres, G. F., Finkbeiner, H., *J. Am. Chem. Soc.* **87**, 3996 (1965).
(6) Cooper, G. D., Gilbert, A. R., Finkbeiner, H., *Am. Chem. Soc., Div. Polymer Chem., Polymer Preprints* **7**, 166 (1966).
(7) Cooper, G. D., Katchman, A., ADVAN. CHEM. SER. **91**, 660 (1969).
(8) Endres, G. F., Kwiatek, J., *J. Polymer Sci.* **58**, 593 (1962).
(9) Factor, A., Heinsohn, G. E., Vogt, L. H., Jr., *J. Polymer Sci., Pt. B* **7**, 205 (1969).
(10) Finkbeiner, H. L., Hay, A. S., "Newer Polymerization Reactions," in press.
(11) Hay, A. S., *Macromolecules* **2**, 107 (1969).
(12) Hay, A. S., *Advan. Polymer Sci.* **4**, 496 (1967).
(13) Hay, A. S., *J. Polymer Sci.* **58**, 581 (1962).

(14) Hay, A. S., Blanchard, H. S., Endres, G. F., Eustance, J. W., *J. Am. Chem. Soc.* **81**, 6335 (1959).
(15) Hay, A. S., Endres, G. F., *Polymer Letters* **3**, 387 (1965).
(16) Hay, A. S., White, D. M., *ACS, Div. Polymer Chem., Polymer Preprints* **10**, 92 (1969).
(17) Karasz, F. E., O'Reilly, J. M., *Polymer Letters* **3**, 561 (1965).
(18) White, D. M., *J. Org. Chem.* **34**, 297 (1969).
(19) White, D. M., "Abstracts of Papers," 155th Meeting, ACS, April, 1968, p. T103.

RECEIVED February 19, 1970.

Composites

28

Polymer–Fibrous Glass Composites: Advances and Potential Properties

FRED G. KRAUTZ

Owens-Corning Fiberglas Corp., Technical Center, Granville, Ohio 43023

> *The 1960–1970 decade marked the "start of something new" for fibrous glass composites. A remarkable growth in RTP (reinforced thermoplastics) occurred. Chemically tailoring a polymer with the sole intent of combining it with fibrous glass is now a reality. In the thermoset area, SMC (sheet molding compound) and BMC (bulk molding compound), including low profile systems, are getting general industry acceptance. A partial explanation for all the activity with fibrous glass composites is stated in terms of polymer property improvements resulting from the addition of glass. Consideration is also given to the importance of polymer chemistry in obtaining adhesion to glass. Potential strengths are given for a selected system based on a theoretical model.*

It always seems fitting at the start of a new decade to look back on the previous one. The topic of fibrous glass composites is no exception. As a clarification, the term fibrous glass composites as used here refers to thermosetting and thermoplastic molding resins reinforced with fibrous glass.

Fiber glass reinforced plastics have reached an important milestone. This industry has grown from relative obscurity to a market which has reached the 1 billion pounds of laminate per year level. Also significant is the milestone reached by RTP. This segment of fibrous glass composites is nearing 10% of the entire FRP (fiber glass reinforced plastic) volume. More importantly, essentially all this volume in RTP was generated in the last decade.

The difficult task in treating this type of subject matter is in the selection of material. It is impossible to cover everything significant. Omissions are not intentional but unavoidable. The technical highlights are stressed.

Two specific areas are covered more comprehensively: fibrous glass polyesters and fibrous glass thermoplastics. The advances in reinforced polyesters have been in the materials and process areas. These changes are more profound than those related to mechanical properties. Hence, a treatment of mechanical properties of reinforced polyesters is not attempted. On the other hand, the reinforced thermoplastics advances are essentially in terms of many new reinforced polymers as well as a greater variety of compounds. For these reasons, it seems appropriate to discuss advances in this area in terms of properties.

To give perspective to these technical developments, end-use applications are used to illustrate them. The use of actual parts is hopefully an effective aid in covering such a broad subject.

Advances in Terms of Applications

One of the most effective ways to gage the advances in a particular field is through the commercial realities based on the new technology. The applications listed below illustrate the technical growth of a diversified industry which has fibrous glass as its common denominator.

(1) High appearance exterior polyester parts spanning the width of a car. This is in the form of a combination upper grille and lamp housing as well as lower grille panels.

(2) High appearance polyester automotive headlamp housings.

(3) High appearance polyester automotive air scoops.

(4) Complex structural polyester parts of various appliance applications such as air conditioners and humidifiers.

(5) Underground gasoline storage tanks.

(6) Polyester bathroom components.

(7) Marine applications reflect a remarkable penetration for FRP.

(8) Injection molded styrenics span the width of the car in the form of instrument panels and crash pad retainers. Shot weights approach 10 lbs in some cases.

(9) Injection molded automotive shift consoles.

(10) Injection molded polypropylene automotive tail lamp housings.

(11) Injection molded olefin fender liners for automobiles.

(12) Injection molded complex appliance fans.

(13) Injection molded polypropylene parts which comprise the complete hot water distribution system for dishwashers.

(14) Injection molded washing machine housings.

(15) Injection molded calculator housings.

Advances in Fibrous Glass Thermosets

The materials and process development relating to sheet molding compounds (SMC) and bulk molding compounds (BMC) represent the most significant breakthroughs in this broad field. The development of SMC and BMC in conjunction with the development of low profile polyester resin systems are contributing to a great amount of interest in glass-reinforced polyesters.

The addition of SMC and BMC to reinforced polyesters has changed the basic mechanical property picture very little. The same is true for low profile polyesters. The mechanical performance is at the same general level that is attained with the older wet systems, such as preform and mat, and premix molding.

Sheet and Bulk Molding Compounds. The activity with SMC and and BMC is based on resin technology which was known more than 20 years ago (3). The key concept involves increasing the viscosity of the unsaturated polyester resin. The polyester resin with an initial viscosity of several thousand centipoise is advanced chemically to a viscosity of several million centipoise.

This thickening reaction is accomplished by using Group 2a metal oxides and hydroxides, such as calcium hydroxide. Although all the variables affecting this mechanism are not fully understood, it is generally accepted that the metal ion acts as a bridge between residual acid groups on the polyester chain.

At first, it may not seem significant to be able to increase the viscosity of a polyester resin without crosslinking it. The following points will illustrate the advantages that such a system brings to molding reinforced polyesters:

(1) The viscosity of the resin can be maintained at a much higher level in the heated mold. This allows the material to hold molding pressure, which results in much improved surface finish.

(2) There is little separation of fiber and resin during flow in the mold. This is also attributable to maintaining viscosity.

(3) Material efficiencies are high since the molding compound is charged within the confines of the mold. This results in no flash waste.

(4) The "mess factor" or dispensing a liquid resin onto a preform or mat is eliminated.

(5) The handleability of the molding compound makes it more suitable for an automated molding process. The compounds are tack-free as ready to mold.

The development of low profile polyester resin adds an additional dimension to SMC and BMC. The post-molding finishing steps with FRP have been a strong deterrent in high appearance requirement applications. The low profile resins utilize thermoplastic additives to obtain

Table I. Typical Mechanical Properties of Polyester BMC and SMC

	BMC 20% Glass	SMC 30% Glass
Flexural strength, psig	19,000	33,000
Flexural modulus, psig	1.5×10^6	1.4×10^6
Tensile strength, psig	7000	17,000
Notched Izod, ft lbs/inch	6	13

the very high quality as molded surfaces, and post-molding operations are essentially eliminated. For example, one of the commercial systems developed employs an acrylic additive in the polyester resin (7). This additive reduces polymerization shrinkage to the point where the fibrous glass pattern is essentially eliminated from the surface.

The process by which SMC is made is usually based on fibrous glass in roving form. The polyester resin containing the catalyst system, mineral fillers, and the Group 2a metal oxide or hydroxide is dispensed onto polyethylene film. The roving is then cut onto the liquid resin. Since the thickening reaction has not started, it is easy to impregnate the cut glass. Compaction rolls then remove air while aiding in the wetting. After a relatively short aging time, the thickened SMC is ready for molding.

BMC is processed by premix techniques, using the standard sigma-blade mixer. The only change is the use of Group 2a metal oxides and hydroxides.

Table I lists some properties of SMC and BMC. These are a function of resin composition, reinforcement, and molding conditions and may be regarded as typical. This will serve as a frame of reference as to the property levels obtained with SMC and BMC. The differences which exist between SMC and BMC in tensile, flexural, and impact strengths are attributable to more than just the difference in glass loading. Fiber attrition arising from the compounding techniques for BMC as well as the shorter input fiber length account for the lower strengths.

Advances in Fibrous Glass Thermoplastics

The growth of these materials is reflected in the number of polymers which are being glass reinforced. These include polypropylene, polystyrene, styrene acrylonitrile, nylon, polyethylene, acrylonitrile–butadiene–styrene, modified polyphenylene oxide, polycarbonate, acetal, polysulfone, polyurethane, poly(vinyl chloride), and polyester. In addition, the reinforced thermoplastics available now include long-fiber compounds, short-fiber compounds, super concentrates for economy, a combination of long and short fibers, and blends of polymer and fibrous glass.

The acceptance of these materials largely arises from the property improvements obtained on adding glass. Table II shows some typical property improvements on adding 30% glass to polystyrene, nylon 66, and polypropylene. Both tensile and flexural strengths can be approximately doubled by adding glass. The modulus is improved by a factor of 3. The heat distortion improvements range from a modest 20°F for polystyrene to 140°F for polypropylene, to 280°F for nylon 66.

Tensile and Flexural Properties of Thermoplastics. It is interesting to show graphically the effect of introducing reinforced thermoplastics on the properties available in injection molding materials. In Figure 1 tensile

Table II. Effect of Glass on Base Polymer Properties at 30% Glass Level

	Polystyrene	*Nylon 66*	*Polypropylene*
Tensile strength[a]	1.9	2.1	1.8
Flexural strength[a]	1.5	2.1	2.1
Modulus	2.8	3.3	3.1
Heat distortion	+20°F	+280°F	+140°F

[a]Strength numbers are ratios of properties of reinforced polymer to unreinforced polymer.

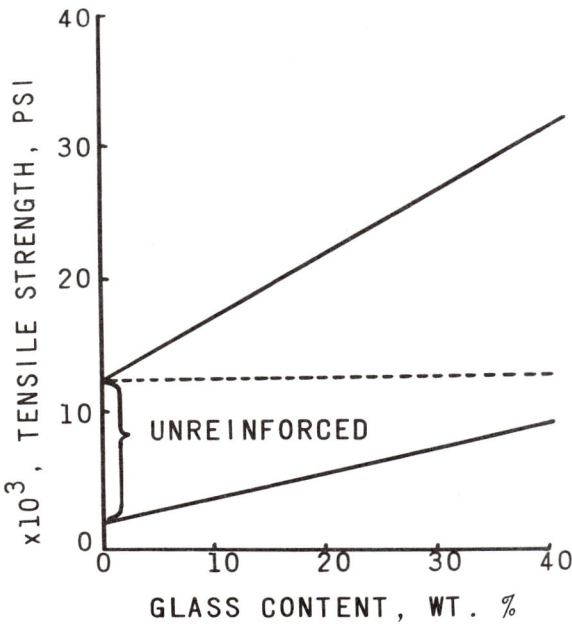

Figure 1. Composite tensile strength of thermoplastics

strength is used to illustrate this point. The range of tensile strengths for unreinforced polymers is *ca.* 2000–12,000 psig. The addition of 40% glass extends this to around 30,000 psig for the highest strength polymer (upper line). The lower line shows the improvement in the lowest strength polymer. The portion in the figure above the dashed line then represents properties not attainable without glass reinforcement.

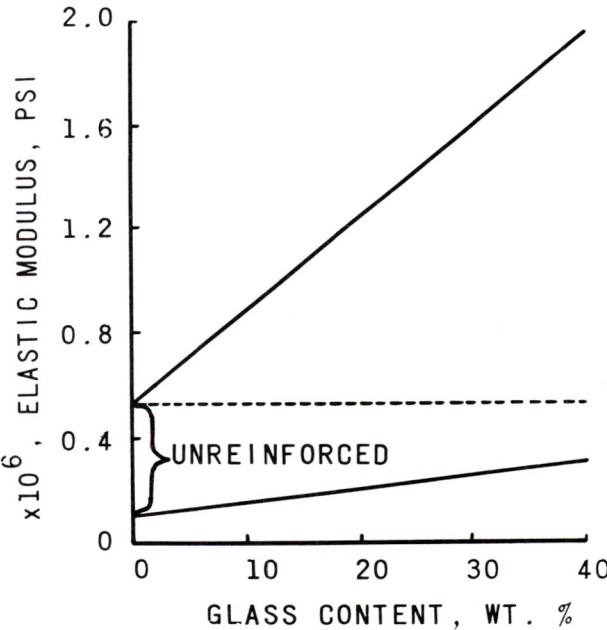

Figure 2. Composite elastic modulus of thermoplastics

A similar situation is illustrated in Figure 2 for elastic modulus. Thermoplastics fall in a rather narrow range of 100,000–500,000 psig. This is extended to approximately 2,000,000 psig by using 40% glass (upper line). The lower line shows the effect for the lowest modulus polymer. The area above the dashed line is the contribution made by fibrous glass to available properties in injection molding.

Figure 3 shows the effect on the available flexural strengths in injection molding. The upper limit of approximately 20,000 psig is more than doubled by using fibrous glass. The area above the dashed line again represents an extension of available properties in injection molding compounds.

Impact Strength of Reinforced Thermoplastics. Impact strength is a difficult topic to treat rigorously mainly because of the shortcomings of the most common test—notched Izod (12). The most serious shortcoming is the extreme complexity of the stress field which does not allow an accurate calculation of maximum local strain rates. The effect on impact strength of specimen width and the method of introducing the notch are two other factors which make it difficult to analyze data. It is important not to use notched Izod data as an absolute indicator of end-use impact performance. Special precaution must be taken in comparing notched Izod impact strength of an unreinforced polymer to its reinforced analog.

One of the important considerations in describing impact behavior of reinforced thermoplastics is the markedly different stress–strain rela-

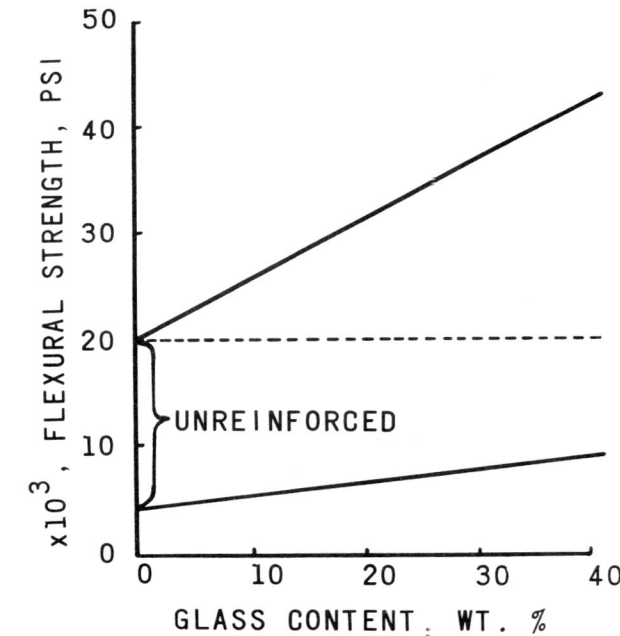

Figure 3. Composite flexural strength of thermoplastics

tionship. Figure 4 illustrates this for a linear polyethylene. The ultimate elongation of approximately 200% is reduced to approximately 3%. Actually, almost all reinforced thermoplastics with 20% or more glass will fall in a 1–5% range. Polymers that have a yield point, such as nylon and ABS, do not exhibit a yield point upon adding glass. Common to all reinforced thermoplastics is low ultimate elongation regardless of polymer type.

Another important factor in relation to impact strength is illustrated in Figure 5. A brittle polymer and a tough polymer, polystyrene and ABS respectively, are used as an example. Their impact strengths differ by a factor of 32, with impact strengths of 0.25 ft lb/inch for polystyrene and 8 ft lbs/inch for ABS. The addition of fibrous glass is the great

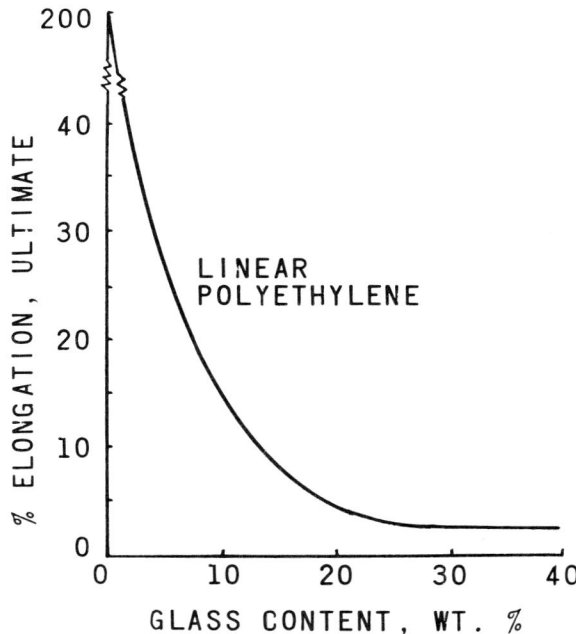

Figure 4. Effect of glass on elongation of thermoplastics

equalizer. At the 20% level, the impact behavior of the vastly different matrix resins is shown to be equivalent. This behavior is also seen in all other brittle and tough thermoplastics. Generally, irrespective of the polymer, the addition of fibrous glass will result in a notched Izod impact strength ranging from 1.0 to 4.5 ft lbs/inch.

An important corollary to impact behavior of reinforced thermoplastics is shown in Figure 6. Two different ABS resins are used for this illustration relating to low temperature impact strength retention. The unreinforced ABS will retain from 20–30% of its room temperature impact strength in going to $-20°F$. The graphs for both the low and high toughness ABS show the benefits obtained from fibrous glass. The low toughness ABS with 20% glass actually retains 100% of its room temperature impact strength in going to −20°F. Further, the low temperature impact strength of reinforced thermoplastics is equal to the room temperature impact strength. This is important since all unreinforced thermoplastics lose impact strength in going to lower temperatures.

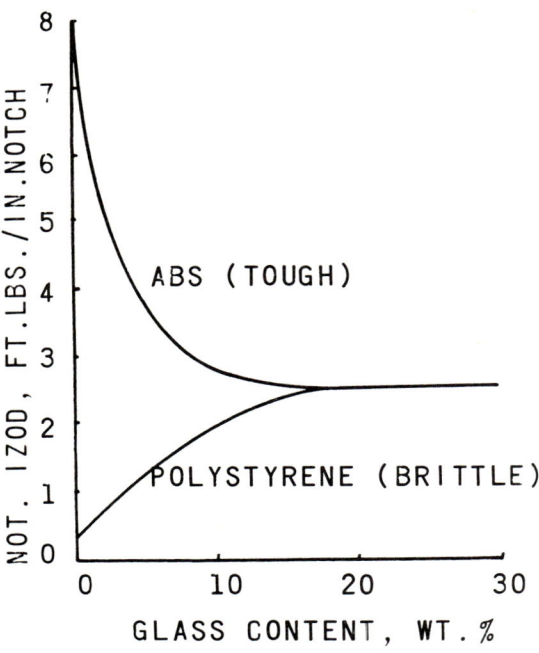

Figure 5. Fibrous glass and impact strength of thermoplastics

DTUL at 264 psig (Heat Distortion) for Reinforced Thermoplastics. By definition, thermoplastics have limitations at elevated temperatures. It is in this particular property that fibrous glass can lead to remarkable improvements. However, a sharp division exists for reinforced thermoplastics. The various reinforced thermoplastics can be put in two groups relative to DTUL. These consist of amorphous and crystalline or semicrystalline polymers. The amorphous polymers such as styrene–acrylonitrile, polystyrene, polycarbonate, poly(vinyl chloride), and acrylonitrile–butadiene–styrene are generally limited to modest DTUL improvements, usually on the order of 20°F with 20% glass. However, crystalline polymers such as the nylons, linear polyethylene, polypropylene, and polyethylene terephthalate, can be increased remarkably. The case for nylon 66 and linear polyethylene is illustrated in Figure 7. Nylon 66 is raised from a DTUL of approximately 160°F to *ca.* 500°F, which is near its crystalline melting point. Linear polyethylene is improved from *ca.* 120° to *ca.* 260°F, which is also within a few degrees of its crystalline melting point. Although the DTUL temperatures do not translate directly to end-use temperatures, they serve as an indicator of much improved elevated temperature performance for reinforced thermoplastics.

Dimensional Stability of Reinforced Thermoplastics. Thermoplastics have relatively large linear coefficients of thermal expansion. This can

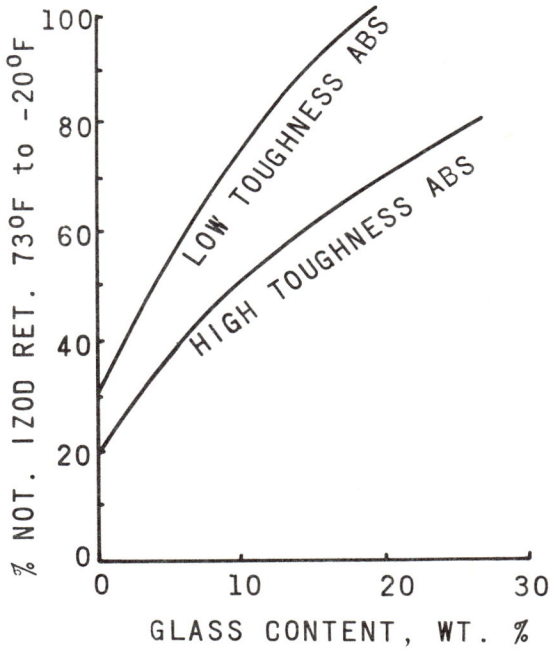

Figure 6. Impact strength retention at low temperatures

lead to difficulties when plastic parts are used in combination with metal and where temperature extremes are encountered. This is, of course, magnified when large parts, such as a 5-ft wide instrument panel, are exposed to a surface temperature range of approximately 200°F.

Figure 8 shows the effect of fibrous glass on linear coefficient of thermal expansion for ABS, one of the more dimensionally stable thermoplastics. The improvement in dimensional stability is on the order of

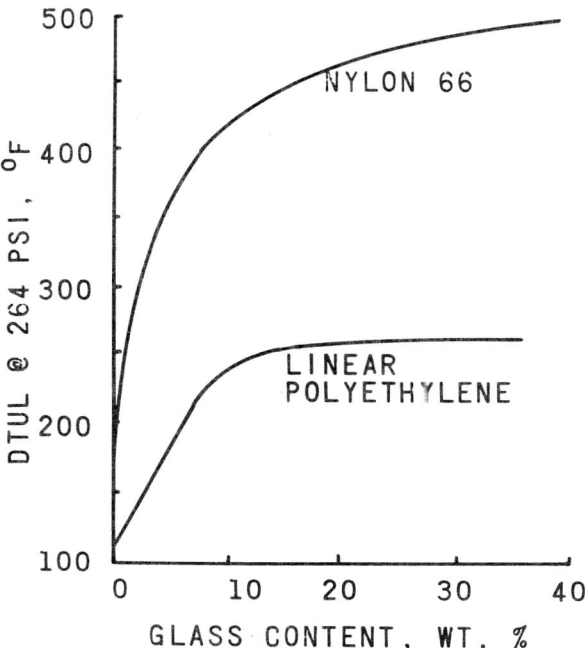

Figure 7. Composite DTUL at 264 psig

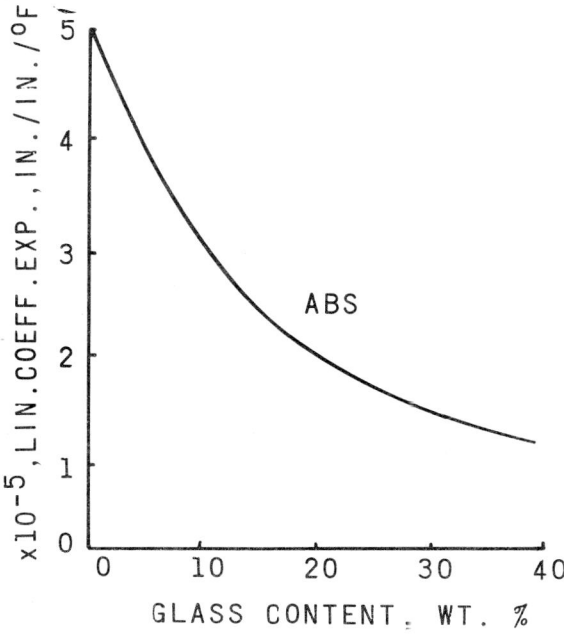

Figure 8. Composite thermal expansion

three-fold with 30% reinforcement. This improvement is characteristic for all reinforced thermoplastics, with the effect of glass being more pronounced in some polymers than others.

Physical and Chemical Concepts of Fibrous Glass Composites

In a discussion of why fibrous glass is used in combination with various polymers, two important questions must be answered. First, what properties of fibrous glass are responsible for its use in combination with plastics? Secondly, why does it work?

Properties of Glass. Some important properties of fibrous glass are listed in Table III. The 400,000-psig tensile strength and 10,000,000-psig modulus are the two key mechanical properties. When these are com-

Table III. Properties of "E" Glass

Tensile strength, psig	400,000
Elastic modulus, psig	10,000,000
Coefficient of thermal expansion, /°C	5×10^{-6}
Softening point, °F	$\sim$1500°F
Material class	Elastic

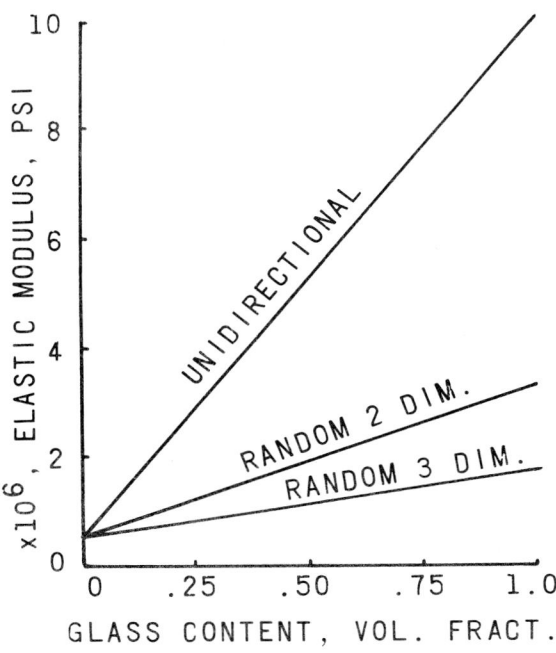

Figure 9. Composite elastic modulus (orientation considerations)

pared with typical tensile strengths and modulus numbers of 8000 psig and 500,000 psig, respectively for unreinforced resins, it is easy to develop a case for composite systems.

Orientation effects are important in utilizing fibrous composites. The theoretical relationships are shown in Figure 9. The greatest benefit is obtained with unidirectional fibers. The more common case in practice is closer to the random orientation in two dimensions. The theoretical modulus that is obtained in this fiber configuration is one-third of the fully oriented composite.

Of interest is the fact that glass is an elastic material. Thermoplastics are viscoelastic and, therefore, exhibit creep or cold flow. The use of glass is an efficient means of reducing creep.

One of the shortcomings of thermoplastic materials is dimensional stability owing to the high coefficient of thermal expansion. Typically, this can be on the order of 50×10^{-6} cm/cm/°C. The coefficient for glass is a much lower 5×10^{-6} cm/cm/°C, and it is, therefore, effective in reducing dimensional changes.

A typical unsaturated unreinforced polyester resin has an extremely low notched Izod impact strength. The addition of fibrous glass can change this extremely brittle material into a high impact strength composite. The same phenomenon occurs with some brittle thermoplastics, such as polystyrene and styrene–acrylonitrile.

In an oversimplification, it is the combination of high strength, high modulus, creep resistance, dimensional stability, and impact modification of the brittle matrices which leads to the selection of fibrous glass in composites.

Achieving Glass-to-Matrix Adhesion. This subject has been expanded upon in several papers and is summarized well by Sterman and Marsden (13). A review of some of the key concepts in glass-to-polymer adhesion as it relates to thermosets and thermoplastics will serve the purposes of this paper.

The approach to obtaining adhesion is to use an intermediate chemical compound between the glass and the matrix. The bridging agents used are organofunctional silanes. These materials bond to the glass surface as well as to the matrix.

The organosilane chemicals utilized in glass composites are trifunctional silanes—i.e., they contain three hydrolyzable groups per silicon atom. Upon hydrolysis, the silanol group adheres strongly to the glass surface. The mechanism by which this takes place is inherently difficult. The indirect confirmation is much better documented through mechanical property data.

Table IV. Ambifunctional Silanes

$$R-Si(OR')_3$$

$NH_2(CH_2)_2NH(CH_2)_3-$
$NH_2(CH_2)_3-$
$CH_2=CH-$
$HS(CH_2)_3-$
$Cl(CH_2)_3-$

$$CH_2=\underset{\underset{O}{\|}}{C}-\overset{\overset{CH_3}{|}}{C}-(CH_2)_3-$$

$$CH_2\underset{\diagdown\;O\;\diagup}{-}CHCH_2O(CH_2)_3-$$

Adhesion to the glass surface completes only one-half of the bridge. The other half must adhere to the matrix. This mechanism is even more difficult to study. Again, direct confirmation is much easier to show through property improvements. Table IV shows some of the common functionalities for various commercial silanes.

The adhesion of thermoplastic to glass cannot be explained by copolymerization of the coupling agent with the matrix. The polymers have essentially fully reacted. Chemical reactivity on the polymer backbone is now of interest. The picture is also more complex in that the number of polymers to be reinforced is much larger than for the thermosetting materials. Virtually every polymer which is being injection molded is also being used in its reinforced form.

Several papers have been published on the effects of silane coupling agents in thermoplastics (4, 9, 10, 11, 14, 15). Silanes are very effective in improving the strengths of fibrous glass thermoplastics. The functionality of the silane as well as time–temperature conditions in the molding cycles are of great importance. The area of question is on how much of this increase can be realized in an injection molding environment.

The work of Sterman and Marsden and Plueddeman was carried out with glasscloth polymer laminates (10, 11, 14, 15). In this way the effect of the silane could be optimized as to time and temperature of the molding cycle. Additionally, the glass was being used in its maximum strength form in that it was continuous. An additional factor is the absence of any shear forces. As an example, for a polystyrene laminate, the use of an epoxyfunctional silane increased the flexural strength 90% over the bare glass control. The wet flexural strength improvement was an even greater 140%.

To a producer of fibrous glass reinforcements it is distressing that the results demonstrated with glass fabrics could not be translated to glass subjected to an injection molding environment. To account for this apparent anamoly, an investigation was undertaken (2). The earlier experiments with fabric laminates were essentially duplicated with respect to time and temperature, and compression molding was used. The difference was in the form of the reinforcement used. For this work, the glass and polymer were combined in a compounding extruder, and this compound was plasticized in an injection molding machine. The molten compound was compression molded, and the mechanical properties were determined.

Table V. Shear Effects for Silane-Only Systems[a]

Property Improvements over Bare Glass

	Fabric Laminate, Continuous Fibers	High Shear History, Discontinuous Fibers
Flexural strength	+90%	+10%
Wet flexural strength	+140%	+10%

[a]Polystyrene, 20% by weight glass, 32% for fabric laminate, epoxy silane compression molded at 500°F.

Table V illustrates the results obtained by the experimental technique which essentially simulates injection molding while satisfying the time–temperature conditions effective in the fabric laminate work. The large strength increases were reduced to only approximately 10% over the bare glass controls. The conclusion is that the high shear environment of injection molding which yields discontinuous fibers does not allow the same strength increases obtained in earlier experimental work.

Table VI. Reducing Shear Effects on Glass[a]

Property Improvements over Bare Glass

	Epoxy Silane Only	No Silane Film Former	Epoxy Silane Film Former
Flexural strength	+10%	+15%	+25%
Wet flexural strength	+10%	−10%	+30%
Tensile strength	+10%	−30%	+50%
Notched Izod	+40%	−90%	+100%

[a]Polystyrene, 20% by weight glass, compression molded 15 min at 500°F.

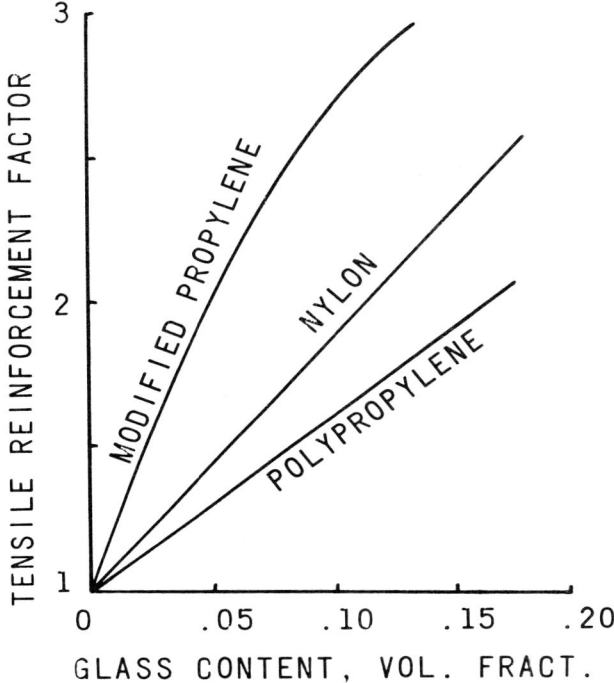

Figure 10. Effects of polymer chemistry in reinforced thermoplastics

In addition to the work using silanes only, the presence of polymeric material at the glass surface was investigated. Table VI illustrates the beneficial effects obtained by protecting the glass filaments with a polymer, commonly referred to as a film former. The combination of silanes with appropriate film formers was shown to be essential in developing satisfactory strength levels. The improvements still did not equal the fabric laminate results. However, the importance of protecting the filaments was illustrated when an injection molding environment is involved.

Polymer chemistry is important in obtaining adhesion to the glass surface (Figure 10). The tensile reinforcement factor—the ratio of tensile strengths of the reinforced system to the matrix resin—is used as a measure of adhesion. Two dissimilar polymers, polypropylene and nylon, are used to illustrate the importance of polymer chemistry. Polypropylene is an inherently difficult polymer to reinforce because of its nonpolar nature and lack of reactivity. Nylon, on the other hand, is highly polar and is one of the easiest thermoplastics to reinforce. The modified polypropylene refers to a polymer specifically developed to be used with glass (1). It can be considered a second generation reinforced thermoplastic. The remarkable difference in the ease of reinforcement can be attributed entirely to a chemical modification of the polypropylene.

Potential of Fibrous Glass Composites

Thermosets. The potential of thermosets is not solely a function of mechanical performance. These materials have established themselves as suitable for many demanding environments. The potential must be measured in terms of their ease of fabrication into parts. This is where limitations have existed for reinforced thermosets, specifically polyesters; this, in combination with the limitation related to the necessity of post-molding operations to obtain high appearance surfaces.

With the introduction of SMC and BMC which give much improved process techniques, the fabrication shortcomings have been largely overcome. The introduction of low profile resin systems has eliminated the costly post-molding operations for high appearance parts.

Table VII. Theoretical Strength Model for Fibrous Glass Thermoplastics

$$\sigma_c = \sigma_f V_f \left[1 - \frac{1}{2\alpha} \right] - \sigma_m (1 - V_f)$$

$$l_c = \frac{d\, \sigma_f}{2\, \tau}$$

$$\alpha = l_a / l_c$$

V_f = volume fraction of glass fiber
l_c = critical fiber length
l_a = actual fiber length
d = fiber diameter
τ = shear strength of matrix resin
α = ratio of actual to critical fiber length
σ_c = composite tensile strength
σ_f = fiber tensile strength
σ_m = matrix tensile stress at failure strain of the composite

The potential of these materials should then be a function of these two technical advances. It is entirely related to how well this technology is converted to general practice. Although additional progress in materials will be a factor, the existing technology can take reinforced polyesters a long way.

Thermoplastics. This segment of fibrous glass composites came out of its infancy in the 1960's. The potential of these materials is tremendous because no new molding process technology is required. Advances will be in the area of materials performance.

If it were possible to develop a theoretical strength model for reinforced thermoplastics, it would be much easier to measure their poten-

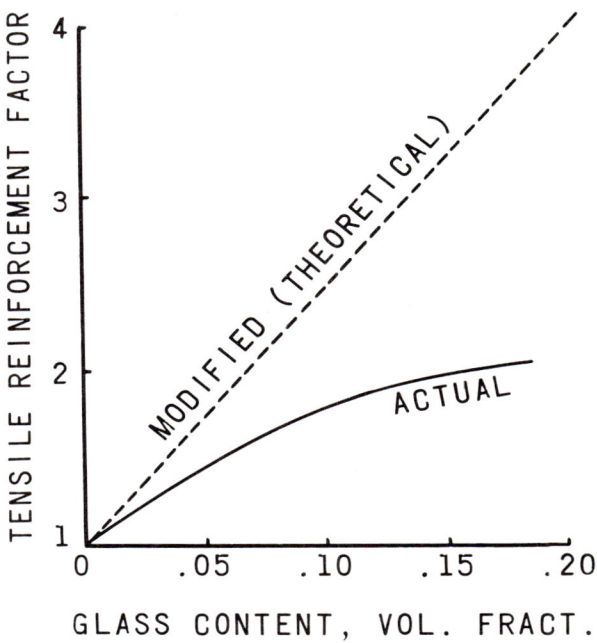

Figure 11. Effects of polymer chemistry—potential strengths of styrene

tial. It would then be possible to predict optimum strength levels. It appears that such a theoretical strength model exists for reinforced thermoplastics, based on the plastic stress transfer model for fiber reinforced metals (5, 6) (Table VII). It is supported by experimental data for metals as well as for fibrous glass reinforced thermoplastics (8).

Using this strength model it is interesting to predict strengths of a composite, assuming satisfactory adhesion of matrix to glass. This was done with polystyrene, which is regarded as difficult to bond to the glass surface. Figure 11 shows the property improvements which could be obtained in glass-reinforced polystyrene if improved adhesion could be achieved. It is suggested that a chemical polymer modification must be made to attain the required adhesion. Since the feasibility of this has been demonstrated for polypropylene, the predicted strength levels could well be within reach.

Glass-reinforced polystyrene points out that some reinforced thermoplastics have not reached their performance potential. The continuing activity in developing improved reinforcements coupled with development work in the area of designing polymers to be chemically compatible with glass will determine the true potential of these composites. Since less than 2% of the thermoplastics injection molded in the United States contain fibrous glass at this time, the growth curve of these materials is highly unpredictable. It is estimated that these materials are growing at a current annual rate exceeding 25%. There is little doubt that a tremendous potential exists for these composites.

Literature Cited

(1) Cessna, L. C., Thomson, J. B., Hanna, R. D., *SPE J.* **25**, 35 (1969).
(2) Englehardt, J. T., Krautz, F. G., Philipps, T. E., Preston, J. A., Wood, R. P., *SPI Reinforced Plastics Div. Ann. Tech. Conf., 22nd, 1967.*
(3) Frilette, Vincent J., U. S. Patent **2,568,331**.
(4) Hall, N. T., *SPI Reinforced Plastics Div. Ann. Tech. Conf., 21st, 1966.*
(5) Kelly, A., Davies, G. J., *Metallurgical Rev.* **10** (37) 1.
(6) Kelly, A., Tyson, W. R., *J. Mech. Phys. Solids* **13**, 329 (1965).
(7) Kroekel, C. H., Bartkus, E. J., *SPI Ann. Tech. Conf., Reinforced Plastics/Composites Div., 23rd, 1968.*
(8) Lees, J. K., *Polymer Eng. Sci.* **8** (3), 195 (1968).
(9) Murphy, T. P., "Abstracts of Papers," 150th Meeting, ACS, Sept. 1965, T12.
(10) Plueddeman, E. P., *SPI Reinforced Plastics Div. Ann. Tech. Conf., 20th, 1965.*
(11) Plueddeman, E. P., *SPI Reinforced Plastics Div. Ann. Tech. Conf., 21st, 1966.*
(12) Stephenson, C. E., *British Plastics* **30** (3) 99 (1957).
(13) Sterman, S., Marsden, J. G., *Union Carbide Publ.* **F41920, 10M-368**.
(14) Sterman, S., Marsden, J. G., *SPE Ann. Tech. Conf., 21st, 1965.*
(15) Sterman, S., Marsden, J. G., *SPI Reinforced Plastics Div. Ann. Tech. Conf., 21st, 1966.*

RECEIVED February 25, 1970.

29

Halogenated Epoxy Matrix Plastics in Filament-Wound Composites

JAMES R. GRIFFITH, ARTHUR G. SANDS, and JACK E. COWLING

Naval Research Laboratory, Washington, D. C. 20390

> *Epoxy resins which contain chlorine, bromine, and fluorine in substantial quantities have been evaluated as matrix materials for glass fiber-reinforced composites. NOL rings containing these systems have been filament wound to determine whether or not low water absorption of the matrix would extend fatigue life under water of a composite. Water absorptions of the bulk plastics and winding variables have been measured for each composite system. Fatigue life has been evaluated in a slow, continuous cycle. Standard epoxy resins based upon bisphenol-A yielded composites with fatigue lives in the 5,000–6,000 cycle range. In general, the composites produced from halogenated materials were superior. One system had nearly triple the fatigue life of the standards.*

Structural materials for use in the deep ocean as well as in space require high strength-to-weight ratios. The hull of a deep-submersible vehicle, for example, must be able to withstand water under high pressure while not being excessively heavy. For this reason fiber-reinforced plastics offer potential advantages as structural materials for such a purpose, and filament wound composites in which the fibers are placed in orderly patterns are particularly promising.

Since such composites contain enormous areas of adhesive bonding, and water is generally quite deleterious to the bond strength of an organic adhesive, there is concern regarding the long term reliability under mechanical stress in water. It appears likely that a matrix plastic with minimal response to the effects of water should be more effective in this regard than a more sensitive matrix, but to our knowledge there have been no previous accounts in the literature of efforts to test this idea. Halogenated epoxies are frequently reported in trade publications to absorb less water than the common epoxies, although data on the subject in the chemical literature of epoxy materials are scattered and fragmentary. Consequently, a study was made to evaluate as filament-winding matrices epoxy components containing chlorine, bromine, and fluorine. The chlorinated and brominated materials were obtained from commercial sources, and the fluorinated resins were synthesized locally as part of a continuing study of fluorinated epoxy systems. All of the filament windings, sample preparations, and evaluations were done at the Naval Research Laboratory.

Experimental

Materials. Code designations, chemical compositions, and the commercial sources of the resins, curing agents, and additives containing bromine and chlorine are listed in Table I. The fluorine-containing resins are described below. For convenience in the tables, the hydrocarbon resins without halogen are designated by S; those with chlorine are Cl; with bromine, Br; and with fluorine, F. The curing agents are designated X regardless of halogen content, and the inert additive, which contains chlorine, is A.

The fluorinated resins, F_1 and F_2, are the diglycidyl ethers of 4,4′-dihydroxyoctafluorobiphenyl and of 1,3-bis(hexafluoro-2-hydroxypropyl)

Table I. Identification of Materials

Code	Trade Name	Chemical Composition	Source	Reference
S_1	Epon 828	Diglycidyl ether of bisphenol A	Shell	
S_2	ERL 2256	Combined S_1 & S_3 types	Union Carbide	
S_3	ERLA 0400	Epoxidized cyclopentenyl ether	Union Carbide	
Cl_1	DGEHQ	Chlorinated hydroquinone derivative	Velsicol	2
Br_1	ERX-67	N,N,-diglycidyltribromoaniline	Shell	7
Br_2	DER 542	Brominated S_1 type	Dow	
F_1	—	(see below)	NRL	4
F_2	—	(see below)	NRL	3
X_1	Curing Agent CL	Metaphenylenediamine	Shell	
X_2	NMA	Nadic methyl anhydride	Allied	
X_3	ZZL 0820	Aromatic amines	Union Carbide	
X_4	MOCA	Chlorinated methylenedianiline	Dupont	
X_5	—	Tetrafluorometaphenylenediamine	Narmco, Whittaker	
A	Arochlor 1254	Chlorinated biphenyl	Monsanto	

benzene respectively. They were made from the corresponding dihydroxy compounds by the method of Kelly, Landua, and Marshall (5).

The systems of resin and curing agent for which there were manufacturer's recommended compositions and cure cycles were prepared and cured according to these recommendations. In compositions for which there were no guides, the ratios of resin and curing agent were determined by the type of curing agent. The systems containing amines were blended for exact equivalence of one amino hydrogen atom for each epoxy ring of the resin—*i.e.*, chemically equivalent amounts. Since the stoichiometry of anhydride-cured systems is so complex that trial-and-error determination of resin–curing agent ratios to obtain optimum plastic properties is common practice, the ratio of epoxy to anhydride functional groups which has been found to be best for bisphenol-A resins was used for the new compositions. In all anhydride-cured systems, a catalytic amount of dimethylbenzylamine was added to accelerate the reactions. Final cure temperatures for all systems were near 160°C, and final cure times ranged from 6 to 24 hours. Halogenated aromatic amines are considerably slower in reacting than unhalogenated analogs in general, and these required the longer cure times.

Water Absorption of the Bulk Resins. As filament windings were in progress, samples of the matrix resins were removed and cast in 1-inch diameter cylindrical aluminum cups to form discs 1/8-inch thick. These were cured in the same oven with the filament-wound structures. The discs were then placed into distilled water contained in small beakers located in a room air conditioned to maintain 25°C. Periodically, the

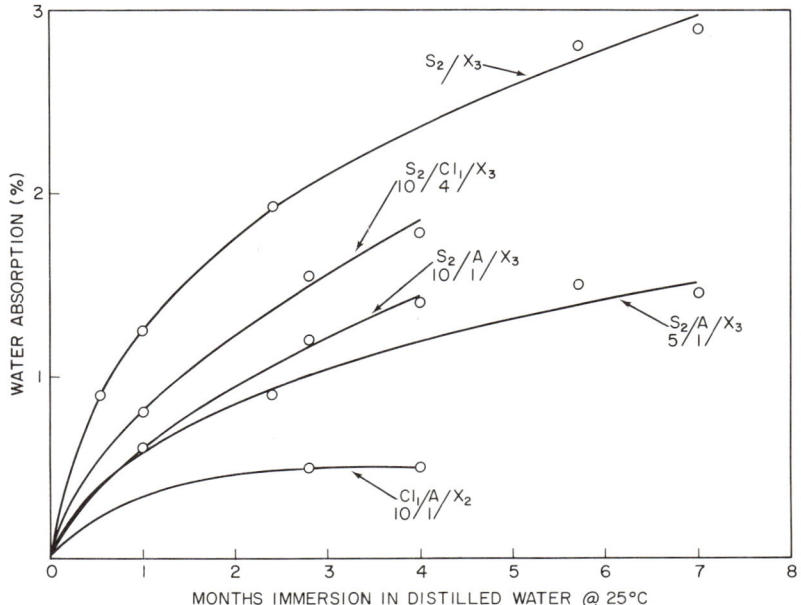

Figure 1. Water absorption of bulk matrix plastics containing chlorine

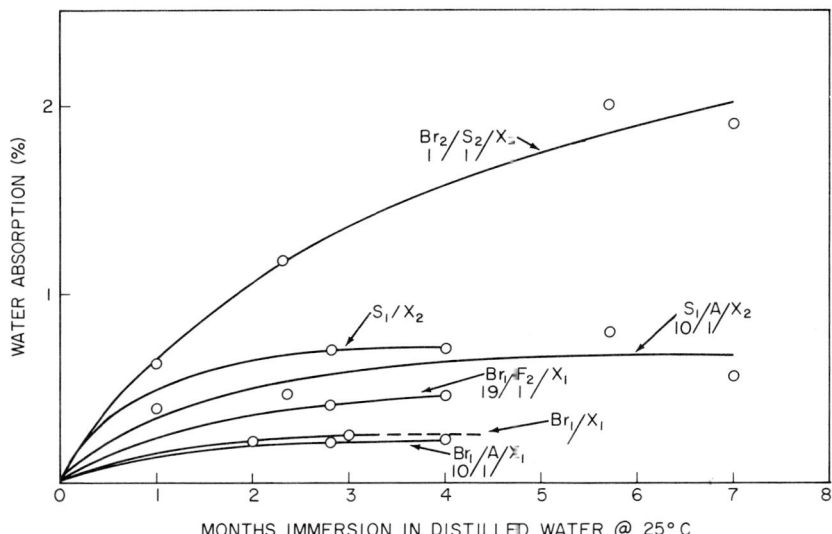

Figure 2. Water absorption of bulk matrix plastics containing bromine

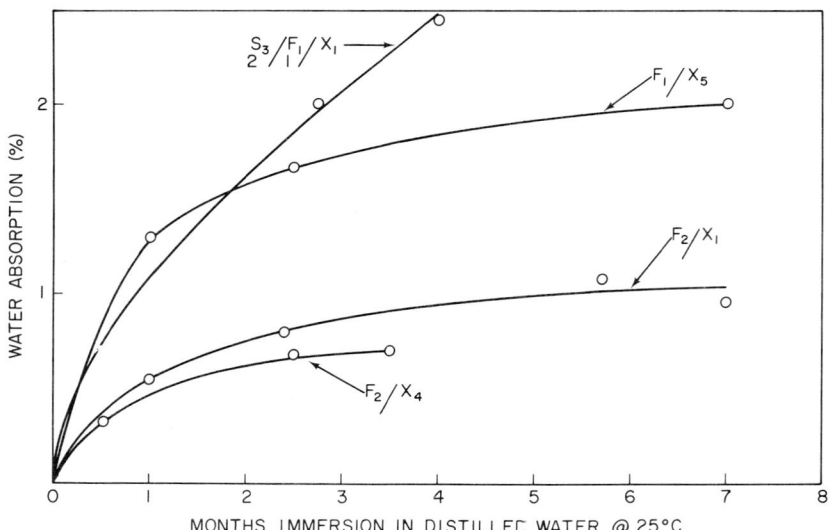

Figure 3. Water absorption of bulk matrix plastics containing fluorine

discs were removed from the beakers, dried quickly with absorbent paper, and weighed to determine water absorptions The results of these tests are shown in Figures 1, 2, and 3.

Fatigue of Glass-Fiber Reinforced Plastics (GRP). NOL rings of 6-inch outer diameter and 0.25-inch width were fabricated from 12 single

Figure 4. Control console and ring fatigue assembly

Figure 5. Close-up of ring fatigue assembly

ends of 204 filaments each of Ferro Corp. S-1014-S24 glass fiber with non-aging finish according to ASTM D2291-67. The thickness of each ring was controlled at 0.125 ± .003 inch by machining the circumference; thus, a fresh surface was exposed to water during fatigue.

The fatigue performance of rings produced from the various resin systems was evaluated in the equipment shown in Figures 4 and 5. The console was programmed to control continuously a hydraulic ram which compressed the rings in the slow cycle illustrated in Figure 6. The rings were held in position by a slotted assembly and restrained from rotation (Figure 5). The force required to compress a ring 2.53 inches along the fatigue diameter was monitored daily (approximately every 1000 cycles). This force produces a strain of 2.2% in the outer fibers of the rings at the

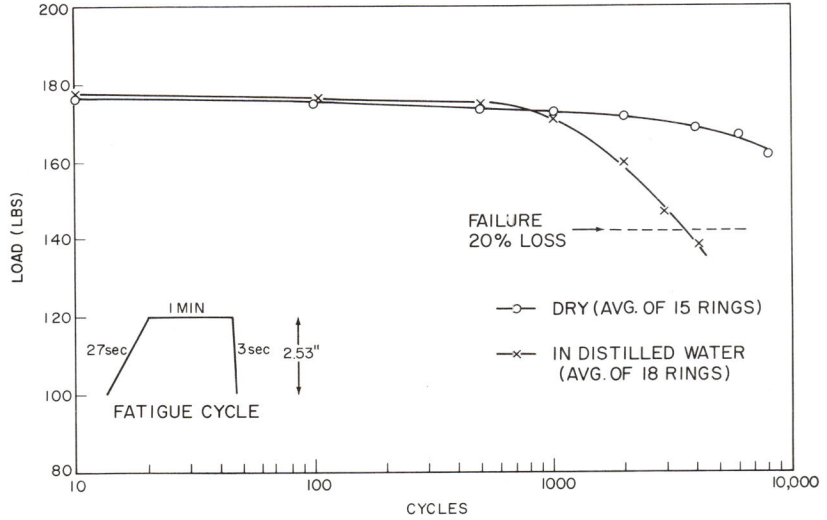

Figure 6. Typical fatigue behavior of identical rings, wet and dry

points of maximum bending, corresponding to a stress of approximately 280,000 psig for Type S glass. Failure was arbitrarily defined as the number of cycles at which a ring lost 20% of its original force for the 2.53-inch compression. A plot of force, or load, as ordinate *vs.* number of cycles as abscissa on semi-log paper commonly showed a steep decline in the load-bearing ability of the composite as failure was approached— *e.g.*, as shown in Figure 6. This figure also compares the typical behavior of dry rings with the relatively rapid decay of identical rings immersed in water.

The glass content was determined by resin burn-off of ring segments in crucibles in a muffle furnace. The void content was determined by a method of optical microscopy previously reported (6). Horizontal shear was measured by the Short Beam method (1).

Table II compares fatigue performances of the chlorinated, brominated, and fluorinated systems.

Discussion

The most outstanding results of this study were obtained with the N,N-diglycidyltribromoaniline, Br_1, both in water absorption and in fatigue of the rings produced by curing it with metaphenylenediamine, X_1. After four months immersion in distilled water, discs of the plastic absorbed only 0.25% water by weight, which was the lowest of the materials evaluated. Since this system was derived from glycidylamine, whereas all others contained glycidyl ethers, there is some uncertainty whether the low water absorption is attributable to this, to the presence of bromine in the molecule, or to both factors. The resin also had good filament-winding characteristics since it was a moderate viscosity liquid of relatively slow thickening rate at 55°C, the impregnation temperature used.

Because of the long fatigue lives obtained from rings with the Br_1/X_1 system under water, a more extensive evaluation was made. Ten sets of rings were prepared with glass contents ranging from 63 to 75%. The fatigue results are shown in Figure 7, in which each point represents the average of six rings. A very rough linear relationship between glass content and fatigue life may be seen with the points falling into two groups. Above 69% glass, the over-all average life was about 6,500 cycles, and below 69% glass the over-all average was about 14,000 cycles. Some individual rings of the latter sets survived more than 17,000 cycles.

These results suggest that there may be a critical glass content for this system around 69% at which performance characteristics change abruptly. Such behavior could possibly result from several causes, two of which would be lack of total fiber encapsulation by resin above 69% glass or strain magnification in the resin during stress at higher glass contents. The data of Table II, part B do not eliminate the possibility that glass content is a more significant factor than resin composition insofar as fatigue life is concerned since all the other compositions had higher glass percentage than the Br_1/X_1 system. However, an extensive fatigue evaluation of unhalogenated matrix resins has been performed, and rarely did an individual ring exceed 10,000 cycles regardless of glass content, void content, or horizontal shear.

It can be seen in Figure 1 and Table II that the Arochlor additive, A, which could not react chemically into the resin, was also beneficial in suppressing water absorption and extending fatigue life when incorporated into unhalogenated systems to the extent of 10% by weight. Thus, the water absorption during four months of immersion of the S_2/X_3 plastic was reduced from 2.6 to 1.8% by 10% of A. The average fatigue life of rings produced was extended from 5,300 cycles to 7,900 cycles while all other factors remained essentially constant.

Table II. Fatigue Life under Water of GRP

Matrix Resin Composition	Glass Content, wt %	Void Content, vol %	Horizontal Shear Strength, psig	Shear Std. Deviation, psig
A. Chlorine-Containing Resins				
S_1/X_2[a]	74.0	0.2	14,000	280
$S_1/A/X_2$ 10/1	74.0	0.2	13,800	225
$S_2/A/X_3$ 10/1	72.0	1.6	16,000	176
$S_2/A/X_3$ 4/1	74.1	2.5	15,700	395
Cl_1/X_2	64.0	2.3	12,200	175
$Cl_1/A/X_2$ 10/1	71.2	1.5	12,000	95
S_1/X_4	73.5	1.1	12,400	278
B. Bromine-Containing Resins				
S_1/X_2[a]	74.0	0.2	14,000	280
$Br_2/S_2/X_2$ 1/1	81.8	5.6	12,500	221
Br_2/X_2	75.3	2.5	12,100	227
$Br_2/S_2/S_3/A/X_3$ 5/5/1/2	77.5	4.0	14,600	235
Br_1/X_1	72.2	2.0	13,600	114
Br_1/X_1	63.8	1.4	14,300	99
C. Fluorine-Containing Resins				
F_2/X_5	—	—	—	—
F_2/X_1	—	—	—	—
$F_2/S_2/X_3$ 1/1	—	—	—	—
$F_2/S_3/X_3$ 1/2	69.6	5.6	12,100	1,044
$F_2/Br_1/X_1$ 1/19	73.7	4.0	12,600	393
F_2/X_4	—	—	—	—
$F_1/S_1/X_2$ 1/2	81.1	2.7	10,700	93

[a] Unhalogenated standard for comparison.

with Halogen-Containing Matrix Resins

Coeff. of Variation, psig	Initial Load, lbs	No. of Rings	Average Fatigue Life		
			Cycles	High	Low
A. Chlorine-Containing Resins					
2.0	160	27	5,300	6,100	4,100
1.6	160	5	7,900	8,400	6,900
1.1	141	7	10,900	11,800	9,600
2.5	155	7	7,900	8,600	6,900
1.4	130	7	8,300	13,200	5,100
0.8	153	7	5,600	6,400	3,700
2.2	161	6	5,100	6,000	4,000
B. Bromine-Containing Resins					
2.0	160	27	5,300	6,100	4,100
1.8	177	7	3,900	4,300	3,600
1.9	153	6	5,000	5,600	3,600
1.6	144	7	6,440	7,200	5,400
0.8	166	6	7,000	7,400	6,000
0.7	150	7	15,700	17,100	13,200
C. Fluorine-Containing Resins					
—	130	3	1	(sheared)	
—	134.5	2	3,850	4,900	2,800
—	159	1	6,700	—	—
8.6	135	7	2,500	4,000	900
3.1	155	7	4,900	6,800	3,700
—	129	8	1,000	(sheared)	
0.9	178	6	1,800	2,600	1,400

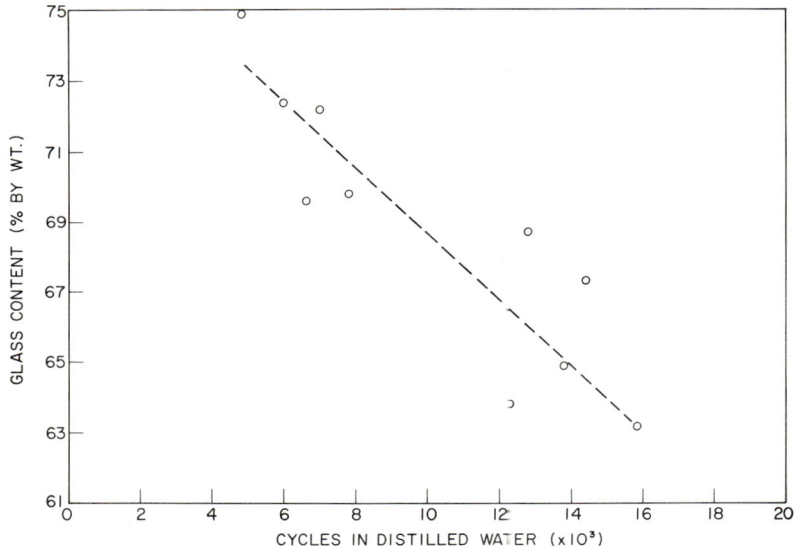

Figure 7. Effect of glass content upon the fatigue life of rings produced with the Br_1/X_1 matrix

The glycidyl ether of the chlorinated hydroquinone derivative, Cl_1, also demonstrated low water absorption (Figure 1). The resin was somewhat difficult to use for filament winding because of a relatively high melt viscosity, but the system Cl_1/X_2 had a respectable fatigue life of 8,300 cycles, with one ring surviving over 13,000 cycles. The addition of Arochlor to this heavily chlorinated resin detracted from the fatigue life of the rings produced therefrom.

The brominated bisphenol-A resin, Br_2, was also too viscous for convenient filament winding, and in this case the halogen appeared to have small effect upon water absorption or fatigue life.

The fluorinated diglycidyl ethers were available only in small quantities which could be produced in glassware. Fluorine was fairly effective in suppressing the water absorption of the cast plastics (Figure 3), but the fatigue lives of rings produced from F_1 and F_2 were less than those of conventional materials (Table II, part C). Rings which contained fluorine were usually more opaque than any of the others, and it appeared that the glass finish was not optimum for these resins. Shearing of rings during cycling occurred more frequently than normal, and this also indicated poor bonding between resin and glass. New finishes which contain fluorine may be required to realize the full potential of these resins.

Conclusions

Epoxy resins which contain appreciable amounts of halogen absorb less water on a weight percentage basis than unhalogenated resins. Resins which contain bromine or chlorine, and are suitable for filament winding, often yield composites more resistant to fatigue under water than similar composites produced from common resins. Other than resin composition, glass content in GRP appears to be the most significant factor which influences underwater fatigue life. Halogenated glycidylamines appear particularly promising as matrix resins for GRP subject to stress under water.

Literature Cited

(1) American Society for Testing and Materials Designation D 2344-65 T, ASTM Standards, Part 26, p. 499, 1967.
(2) Dissen, I. J., U. S. Patents **3,235,569** (Feb. 15, 1966) and **3,366,602** (Jan. 30, 1968).
(3) Farah, B. S., Gilbert, E. E., Sibilia, J. P., *J. Org. Chem.* **30**, 998 (1965).
(4) Griffith, J. R., Quick, J. E., ADVAN. CHEM. SER. **92**, 8 (1970).
(5) Kelly, P. B., Landua, A. J., Marshall, C. D., *J. Appl. Polymer Sci.* **6** (22), 425–532 (1962).
(6) Kohn, E. J., Sands, A. G., Clark, R. C., *Ind. Eng. Chem., Prod. Res. Develop.* **7**, 179 (1968).
(7) Newey, H. A., U. S. Patent **3,449,375** (June 10, 1969).

RECEIVED January 16, 1970.

30

Inorganic Reinforcements for High-Performance Structural Composites

WALTER H. GLOOR

Air Force Materials Laboratory (AFSC), Wright-Patterson Air Force Base, Ohio 45433

> *The range of inorganic fibrous reinforcements available to the composite materials designer is an ever-broadening one. It stretches from E-glass to graphite, from boron to whiskers. For continuous-filament materials, the high strength glasses have yet to be exceeded in tensile strength but suffer in comparison with many of the newer reinforcements in the area of elastic modulus. A large portion of the recent development effort on improved reinforcement fibers has been devoted to attainment of higher modulus while retaining an acceptable strength level. In addition to the graphite fiber development this includes effort on boron, silicon carbide, sapphire, various whisker materials, and a number of polycrystalline ceramic fibers. Information is presented on the formation, properties, and performance in composites for these emerging materials, together with a discussion of the major advantages and disadvantages for each class of reinforcement.*

Much has been said and written recently about the beginning of a new decade (3). The 70's have been referred to as "soaring," "sizzling," "scintillating" and any number of other alliterative adjectives, but this is a good time to look backward too. In 1960 the structural composites designer had a choice of reinforcements from which to select—he could choose E-glass strand, E-glass fabric, or E-glass mats. Graphite was a material used to fill pencils and form the anode in dry cells. Boron was a catchy word to sell gasoline and the last word in exotic rocket fuels. Whiskers were what we faithfully scraped off our faces each morning, and a few of us still do. Sapphire was an expensive adornment.

Indeed, there has been much progress in high performance inorganic reinforcement materials in the past decade (2). The intent of this paper is not to review in detail these developments but rather to summarize where we now stand, to some extent how we got there, and attempt to predict where the 70's will take us.

First, let us consider the glass fiber area (7). E-glass was for years the primary reinforcement material for resin matrix structural composites. From the standpoint of pounds used it is still the leader. In the early 60's, two new glasses were developed by Owens-Corning under sponsorship of the Air Force Materials Laboratory. These were S-glass, originally known as AF-994, and YM-31A, a beryllia-containing glass. A summary of fiber properties of these two materials, plus E-glass, is shown in Table I. S-glass provides improvements in both tensile strength and modulus of the order of 20% when compared with E-glass and has become the leading glass fiber for reinforcement where maximum performance is required. YM-31A, although displaying a high modulus in psi, suffers from high density, low strength, and the stigma of beryllia. Therefore, it was never used widely.

Table I. Glass Fiber Comparison

Type	Virgin Fiber Tensile Strength, kpsi	Elastic Modulus, $psi \times 10^6$	Density, $grams/cm^3$	Specific Modulus, $inches \times 10^5$
E	450–500	10.5	2.54	115
S	650–700	12.5	2.49	139
YM31A	500	15.8	2.89	151

Table II. Glass Comparison

Type	Fiber Density, $grams/cm^3$	Tensile Strength, Kpsi	Modulus $psi \times 10^6$	Modulus $Inches \times 10^6$
E	2.54	450–500	10.5	115
S	2.49	650–700	12.5	139
970S	2.41	700+	14.9	171
2285	2.55	750–800	15.2	166

Recently, two more high modulus, low density glasses (4) were developed—composition 2285 by Owens-Corning and 970S by Aerojet-General's Fiberglass group, now Glass Fibers Products, Inc. Table II compares these two, both of which contain beryllia, with E- and S-glass. Both exhibit approximately a 20% improvement in specific modulus over S-glass, with the 970S having the edge in strength although at a somewhat higher density. Over-all, the properties of these two glasses are about the same. In attempts to scale up from laboratory operation to

204-hole production-type bushings, difficulties with reproducibility in continuous operation were encountered with the 2285 glass, while with 970S it was possible, with minor compositional changes, to obtain essentially the same mechanical properties as lab operation had produced. Commercialization of 970 is now under consideration, although extensive testing in composite specimens by industry and government will be required prior to such a decision. Additionally, some finish or coupling agent development will probably be required.

One or two other developments in the glass fiber area warrant mention. One is the work that has been done on hollow glass fibers of conventional glass formulations (1). While offering no advantage in tensile properties, such fibers provide a method of achieving reduced weight for a given size structure, as well as offering a route to improved flexural and compressive performance.

Finally, a current effort in glass compositional research is worthy of mention. This is a United Aircraft program under NASA sponsorship which originated with intensive studies of the thermodynamics of crystallization. Low silica compositions which would be expected to crystallize are urged into the glassy state by adding rare earth oxides such as yttria and lanthana. These additives reduce the rate of crystallization sufficiently so that glass fibers may be drawn. Although most of these glasses have fairly high densities, several of them demonstrate moduli sufficiently high (18 to 22 million psi) so that their specific moduli approach 200 million inches. Accurate tensile strengths have not yet been determined, but one of these glasses, at about 18.5 million psi and a density of about 3.3, has undergone preliminary evaluation in epoxy matrix composites and yielded good tensile and flexural performance. Drastic deterioration of composite properties after a water boil indicates the needs for coupling agent studies. Typical dry properties obtained at approximately 60% fiber by volume are shear strength, 15,000 psi, flexural strength 250,000 psi, and flexural modulus of 11 million.

Despite the significant improvements in glass fibers mentioned, it is apparent that if reinforcement requirements are for specific moduli much above about 200 million inches, materials other than glass, or at least glass as we know it, will be necessary. The search for materials which can be put into fibrous form and possessing high Young's modulus, low density, and the potential for acceptable levels of tensile strength has been active over these last 10 years.

A large portion of such effort has been expended in the carbon and graphite area, materials of tremendous and exciting potential. Costs will certainly be reduced drastically in the years ahead. Specific cost projections will be left to the producers who will have to sell their product profitably, but there seems to be general agreement that graphite fibers and yarns will be at least competitive pricewise with other high performance reinforcement materials. Looking ahead on tensile properties, strengths greater than 600,000 psig and elastic moduli of about 100 million on graphite in production quantities can probably be expected. At a density of around 2 grams/cm^3, this translates to a specific modulus of about 1.4 billion inches, or about twice that available with boron, and seven times the best glasses.

Probably no fiber since the advent of nylon has received the widespread publicity afforded boron. This space-age super-fiber captured the imagination of both the professional and the layman, opening the door to serious consideration of resin–matrix composite materials for structural applications previously the exclusive domain of metals.

Table III. Typical Epoxy Composite Properties

	S-Glass	T-50 S	Boron
Tensile strength, psi	300,000	120,000	200,000
Tensile modulus, psi	8×10^6	25×10^6	35×10^6
Compressive strength, psi	250,000	90,000	250,000
Interlaminar shear, psi	15,000	8,000	14,000
Density, grams/cm^3	2.0	1.5	2.0

Uniaxial specimens, 60 vol %

It is not necessary to dwell on the properties of boron that make it attractive, especially in prepreg tape form, for both filament-wound and laid-up structures. Table III compares boron with both S-glass and Thornel 5-S in epoxy composite form. The numbers shown are typical for uniaxial specimens at about 60 vol % fiber. Specific points worthy of mention for their merit, or lack thereof, are

(1) boron tensile modulus
(2) glass strength
(3) graphite compressive and shear strengths

Boron fiber is formed by vapor-phase decomposition of boron trichloride and deposition on an electrically heated tungsten filament substrate. Typically, the substrate is 1/2-mil in diameter, and the final boron filament is 4 mils. Considerable work has gone into development of substrate materials other than tungsten, which imposes a rather severe density penalty. This has included both electrically conductive glass and quartz fibers and carbon. To date, no commercialization has taken place of a boron process not utilizing tungsten as the substrate material. There is currently some interest in larger diameter boron, primarily as a means to improved economics. For some applications the larger diameter would not be a disadvantage, and going to an 8-mil on a 1-mil tungsten substrate could mean a significant decrease in cost per pound, both of the fiber and in associated handling costs.

The 60's also saw the development of silicon carbide fibers by vapor deposition on tungsten. Although these fibers exhibit a higher modulus than boron, lower strength and higher density nullify any advantage they might have in resin–matrix composites. Silicon carbide enjoys an advantage over boron in compatibility with metals, but this is also available in "Borsic," boron filament with a thin silicon carbide layer deposited over the boron. Currently, no manufacturer produces silicon carbide filament.

Three of the fiber areas just discussed—glass, graphite, and boron—have one thing in common. Either they are currently being used in structural composite materials or they are now undergoing extensive testing, in hardware, for such applications. They offer the designer a wide choice of properties, at a wide range of costs and, particularly for graphite, promise continued improvement in mechanical properties. It is reasonable to ask, therefore, why develop additional fiber types? A major portion of the answer is concerned with fiber properties other than simply mechanical. Compatibility with the matrix, especially important in metal–matrix composites, has already been mentioned. An important consideration for composites intended for radomes and antenna windows is the fiber's transparency to electromagnetic radiation. Continuous filament, single-crystal alumina, or sapphire, and some of the whisker materials, are candidates for these applications. Naturally, if high levels of mechanical properties are not needed, glass and silica fibers can be used.

Sapphire fibers (5) are grown continuously from an alumina melt by pulling an oriented seed crystal from the melt through a controlled temperature gradient. The resulting filament is nominally a c-axis crystal, whose length is limited only by the take-up and melt replenishing systems. Orientation of the fiber axis relative to the crystal c-axis is normally found to be within 4°. Sapphire fibers have been grown as small as 2-mil diameter and up to about 15 mils. Typically, diameters are in the 8–10 mil range. Average mechanical properties are: strength, about 400,000 psi; modulus, about 70 million psi; density 3.94 grams/cm^3. Strengths over 600,000 psi have been measured on selected samples. Recent resin–matrix composite specimens with sapphire reinforcement have shown excellent compressive strength. Because of their high temperature capabilities, high level of mechanical properties, and chemical compatibility, sapphire fibers are also of interest in the metal–matrix composite field. Costs are currently relatively high; however, improved production methods, including multi-filament capabilities and increased growth rates should effect sizeable reductions in price in the near future.

To this point, the discussion has centered on continuous filament materials, either multi-filament or large monofil types. There has also been considerable research and development expended on short fiber types. Outstanding in this category are the inorganic whisker materials (6, 8). These fiber-like single crystals, normally formed by vapor-phase growth, have demonstrated exceptionally high strengths and the theoretical modulus for the material in question. Typically, whiskers range in size from submicron to a mil or larger in diameter and with aspect ratios of from less than 50 to over 10,000. In fact, nonuniformity, not only in size but also in mechanical properties, is a major shortcoming of whisker materials today. Before their potential composite materials can be evaluated fully, it will be necessary to grow whiskers in a narrower range of dimensions and exhibiting a much smaller spread in measured strength. Currently this spread in one whisker batch (alumina, for example) may be from under 100,000 psi to 2 million psi or higher. Also, in some whisker batches, a large percentage of the weight is made up of low aspect ratio, essentially unusable material.

These comments on whiskers apply roughly equally to alumina, silicon carbide, and silicon nitride, the three types on which the most work has been done.

When and if uniform size and property whisker materials become available, techniques will be required for aligning and incorporating them in composite structures. Limited work in this area has been accomplished. Since costs are likely to remain fairly high, efficient use of whiskers will require good alignment and minimum handling. The need for alignment also stems from the necessity for achieving high volume fraction of reinforcement in the matrix. Non-aligned whiskers pack poorly, restricting loading to 20 vol % or less. In short, the potential for inorganic whisker materials in high performance resin–matrix composite materials is great, but most of the development effort remains to be done to realize this potential.

Several fiber types have been mentioned so far, and several other types have been neglected that have been worked on over the past few years. Some of those not discussed may become important fibers for reinforcement in the years ahead. To date though, they have not been available in sufficient quantity for thorough evaluation in composite specimens. Included in this group are boron carbide, spinel, polycrystalline alumina and silica, titanium diboride, and miscellaneous silicides and intermetallics. Ten years from now as we look back on the 70's we no doubt will have an entirely different view of some of these materials.

Another fiber available in heavy yarn form is boron nitride. Tensile strengths are reported to be of the order of 150–200,000 psi and modulus around 20 million.

Table IV lists the main fiber types discussed as well as their form and method of formation. Graphite is included even though not discussed in detail since it belongs in such a listing.

Table IV. Inorganic Reinforcement Fibers

Type	Available Forms	Method of Formation
Glass	Yarn, strand, mats, etc.	Melt draw
Carbon/graphite	Yarn, monofil	Pyrolysis
Boron	Monofil	Vapor deposition
Sapphire	Monofil	Growth from melt
Whiskers	Short fibers	Growth from vapor

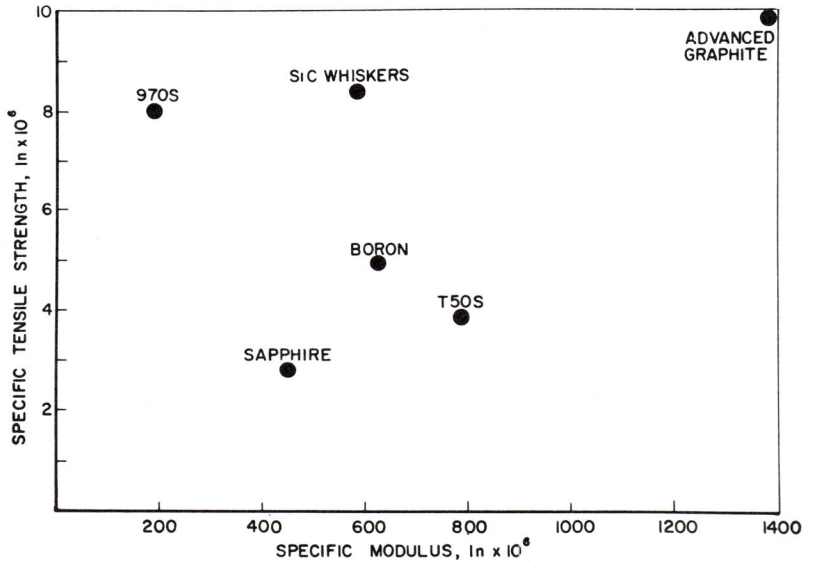

Figure 1. *Specific strength vs. specific modulus for various reinforcement fibers*

Figure 1 presents a familiar type of plot used to compare reinforcement fibers—specific strength *vs.* specific modulus. The "advanced graphite" point shown assumes 700,000 psi tensile, 100 million psi modulus, and a density of 2 grams/cm³. Its position on the chart indicates clearly why there is a future in working on graphite. This type of plot, however, emphasizes tensile performance. It gives no clues as to how a particular reinforcement will perform with respect to shear or compression. Nor does it provide information on such properties as adhesion, processability, or response to electromagnetic radiation.

In summary, a busy decade in improvements in properties of inorganic reinforcement materials has just ended. The rate of advancement in the future will probably be somewhat slower although it is unlikely that a plateau is being approached. Hopefully, the price curve can be reversed now and the rate of cost reduction accelerated. In any event, this will remain a tremendously active and interesting field.

Literature Cited

(1) Bergman, J. A., *Tech. Conf. Soc. Plastics Ind.*, 20th, Chicago, Ill., Feb. 1965.
(2) Bethune, A. W., Davis, R. A., *Space/Aeronautics* 64 (July 31, 1967).
(3) Fabian, R. J., *Materials Eng.* 20 (Oct. 1969).
(4) Gloor, W. H., *Ann. Tech. Conf., Soc. Plastics Ind.*, 24th, Washington, D. C., Feb. 1969.
(5) Hurley, G. F., LaBelle, H. E., Jr., *SAMPE J.* 17 (Dec./Jan. 1970).
(6) Krack, R. H., Kelsey, R. H., *Ind. Res.* 7 (2), 46 (1965).
(7) McMarlin, R. M., *Ind. Eng. Chem.* 58 (3), 21 (1966).
(8) Parratt, N. J., *Chem. Eng. Progr.* 62 (3), 61 (1966).

RECEIVED May 7, 1970.

31

Mechanisms of Reinforcement of Elastomers by Polymeric Fillers

MAURICE MORTON

Institute of Polymer Science, The University of Akron, Akron, Ohio 44304

> *Two systems have recently been studied involving elastomeric networks reinforced by finely divided glassy polymeric fillers. One of these was prepared by blending latices of polystyrene, or other glassy polymers, with SBR, and vulcanizing the resulting elastomer. It was found that the tensile strength of the SBR increased with higher filler content, lower particle size, and increase in rigidity of the filler and was virtually unaffected by filler–elastomer bonds. The other system studied consisted of the "thermoplastic elastomers" prepared from styrene–diene–styrene triblock polymers. Here the polystyrene forms small "domains" ($\sim$200 A) which act both as network junctions and reinforcing filler. These glassy particles yield at high stresses, thus relieving high stress concentrations and resulting in high tensile strength but large inelastic deformations.*

During the last few years new light has been thrown on the mechanism of reinforcement of elastomers by particulate fillers, based on studies of the effect of polymeric fillers on elastomers. These studies have involved two different systems, *i.e.*, (1) the use of model polymeric fillers (8, 9), such as polystyrene prepared by emulsion polymerization, in SBR vulcanizates, and (2) the "thermoplastic elastomers" obtained from ABA block copolymers (10, 11) where A is a polystyrene block and B is a polydiene block.

The morphology of these two systems is shown in Figures 1 and 2, respectively. Figure 1 shows electron photomicrographs of fracture replicas of SBR vulcanizates containing polystyrene fillers of two different particles sizes, and the existence of the individual polystyrene particles is easily confirmed. Figure 2 shows a schematic of the morphology of a styrene–diene–styrene block copolymers, in which the formation of a

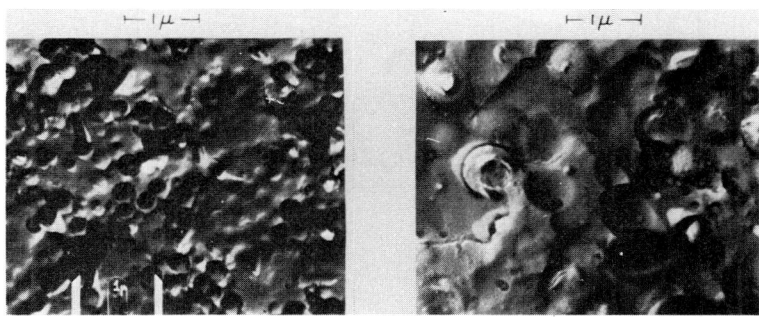

Figure 1. Morphology of polystyrene-filled SBR vulcanizates (× 10,000)

Left: R–polystyrene (2200 A) vulcanizate ($V_f = 0.25$)
Right: SBR–polystyrene (6500 A) vulcanizate ($V_f = 0.25$)

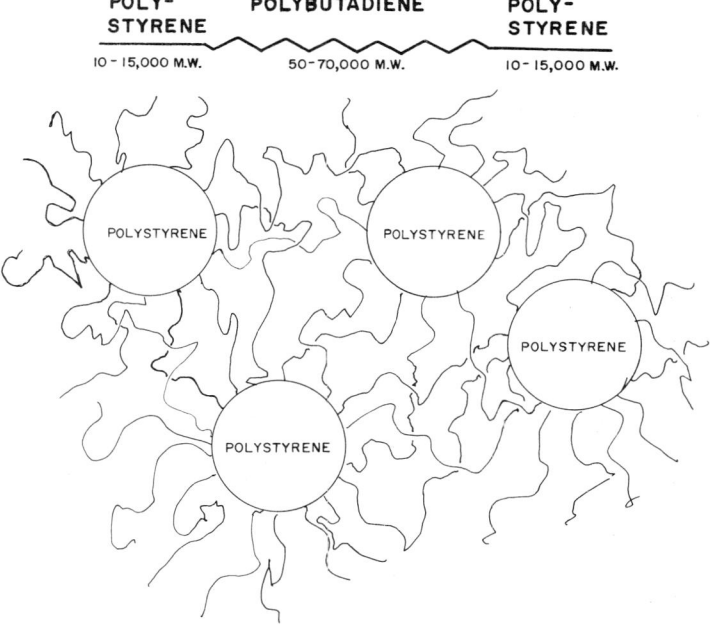

Figure 2. Morphology of SBS block polymers

separate polystyrene phase is shown in the form of spherical "domains." These have actually been observed by electron microscopy and are quite small (10), approximately 300 A in diameter. Hence, these polystyrene domains virtually constitute a finely divided filler dispersed within an elastic medium, similar, in principle, to the model polystyrene fillers of the first system described above. The only distinction between the two systems is that the "network" in the case of the block copolymers does not involve chemical crosslinking of the elastomer chains, the polystyrene domains themselves acting as network junctures for the elastic chains. The striking similarity in the morphology of the two systems makes it interesting to compare their mechanical properties which have already been studied extensively (8, 9, 10, 11).

Model Filler Studies

As previously reported (8), the effect of incorporating a finely divided polystyrene filler into an SBR vulcanizate is to raise the tensile strength by increasing the stiffness or modulus of the material. This is shown clearly in Figures 3 and 4 where it is seen that the increase in tensile strength, at any given temperature, depends directly on the amount of polystyrene filler. The polystyrene fillers in these figures are presumably not bonded chemically to the rubber. However Figure 4 shows the effect of a chemically bonded filler (SB-10, a styrene–butadiene copolymer) on the tensile strength, and it appears to have a somewhat smaller reinforcing effect than the equivalent polystyrene.

The interesting conclusion derived from this work is that the effect of the filler on the vulcanizate strength can still be explained in terms of the over-all viscoelastic response of the material. This is demonstrated in Figure 5 where it can be seen that all the vulcanizates, filled and un-

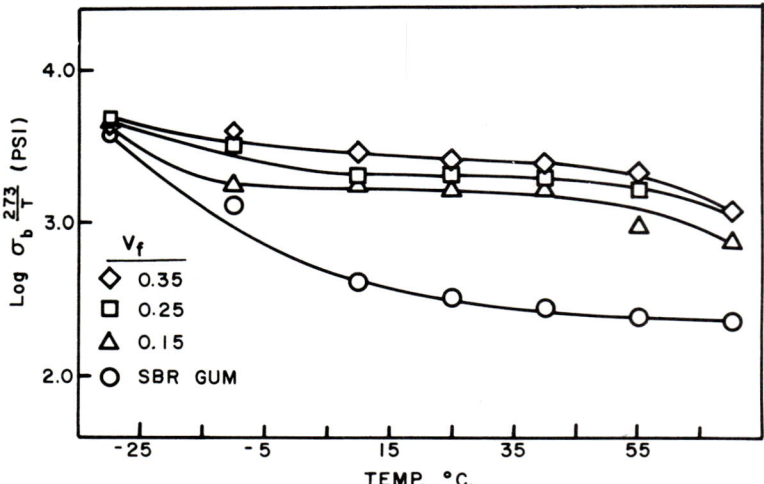

Figure 3. Effect of 485-A polystyrene filler on tensile strength of SBR (strain rate 20 inches/min)

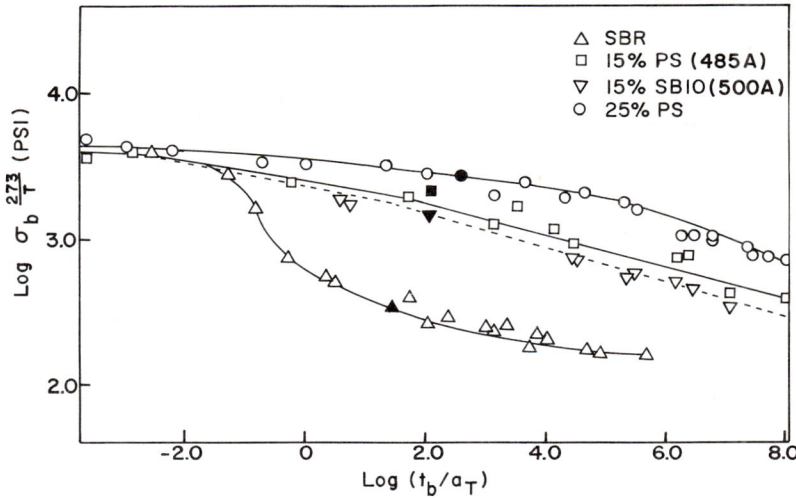

Figure 4. Effect of fillers on tensile strength of SBR. Filled points refer to 25°C and 20 inches/min.

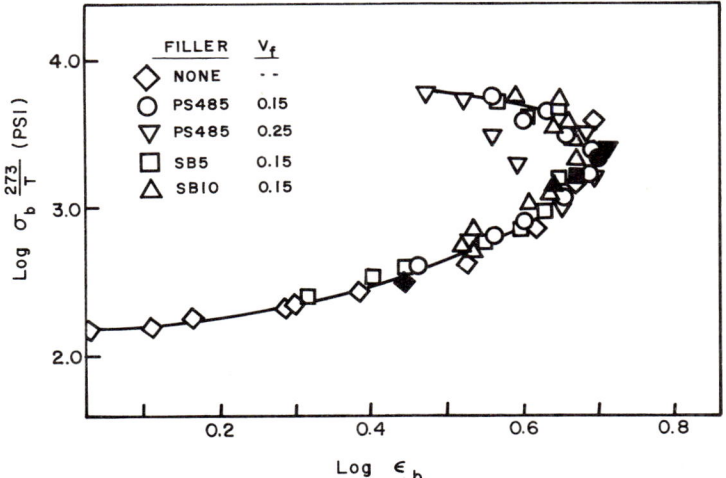

Figure 5. Failure envelope for filled SBR vulcanizates. Filled points refer to 25°C and 20 inches/min.

filled, can be plotted on the same "failure envelope," in accordance with the treatment proposed by Smith (*12*). Since all the vulcanizates were prepared with the same crosslink density, the effect of the fillers is to increase the viscous component of the network—*i.e.*, to move the failure

Table I. Properties

Filler	Type	$M_n \times 10^{-5}$
TFE	Teflon	—
SB-10	Styrene/Butadiene (90-10)	3
DCB	2,6-Dichlorostyrene/Butadiene (78-22)	Gel
PS	Styrene	2.8
DC	2,6-Dichlorostyrene	1.7
DCA	2,6-Dichlorostyrene/Ethyl acrylate (83-17)	3.0
PAB	Acenaphthylene/Butadiene (81-19)	Gel
PA	Acenaphthylene	0.9

points up the envelope. This effect is, of course, analogous to that of decreasing the temperature and/or increasing the rate of the test. Hence, the presence of these fillers simply shifts the time–temperature response of these vulcanizates to higher stress values.

The apparent effect of chemical bonding between filler and rubber, as demonstrated by the behavior of the SB-10 filler in Figures 4 and 5, is to decrease slightly the tensile strength. However, since this chemical bonding was accomplished by introducing some 10% of butadiene by copolymerization with the styrene, the resultant SB-10 filler had a substantially lower glass transition temperature than polystyrene (69° vs. 105°C). T. L. Smith brought to our attention the fact that such a drop in T_g might actually decrease the rigidity of the filler and thus possibly alter the viscoelastic response of the filled vulcanizate—i.e., decrease the viscous component and hence the strength.

Hence, it was of interest to investigate any possible effects of the T_g and modulus of polymeric fillers on the tensile strength of vulcanizates. For this purpose a series of polymeric fillers was prepared by emulsion polymerization, using monomers or mixtures of monomers designed to yield polymers or copolymers of varying T_g and modulus. The experimental details of the polymerization will be described in a forthcoming publication. The characteristics of these polymeric fillers are given in Table I.

The original objective in preparing emulsion polymers from the 2,6-dichlorostyrene and acenaphthylene was to obtain polymeric fillers of higher T_g than that of polystyrene. It was also presumed that these fillers would not be bonded chemically to the rubber during vulcanization and that the copolymers with butadiene would enable such bonding to be effected. Actually, the polydichlorostyrene and polyacenaphthylene did become bonded to the rubber, as indicated by the inability to extract most of the filler by solvents. The final result was that the copolymers with butadiene served merely as fillers of lower T_g than the above homo-

of Polymeric Fillers

Diameter D_n, A	T_g, °C	Young's Modulus, $kg\ cm^{-2} \times 10^{-4}$
1700	—	0.33
530	70	1.2
460	100	1.6
570	105	3.5
560	165	4.0
470	120	4.0
550	155	3.4
380	250	>10

polymers. The ethyl acrylate copolymer also served the analogous purpose. In this way a series of fillers was prepared having T_g values varying from 70° to 250°C.

The Teflon filler was obtained as a latex kindly supplied by the E. I. du Pont de Nemours Co. It had a much larger particle size than the fillers prepared in this laboratory, the latter having been designed to fall in the particle size range 400–600 A. However, it was still of interest to study the effect of the Teflon filler in view of the low rubber–filler interfacial adhesion that could be expected.

The modulus of the fillers was determined by tensile measurements on films cast from suitable solvents, while the T_g values were determined by differential scan calorimetry. The T_g and modulus values agreed well with available published values. There is no direct correlation between T_g and Young's modulus of these glassy polymers. The latter appear to fall into three ranges of modulus and can be grouped as follows: (1) SB-10 and DCB, (2) PS, DC, DCA, and PAB, and (3) PA. The modulus of the polyacenaphthylene (PA) could not be determined accurately because of the extreme brittleness of the film, but it appeared to be greater than 10^5 kg cm^{-2}. The other two groups contained polymers having similar modulus values. The small difference in modulus between the polystyrene and the poly-2,6-dichlorostyrene was quite unexpected in view of the wide dispersity in T_g values.

The characterization of the SBR vulcanizates containing these fillers is shown in Table II. The low values of sol content indicate the presence

Table II. Network Density of Filled SBR Vulcanizates

Filler, 10 vol %	None	PS	DC	DCA	DCB	PA	PAB
% Sol, benzene	3.8	7.5	6.6	7.6	5.4	7.1	6.8
Swelling Ratio, q_e							
Benzene	5.4	5.73	5.54	5.35	5.39	5.70	5.85
Decane	2.5	2.44	2.35	2.30	2.35	2.42	2.44

of chemical bonding between filler and rubber in all cases. The swelling measurements in benzene cannot be taken as a reliable index of the network density owing to the swelling of the bound fillers in this solvent. However, the values in n-decane are valid since the latter is not a solvent for the fillers. On this basis, the crosslink densities can be taken as reasonably constant for all the vulcanizates.

The comparative effect of the polystyrene and poly-2,6-dichlorostyrene fillers on the tensile strength of a polybutadiene vulcanizate is shown in Figure 6. Despite the large difference in T_g values for these fillers, there is no difference in their effect on the vulcanizate. This is illustrated further by the failure envelope plot shown in Figure 7, where the data points for the two fillers, at equal volume fraction, appear to coincide quite well. The fact that all the points fall on the same envelope is a good indication of the constant crosslink density for these vulcanizates. Thus, the similarity in effect of these two fillers appears to be more related to their similar modulus values.

Figure 6 is also useful in demonstrating the difference in viscoelastic response of polybutadiene and SBR vulcanizates. The higher values of tensile strength of the latter, at any given temperature, can obviously be ascribed to the substantially higher T_g of the SBR since the crosslink densities of the two vulcanizates are similar.

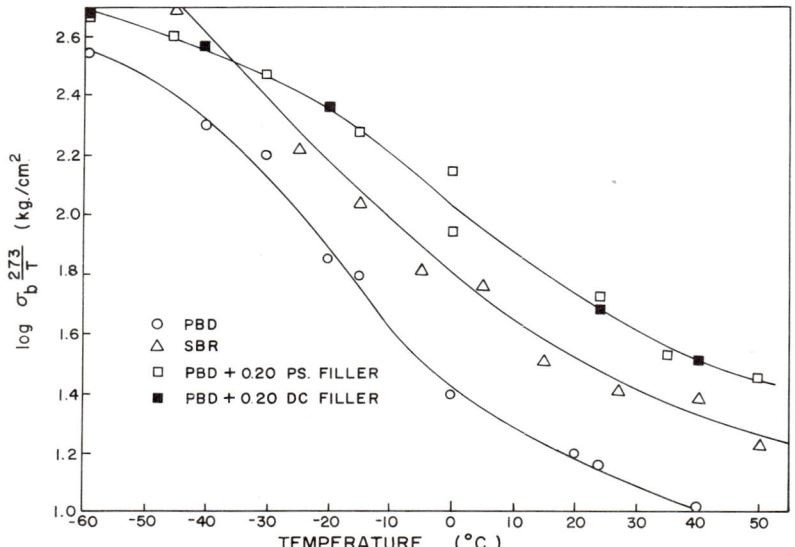

Figure 6. Tensile strength of polybutadiene with poly-2,6-dichlorostyrene (DC) filler. Strain rate, 2 inches/min.

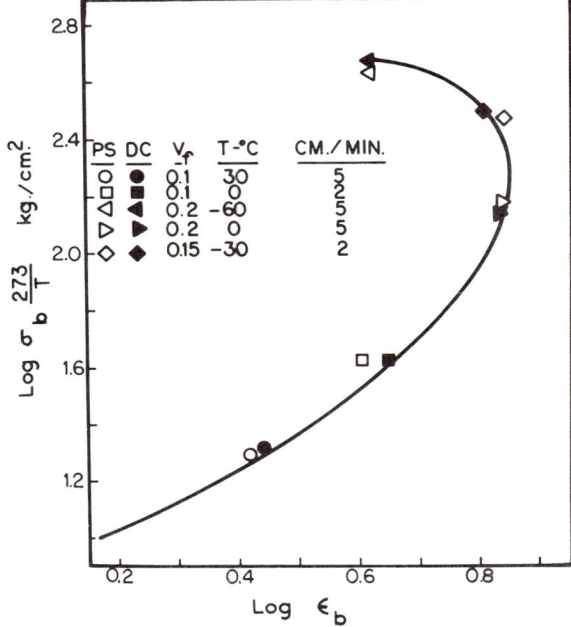

Figure 7. Failure envelope for PBD with poly-2,6-dichlorostyrene filler

The behavior of the SBR vulcanizates containing the series of fillers shown in Table I can be seen in Figure 8. Thus, it appears that the three groups of fillers having different modulus values lead to three different tensile strength curves and that the tensile strengths are directly related to the magnitude of the modulus. Hence the original results (8) reported for the SB-10 "bonded" filler can now be explained adequately by the lower modulus of this filler, and this effect also appears to be corroborated by the higher tensile strength of the polyacenaphthylene-filled SBR. In other words, the rigidity of the filler appears to be sufficiently important to affect the viscoelasticity of the matrix and hence the tensile strength. These findings must be considered tentative, pending corroboration by more extensive studies.

The Teflon-filled vulcanizates have not been included until now since this filler must be considered as a special case, involving poor adhesion at the filler–rubber interface. The marked difference between Teflon and the other fillers is seen in Figure 9, which shows that the Teflon filler exerts only a slight effect on the tensile strength of the polybutadiene vulcanizate. As a matter of fact, although this filler does increase the strength slightly at temperatures above 0°C, it actually appears to de-

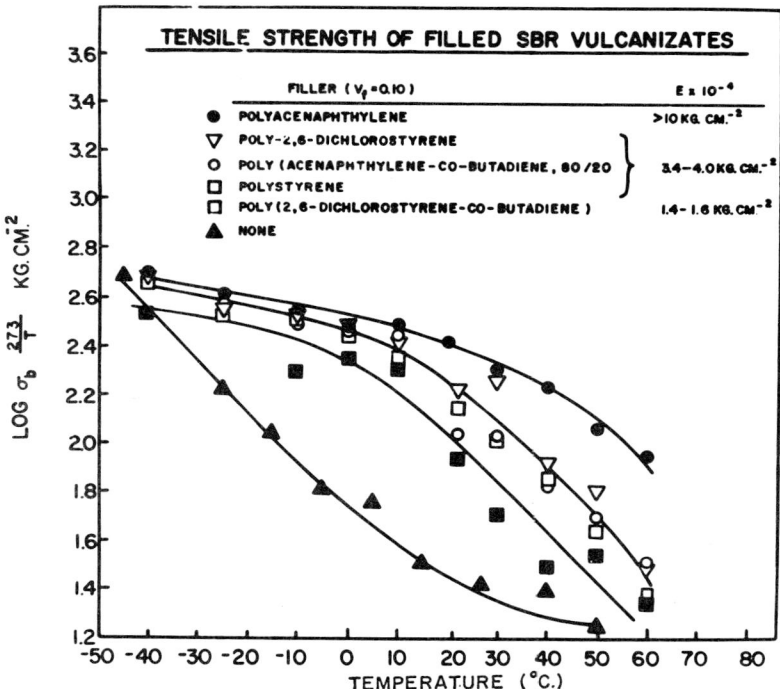

Figure 8. Effect of various fillers on tensile strength of SBR

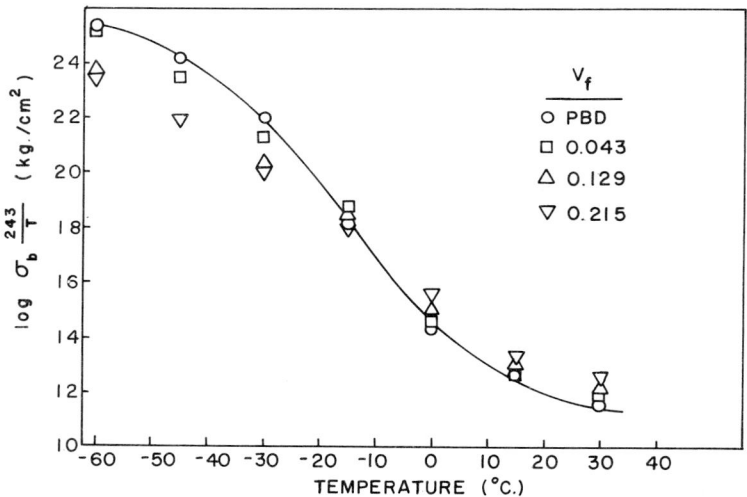

Figure 9. Tensile strength of Teflon-filled PBD. Strain rate, 2 inches/min.

crease the strength, to some extent, at lower temperatures. No explanation of this effect can be advanced at this time. However, electron photomicrographs of this filler, in latex form, show that the particles are generally nonspherical and have axial ratios greater than 1. Furthermore, it is assumed that this filler does not reinforce the rubber because its poor adhesion leads to early dewetting of the particle surface. Thus, it is difficult to predict what effect low temperature will have on the dewetting and subsequent stress concentration in the vicinity of such nonspherical particles.

The nonreinforcing character of the Teflon filler is also shown in Figure 10, where it is seen that the points for the filled and gum vulcanizates coincide at any given temperature or strain rate. This filler shows no reinforcement of the strength of the vulcanizate, and this is most readily ascribed to the poor filler–elastomer adhesion, which prevents the filler particle from exerting viscous, energy dissipating forces on the vulcanizate in its vicinity. However, this filler also has a significantly lower Young's modulus than any of the others, so it might be expected to show the lowest reinforcing effect. Even so, its modulus of 3×10^3 kg cm^{-2} is several orders of magnitude greater than that of the polybutadiene vulcanizate at ambient temperatures (~ 10 kg cm^{-2}), and it would hardly

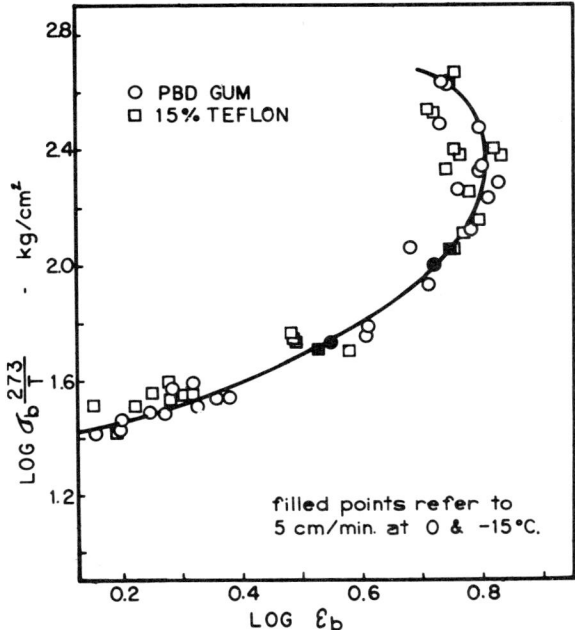

Figure 10. Failure envelope for Teflon-filled PBD. Filled points refer to 2 inches/min at 0° and −15°C.

be expected that such a difference in rigidity of the two phases would not exert some effect on the rupture strength if the adhesion were adequate.

Strength Mechanisms in "Thermoplastic Elastomers" (ABA Block Copolymers)

The synthesis and characterization of these ABA block copolymers of styrene and dienes have been described elsewhere (10, 11). Since the polystyrene end blocks aggregate into glassy domains which act as network junctions, the elastic center blocks must virtually represent the "network chains." The polystyrene domains should also act as a finely divided filler. Hence it might be expected that the mechanical properties of these materials could depend on the two basic parameters: polystyrene content and length of center block ("molecular weight between crosslinks").

Some interesting results have already been obtained (10, 11) on these polymers, where the effect of the above molecular parameters on the mechanical properties has been studied. Thus, Figure 11 shows the effect of variations in block length and styrene content on the stress–strain behavior of styrene–butadiene–styrene (SBS) polymers. As expected, the stress levels increase with increasing styrene ("filler") content but are independent of the block lengths. In other words, the center block size does not exert the same influence as the "molecular weight between cross-

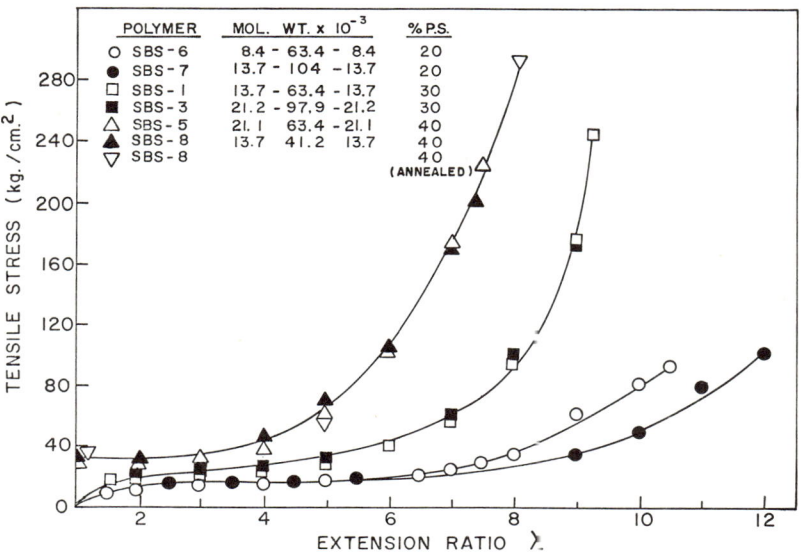

Figure 11. Stress–strain properties of SBS elastomers

links $(M_c)''$ in a normal vulcanizate. This effect (or lack of effect) has already been explained (7) as being caused by the fact that the size of the center block ($MW \sim 60,000$) is such as to include a sizeable number (~ 10) of chain entanglements, which really act as network restraints.

The tensile strengths of the polymers in Figure 11 are shown as the last points in the curves. These, too, show an increase with increasing styrene content, and this could also possibly be explained as a "filler" effect. However, these samples were made by heat molding in a press, and the high styrene polymer (40%) did not exhibit its maximum strength unless it was "annealed" (slow cooling, at 1°C/min instead of 20°C/min). This indicates a considerable influence of thermal treatment, presumably on the degree of phase separation. In view of this, further samples were always prepared by casting films from solution (11).

Two other aspects of the behavior of these block polymers are noteworthy. The first is demonstrated in Figure 11 by the curve for the 40% styrene polymers. These exhibit an unusual "yield point" at very low strain, after which they show a typical curve for an elastomer. This yield point occurs only during the first draw and is not reproduced unless the sample is remolded or reheated. This behavior has been ascribed to the presence of a "continuous" polystyrene phase at such high styrene levels, and this has actually been observed by electron microscopy (1, 6). The interdomain contacts are, of course, disrupted on the first extension and cannot reform unless the material is reheated.

The second aspect that bears mention is the severe distortion of the polystyrene domains which occurs, especially at high strain. This is reflected in an unusually high "set," or unrecoverable deformation, that these materials show at the breaking point or even before. Figure 12 shows this deformation as "% set"—*i.e.*, percent increase over original length, as a function of styrene content and strain. Thus the 40% styrene polymers show as much as 50–60% set near the breaking point. This irreversible extension is completely recoverable upon heating the samples near the T_g of polystyrene ($\sim 100°C$), and it is thus reasonable to ascribe it to an actual distortion, or "cold drawing" of the polystyrene domains. Such distortions have been demonstrated by electron photomicrographs (1, 6) of strained polymers, which show the polystyrene domains capable of assuming shapes of high axial ratio.

It is this ability of the glassy polystyrene domains to yield under stress that has been invoked (10, 12) to account for the high tensile strength exhibited by these materials. In other words, the polystyrene domains do not act in a similar manner to the above model fillers in reinforcing vulcanizates—*i.e.*, by raising the viscosity of the matrix, but instead they offer "energy sinks" for the dissipation of the strain energy by being capable of yielding. Hence, the strength of these block polymers

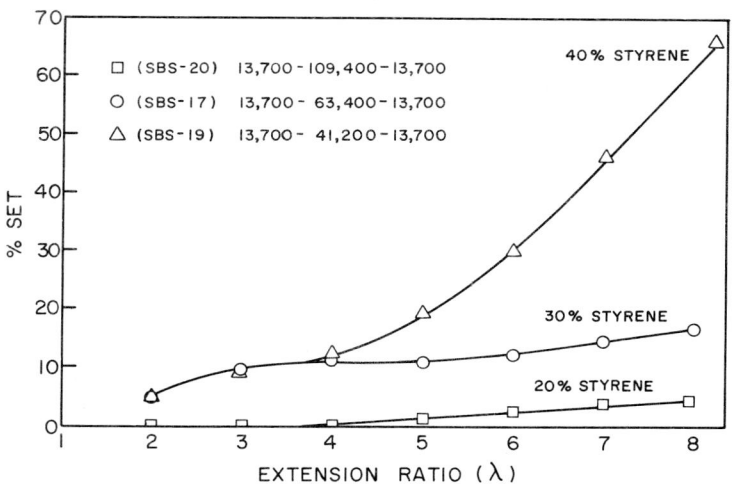

Figure 12. Percent set vs. extension for SBS elastomers

should depend on the extent to which the domains can retain their integrity and still absorb energy.

In this regard, the behavior of the styrene–isoprene–styrene (SIS) block polymers is illustrative. An SIS series of polymers is shown in Figure 13, and these can be compared with the analogous SBS polymers shown in Figure 11. Although the stress levels of the SIS polymers depend mainly on the styrene content and not on block size (as expected), the tensile strengths appear to be independent of both, with one exception. Unlike the SBS polymers, the tensile strengths are very similar for all the SIS polymers except for the one case involving the shortest end block ($MW = 8400$).

The constancy of tensile strength of the SIS polymers, above a certain minimum end-block size, can be explained best on the basis of efficiency of phase separation. Since the latter depends on the three parameters—*i.e.*, incompatibility of the two blocks, composition ratios, and block size—the SIS polymers must undergo a much better phase separation at equivalent composition and block size than the SBS polymers, and the latter require a higher styrene content and a longer polystyrene end block to accomplish good phase separation. This is in accord with the generally accepted fact that polyisoprene is more incompatible with polystyrene than is the case for polybutadiene.

The importance of efficient phase separation on the "purity" of the domains is illustrated even better in Table III, which shows the tensile strength of the SIS polymers as a function of composition and end-block length. The values at higher styrene content were taken from Figure 13, but the striking results are those in which the polystyrene end block drops from a MW of 8400 to 7000 and finally to 5000, where it can be assumed that a separate polystyrene phase no longer exists. Hence the "critical" value for the molecular weight of the polystyrene is in the vicinity of 6000.

If the strength of these block polymers depends on the ability of the polystyrene domains to absorb strain energy by a yielding process, it should be possible to design a material of higher strength by creating domains capable of absorbing more energy—*i.e.*, having a higher modulus. Such domains could possibly be formed, for example, from end blocks having a higher T_g. In accord with this approach, such a block polymer has been synthesized (5) in which polyisoprene is the center block and

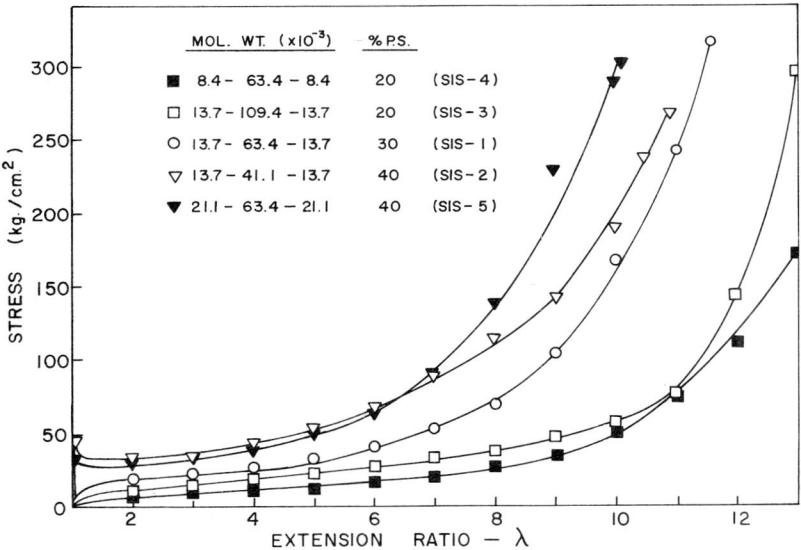

Figure 13. Stress–strain properties of SIS elastomers

Table III. Tensile Strength of SIS Block Polymers

Polymer	Styrene, wt %	MW, × 10⁻³	Tensile Strength, kg cm⁻²
SIS–5	40	21.1– 63.4–21.1	310
SIS–2	40	13.7– 41.1–13.7	306
SIS–1	30	13.7– 63.4–13.7	321
SIS–3	20	13.7–109.4–13.7	270
SIS–4	20	8.4– 63.4– 8.4	160
SIS–9	19	7.0– 60.0– 7.0	22
SIS–10	11	5.0– 80.0– 5.0	0

poly-α-methylstyrene constitutes the two end blocks. The stress–strain properties of this polymer are compared with those of an equivalent SIS polymer in Figure 14; the stress levels of the α-methylstyrene polymer are consistently higher than those of the SIS polymer, up to and including the point of tensile rupture. Furthermore, since the poly-α-methylstyrene also has a much higher T_g (170°C) than polystyrene (105°C), the corresponding block polymer shows a much better temperature–strength retention, as shown in Figure 15.

These results with the α-methylstyrene block polymers are especially interesting in that they indicate that end blocks of higher T_g and rigidity are still capable of yielding under stress and thus absorbing energy rather than undergoing brittle fracture. It would, of course, be instructive to determine to what extent this yielding behavior prevails as the T_g and rigidity of the domains increase.

The above data on these block polymers have thrown great emphasis on the critical role of the "rigid" domains in the strength mechanisms of these elastomers. It is interesting, therefore, to explore the effect of making "composite" domains by adding, for example, free polystyrene to these block polymers. The techniques will be described in detail in a forthcoming publication, but suffice it to say that all blending of these additive polymers was done in solution prior to film casting.

The properties of the blended polymers are shown in Table IV for both SIS and SBS block polymers. The added polystyrene was designed

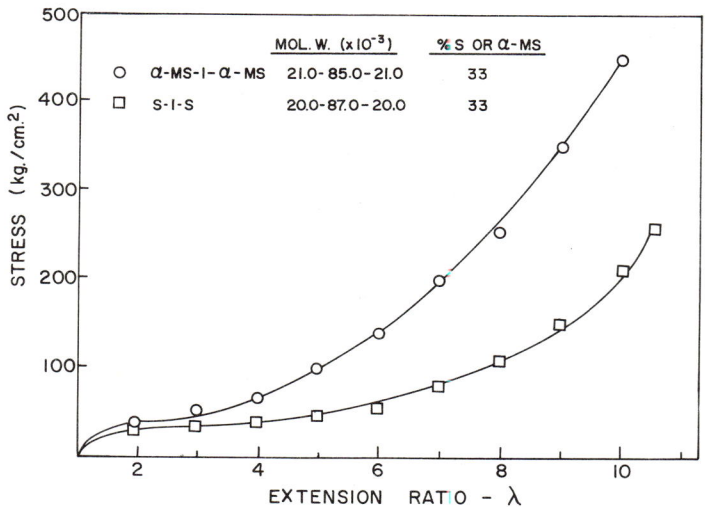

Figure 14. Stress–strain properties of α-methylstyrene block polymers

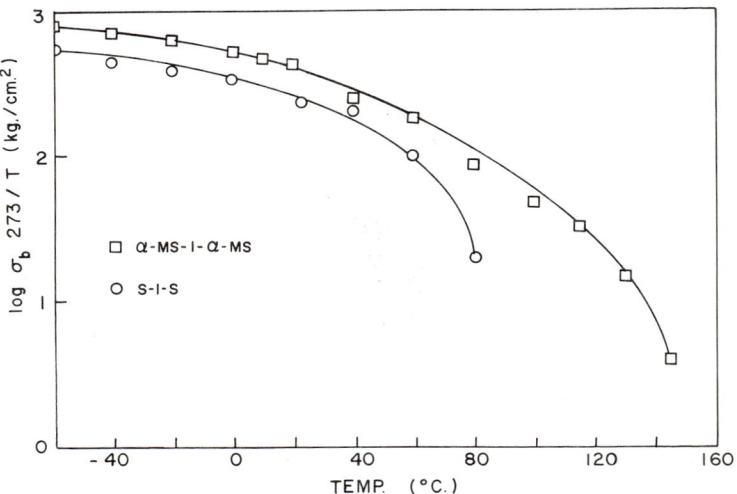

Figure 15. Tensile strength vs. temperature of α-methylstyrene block polymers

Table IV. Effect of Added Homopolymers

Polymer Added	Wt. %	Total Styrene Content, %	Stress at $\lambda = 4$, kg cm^{-2}	Tensile Strength, kg cm^{-2}
SIS–3: 13,700–109,400–13,700				
None	—	20.0	17.5	296
PS–15,000	5	22.8	24.0	306
PS–15,000	10	25.5	25.4	302
PS–15,000	20	30.2	33.5	288
SIS–20: 13,700–109,400–13,700				
None	—	20.0	—	127
PS–12,500	14.9	30.0	—	342
SBS–17: 13,700–63,000–13,700				
None	—	30.0	—	313

to have approximately the same molecular weight as the polystyrene in the end blocks. All of the resulting cast films were optically clear, except for the blend having 20% added polystyrene, which showed some haze. This can be taken to indicate that the added polymer was able to be included in the polystyrene domains, which are, of course, too small to cause noticeable light scattering.

It is not surprising, therefore, that the added polystyrene causes no change in the strength of the SIS polymer since, at these styrene contents and molecular weights, the tensile strength has been shown (Figure 13) to be independent of these parameters. The moduli, however, do show an increase with increasing styrene content, as expected. The significant aspect of these results is the fact that the added polystyrene does apparently enter the domains and does not form a separate polystyrene phase, as would be the case in any attempted blend of the two homopolymers, i.e., polystyrene and polyisoprene. The same "compatibility" of added and block polystyrene is demonstrated in the case of the SBS polymer, only here there *is* an effect on the tensile strength caused by the increased styrene content, regardless of whether the polystyrene is present as an end block or as added homopolymer. Again the evidence suggests that the added polystyrene enters the end-block domains, and further that it actually improves phase separation, presumably by virtue of the higher styrene content of the mixture.

The above experiments, besides demonstrating the phase separations which occur in these blends, can also be used to show the possible effect of synthesis problems on the properties of these block polymers. In other words, it is obvious that any homopolystyrene that is formed during the block polymerization, e.g., by fortuitous termination arising from impurities, would have little effect on the tensile strength (provided it is of a suitable molecular weight (*11*) to be "compatible" with the polystyrene end blocks in the domains). In this connection it is of interest to explore this approach one step further—i.e., to determine the effect of the presence of diblocks since these also could be the result of adventitious termination in the synthesis process. Such a study has been carried out by preparing solution blends of SB diblocks and SBS block polymers, having similar block sizes, and the results are shown in Table V. Even minor amounts of the diblock polymer have a profound effect on the tensile strength. The fact that the modulus is not influenced to any extent again indicates that the latter depends primarily on the styrene (or "filler") content. However, the presence of the SB diblocks apparently introduces "network defects" which can act as flaws to initiate rupture.

Table V. Effect of Added Diblock Polymer[a]

Added Diblock, %	Stress at $\lambda = 4$, kg cm^{-2}	Tensile Strength kg cm^{-2}	% Set
0	46.3	319	50
1	50.3	308	50
2	47.5	264	50
5	48.3	244	50
67	14.6	49	200

[a] SBS-5: 21,100–63,400–21,100; SB diblock: 21,100–63,400.

The network structure of these block polymers can also be altered by actually crosslinking the elastomeric polydiene chains, thus introducing a "chemical" network. In this approach it was necessary to use a crosslinking method which would not result in any measurable chain scission. Dicumyl peroxide (Dicup) was chosen for crosslinking an SIS polymer since this peroxide is known to accomplish exclusive crosslinking without any observable chain scission (3). The Dicup was dissolved in a THF solution of the polymer, and a cast film was prepared which was then vulcanized in a press at 155°C for 35 min. A control sample, without Dicup was treated in the same way.

Table VI. Properties of SIS Polymers Crosslinked by Dicumyl Peroxide

Polymer	Swelling Ratio[b] (q_m)	Stress at $\lambda = 4$, kg cm^{-2}	Tensile Strength, kg cm^{-2}	λ at Break
SIS–2A[a]	9.8	50	225	11
Vulcanized SIS–2A	5.4	53	160	10

[a] $MW = (13.7–41.1–13.7) \times 10^3$
[b] Isooctane, 48 hour at 25°C.

The effect of such crosslinks is shown in Table VI. Whereas the modulus was affected only slightly by the number of crosslinks introduced, the tensile strength was decreased markedly. Hence, although the actual number of crosslinks produced by the peroxide is not very great relative to the number of chain entanglements already present, these "fixed" crosslinks must introduce sites of stress concentration which cannot transmit the stress to the polystyrene domains as efficiently as the entanglements do. This conclusion is, of course, based on the premise that there is a negligible amount of chain scission occurring during the crosslinking reaction. Note should be taken of the fact that although the crosslinking did not raise the modulus much, it did reduce the swelling markedly. However, this anomalous lack of agreement between modulus and swelling may be ascribed to the unusual swelling behavior of these block polymers. This has already been noted (2) as "swelling creep" which occurs in presence of a specific solvent for the elastomer phase—e.g., isooctane—and renders swelling equilibrium difficult to obtain.

Finally, it is instructive to compare the temperature effect on the tensile strength of the SBS and SIS block polymers. As noted previously (Figure 6) the tensile strength of an elastomer vulcanizate can be related to the difference between the test temperature and the T_g of the elastomer, in accordance with the viscoelastic theory of tensile strength. Since the T_g values for polyisoprene ($-65°C$) and polybutadiene ($-95°C$) differ

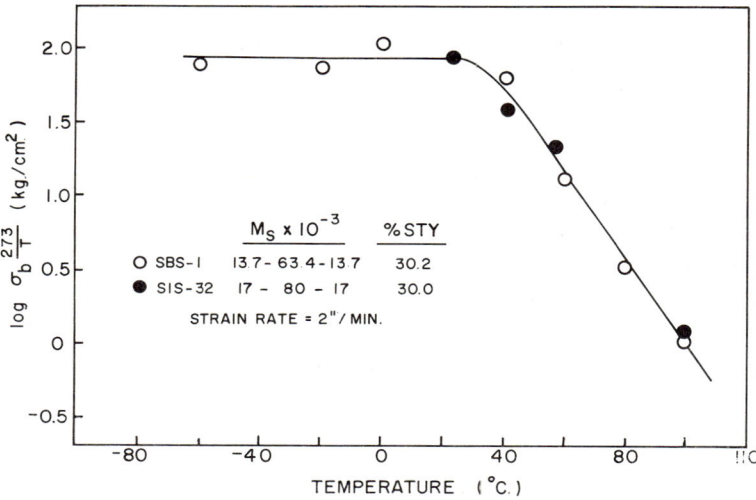

Figure 16. Tensile strength vs. temperature for SBS and SIS polymers. Strain rate, 2 inches/min.

by 30°, one would expect the SIS polymers to have a consistently higher tensile strength at any given temperature and strain rate. However, this has already been shown not to be true at room temperature. Figure 16 demonstrates that the strength–temperature relations of the SIS and SBS polymers are virtually identical, again indicating that it is the response of the polystyrene domains in relieving the stress concentration which governs the rupture of these materials. Since the modulus of these domains is several orders of magnitude higher than that of the elastic matrix surrounding them, the viscosity of the matrix contributes little, if anything, to the tensile strength.

Summary and Conclusions

The two systems discussed above demonstrate two mechanisms whereby the tensile strength of elastomers can be reinforced by the presence of "rigid" fillers. For the polymeric fillers dispersed within a vulcanizate, the filler operates by raising the viscosity of the matrix, analogous to a decrease in temperature, but without affecting the dynamic, high frequency response (there is ample experimental evidence of the independence of T_g on presence of filler). There is also some indication that the rigidity of the filler affects the extent of reinforcement.

In the case of the heterophase systems resulting from the ABA block polymers, the strength is reinforced because of the ability of the plastic, glassy domains to yield and absorb energy, thus delaying the onset of crack propagation. This is undoubtedly caused by the efficient transmittal of the applied stress through the elastic chains which are bonded to the domains, as well as by the presence of some phase mixing at the interface. In this way, the strength reinforcement increases if the energy of distortion of the domains is greater—e.g., for domains of higher T_g. The absorption of energy by the elastic matrix does not seem to be significant in these systems.

These studies of the ABA block polymers also help to explain the results obtained by many investigators, as described in a recent comprehensive review (4).

Acknowledgments

The results described here are based on the excellent experimental work carried out during the past several years by the following contributors: J. C. Healy, R. L. Denecour, J. Trout, J. E. McGrath, P. C. Juliano, F. C. Schwab, and C. R. Strauss. The interest and collaboration of L. J. Fetters is also gratefully acknowledged. Various aspects of this work were supported by the Cabot Corp., B. F. Goodrich Co., and the Materials Laboratory, U. S. Air Force.

Literature Cited

(1) Beecher, J. F., Marker, L., Bradford, R. D., Aggarwal, S. L., *J. Polymer Sci., Pt. C* **26**, 117 (1969).
(2) Bishop, E. T., Davison, S., *J. Polymer Sci., Pt. C* **26**, 59 (1969).
(3) Calderon, N., Scott, K. W., *J. Polymer Sci., Pt. A* **3**, 551 (1965).
(4) Estes, G. M., Cooper, S. L., Tobolsky, A. V., *J. Macromol. Sci., Rev. Macromol. Chem.* **C4** (2), 313–366 (1970).
(5) Fetters, L. J., Morton, M., *Macromolecules* **2**, 453 (1969).
(6) Hendus, H., Illers, K. H., Ropte, E., *Kolloid-Z.Z., Polymere* **216–217**, 110 1967, MacLaren, London, 1968, p. 175.
(7) Holden, G., Bishop, E. T., Legge, N. R., *J. Polymer Sci., Pt. C* **26**, 37 (1969).
(8) Morton, M., Healy, J. C., Denecour, R. L., *Proc. Intern. Rubber Conf. 1967*, 175 (1968).
(9) Morton, M., Healy, J. C., *Appl. Polymer Symp.* **7**, 155 (1968).
(10) Morton, M., McGrath, J. E., Juliano, P. C., *J. Polymer Sci., Pt. C* **26**, 99 (1969).
(11) Morton, M., Fetters, L. J., Schwab, F. C., Strauss, C. R., Kammereck, R. F., "SRS-4," p. 1, Rubber and Technical Press, London, 1969.
(12) Smith, T. L., Dickie, R. A., *J. Polymer Sci., Pt. C* **26**, 163 (1969).

RECEIVED August 3, 1970.

32

Short Fiber–Elastomer Composites

GEORGE C. DERRINGER

PPG Industries, Inc., Box 31, Barberton, Ohio 44203

> *Short fibers have typically failed to reinforce elastomers to the extent that they reinforce plastics. This is the result of the relatively poor adhesion of most fibers to elastomeric matrices. This poor adhesion, however, can be overcome by using a tricomponent resin forming system consisting of hexamethylenetetramine, resorcinol, and high surface area hydrated silica (i.e., HRH). When this system is incorporated into the rubber compound along with the fiber, a high degree of fiber–matrix adhesion develops upon vulcanization. The important properties of the resultant composites are extremely high modulus with retention of a high level of resilience and a high degree of anisotropy, especially with regard to modulus. Such properties are presently unobtainable with conventional reinforcing fillers.*

Short fibers have long been used as fillers for elastomer compounds. Unlike their function in plastics, however, they have not functioned as reinforcing materials in rubber mainly because of the absence of adequate fiber–matrix adhesion. Good adhesion between fiber and matrix is necessary for the development of fiber–matrix interfacial shear stress through which load is transferred to the fiber. To obtain good adhesion in plastic composites, multifunctional silanes are used chemically to bond fibers, such as glass fibers, to the polymer matrix. Little has been done along these lines with elastomers and chopped organic fibers, however, probably because of the high cost of the silanes relative to the cost of commercial rubber compounds.

Several years ago it was discovered that a tricomponent system (*1, 3*) consisting of hexamethylenetetramine, resorcinol, and high surface area hydrated silica (HRH system) when incorporated into a rubber compound, brought about extremely good adhesion to many types of fabrics. Subsequently, the system was also evaluated in elastomer composites for

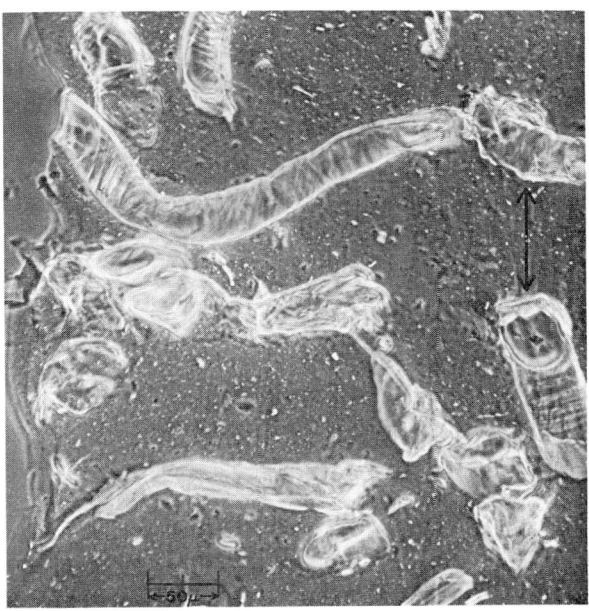

Figure 1. 50-Volume loading of nylon fibers in standard natural rubber formulation

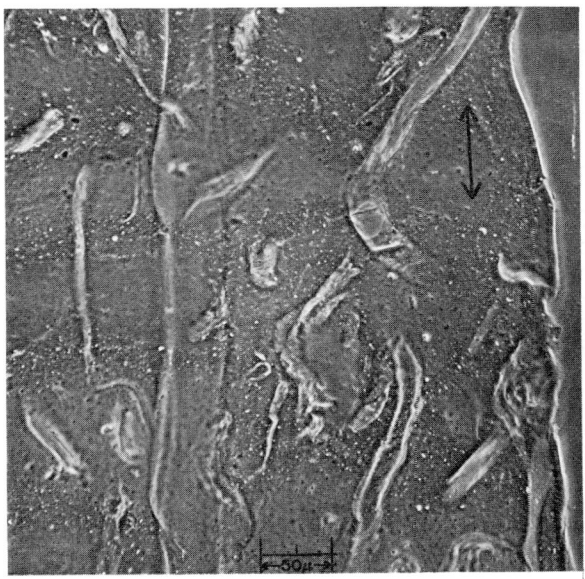

Figure 2. 50-Volume loading of acrylic fibers in standard natural rubber formulation

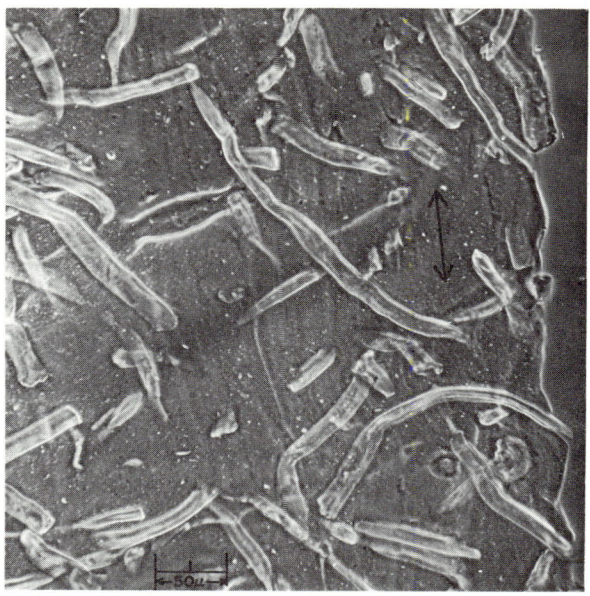

Figure 3. 50-Volume loading of rayon fibers in standard natural rubber formulation

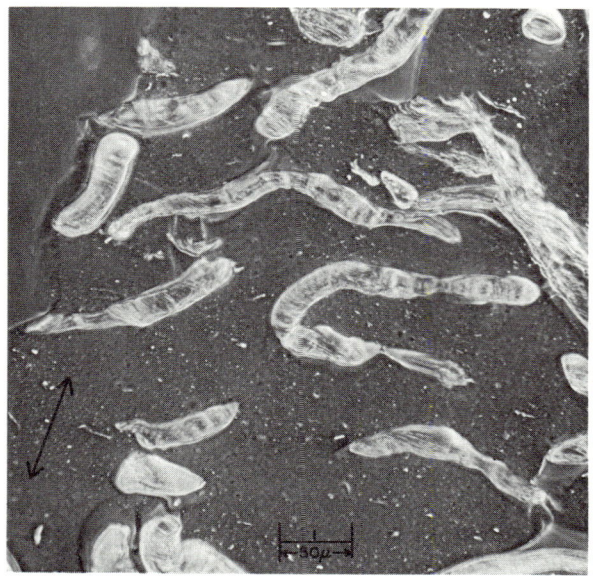

Figure 4. 50-Volume loading of polyester fibers in standard natural rubber formulation

bonding discontinuous fibers to elastomeric matrices. This system results in extremely good adhesion between many types of commercial fibers and most commercially important elastomers.

Fiber Characteristics

The organic fiber reinforcement used in this study consisted of commercially available nylon, rayon, polyester, and acrylic flocks. They were cut to lengths of approximately 0.4 mm, the short length being required for optimum dispersion. The actual length, however, varied considerably as did the fiber diameters. The micrographs shown in Figures 1–4 show these four types of fiber dispersed in a standard natural rubber formulation give in Table I. The length and diameter distribution varied considerably not only between fiber types but also among fibers of the same type. The fiber lengths ranged from approximately 300μ to fragments barely longer than the fiber diameters. The fiber diameters varied as follows: nylon (20–35μ) > polyester (15–25μ) > rayon (10–20μ) > acrylic (2–15μ).

The fiber glass used was also commercial material cut to 1-inch lengths with a diameter of 10μ. Because the glass fibers, unlike the organic fibers, are very brittle, the length in rubber was considerably smaller than the cut length. For example, the longest fiber in Figure 5, which shows the glass fibers dispersed in the standard rubber formula-

Table I. Base Formulation for Fiber Composites

First stage of mix: banbury. Mixing time 6 minutes; maximum temperature $300° \pm 5°F$

	Parts by weight
Natural rubber (SMR–H5L)	100
Hydrated silica[a,d]	25
Resorcinol[b,d]	2.5

Second stage of mix: mill. Mixing time, 4–6 minutes; temperature $200° \pm 10°F$.

Hexamethylenetetramine[c,d]	1.6
Zinc oxide	5
Stearic acid	2
2,2'-Benzothiazyl disulfide	0.7
Diphenylguanadine	0.3
Sulfur	2.5

[a]Hi-Sil 233, PPG Industries, Inc.
[b]Technical grade
[c]Hexa Flo Powder, Hyden Chem. Div., Tenneco, Inc.
[d]Constituents of HRH system

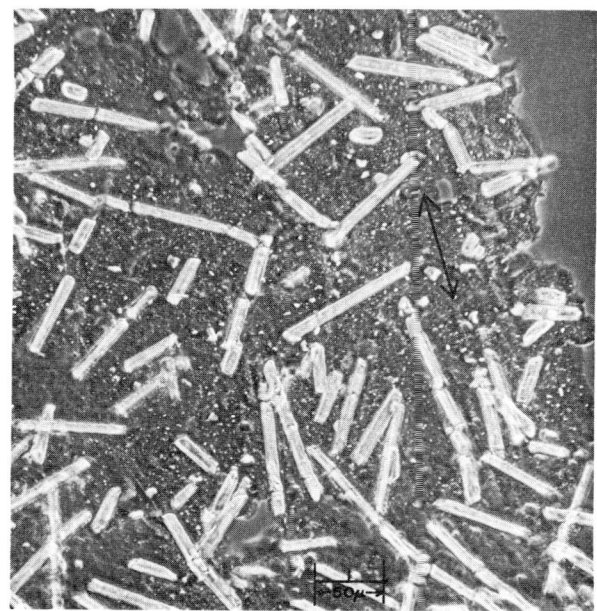

Figure 5. 50-Volume loading of fiber glass in standard natural rubber formulation

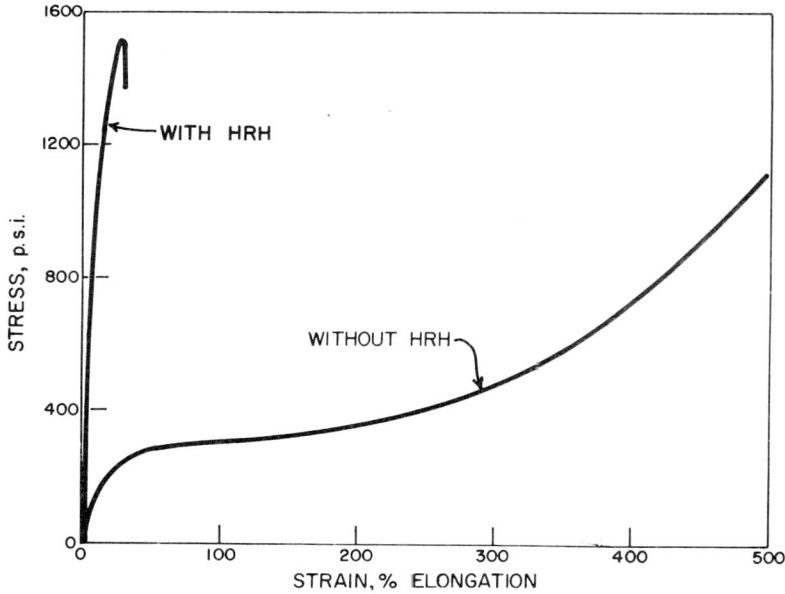

Figure 6. Stress–strain curves for 50-volume rayon composite with and without HRH system

tion, is only approximately 100μ in length. Here again, however, the length distribution is very wide including many fragments. Also, unlike the organic fibers, the glass fibers had very uniform diameters.

Because of the nonuniform nature of the fibers studied, this work was necessarily empirical in nature. Its main purpose is to illustrate the unique properties and property trends of these elastomeric composites in comparison with conventional rubber compounds.

Effect of HRH System on Fiber–Matrix Adhesion

In the absence of fiber-matrix adhesion, fibers add very little to the strength of rubber compounds. This is illustrated in Figure 6 which shows stress–strain curves for a 50-volume loading of rayon fibers with and without HRH in the standard natural rubber formulation shown in Table I. For the composite without HRH, a region of large yielding occurs at a stress level of approximately 200 psi and extends over a large portion of the strain range. This yielding behavior accompanied by low tensile strength and high permanent set is characteristic of elastomer composites which lack adequate fiber–matrix adhesion. On the other hand, the stress–strain curve for the composite containing the HRH

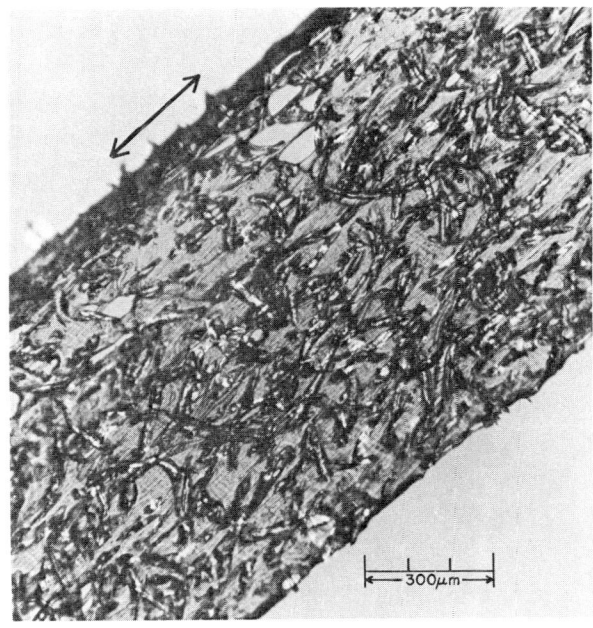

Figure 7. Thin section of 50-volume rayon composite without HRH system stretched to 25% elongation

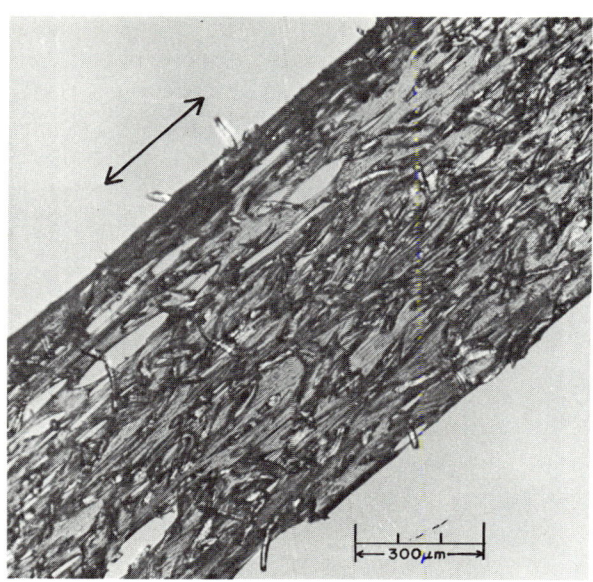

Figure 8. Thin section of 50-volume rayon composite without HRH system stretched to 100% elongation

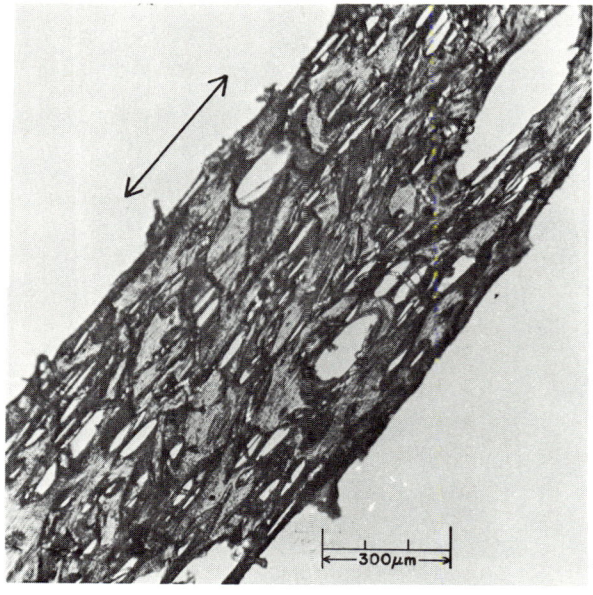

Figure 9. Thin section of 50-volume rayon composite without HRH system stretched to 150% elongation

system is nearly linear and demonstrates high modulus and low elongation at break—features expected for good fiber-matrix adhesion.

The effect of the HRH system on adhesion is further illustrated by the micrographs (Figures 7–11) of the same rayon–natural rubber composite with and without HRH. Figures 7–9 show a thin section of the composite without HRH stretched to various elongations with the force applied parallel to the direction of orientation. Many voids form as the strain is increased owing to fiber–matrix bond failures. Both the number and size of voids increase with increasing strain.

Figures 10 and 11 show the identical composite with HRH stretched to 25 and 100% elongation. In this case only two relatively small voids were formed and not until a strain of 100% elongation was reached. This is a clear demonstration of the effectiveness of the HRH system, especially when it is noted that at any given strain level, the HRH composite was under considerably more stress than the non-HRH composite owing to its considerably higher modulus. Unfortunately, it was impossible to determine the actual stress levels incurred for each micrograph since the rate of deformation could not be controlled and the samples were considerably thinner than those used in Figure 6.

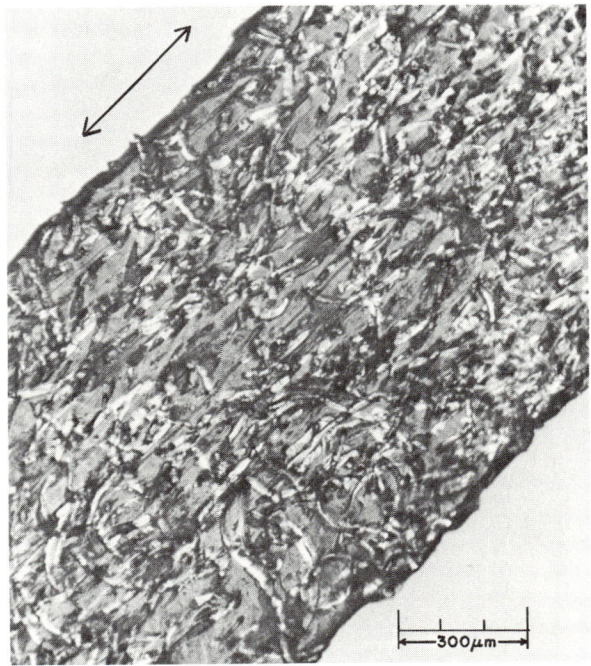

Figure 10. Thin section of 50-volume rayon composite with HRH system stretched to 25% elongation

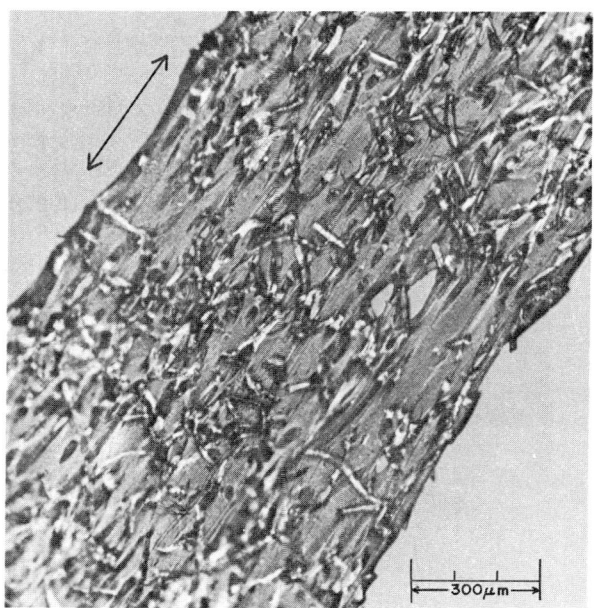

Figure 11. Thin section of 50-volume rayon composite with HRH system stretched to 100% elongation

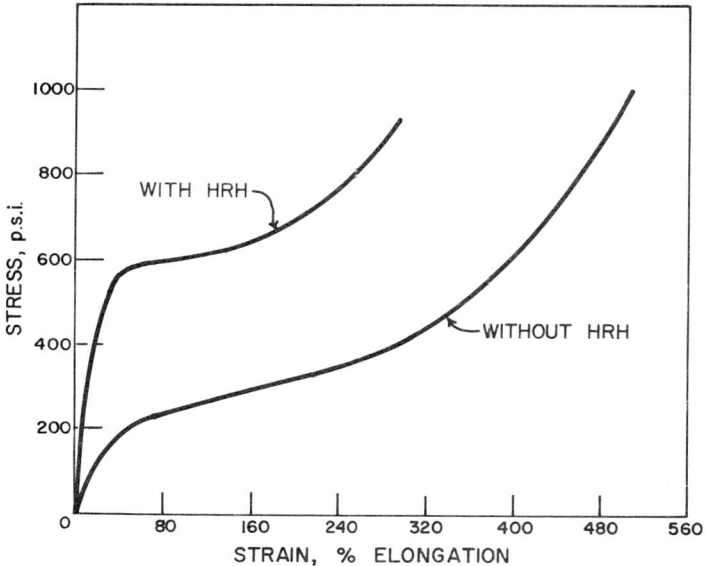

Figure 12. Stress–strain curves for 50-volume polyester composite with and without HRH system

Stress–strain curves were also obtained for 50-volume acrylic, nylon, polyester, and fiber glass composites, with and without HRH. All but the polyester showed the same general behavior as rayon illustrated in Figure 6, indicating good adhesion with the HRH system. The polyester stress–strain curves are illustrated in Figure 12 where the curve without HRH is similar to those for the other non-HRH composites. The curve with HRH, however, shows a yielding region at approximately 600 psi which is believed to be the stress level at which the polyester–rubber bond breaks down. The micrograph of the unstrained polyester composite containing HRH shown in Figure 4 shows several voids between fiber and matrix. These voids, which are not present for the other fibers, are either the cause or the result of poor polyester–rubber adhesion. If they indicate poor wetting, they are the cause. On the other hand, if they resulted from stresses within the sample upon cooling or during sample preparation, they are probably the result of poor adhesion. In any case, adhesion of polyester fabric to rubber even in the presence of the HRH system has been typically poor.

Stress–Strain Properties

The longitudinal stress–strain properties of all of the composites can be represented by the following empirical equations which have been found to represent the data satisfactorily:

Ultimate tensile strength

$$T = T_o + aV - bV^{1/2} \tag{1}$$

Elongation at break

$$W = W_\infty + (W_o - W_\infty) \exp(aV^b) \tag{2}$$

Young's modulus

$$E = E_o - 1 + \exp(aV^b) \tag{3}$$

where

T = ultimate tensile strength of composite, psi
T_o = ultimate tensile strength of matrix compound
W = elongation at break of composite, %
W_o = elongation at break of matrix compound
W_∞ = limiting elongation at break of highly loaded composite
E = Young's modulus of composite at zero strain, psi
E_o = Young's modulus of matrix compound at zero strain
V = volume loading of fibers, cm³ per 100 gram rubber
a and b = empirical constants

Values of E_o, T_o, W_o, W_∞, a, and b are summarized in Tables II–IV.

Table II. Equation Parameters for Composite Tensile Strength

Fiber	a	b	T_o
Nylon	51.455	726.57	3673
Acrylic	85.917	741.81	3530
Rayon	101.33	929.02	3647
Fiber glass	94.704	819.38	3437

Table III. Equation Parameters for Composite Elongation at Break

Fiber	a	b	E_∞
Nylon	−0.0154586	1.3619	25
Acrylic	−0.0766899	1.09834	10
Rayon	−0.0405033	1.37792	10
Fiber glass	−0.129811	1.0919	36

$E_o = 607$

Table IV. Equation Parameters for Composite Young's Modulus at Zero Strain

Fiber	a	b
Nylon	4.16637	0.172268
Acrylic	6.71147	0.10000
Rayon	4.54637	0.196473
Fiber glass	3.88161	0.201548

$M_o = 179$

These properties were determined at an extension rate of 5 inches/minute with force applied parallel to the rubber grain direction. Young's modulus at zero strain was found by taking the first derivative of polynomial equations fitted to the low extension portion of the stress–strain curves.

The equations are plotted in Figures 13–15. The standard errors for the tensile and modulus curves ranged between 10–15% of the mean response levels. For elongation at break, the standard error was 6–10%. The equations were chosen from many candidates, all of which were evaluated by linear or nonlinear least squares. They are considered to give a good description of the important features of the data for each response.

The tensile curves are characterized by a sharp decrease in tensile with loading at low loadings. For rayon, acrylic, and fiber glass the tensile was minimized at approximately 20 volumes. For nylon the minimum was reached at approximately 50 volumes. Beyond the minimum points in the curves the tensile increased but at a slower rate than the initial rate of decrease. Eventually the tensile reached that of the natural rubber matrix for all but the nylon fibers. Since the loading resulting in

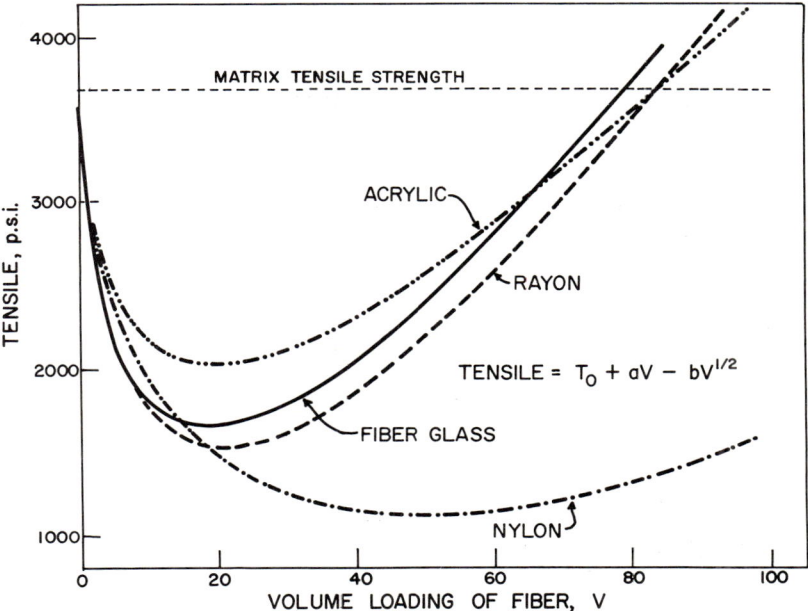

Figure 13. Fiber loading vs. longitudinal tensile strength for various fibers

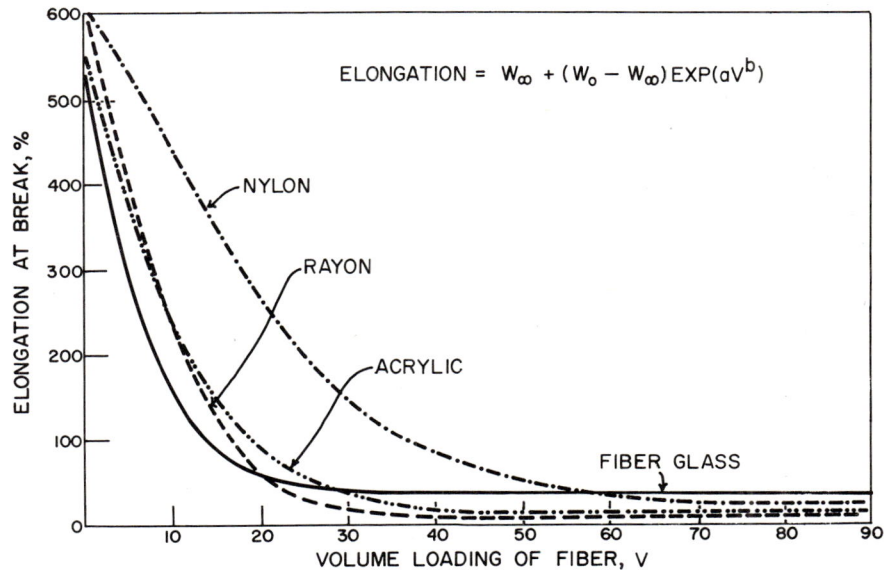

Figure 14. Fiber loading vs. longitudinal elongation at break for various fibers

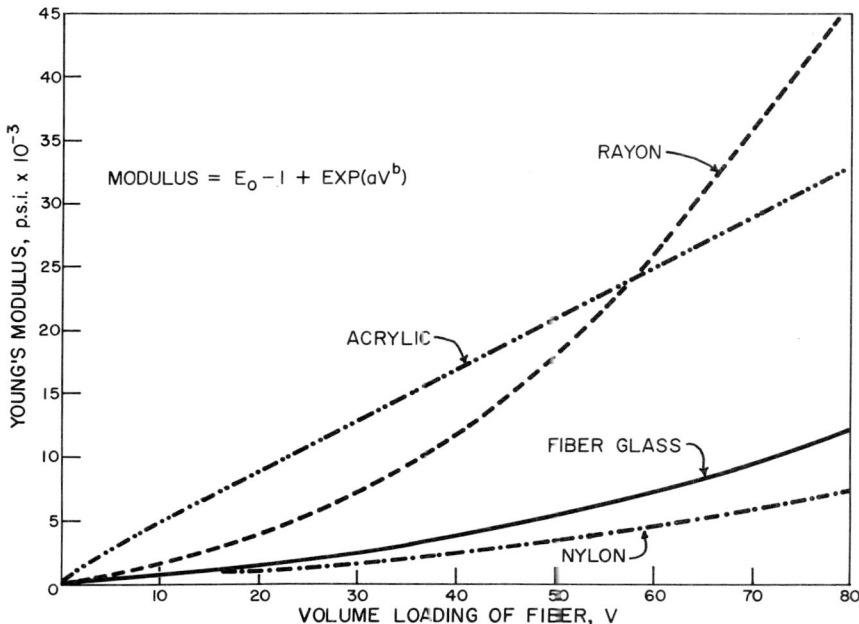

Figure 15. Fiber loading vs. longitudinal Young's modulus for various fibers

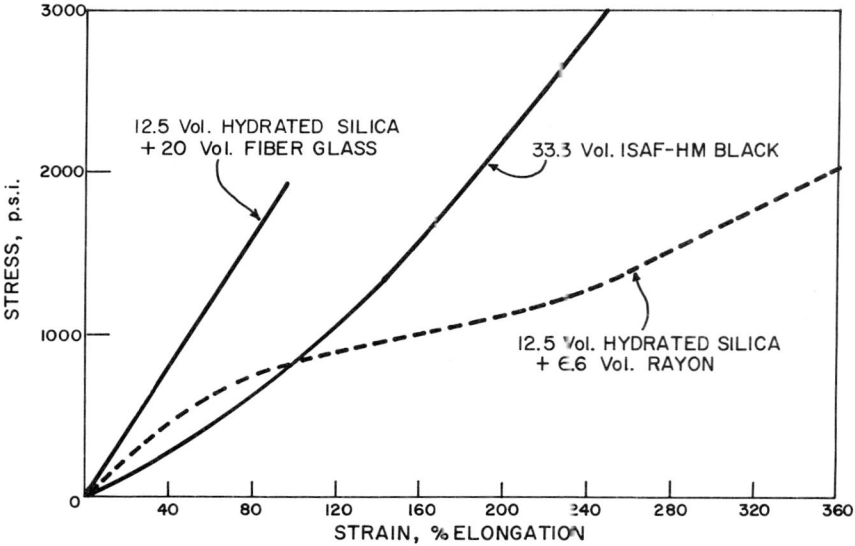

Figure 16. Comparison of effects of fiber reinforcement vs. high modulus carbon black on stress–strain behavior of rubber

minimum tensile strength is equal to $1/4(b/a)^2$ and that giving the matrix strength is $(b/a)^2$, the matrix tensile strength is obtained at a loading four times that giving the minimum tensile.

The elongation at break curves are characterized by a high rate of elongation loss with fiber loading at low fiber volumes. The elongation for all fibers is minimized at levels lower than 50%. Rayon, acrylic, and fiber glass reach minimum elongation at approximately 30 volumes and nylon at approximately 50 volumes. It is interesting that tensile and elongation are minimized at roughly the same loadings.

The modulus plots show wide variations among fiber types, especially at high fiber loadings. Rayon is the most reinforcing fiber with respect to modulus at high loadings and acrylic at lower loadings. Nylon imparted the least modulus reinforcement at all loadings, followed by fiber glass.

However, even the poorer reinforcing fibers—nylon and fiber glass—impart higher low strain moduli than high modulus carbon blacks. This is illustrated in Figure 16 where stress–strain curves are compared for the standard compound filled with ISAF–HM black, and rayon and glass fibers. The black-filled compound did not contain the HRH system. The composite containing only 6.6 volumes of rayon along with the 12.5 volumes of hydrated silica from the HRH system exhibited significantly higher modulus than the black-filled compound at strains up to 100% elongation. The composite containing 20 volumes of fiber glass along with hydrated silica exhibited considerably higher modulus at all elongation levels below the breaking strain. High modulus at low extensions is very important because most rubber articles do not encounter strains larger than 100%.

Comparison of Elastomer Composites with Metal and Plastic Composites

Relatively little theoretical and experimental work has been done on discontinuous fiber composites, and most of this work has been confined to metal–metal composites (4, 5, 8). For the metal–metal systems the law of mixtures is applicable for composite modulus as follows:

$$E_c = V_f E_f + V_m E_m = V_f(E_f - E_m) + E_m \qquad (4)$$

where

V_f = volume fraction of fibers
V_m = matrix volume fraction
E_f = fiber tensile modulus
E_m = matrix tensile modulus

The basic assumption of this equation is that fiber strain, ϵ_f is equal to the matrix strain, ϵ_m for any given stress applied to the composite. This assumption is reasonable for metal–metal systems but not for plastic or elastomer systems where the matrix has significantly lower modulus than the fiber.

To account for this, Lees (6) proposed Equation 5 where the unequality of ϵ_f and ϵ_m is taken into account:

$$E_c = E_f V_f \alpha^2 - \frac{E_m V_f}{15}[1 + 3\alpha + 11\alpha^2]$$
$$- \frac{2E_m}{15} V_f^{1/2}[2 + \alpha - 3\alpha^2] + \frac{E_m}{3}[1 + \alpha + \alpha^2] \quad (5)$$

where

$$\alpha = \frac{\varepsilon_f}{\varepsilon_m}$$

Applying this equation to elastomer composites, considerable simplification is possible. Because the elastomeric matrix has very low modulus and high elasticity, α can be assumed to be nearly zero. The resulting equation is then:

$$E_c = \frac{E_m}{3} - \frac{E_m}{15} V_f - \frac{4}{15} E_m V_f^{1/2} \quad (6)$$

Because this equation predicts decreasing modulus with increasing fiber loading, it is obvious that Lees' model for modulus is not applicable to elastomer systems.

Lees' results for elongation at break for polyethylene–glass composites agrees more favorably with the elastomer composites. He shows that elongation at break is a function of $V_f^{-1/2}$, indicating dependence of high stress flow on the distance between fibers. This relationship holds relatively well for the elastomer composites. This can be seen by transforming Lees' linear, elongation vs. $V_f^{-1/2}$ plot to an elongation vs. V_f plot. The latter is of the same form as the elastomer composite plots of Figure 14.

The tensile strength curves shown in Figure 13 have no theoretical justification. In fact, the existing theories for metal and plastic systems predict continual increase in longitudinal tensile strength with increased fiber loading. It is interesting, however, that the transverse tensile vs. V_f curve for fiber glass in poly(methyl methacrylate) shown by Lees (7) is very similar to those for elastomer composites. It shows the same sharp tensile loss with initial fiber loading followed by the gradual increase, eventually reaching the matrix tensile strength. Lees attributed this to

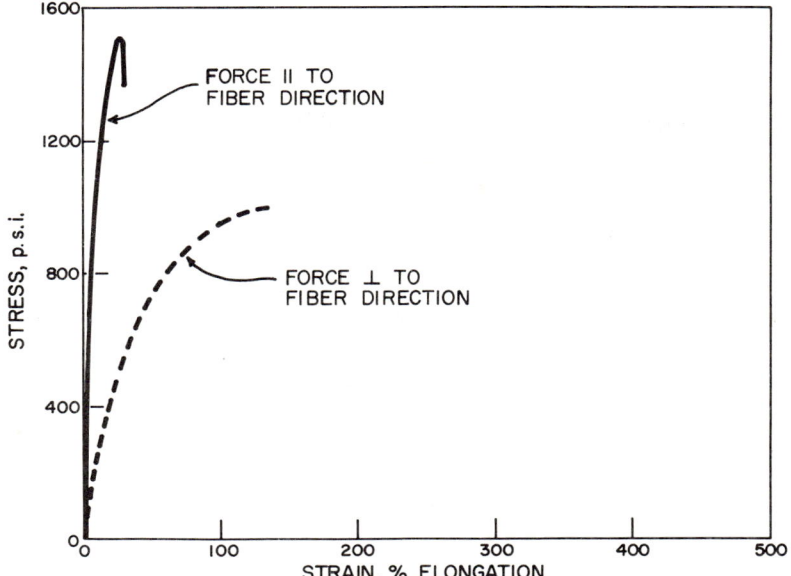

Figure 17. Longitudinal vs. transverse stress–strain behavior for 50-volume rayon composite

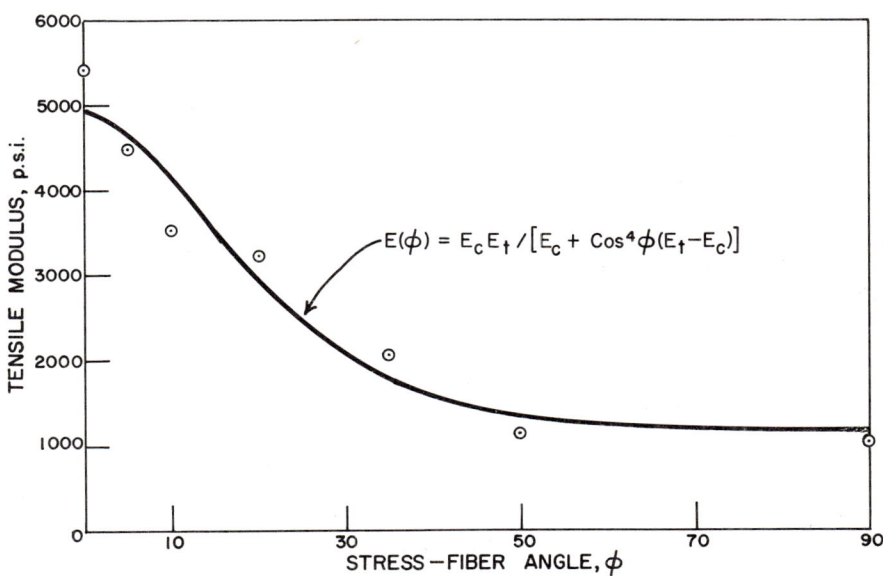

Figure 18. Angular dependence of Young's modulus for 30-volume fiber glass composite

either crack sensitivity of the matrix or a transverse critical volume effect. It is possible that the rubber composite behavior may result from the same factors because a significant fraction of the fibers are oriented in a transverse rather than longitudinal direction (discussed next).

Orientation

All of the composites in this study were prepared in a Banbury internal mixer. After mixing, the curing ingredients and hexamethylenetetramine were added on a 10-inch rubber mill. This milling step causes molecular orientation or "grain" even in unfilled rubber although the effect is relatively small.

The orientation effect is magnified when the compound contains fibrous reinforcement which tends to align itself with the direction of milling. The amount of orientation depends upon the viscosity of the stock as well as the type of rubber and type of fiber.

The amount of orientation achieved for the composites of this study was relatively small but significant as shown in Figures 7–11. This is also illustrated in Figures 1–5 where the arrow indicates the grain direction. Owing to the high magnification, however, the orientation is not as obvious.

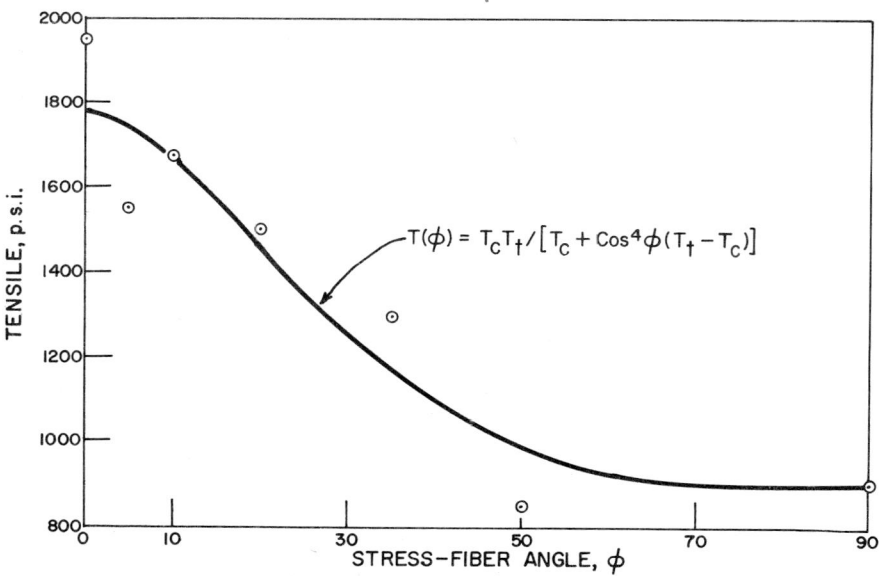

$T(\phi) = T_c T_t / [T_c + \cos^4\phi(T_t - T_c)]$

Figure 19. Angular dependence of tensile strength for 30-volume fiber glass composite

Figure 17 illustrates the effect of orientation on the stress–strain properties of the rayon composite shown in Figures 10 and 11. The upper curve represents stress–strain behavior for stress applied parallel to the fiber orientation direction. In the lower curve the force is applied perpendicularly. Even a small degree of orientation has a large effect on the anisotropy of the composite. The differences in tensile strength, modulus and elongation at break in the two directions are considerable.

Even though the initial degree of orientation is low, Figures 7–11 show that it increases with increasing strain on the sample. This phenomenon is unique to elastomer composites since metal and plastic composites do not have large enough breaking elongations to make this effect significant.

The effect of angle of applied stress on modulus, tensile strength, and elongation at break was further studied for a 30-volume fiber glass composite in a formulation identical to the standard formulation, except that the silica was reduced to 15 parts per hundred. The results for modulus are illustrated in Figure 18 where the data show good agreement with the theoretical relationship developed by Lees (6).

$$E(\phi) = E_c E_t / [E_c + \cos^4 \phi (E_t - E_c)] \tag{7}$$

where

E_c = modulus in the longitudinal direction relative to the direction of orientation
E_t = modulus in the transverse direction
ϕ = angle between the applied stress and the orientation axis

The same equation form also fits the tensile strength data quite well as shown in Figure 19. In this case

$$T(\phi) = T_c T_t / [T_c + \cos^4 \phi (T_t - T_c)] \tag{8}$$

where

T_c = longitudinal tensile
T_t = transverse tensile

Equation 8 has no theoretical basis but is a good empirical representation of the data. Elongation at break, shown in Figure 20, appears to be a simple linear function of the orientation angle as follows:

$$W(\phi) = W_c + b\phi \tag{9}$$

where

$W(\phi)$ = breaking elongation of composite
W_c = elongation at break in longitudinal direction
b = slope of the line

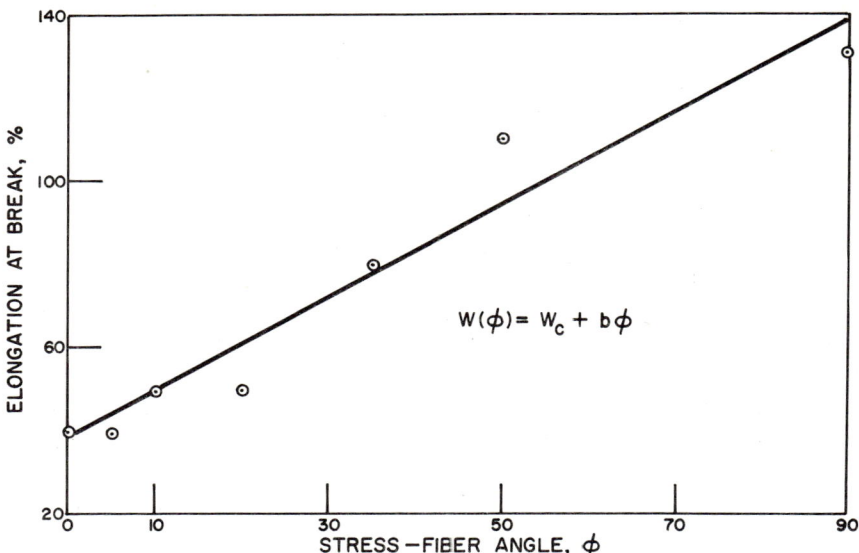

Figure 20. Angular dependence of elongation at break for 30-volume fiber glass composite

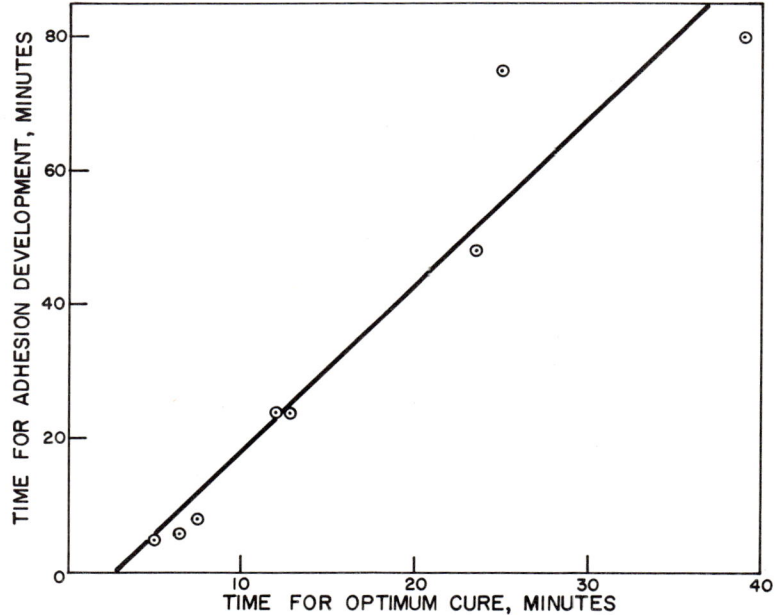

Figure 21. Rate of crosslink development vs. rate of fiber–matrix adhesion development for 12-volume fiber glass composite

The degree of orientation can be increased by repeated milling in the same direction with a tight mill setting or by increasing the viscosity of the rubber. In this study no extra milling was used, and the compound viscosity was relatively low. Previous results not reported here have shown that almost complete orientation can be obtained with high fiber glass loadings in formulations of relatively high viscosity.

The effects of other variables such as polymer type, processing temperature, extrusion *vs.* milling, etc., have not been investigated but may be fruitful areas of exploration.

Rate of Adhesion Development

Earlier work (2) has shown that adhesion of rubber to fabric with the aid of the HRH system develops at a slower rate than the rate of crosslinking as measured by the Monsanto oscillating disc rheometer. To determine if this was true of elastomer–fiber composites, the standard compound containing 12 volumes of fiber glass was cured at four different temperatures: 280°, 300°, 320°, and 340°F. Rate of adhesion development was measured by curing a series of sheets over an extended time and determining the time at which elongation was minimized for each temperature. The process was repeated on the same compound with the vulcanization accelerator levels reduced by 50% to reduce the rate of cure. Rate of adhesion development was not independent of time to reach optimum cure. At a given temperature the time to optimum adhesion varied with the optimum cure time. The relationship is shown in Figure 21 where time to optimum cure is plotted *vs.* time to optimum adhesion. The latter appears to be a linear function of optimum cure time. The slope of the line is 2.47, indicating that for an increase in optimum cure time, Δt, time to optimum adhesion is increased $2.47(\Delta t)$. Since natural rubber networks degrade as optimum cure time is surpassed, fast cure times are desirable as the lower the optimum cure time, the less overcure is necessary for adequate adhesion development. At $t = 5$ minutes, optimum cure and adhesion time are equal.

Conclusions

The HRH system has made it possible to study many types of fiber–elastomer composites with good fiber–matrix adhesion. The present study leaves many questions unanswered, such as the effect of fiber length, of higher degree of orientation, of different elastomers and fibers as well as various changes in other compounding materials. In view of the unique properties of these systems, continued investigation along these lines will be profitable.

Acknowledgments

The author thanks M. P. Wagner for his many helpful suggestions, James Borgen for the excellent micrographs, and Gerald Humes for the excellent drawings.

Literature Cited

(1) Creasey, J. R., Wagner, M. P., *Rubber Age* **1968**, 100 (10), 72.
(2) Creasey, J. R., Russell, D. B., Wagner, M. P., *Rubber Chem. Technol.* **1968**, 41, 1300.
(3) Degussa Technische Berichte, "Adhesion of Rubber Compounds on Textiles and Metals," 1967.
(4) Hashin, Z., Rosen, B. W., *J. Appl. Mech.* **1964**, 223, 32.
(5) Kelly, A., Tyson, W. R., *J. Mech. Phys. Solids* **1965**, 13, 329.
(6) Lees, J. K., Polymer Eng. Sci. **1966**, 8, 186.
(7) *Ibid.*, p. 195.
(8) Piggot, M. R., *Acta Metallurgica* **1966**, 14, 1429.

RECEIVED February 5, 1970.

Reinforcement of Thermosetting Cycloaliphatic Epoxy Systems with Elastomers

A. C. SOLDATOS and A. S. BURHANS

Research and Development Department, Chemicals and Plastics, Union Carbide Corp., Bound Brook, N. J. 08805

> *The toughness of crosslinked cycloaliphatic epoxy systems can be improved significantly by modification with elastomeric materials containing functional groups of varied molecular configuration. The nature of the reactive groups and the polarity of the elastomer affect rate of reaction and compatibility with epoxy. These variations, in turn, influence the molecular structure of the matrix resin. Two to 10-fold improvements in toughness, both at room temperature and $-70°C$, have been obtained at the 4–33% weight elastomer modification level. Similar improvements in tensile strength and elongation were realized. A correlation between the elastomer concentration and mechanical and optical properties of cured polymers suggests changes in morphology of the polymer at 20–25% elastomer level. Electron microscopy demonstrates that in the two-phase system the phase inversion takes place at about 20% rubber.*

To maximize the utility of crosslinked cycloaliphatic epoxy resins in some of the more critical application areas, improved toughness is required. Such improvements can often be made through modification with various flexibilizing agents, but as a rule this improvement is accompanied by a severe degradation of the strength and heat distortion temperature of the cured system.

For polystyrene and related thermoplastic copolymers, this inherent brittleness has been corrected effectively by including rubbery particles, properly modified by grafting with thermoplastic chains, for energy dissipation (6). With crosslinked epoxy resins, chemical attachment of the ends of low molecular weight rubbery chains to the epoxy resin through functional end groups provides the necessary means of translating the energy to the rubbery particle formed by precipitation during cure.

Previous work (5) has been limited to the glycidyl ether type epoxies, requiring catalytic curing rather than the conventional hardener type curing to facilitate formation of the desired rubbery particles.

The present study, by contrast, deals with toughening of cycloaliphatic type epoxies with anhydride curing agents normally used in industrial applications. In addition to developing mechanical property data, the morphological characteristics were also studied.

Discussion

This investigation shows that the toughness of cycloaliphatic epoxy systems can be improved significantly by modification with several elastomeric materials containing functional groups and varying in molecular structure. Liquid rubbers which are soluble in the uncured epoxy resin–hardner system, through reaction (copolymerization) with the epoxy resin, are crosslinked to elastomeric networks and partially precipitate as distinct rubber particles which are chemically bonded to the matrix. Furthermore, this investigation shows that the nature of the reactive groups and the polarity of the elastomer affect the rate of reaction as well as the compatibility with the epoxy resin. These variations, in turn, influence the molecular structure of the matrix resin.

For simplification, the same cycloaliphatic epoxide was used throughout this investigation. The resin was the 3,4-epoxycyclohexylmethyl-3,4-epoxycyclohexane carboxylate, designated ERL-4221 (Figure 1). Hexa-

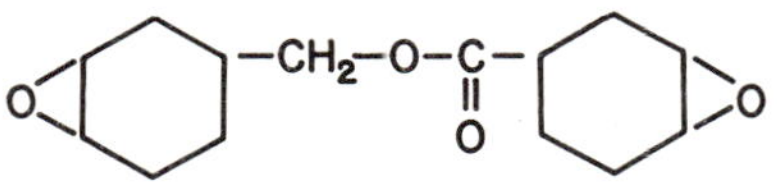

3,4 — EPOXYCYCLOHEXYLMETHYL
3,4 — EPOXYCYCLOHEXANE CARBOXYLATE

Figure 1. Structure of ERL-4221

hydrophthalic anhydride (HHPA) was chosen as the hardener and benzyldimethylamine (BDMA) as the catalyst. The small amount of ethylene glycol added serves to open some of the anhydride rings; the resultant acid groups initiate the curing reaction.

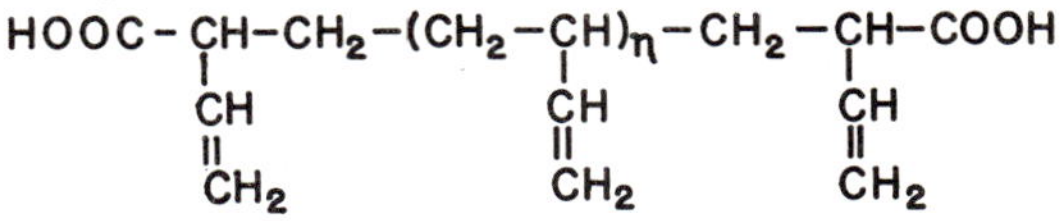

$$\text{HOOC-CH-CH}_2\text{-(CH}_2\text{-CH)}_n\text{-CH}_2\text{-CH-COOH}$$
$$\underset{\underset{\text{CH}_2}{\overset{\|}{\text{CH}}}}{|} \qquad \underset{\underset{\text{CH}_2}{\overset{\|}{\text{CH}}}}{|} \qquad \underset{\underset{\text{CH}_2}{\overset{\|}{\text{CH}}}}{|}$$

Figure 2. Structure of α,ω-polybutadiene dicarboxylic acid

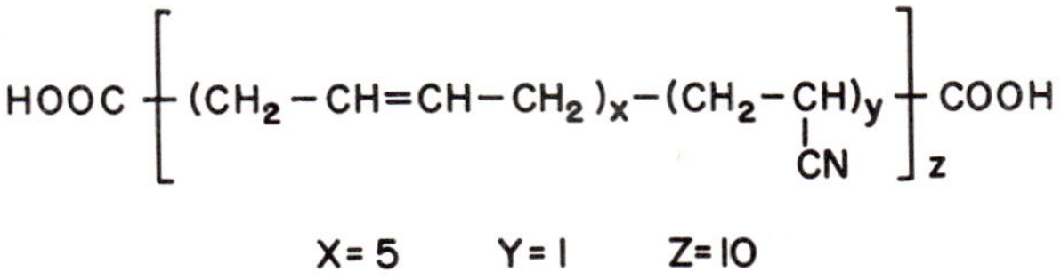

$$\text{HOOC}\left[(\text{CH}_2\text{-CH=CH-CH}_2)_x\text{-}(\text{CH}_2\text{-}\underset{\text{CN}}{\text{CH}})_y\right]_z\text{COOH}$$

X = 5   Y = 1   Z = 10

Figure 3. Structure of CTBN

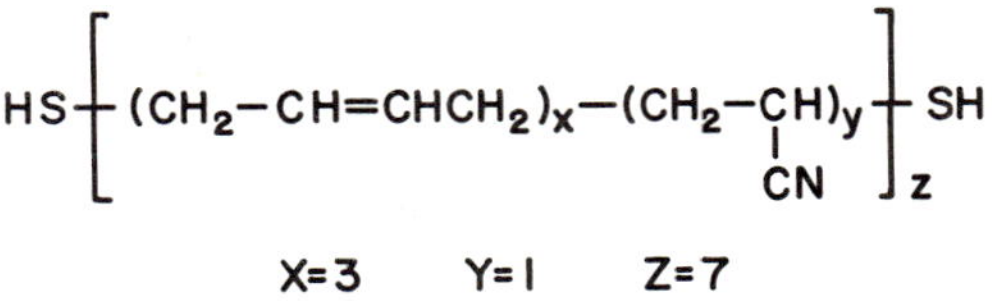

$$\text{HS}\left[(\text{CH}_2\text{-CH=CHCH}_2)_x\text{-}(\text{CH}_2\text{-}\underset{\text{CN}}{\text{CH}})_y\right]_z\text{SH}$$

X = 3   Y = 1   Z = 7

Figure 4. Structure of MTBN

The elastomeric materials which were tested and found effective in improving the toughness of ERL-4221 were the following:

(1) A carboxyl terminated, 3000 molecular weight polybutadiene designated C-3000. This polymer is of 1,2 configuration (2) as shown in Figure 2.

(2) A 3300 molecular weight carboxyl terminated 80–20 butadiene–acrylonitrile random copolymer designated CTBN (1) (Figure 3).

(3) A 1700 molecular weight mercaptan terminated random copolymer of 70% butadiene and 30% acrylonitrile designated MTBN (5) (Figure 4).

Mechanical Properties of Cast Resins. The toughness of the crosslinked polymers was determined by the area under the stress–strain curve and by the energy required to fracture. The fracture or impact energy, expressed in inch-pounds, was measured by a simple but reliable test— the Gardner impact—which consists of striking cured specimens with a 2-lb round-nose rod, ½-inch in diameter, from various distances; the test specimens were discs 0.1 inch thick by 2.0 inches in diameter.

The impact of ERL-4221 was essentially doubled when modified with 10 parts per 100 parts resin (phr) of each of the above elastomers (Table I) which is actually only approximately 5% by weight based on the total formulation. The tensile strength and elongation were

Table I. Cast Resin Properties of ERL-4221/HHPA Modified with Various Elastomers

	ERL-4221/HHPA			
	Elastomer 10 pts./100 pts. Resin			
	None	CTBN	MTBN	C-3000
Gardner impact, inch-lbs.	40	80	70	75
Heat distortion temperature, °C	191	187	169	188
Tensile strength, psig	7,500	12,000	9,600	4,300
Tensile modulus, psig	435,000	401,000	412,000	401,000
Elongation, %	1.9	5.0	3.0	1.5

Formulation	Parts
ERL-4221	100
Hexahydrophthalic anhydride (HHPA)	100
Ethylene glycol	1.5
Benzyldimethylamine (BDMA)	1.0
Modifier (as indicated)	10

also increased without severely degrading the modulus and heat distortion temperature. The only exception was the carboxyl terminated polybutadiene, which lowered the tensile value.

The tensile strength and elongation were increased in the case of CTBN from 7500 psi and 1.9% to 12,000 psi and 5% which represents improvements of 60% and 250%, respectively. The maximum drop in modulus was only 7% (from 435,000 psi to 401,000 psi), and the heat distortion temperature was decreased by a few degrees (from 191° to 187°C) with the exception of the MTBN, which degraded this property more excessively, by 22°.

It is obvious from these data that CTBN (the carboxyl terminated butadiene–acrylonitrile copolymer) is the most effective modifier, and therefore it was selected for further study. As the concentration of the elastomer was increased to levels up to 100 parts, the impact also increased. The data in Table II show that the room temperature impact of ERL-4221 increased from 40 inch-lbs to greater than 320 inch-lbs by adding 100 phr or 33 wt % of CTBN. At very low temperatures ($-160°F$) the impact of the system modified with 60 parts, or 23 wt %, of CTBN was 120 inch-lbs. These impact improvements appear to be directly proportional to the concentration of the elastomer modifier.

A parallel increase in tensile strength was also observed at the 10–50 phr level of CTBN modification (Table III). The most significant improvement was obtained at the 10 parts level, where the tensile strength was increased to 12,000 psi; the elongation was also increased to 5%, whereas the modulus and heat distortion temperature were reduced only slightly.

Higher concentrations of elastomer failed to increase the tensile any further, and at the 50 parts level the strength was only slightly higher than the unmodified system. At 80 parts CTBN, the tensile strength was already lower than the control. As the elastomer was increased, the heat distortion temperatures and moduli were decreased, but the elongation continued to increase, as might have been expected.

Table II. Gardner Impact of ERL-4221/HHPA Modified with Carboxyl Terminated Elastomer (CTBN)

ERL-4221/HHPA

	\multicolumn{9}{c}{CTBN pts./100 pts. Resin}
	0	10	20	30	40	50	60	80	100
Gardner impact at room temp., inch-lbs.	40	80	110	90	100	120	130	168	>320
Gardner impact at $-160°F$, inch-lbs.					120				

Formulation	Parts
ERL-4221	100
HHPA	100
Ethylene glycol	1.5
BDMA	1.0
CTBN	as indicated

This substantial toughening effect of CTBN on the cycloaliphatic epoxide ERL-4221, coupled with significant increase of the strength of the resin without seriously lowering the heat distortion temperature, is quite unusual for thermosetting systems.

Further evidence of the reinforcing effect of this elastomer on the epoxide ERL-4221 was obtained from measurements of the area under the stress–strain curve. Epoxy resin systems with high elongation and high tensile strength, in general, have greater toughness as measured by the area under the curve (8). The tensile stress–strain curves for the cast resin system ERL-4221–HHPA modified with zero, 10, 20, and 40 parts of CTBN are shown in Figure 5. The areas under the curves of the modified systems are considerably larger (382 to 458 inch-lb/inch^3) than that of the unmodified system which is only 86 inch-lb/inch^3. These improvements are the result of the increased elongation and tensile strength.

Morphology of Polymers. In conjunction with the measurement of various mechanical properties, the morphology of the rubber modified

Table III. Cast Resin Properties of ERL-4221/HHPA

	0	10
Heat distortion temp., °C	191	187
Tensile strength, psig	7,500	12,000
Tensile modulus, psig	435,000	401,000
Elongation, %	1.9	5

Formulation	Parts
ERLA-4221	100
HHPA	100
Ethylene glycol	1.5
BDMA	1.0
CTBN	as indicated

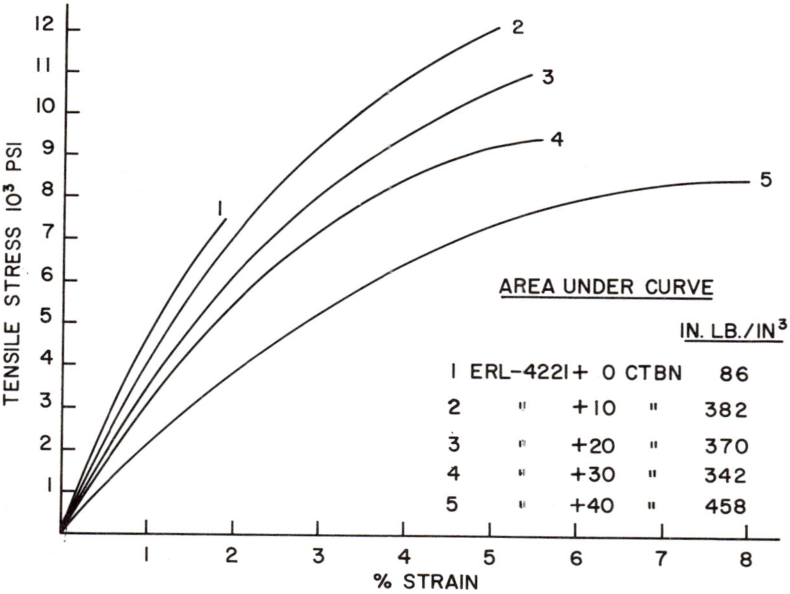

AREA UNDER CURVE

IN. LB./IN^3

1	ERL-4221+	0 CTBN		86
2	"	+10	"	382
3	"	+20	"	370
4	"	+30	"	342
5	"	+40	"	458

Figure 5. Cast resin tensile stress–strain curves

cycloaliphatic epoxides was studied with optical methods and electron microscopy.

The carboxyl terminated butadiene–acrylonitrile rubber, which is soluble initially in the liquid phase, precipitates out as a second phase during the crosslinking reaction of the epoxy. The cured specimens are

Modified with Carboxyl Terminated Elastomer (CTBN)

ERL-4221/HHPA

CTBN pts./100 pts. Resin					
20	30	40	50	80	100
188	168	159	158	127	92
10,600	9,390	8,500	8,050	5,850	4,980
354,000	308,000	276,000	270,000	189,000	142,000
5.3	5.5	8	8.5	20	30

Cure Cycle

2 hrs at 120°C
4 hrs at 160°C

opaque, in contrast to the liquid solutions which are clear, and the degree of opacity depends on the concentration of the elastomer and the degree of epoxy–elastomer incompatibility. This was proved experimentally by measuring the percent haze of cured films using the pivotable-sphere hazemeter (ASTM D-1003-59T).

Haze is the percentage of transmitted light which, in passing through the specimen, deviates from the incident beam by forward scattering. The various resin systems were applied to clean glass panels with a drawdown blade having a gap clearance of 30 mils. The panels were tested after they were cured properly. The percent haze of ERL-4221–HHPA modified with zero to 100 phr CTBN was plotted against the elastomer

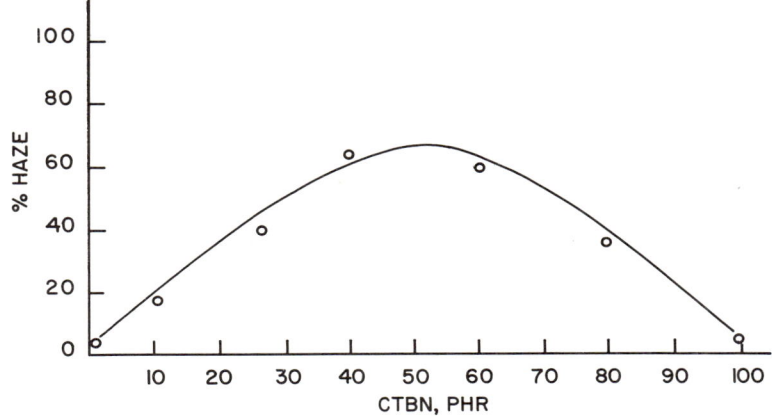

Figure 6. Percent haze of ERL-4221 modified with CTBN

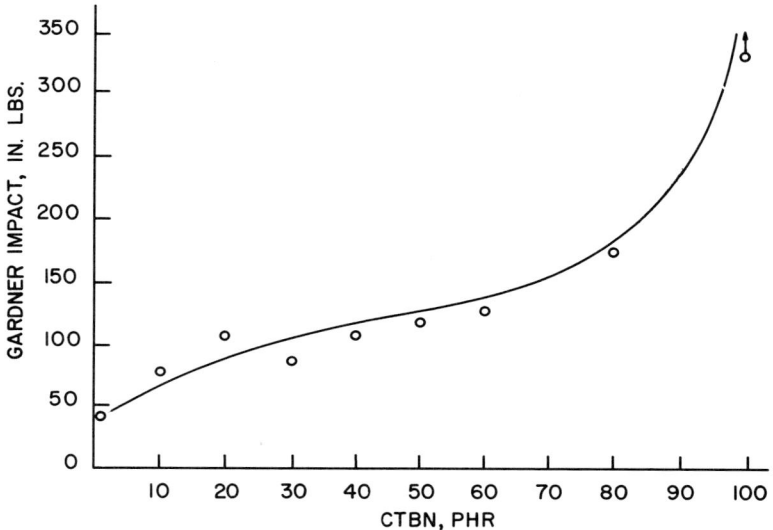

Figure 7. Gardner impact of ERL-4221–CTBN systems

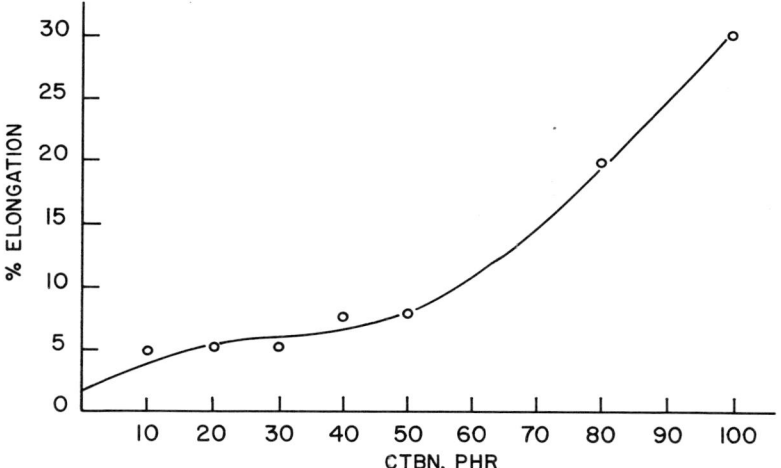

Figure 8. Percent elongation of ERL-4221–CTBN systems

concentration (Figure 6). These data show clearly that the haze increases by increasing the CTBN elastomer concentration from zero to approximately 50–60 phr. Beyond this concentration the haze decreases, and at the 100 phr modification level the haze is 8% only.

This drastic change in the slope of the haze curve at the 50–60 phr CTBN suggests that some morphological changes in the cured polymer have taken place. An inspection of several mechanical properties tends to support this hypothesis. These data indicate that there is a correlation between the elastomer concentration and macroscopic properties of the polymer. The Gardner impact values of the various resin system compositions, shown in Figure 7, indicate that the impact starts to increase more rapidly at about the 60 phr CTBN concentration. Also, at the 50–60 phr level, the elongation increased sharply, as shown in Figure 8.

Finally, the curve of Figure 9 indicates that the heat distortion temperature drops somewhat more sharply as the CTBN concentration increases beyond the 50 phr level.

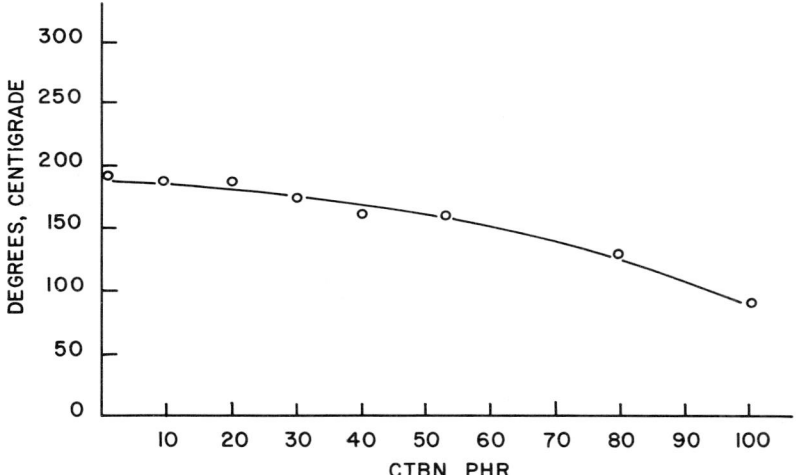

Figure 9. Heat distortion temperature of ERL-4221–CTBN systems

Electron microscopy further supported the hypothesis or morphological changes by demonstrating graphically the two-phase system. Electron microscopy also provided correlation of the mechanical properties with the second phase microstructure, distribution of particle size, and second phase content.

The electron micrographs of the various ERL-4221–CTBN systems were prepared with the osmium tetroxide technique (3). The castings were stained by reaction with osmium tetroxide vapors for 24 hours. Ultrathin specimens, approximately 1000 A thick, were cut with the Reichert OMU2, ultramicrotome equipped with a diamond knife and were stained again for one hour. The osmium tetroxide selectivity stained the rubber phase, while the epoxy remained unaffected, revealing many structural features in the polyphase systems with excellent contrast and

Figure 10. Electron micrograph of ERL-4221–HHPA system

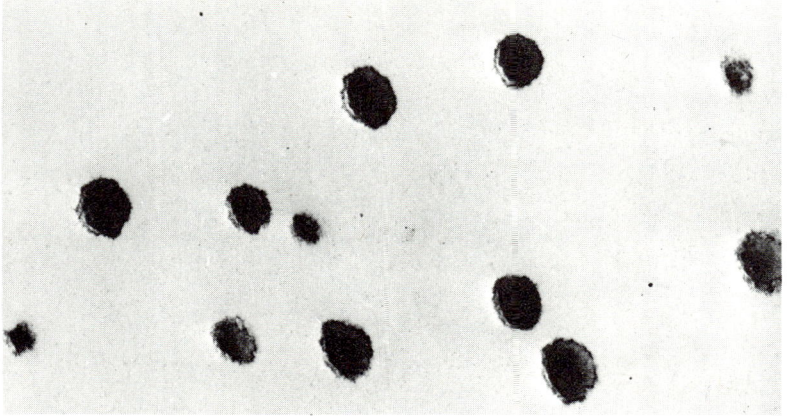

Figure 11. Electron micrograph of ERL-4221–HHPA modified with 10 phr CTBN

definition. An additional advantage of this technique is that osmium tetroxide hardens the soft rubber which facilitates the cutting of ultrathin sections. The RCA electron microscope, model EMU-3B, was used, which produced a magnification of 25,000.

The electron micrographs shown in Figures 10 through 16 demonstrate the multiphase structure of the ERL-4221–HHPA system modified with various concentrations of CTBN. Figure 10 illustrates the control which contains no rubber, and it is clearly a monophase system. The system containing 10 parts of CTBN is depicted in Figure 11 where the two-phase structure is evident. The continuous phase (light color) consists mainly of epoxy resin, and the disperse phase (black color) consists of rubber with an average particle size of approximately 5000 A. The rubber particles are dense spheroids with well defined boundaries. A careful examination of the topography of this micrograph shows that the

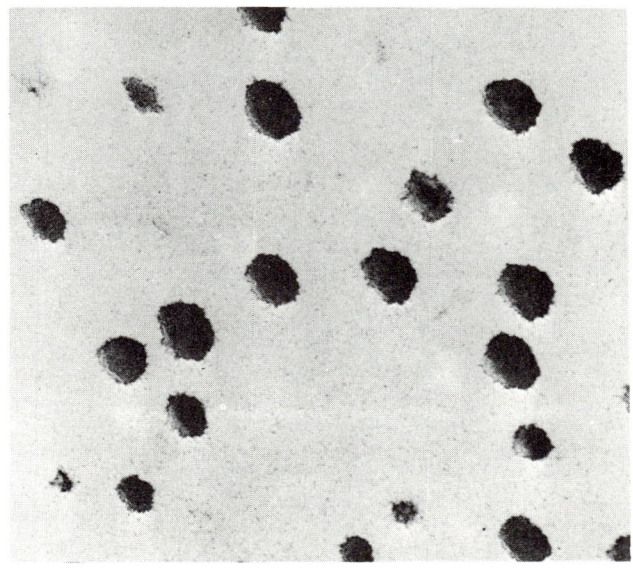

Figure 12. Electron micrograph of ERL-4221–HHPA modified with 20 phr of CTBN

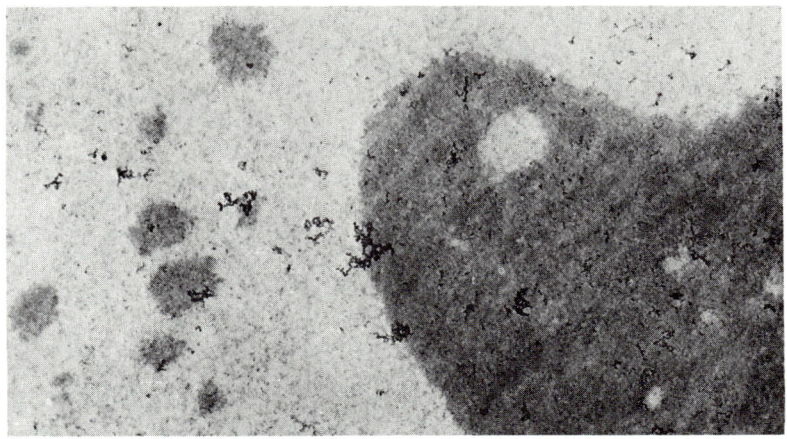

Figure 13. Electron micrograph of ERL-4221–HHPA modified with 40 phr CTBN

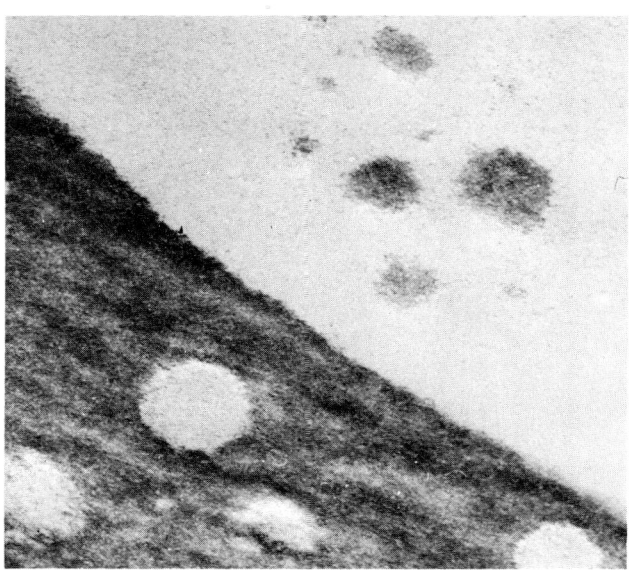

Figure 14. Electron micrograph of ERL-4221–HHPA modified with 50 phr of CTBN

continuous resinous phase contains numerous small rubber particles, less than 100–200 A in diameter. At the 20 parts CTBN level (Figure 12) the average particle size and the volume content of precipitated rubber particles are increased somewhat. The boundaries of the particles are less well defined and contain some epoxy resin (light regions). Further, the number of small rubber particles, less than 100–200 A, dispersed through the resinous phase has increased. As the concentration of CTBN increases to 40 parts (Figure 13) the rubber particles start segregating into large islands, and at the 50 phr CTBN (Figure 14), it is evident that a phase inversion is beginning. Further increase of CTBN to 70 parts (Figure 15) reveals that apparently complete inversion has taken place. The continuous phase is now dark and consists of rubber saturated with resin, whereas the disperse phase is white, consisting of large particles of resin (20,000–25,000 A in diameter) and finely dispersed rubber particles. The 100 parts CTBN system illustrated in Figure 16 has a topography similar to the 70 parts, but the size of the rubber particles has now decreased substantially. A close examination of the texture of this micrograph clearly shows that both phases are contaminated heavily with material of other phase.

The heterophase and phase inversion phenomena have been observed previously in rubber systems modified with hydrocarbon resins (*10*), but the inversion has taken place only at high (50–70%) resin concentra-

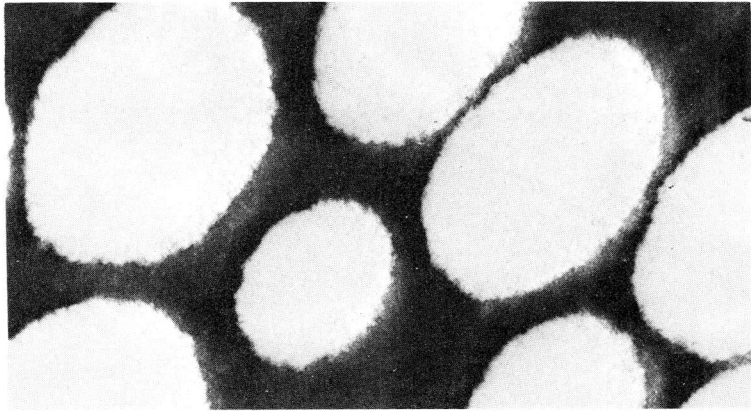

Figure 15.   Electron micrograph of ERL-4221–HHPA modified with 70 phr of CTBN

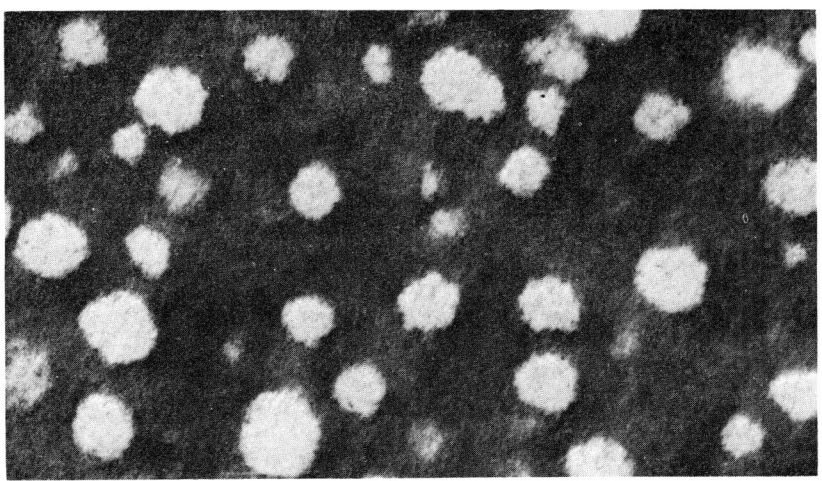

Figure 16.   Electron micrograph of ERL-4221–HHPA modified with 100 phr of CTBN

tions. The interesting and somewhat puzzling observation is that in our case the phase inversion has started at the very low rubber concentration of 17–25% (40–70 phr resin).

Although the micrographs indicate that at these higher CTBN concentrations the rubber is predominantly the continuous phase, the mechanical properties of the systems do not support this assumption. The experimental tensile modulus of the 100 phr CTBN system is too high (142,000 psi) to support a rubbery continuous phase.

Analysis of the modulus data by Maxwell's equation suggests that at the higher rubber concentrations, both the epoxy and rubber phases are continuous (7), with the epoxy phase slightly predominant.

For tensile modulus, the following equation would hold:

$$E = E_c \left[\frac{E_d + 2E_c - 2X_d(E_c - E_d)}{E_d + 2E_c + X_d(E_c - E_d)} \right] \quad (1)$$

where

E_c = tensile modulus of continuous phase
E_d = tensile modulus of discontinuous phase
X_d = volume fraction of discontinuous phase
E = tensile modulus of composite.

if epoxy is the continuous phase, and since $E_c \gg E_d$, Equation 1 can be written:

$$E = E_c \left[\frac{(2 - 2X_d) E_c}{(2 + X_d) E_c} \right] = E_c \left[\frac{(2 - 2X_d)}{2 + X_d} \right] \quad (2)$$

if rubber is the continuous phase, and since $E_d \gg E_c$, Equation 1 can be written:

$$E = E_c \left[\frac{E_d(1 + 2X_d)}{E_d(1 - X_d)} \right] = E_c \left[\frac{(1 + 2X_d)}{1 - X_d} \right] \quad (3)$$

Applying Equation 2 (epoxy is the continuous phase) for the system containing 33 wt % or approximately 35 vol % rubber

$E = 435,000 (0.552) = 240,120$ psi
$X_d \simeq .35$
$E_c = 435,000$ psi (tensile modulus of pure epoxy resin)

(The experimental tensile modulus for the system is 142,000 psi)

Applying Equation 3 (rubber is the continuous phase) for the same above system where $X_d \simeq 0.65$ and E_c, the tensile modulus of the cross-linked rubber, was assumed to be in the range of 100–1000 psi

$$E = E_c [6.58] \ll E_{exp} = 142,000 \text{ psi.}$$

From these results it appears as if both phases are continuous, with epoxy continuity $= \dfrac{142,000}{240,120} \times 100\% = 59\%$.

The haze data of the various systems previously discussed had shown that at the 100 parts rubber concentration the cured castings were very clear, and the percent haze was only 8%. However, the electron micrograph of this sample (Figure 16) has shown clearly that the dispersed phase is large enough to scatter light, and therefore the haze and opacity should have been much higher. This phenomenon can be explained possibly by the similarity of the refractive indices of the two phases.

The refractive index of the cured ERL-4221–HHPA system is 1.5070, which is significantly different from the refractive index of CTBN (1.5142). This difference can possibly account for the considerable haze of the systems containing the low concentration (10 to 70 phr) of rubber. As the concentration of CTBN is increased, instead of two distinctly separated phases, two continuous phases are formed, along with an inversion of the phases plus considerable intermixing of the rubber and epoxy components, the differences in the refractive indices are gradually diminished. More specifically, the refractive index of the ERL-4221–HHPA modified with 100 parts CTBN is 1.5110 which is obviously between those of the two pure phases.

Effect of Molecular Configuration of Elastomer. The extent of the impact and strength improvements of ERL-4221 depends on the chemical structure and composition of the elastomer modifier. The data shown in Table I indicate that the carboxyl terminated 80–20 butadiene–acrylonitrile copolymer (CTBN) is the most effective toughening and reinforcing agent. The mercaptan terminated copolymer (MTBN) is considerably less effective as far as tensile strength and heat distortion temperature are concerned. The mercaptan groups are considerably less reactive with epoxides than carboxyls (4), and this difference in the rate of reaction may influence the extent of the epoxy–elastomer copolymerization and therefore the precipitation of the rubber as distinct particles.

The carboxyl terminated polybutadiene (C-3000) is about equally effective to CTBN in heat distortion temperature and impact but considerably less effective in strength. From the haze data (the percent haze of ERL-4221 modified with 10 phr of CTBN and C-3000 were 17 and 85% respectively) it is quite clear that this elastomer (C-3000) is highly incompatible with the epoxy–hardener system in the cured state. A 2000 molecular weight polybutadiene elastomer, containing no carboxyl groups, was completely incompatible with the epoxy system and segregated in the cured state.

These findings suggest strongly that the composition of the elastomeric molecule and the nature of the functional groups affect its compatibility and rate of reaction with the epoxy resin, which in turn affect the molecular and morphological structure of the heterophase system. These data indicate the importance of the acrylonitrile comonomer and the carboxyl groups in controlling the polarity of the rubber, and subsequently its compatibility characteristics with the epoxy. We could also postulate further that highly polar materials, such as the carboxyl terminated butadiene–acrylonitrile copolymers significantly improve both the impact and strength of the cycloaliphatic epoxy resins, whereas similar elastomers of reduced polarity, such as the carboxyl terminated polybutadiene, improve the impact but severely degrade the strength.

The practical application of this study lies in the beneficial effect on composite properties observed as a result of toughening these resin systems. These data were reported at the 1970 SPI Conference (9). Current work on the effect of this type of modification on adhesives will be the subject of a future report.

Acknowledgments

The authors express their appreciation to W. D. Niegisch who carried out the electron microscopy studies and L. M. Robeson for his mathematical treatment of the Maxwell equation.

Literature Cited

(1) Drake, R. S., McCarthy, W. J., *Rubber World* (October, 1968).
(2) Fukushima, Masatoshi, *Japan Chem. Quart.* 111-11.
(3) Kato, K., *Poly. Eng. Sci.* 38 (Jan. 1967).
(4) Lee, H., Neville, K., "Handbook of Epoxy Resins," McGraw-Hill, New York, 1967.
(5) McGarry, F. J. et al., "Relationships Between Resin Fracture and Composite Properties," A.F.M.L. TR-67-381 (1967).
(6) Molan, G. E., Keskkula, H., *J. Polymer Sci.* 4, 1595 (1966).
(7) Robeson, L. M., Union Carbide Corp., personal communication.
(8) Soldatos, A. C., Burhans, A. S., Cole, L. F., *S.P.I., 24th Ann. Tech. Mgt. Conf.*, Feb. 1969.
(9) Soldatos, A. C., Burhans, A. S., *S.P.I., 25th Ann. Tech. Mgt. Conf.*, Feb. 1970.
(10) Wetzel, F. H., *Rubber Age* 291 (Nov. 1957).

RECEIVED March 10, 1970.

34

The Preparation and Characteristics of Concrete–Polymer Composites

M. STEINBERG, L. E. KUKACKA, P. COLOMBO, and B. MANOWITZ

Brookhaven National Laboratory, Department of Applied Science, Upton, N. Y. 11973

Preformed mortar and concrete can be impregnated with monomer and polymerized by radiation or thermal-catalytic techniques. The resulting concrete–polymer has substantially improved structural properties and durability. Maximum increases in strength are obtained by forming and curing the concrete, subsequently drying, evacuating, and soaking the concrete in monomer, and initiating polymerization by radiation. Crosslinked and thermosetting monomers allow the maintenance of strength and durability up to temperatures of ca. 150°C. Radiation treated concrete–polymer displays strength properties from 12 to 38% higher than thermal catalytically treated material. Concrete–polymer significantly increases the bond strength and flexural strength of steel-reinforced and fiber-reinforced concrete. There appears to be little difference between the effects of radiation and thermal treatment on the durability of concrete–polymers.

Concrete is used in nearly all types of construction throughout the world because of its capacity to be formed into various sizes and shapes, the ready availability of the raw material from which it is made, and its relatively low cost. Although concrete is an excellent building material, certain limitations to its use are recognized. These relate mainly to its relatively low tensile strength, its tendency to crack with changes in temperature and moisture, and its deterioration because of permeability, absorption, and chemical attack under various environmental conditions. Improvements in strength and durability could extend significantly the usefulness of concrete.

Based on the concept of diffusing a monomer through the porous mass of concrete and *in situ* polymerization of the monomer with ^{60}Co gamma radiation, initial experiments were carried out with methyl methacrylate and styrene-impregnated mortar bars. A hard glossy polymer was formed throughout the cross section of the concrete, and the compressive strength was improved by as much as a factor of 2.4, the absorption was decreased by 98%, and the hardness was increased by a factor of 1.9. Based on these remarkable preliminary results a cooperative research effort was undertaken by the Brookhaven National Laboratory and the U. S. Bureau of Reclamation, supported jointly by the U. S. Atomic Energy Commission and the U. S. Department of the Interior (5).

The long range objectives of the joint program are the investigation and development of a concrete–polymer composite as a new construction material. The program includes the development of techniques for preparing concrete–polymer material, measurement of the physical and chemical properties, preparation of full-scale concrete products, and the conceptual design and evaluation of various specific applications.

Concrete–Polymer Preparation Techniques

Concrete Formation. Concrete is usually prepared by mixing thoroughly hydraulic cement and water with sand, gravel, crushed rock, or some other aggregates, casting the mixture in a mold, and allowing it to hydrate or cure in a humid atmosphere (3). After a 28-day cure the concrete reaches about 80% of its ultimate strength. Concrete at this point is ∼3.5% by weight of bound or hydrated water and ∼3.5% by weight of free water. When an air-entraining agent is used, the total available void volume in the concrete is ∼14%. The amount of free water and the void volume in the concrete determine the quantity of monomer and polymer that can be added or loaded into a precast concrete.

Drying. To obtain maximum loading of the concrete with the monomer, the free water is removed by drying. This can be done by various techniques such as oven drying, vacuum drying in a heated air stream, or infrared or microwave heating. To dry thermally 3-inch diameter $\times$ 6-inch long cylindrical concrete specimens to constant weight in a muffle furnace at 221°F (105°C), a drying time of 100 hours is required. This time is decreased considerably at higher temperatures —*e.g.*, 8 hours at 302°F (150°C). For undried specimens the monomer loadings were usually <3.5%, and for dried specimens >3.5% of the unimpregnated specimen.

Evacuation, Soaking, and Pressurizing. The impregnation of concrete with monomer is essentially a displacement of the air in the void volume of the concrete. This can be accomplished by lowering the specimen slowly into the liquid monomer and allowing the liquid pressure and capillary forces to displace the air volume. The air can also be displaced by forcing the monomer through the concrete under a differential pressure. Another method is to evacuate the air and submerge the specimen in the liquid monomer. This eliminates the possibility of entrapment of air toward the center of the specimen, which would prevent full loading. It was also found that evacuation of specimens significantly decreased the subsequent soaking time needed to obtain comparable monomer loading. Application of pressure during soaking in monomer is another technique that can be used to reduce loading time.

Table I. Techniques for Monomer Impregnation of Concrete

Sample[a]	Evacuation Time, hours[d]	Soak Time in Monomer,[c] hours	Nitrogen Pressure, psig	Polymer[e] Loading, % wt
1[d]	0	3	0	1.89
2	0	3	0	4.61
3	0	1	5	3.62
4	0	3	5	4.62
5	0	5	5	4.81
6	2	1	0	4.54
7	2	3	0	5.44
8	2	5	0	6.32
9	2	1	5	6.64
10	2	2	5	6.15
11	2	3	5	6.43
12	1	1	5	6.51

[a] All samples are 3-inch diameter × 6-inch long cylinders.
[b] Vacuum pump to ≈1 inch Hg.
[c] Methyl methacrylate monomer.
[d] All samples except No. 1 were dried to equilibrium before soaking.
[e] Uptake of monomer measured by polymer loading. Complete polymerization of monomer obtained by ^{60}Co gamma radiation.

Table I indicates the effect of evacuation, soaking, and pressure on the time required to impregnate a 3-inch diameter × 6-inch long cylinder of concrete and the polymer loadings achieved with a monomer of relatively low viscosity, such as methyl methacrylate. For similar soaking times the polymer loading values were 20–30% higher for evacuated than for non-evacuated samples. The effect of 5 psig nitrogen gas pressure over the monomer in which the sample was soaking indicates that for the non-evacuated specimens there is little change in polymer loading. However, for evacuated samples the soaking time required to reach a loading of 6.5% was reduced from 5 to 1 hour (cf. specimens No. 8 and 12).

Monomer Selection. The selection of a monomer to form a concrete–polymer material depends on the end application of the material, which in turn depends on the properties of the polymer, and on the cost, process requirements, and properties of the monomer.

For ambient or low temperature application, thermoplastic polymers can be used. Low cost monomers that have been used in this category include ethylene, ethylene–SO_2, vinyl acetate, methyl methacrylate, styrene, styrene–acrylonitrile, and chlorostyrene. Others awaiting test are vinyl chloride, vinylidene chloride, and *tert*-butylstyrene. These monomers are limited for use at temperatures below $\sim 100°C$ because of their softening points.

For higher temperature applications thermosetting monomers can be used, and crosslinking agents can be added to the thermoplastic-type monomers. Monomers selected for studying high temperature properties of concrete–polymer materials include 90% methyl methacrylate–10% trimethylolpropane trimethacrylate (the latter, TMPTMA, is a polyfunctional crosslinking agent), 60% styrene–40% TMPTMA, acrylonitrile, diallyl phthalate, polyester–styrene mixtures, and epoxy–styrene mixtures. These polymers have softening temperatures that permit them to be used above $\sim 150°C$.

For processing purposes the monomer should have a relatively high boiling point or low vapor pressure under polymerization conditions to limit vaporization losses between the time the concrete specimen is impregnated and the time it is polymerized fully.

To prevent evaporation, polymerization can be carried out in a closed system, such as an enclosed vessel, or the concrete can be wrapped in polyethylene sheet. Another possibility is carrying out the polymerization under water, which is especially attractive for decreasing monomer loss. The monomer in the specimen is usually not miscible with water so that it has little effect on the properties of the concrete–polymer.

Gaseous and higher vapor pressure liquid monomers are more difficult to handle in process. Closed pressurized containers are needed, and excess monomer surrounding the specimen is wasted and must be removed after polymerization. For many applications a hard glassy polymer is required to produce a strong bond and provide resistance to chemical attack. The polymers of ethylene formed below their melting point from the gaseous phase yield a powdery material in the interstices of the concrete which does not seem to adhere to the concrete. Poly(methyl methacrylate) forms a hard glassy material which tends to line the voids and capillary structure of the cured concrete.

Most monomers have a low enough viscosity so that they can be diffused easily through relatively thick layers (6 inches) of concrete at reasonable rates. Monomers such as styrene and methyl methacrylate with viscosities <1 cp are typical examples. Other monomers such as polyesters and epoxies have viscosities several hundred times higher, which makes it difficult for them to diffuse through concrete. Dilution with comonomers of low viscosities can reduce the viscosity sufficiently to allow penetration through the concrete. Viscosity effects are also important in partially impregnating a concrete material. With low viscosity monomers it is difficult to form sharply defined partially impregnated layers because of capillary diffusion through the concrete.

Polymerization Technique

Polymerization of monomer-impregnated concrete was initially carried out with ^{60}Co gamma radiation. A principal advantage of using radiation is that free radical chain reactions can be induced at ambient temperature. This limits vaporization loss and may produce a better concrete–polymer bond. The disadvantages are the dose requirements and the relatively high initial investment required for the radiation source and facility.

In the thermal-catalytic method a peroxide catalyst is usually used to initiate the free radical chain reaction. The main disadvantages are the higher temperatures required for carrying out the polymerizations, the potential hazard of explosion on addition of catalyst to the monomer, and disposal of excess catalyzed monomer after impregnating. Combinations of heat, radiation, and catalyst have been experimented with to reduce the radiation and catalyst requirements and to increase the rate of polymerization. In thermal polymerization a muffle furnace, infrared heating, and microwave heating can be used to provide the thermal energy.

Promoters have been used to reduce the temperature at which the catalyst normally decomposes and to effect polymerization at ambient temperature. The use of a combination of monomer, catalyst, and promoter may have some advantage for field impregnation of cast-in-place concrete.

To provide improved control over the polymerization and additional safety, it is possible to add the catalyst to the concrete mix before curing and to add the promoter to the monomer used for impregnating. Benzoyl peroxide when incorporated in premixed concrete will initiate polymerization of methyl methacrylate at room temperature when cobalt naphthenate is added to the monomer used for impregnation.

Some typical ^{60}Co gamma radiation dose requirements for fully polymerizing several concrete–polymer systems are given in Table II. There is some evidence (5) that a lower dose is required to polymerize monomer in concrete than monomer alone. This may be attributed to energy transfer effects in the heterogeneous concrete–monomer system.

Table II. Radiation Requirements for Concrete–Polymer Using ^{60}Co Gamma Radiation

Monomer[a]	Radiation Intensity, rads/hour $\times 10^5$	Radiation Dose, rads $\times 10^6$
Methyl methacrylate	1.6	1.2
	3.4	1.8
Methyl methacrylate + 10% TMPTMA	2.5	1.0
Acrylonitrile	1.9	1.3
Styrene	2.6	25.0
60% Styrene + 40% TMPTMA	5.0	1.0
Chlorostyrene	2.5	6.0

[a] Monomers contain inhibitor.

Table III. Concrete–Polymer Materials: Summary of

(Dried Concrete Specimens

	Control
Compressive strength, psi	5267
Modulus of elasticity, 10^6 psi	3.5
Tensile strength, psi	416
Modulus of rupture, psi	739
Flexural modulus of elasticity, 10^6 psi	4.3
Water absorption, %	5.3[a]
	6.4
Abrasion, inch	0.0497
g	14
Cavitation, inch	0.32
Water permeability, 10^{-4} ft/yr	6.2[a]
	5.3
Thermal conductivity at 73°F (23°C), Btu/ft-hr-°F	1.332
Diffusivity at 73°F (23°C), ft^2/hr	0.0387
Coefficient of expansion, in./in.-°F	4.02×10^{-6}
Creep for 800 psi load after 30 days, 10^{-6} in./in.	−95
Freeze–thaw durability, cycles, % wt loss	590; 26.5[a]
	490; 25.0
Hardness-impact ("L" hammer)	32.0
Corrosion by sulfate solution,	
300-day exposure, % expansion	0.144[a]
Acid corrosion in 15% HCl,	
84-day exposure, % wt loss	10.4
Corrosion by distilled H$_2$O,	
120-day exposure at 206°F (97°C)	Severe attack

[a] From survey PLIC series; all other data from CP-1 test series. Results based on averages of 1 to 3 test measurements for each property.

[b] Difference, % = $\frac{\text{radiation—control}}{\text{control}}$(100).

For thermal-catalytic treatment, benzoyl peroxide was used as a free radical initiator in concentrations varying between 0.2 and 5 wt %. For methyl methacrylate, at a temperature of 75°C, polymerization is complete in about 30 minutes with 5 wt % catalyst and in about 70 minutes when 1% catalyst is present.

For impregnation of concrete specimens the following standard procedure was generally followed.

(1) The dried or undried concrete is evacuated to a pressure of < 80 mm Hg. Drying of concrete is usually accomplished by heating to temperatures of 105° or 150°C to constant weight.

(2) The concrete is soaked in monomer with a nitrogen blanket pressure of 5 psig

Properties of Methyl-Methacrylate Impregnated Concrete

Containing 4.6 to 6.7 wt % PMMA)

Irradiation	Difference[b], %	Thermal-catalytic	Difference, %	Difference[d], %
20,255	285	18,161	244	−10
6.3	80	6.2	77	−2
1,627	292	1,508	262	−7
2,637	256	2,287	210	−13
6.2	44	7.4	65	15
0.29[a]	−95	3.58[a]	−33	1130
1.08	−83	0.34	−95	−69
0.0163	−67	0.0147	−70	−10
4	−71	4	−71	0
0.064	−80	0.020	−94	−69
0[a]	−100	—	—	—
0.8	−85	1.4	−73	75
1.306	−2	1.265	−5	−3
0.0409	6	0.0385	0	−6
5.36 × 10⁻⁶	33	5.25 × 10⁻⁶	31	−2
+38	neg. creep	+56	neg. creep	—
2,420; 0.5[a]	>340	—	—	—
750; 4.0	>53	750; 0.5	>53	−88
55.3	73	52.0	62	−6
0[a]	−100	—	—	—
3.64	−65	3.49	−66	−4
No attack	−100	—	—	—

[c] Difference, % = $\dfrac{\text{thermal}-\text{control}}{\text{control}}(100)$.

[d] Difference, % = $\dfrac{\text{thermal}-\text{radiation}}{\text{radiation}}(100)$.

(3) The monomer is drained.

(4) The specimen is wrapped in polyethylene sheeting to reduce evaporation losses.

(5) Polymerization is carried to completion using either radiation or thermal-catalytic means.

Strength and Durability of Concrete–Polymer

Normal Temperature. Wherever possible, standard test procedures were used in performing strength and durability measurements (3, 5). A summary of the properties for methyl methacrylate impregnated concrete is given in Table III.

Notable improvements in properties compared with those of control specimens of untreated concrete have been obtained as follows.

(a) Compressive strength increases of 285%, or a factor of almost 4, over the compressive strength of the control specimens.
(b) Tensile strength increases of 292%.
(c) Modulus of rupture increases of 256%.
(e) Flexural modulus of elasticity increases of 44%.
(f) Freezing and thawing improved by more than 300%.
(g) Hardness-impact ("L" hammer) increases of 73%.
(h) Water permeability decreases to negligible values.
(i) Water absorption decreases of as much as 95%.

negligible values compared with a severe attack observed on the control specimens.

Methyl methacrylate impregnated concrete–polymer specimens subjected to creep tests exhibited expansion under sustained load (negative creep), in contrast to contraction in length (positive creep) shown by ordinary concrete. If confirmed, this property could influence significantly the design of concrete–polymer structural and prefabricated members.

The effect of polymer loading on compressive strength is shown in Figure 1. The strength generally increases with polymer loading. It is difficult to obtain uniform partial loading of dried concrete so that the partially loaded samples may also show the effect of water content. Attempts to explain the effect of polymer loading on strength can be made by using the theory of particulate filled composite systems and the additive strength rule

$$S_c = S_m V_m + A B S_d V_d$$

S_c is the strength of the concrete–polymer composite, S_m is the cement strength, V_m is the cement volume, S_d is the aggregate strength, V_d is the aggregate volume, A is a theoretically and experimentally determined constant, and B is a factor relating to the bonding between phases. The experimental results on compressive strength correlate well with this theory (1, 2, 4).

High Temperature. The effect of temperature on compressive strength for several concrete–polymer systems is shown in Figure 2 (6).

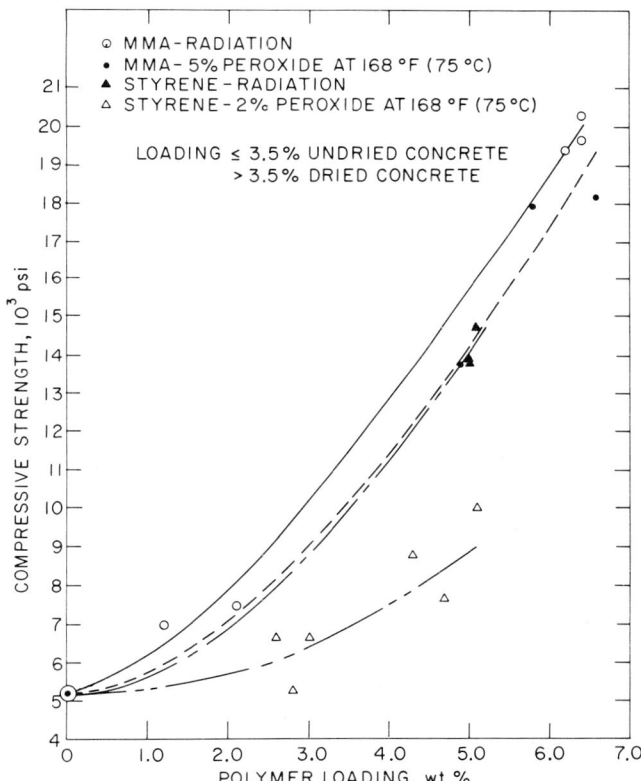

Figure 1. Concrete–polymer material compressive strength vs. polymer loading

Best results were obtained with the 60% styrene–40% TMPTMA monomer mix which gave a 175% increase in strength over the control (high strength concrete containing fly ash and type V cement, 8000 psi) at 25°C. The strength was maintained, decreasing by only 16% up to a temperature of 143°C. The methyl methacrylate +10% TMPTMA lost about 35% strength on increasing the temperature to 143°C. This is probably caused by using an insufficient amount of the TMPTMA crosslinking agent needed to increase the softening point.

Radiation vs. Thermal Polymerization. Table IV compares the compressive strengths of several concrete–polymer systems for each of two polymerization methods. Generally, the radiation polymerized material gave higher strengths than the thermally polymerized material.

For methyl methacrylate plus 10% TMPTMA the strengths were only 12% higher for radiation treatment. This difference increased to 38% for styrene. Other structural properties, such as tensile strength and modulus of elasticity, showed similar trends. A higher bond strength

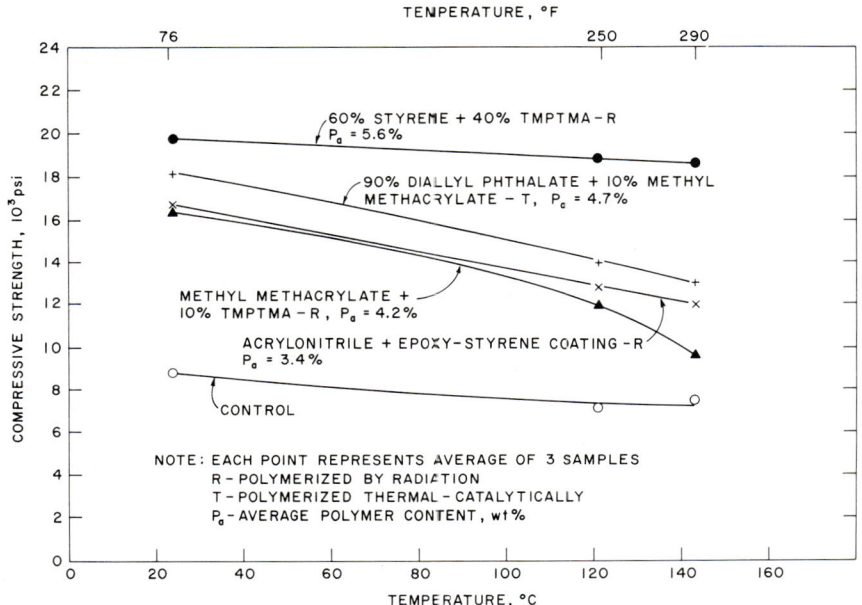

Figure 2. Concrete–polymer materials compressive strength vs. temperature (OSW-type concrete)

between the polymer and the concrete in the case of radiation treatment may account for this result.

Table V shows the differences in the freeze–thaw durability of concrete–polymer specimens receiving radiation and thermal treatment. The data indicate that the thermally treated specimens have either the same or slightly better durability than the radiation treated material. The higher durability, however, may be related to the higher polymer loading present in the thermal samples for this series of tests.

Reinforced Concrete. Because of the low tensile strength of concrete, steel-reinforcing bars are usually incorporated in structures when con-

Table IV. Compressive Strengths for Radiation and Thermally Polymerized Concrete–Polymer

Monomer is Concrete	Method of Initiation[d]	Av. Polymer Content[a], %	Av. Compressive Strength[a], psi	Concrete-Polymer vs Control Difference[b], %	Thermal vs. Radiation Difference[c], %
Control	none	0	5,270	—	—
Methyl methacrylate + 10% TMPTMA (CP-4)	R	6.3	21,593	310	—
	T	7.3	19,000	260	−12
Methyl methacrylate (CP-1)	R	6.3	19,800	276	—
	T	5.8	16,596	215	−16
Chlorostyrene (CP-6)	R	6.1	16,087	205	—
	T	6.2	14,387	173	−10
Acrylonitrile (CP-5)	R	5.4	14,410	173	—
	T	5.5	10,753	104	−25
Styrene (CP-2)	R	5.0	14,135	168	—
	T	4.7	8,788	67	−38

[a] Average of 3 specimens.

[b] % Difference = $\dfrac{\text{test} - \text{control}}{\text{control}} = (100)$.

[c] % Difference = $\dfrac{\text{thermal} - \text{radiation}}{\text{radiation}} = (100)$.

[d] R = radiation polymerized, T = thermally polymerized.

crete is stressed in tension. Although the tensile strength of concrete–polymer is higher by a factor of 4 than that of concrete, it is still relatively low compared with that of steel. Standard tests on reinforcing bars in which a 1-inch diameter steel rod is pulled out of a 6-inch diameter concrete cylinder indicated that the force needed to pull the bar from a concrete–polymer cylinder was three times the force required for a cylinder of ordinary concrete. This greatly improved bond strength between concrete–polymer and steel is important in concrete structural design.

Steel fiber has been used to reinforce concrete. When 10-mil square bar 2 inches long is incorporated into concrete to the extent of 2% by volume, the modulus of rupture (flexural strength) increases by a factor of 2. When impregnated with polymer the fiber reinforced concrete–polymer increases the modulus of rupture by another factor of 2.5. Thus, an over-all increase by a factor of 5 in the flexural strength is achieved

Table V. Freeze–Thaw Durability for Radiation

Monomer in concrete	Initiation[e]	Av. Polymer[a] Content, %
Control	none	0
Methyl methacrylate (CP-1)	R	5.9
	T	6.8
Methyl methacrylate + 10% TMPTMA (CP-4)	R	6.2
	T	7.2
Styrene (CP-2)	R	4.9
	T	4.4
Acrylonitrile (CP-5)	R	5.1
	T	5.5

[a] Average of two specimens.

[b] % Difference = $\dfrac{\text{test cycles} - \text{control}}{\text{control}}(100)$.

Table VI. Comparison of Basic Construction

Material	Strength property (S), psi	Density (W), lb/ft³
Tensile strength, psi		
Concrete	400	156
Concrete-polymer (6% PMMA)	1,600	165
Polymer (PMMA)	7,000	62
Steel	70,000	490
Aluminum	30,000	165
Glass	7,000	137
Compressive strength		
Concrete	5,000	156
Concrete-polymer (6% PMMA)	20,000	165
Polymer (PMMA)	15,000	62
Steel	42,000	490
Aluminum	25,000	165
Glass	300,000	137
Modulus of elasticity (E), psi		
Concrete	3.5×10^6	156
Concrete-polymer (6% PMMA)	6.3×10^6	165
Polymer	0.4×10^6	62
Steel	30×10^6	490
Aluminum	10×10^6	165
Glass	10×10^6	137

and Thermally Polymerized Concrete–Polymer

No. of[a] Freeze–Thaw Cycles	Wt Loss[a], %	Concrete–Polymer vs. Control Difference[b], %	Thermal vs. Radiation Difference[c], %
740	25[d]	—	—
3,450	6.7	>366	—
3,450	1.5	>366	−78
2,560	11.5	>246	—
2,560	0.2	>246	−98
2,635	25	256	—
3,340	10.5	>352	−67
1,690	25	128	—
2,020	2.2	>173	−93

[c] % Difference = $\dfrac{\text{wt loss thermal/cycle} - \text{weight loss rad/cycle}}{\text{wt loss radiation/cycle}}(100)$.

[d] Weight loss of less than 25% is considered durable concrete.

[e] R = radiation polymerized, T = thermally polymerized.

Materials with Concrete–Polymer Material

Strength to weight ratio (S/W), psi-ft³/lb	Unit cost of material in finished product (C), ¢/lb	Cost to volume strength ratio (C/S/W), ¢/psi-ft³
2.6	3	1.2
9.7	6	0.6
112	125	1.1
143	15	0.1
182	60	0.3
51	55	1.1
32	3	0.09
121	6	0.05
241	125	0.5
86	15	0.2
152	60	0.4
2,200	55	0.03
E/W		
1.9×10^4	3	1.6×10^{-4}
3.8×10^4	6	1.6×10^{-4}
0.6×10^4	125	200×10^{-4}
6.1×10^4	15	4.0×10^{-4}
6.1×10^4	60	10×10^{-4}
7.3×10^4	55	7.5×10^{-4}

for a polymer-impregnated fiber-reinforced concrete compared with standard concrete.

Monomer Addition to Premix Concrete

Attempts have been made to add monomers to the concrete mix before curing. Radiation was used to polymerize the mass after 1 day of curing. In many of the tests the concrete did not harden, and for those specimens that did harden the compressive strength was no higher than that of the control without monomer. In one notable case, in which 8% acrylonitrile was added to a standard concrete mix and radiation polymerized, an 87% increase in strength was obtained. Numerous attempts to add various organic constituents to premixed concrete have been made in the past. In most cases the strengths were increased by a relatively small factor, the increase usually being much less than 50%, and improvements in durability were also very limited. These results are not unexpected since organic monomers could interfere with the hydration process of cement. Furthermore, concrete shrinks on setting, and a void volume is formed in the concrete mass. The unique feature of impregnating precast concrete is that a large part of the void volume in the capillaries and pores is filled with polymer and forms a continuous internal reinforcing structure which appears to be the basis for the remarkable improvements in strength and durability.

Value of Concrete–Polymers and Potential Applications

A comparison of the properties of concrete–polymer with those of other more common materials used in construction is shown in Table VI. The tensile strength for concrete–polymer lies between that of the concrete and the polymer alone, while the compressive strength and modulus of elasticity are greater than those of either of these materials. Based on preliminary estimates the cost of prefabricated concrete polymer sewer pipe produced by radiation treatment (6¢/lb) is roughly double the cost of ordinary concrete pipe (3¢/lb). For a 6% methyl methacrylate loading at 20¢/lb of monomer, the cost of monomer is 1.2¢/lb of finished concrete–polymer or about 40% of the cost of manufacturing the impregnated pipe. The structural properties are related to volume of the material; thus, if the compressive strength for concrete–polymer increases by a factor of 4 over that for ordinary concrete and the cost increases by a factor of only 2, then the cost-to-volume strength ratio decreases by a factor of 2. The cost-to-volume strength index is shown in the last column of Table VI. In terms of compressive strength the economic value of concrete–polymer is superior to that of all the other materials except for glass, which is not considered a reliable construction material. Concrete and concrete–polymer are much better than the other materials in terms of modulus of elasticity. However, for tensile strength, steel is the leader, and for this reason concrete is reinforced with steel. The general conclusion is that the use of concrete–polymer for structural purposes appears economically reasonable.

Several possible applications of concrete–polymer material includes sewer and pressure pipe, building material for housing, underwater structures, railroad ties, and chemically resistant material of construction for large desalination plants. The incorporation of colored dyes with the monomer allows concrete–polymer material to be considered for various esthetic applications, such as in decorative wall and floor tile.

Literature Cited

(1) Auskern, A., *Brookhaven Natl. Lab. Rept.* **12890** (Sept. 1968).
(2) Auskern, A., *Brookhaven Natl. Lab. Rept.* **13493R** (March 1969).
(3) "Concrete Manual," 7th ed., U. S. Government Printing Office, Washington, D. C. 1966.
(4) Nielsen, L. E., *J. Composite Mater.* **1**, 100–19 (1967).
(5) Steinberg, M., Kukacka, L. E., Colombo, P., Kelsch, J. J., Manowitz, B., Dikeou, J. T., Backstrom, J. E., Rubinstein, S., *U.S. Dept. Interior and U.S. Bur. Reclamation Brookhaven Natl. Lab. Rept.* **50134 (T-509),** *USBR Gen. Rept.* **41** (Dec. 1968).
(6) Steinberg, M., Kukacka, L. E., Colombo, P., Auskern, A., Manowitz, B., Dikeou, J. T., Backstrom, J. E., Hickey, K. B., Rubinstein, S., Jones, C. W., *U.S. Dept. Interior and U.S. Bur. Reclamation Brookhaven Natl. Lab. Rept.* **50218 (T-560), REC OCE 70-1** (Dec. 1969).

RECEIVED February 9, 1970. Work performed under the auspices of the U. S. Atomic Energy Commission.

35

Carboxylated Polyester Additives for Improving the Adhesion of Coatings

W. J. JACKSON, JR., and J. R. CALDWELL

Research Laboratories, Tennessee Eastman Co., Division of Eastman Kodak Co., Kingsport, Tenn. 37662

> *Carboxylated polyesters were prepared by extending hydroxyl-terminated polyester segments with dianhydrides. Carboxylated polyesters which were soluble in common lacquer solvents were effective in improving the adhesion of coatings on a variety of substrates when 1–10% was blended with cellulose acetate butyrate, poly(vinyl chloride), poly(methyl methacrylate), polystyrene, bisphenol polycarbonates, and other soluble polymers.*

The adhesion of many polymeric coatings on various substrates is poor. Adhesion often can be improved, however, by using a primer, which must be applied and dried before applying the top coat. A more convenient and economical process would be the application of a single coat containing an adhesion-promoting component.

Incorporation of carboxyl groups in vinyl polymers (1) and polyolefins (1, 7) improves the adhesion of these polymers to various materials. However, many of these carboxylated polymers, particularly the carboxylated polyolefins, have limited solubility in volatile, lacquer-type solvents such as butyl acetate or methyl ethyl ketone and thus are limited in their ability to improve the adhesion of coatings applied from solvents. Carboxylated polyesters that are soluble in these solvents can be prepared. We were therefore interested in determining the effects of structure and carboxyl content on the adhesion of coatings of various classes of polymers blended with carboxylated polyesters.

Experimental

Inherent viscosities of the polymers were determined at a concentration of 0.23 gram/100 ml in 60/40 phenol/tetrachloroethane at 25°C. Polyesters for solubility determinations were prepared by conventional

Table I. Polyester Solubilities

Example	Polyester[a] Acid Components[c]	Glycol[d]	Solubility[b] Toluene	Methyl Ethyl Ketone	Butyl Acetate
1	60/40 I/adipic	EG	insol.		
2	60/40 T/phthalic	EG	insol.		
3	I	BD	insol.		
4	60/40 I/T	BD	insol.		
5	60/40 I/H	BD	gel.	gel.	insol.
6	60/40 I/phthalic	BD	insol.	insol.	insol.
7	50/50 I/phthalic	BD	sol.	sol.	insol.
8	70/30 H/T	BD	sol.	gel	gel.
9	70/30 H/I	BD	sol.	sol.	sol.
10	H	NPG	sol.	sol.	sol.
11	60/40 T/H	NPG	sol.	sol.	sol.
12	50/50 T/H	NPG	sol.	sol.	sol.
13	50/50 T/I	NPG	sol.	sol.	sol.
14	50/50 I/adipic	CHDM	sol.	sol.	sol.
15	60/40 I/H	CHDM	gel	insol.	insol.

[a] All inherent viscosities are 0.5–1.0.

[b] Solids content is 20%. Gel indicates separation of polymer as a gel after solution had cooled and stood overnight.

[c] Ratios are molar. I = isophthalic acid, T = terephthalic acid, H = 50/50 cis/trans-hexahydroterephthalic acid.

[d] EG = ethylene glycol, BD = 1,4-butanediol, NPG = neopentyl glycol, CHDM = 30/70 cis/trans-1,4-cyclohexanedimethanol.

procedures (5, 8). The solubilities of these polyesters before introduction of carboxyl groups are listed in Table I. The solubilities were determined at a solids content of 20% in the solvents. If the polymer did not dissolve readily at room temperature, the mixture was heated on a steam bath until the polymer dissolved. The solution was then allowed to cool and stand overnight. Any separation of the polymer as a gel was then noted.

Preparation of Carboxylated Polyesters. Hydroxyl-terminated polyester segments were prepared by conventional procedures and extended with an equimolar amount of dianhydride (2). To avoid the possibility of crosslinking the polymer, reaction with the dianhydride was carried out at 175°C. Depending upon the size of the reaction mixture, about 1 to 3 hours were required for all of the dianhydride to react and for a medium-to-high melt viscosity to be obtained. Inherent viscosities were 0.3 to 0.4.

Adhesion of Coatings. Except for K-1 polycarbonate [4,4'-(2-norbornylidene)diphenol polycarbonate] (4), an experimental polymer (inherent viscosity 0.85), all the coatings were prepared with commercial products: EAB-381-0.5 and EAB-381-20 cellulose acetate butyrates from Eastman Chemical Products, Inc.; VYHH vinyl chloride (87%)/vinyl acetate (13%) copolymer from Union Carbide Corp.; Butvar B76 poly(vinyl butyral) from Shawinigan Resins Corp.; Plexiglas V poly(methyl methacrylate) from Rohm and Haas Co.; Dylene P3I polystyrene from

Sinclair-Koppers Co.; Lexan 125 bisphenol A polycarbonate from General Electric Co.; Polysulfone aromatic poly(sulfone–ether) from Union Carbide Corp.; and PPO poly(2,6-dimethyl-p-phenylene oxide) from General Electric Co.

Initial adhesion tests were made with 3.5- $\times$ 1- $\times$ 0.064-inch specimens of aluminum (Aluminum Associates No. 2024) and cold-rolled steel (polished ASTM A-415). Later tests were made with strips of copper, chrome-coated steel (Weirton Steel Co.), and brass; with molded nylon 66 bars; and with Mylar 300A poly(ethylene terephthalate) film. Before use, the specimens were washed (scrubbed with a soft bristle brush) with a solution of Alconox detergent, rinsed with water, rinsed with acetone, and dried.

The specimens were coated with a blend of the carboxylated polyester and second polymer in a solvent (10% solids), which, when possible, consisted of a conventional lacquer solvent (74.2% toluene, 7.4% butyl alcohol, 7.4% Solvesso 100 solvent, 3.7% ethyl acetate, 3.7% butyl acetate, and 3.6 Cellosolve acetate). If the second polymer was not soluble (polysulfone, polycarbonates, PPO), chloroform was used. The coatings were dried for 1 hour at room temperature and then for 2 hours in an oven at 115°C. The coating thickness was about 0.5 mil.

The coatings were scored (crosshatched) with a razor, and adhesion was determined by the conventional cellophane tape test. The tape was pressed firmly against the coating and then jerked up. If any trace of the coating was removed by the tape on several tests, the coating was considered to have failed the test. The results obtained by the scored cellophane tape test are listed in Tables II and III. All coated specimens used for this test were dried for 2 hours at 115°C to ensure complete removal of all solvents, but a 1-hour drying period at 23°C was sufficient for many of the coatings to pass the cellophane tape test.

Table II. Effect of Carboxyl Content on Adhesion

Polyester			K-1 Polycarbonate Blends	
Segment Mol. Wt. before Carboxylation	Acid No.	Amt. in Blend, %	Steel	Al
		0	—	—
e	1	5	—	—
e	1	10	+	—
5500	26	1	+	—
5500	26	5	+	+
3000	39	1	+	+
3000	39	5	+	+
2100	56	1	+	—
2100	56	5	+	+
1200	126	1	+	—
1200	126	5	+	+

[a] Polyester of neopentyl glycol and equimolar amounts of terephthalic and isophthalic acids extended with pyromellitic dianhydride.

[b] Determined by crosshatched cellophane tape adhesion test (+ is pass; − is fail).

Discussion

Preparation and Solubility of Polyesters. A number of polyesters were prepared from several diols and dicarboxylic acid esters to determine the effect of structure on the solubility in typical solvents used in lacquers. The data in Table I show that solubility in the solvents decreased in the following order: toluene > methyl ethyl ketone > butyl acetate. Polymers that were soluble in all three solvents are examples 9-14.

The carboxylated polyesters were prepared by a two-step process: (1) preparation of a hydroxyl-terminated polyester segment and (2) reaction with a dianhydride to extend the polyester and introduce carboxyl groups. When the dianhydride is pyromellitic (II), the equation is as follows:

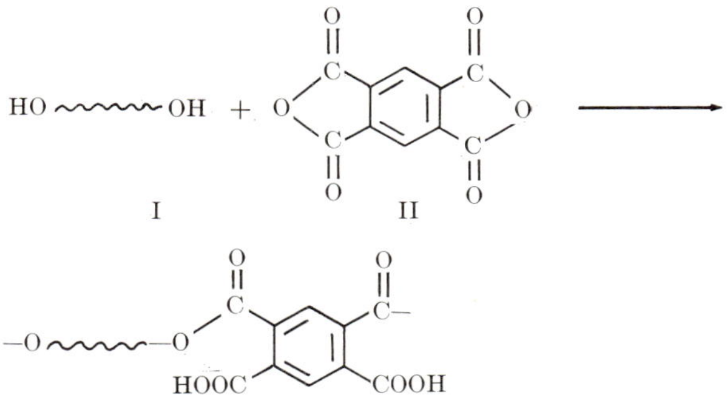

$$HO\sim\sim\sim OH + \text{(pyromellitic dianhydride, II)} \longrightarrow$$

(I)   (II)

$$-O\sim\sim\sim O-\text{C}_6\text{H}_2(\text{COOH})_2-\text{C(=O)}-O-$$

III

of Blends Containing Carboxylated T50I(NPG)[a]

Adhesion[b]			
Cellulose Acetate Butyrate Blends[c]		Vinyl Chloride/Vinyl Acetate Copolymer[d] Blends	
Steel	Al	Steel	Al
−	−	−	−
−	−	+	−
+	−	+	−
+	+	+	−
			+
+	+	+	++
+	+	+	+
+	+	+	+

[c] EAB-381-0.5 from Eastman Chemical Products, Inc.
[d] VYHH from Union Carbide Corp.
[e] Polyester (inherent viscosity 0.46) was not carboxylated.

Table III. Adhesion[a] of Blends Containing Carboxylated T50I(NPG)[b]

Blend		Substrate						
Major Component	Polyester, wt %	Brass	Steel	Cu	Cr[c]	Al	Nylon 66	PET Film[d]
Cellulose acetate butyrate (EAB-381-0.5)[e]	0	—	—	—	—	—	—	—
	5	+	+	+	—	+	+	—
	10	+	+	+	+	+	+	—
Cellulose acetate butyrate (EAB-381-20)[e]	0	—	—	—	—	—	—	—
	5	+	+	+	—	—	—	—
	10	+	+	+	+	+	+	—
Vinyl chloride (87%)/vinyl acetate (13%) copolymer	0	—	—	—	—	—	—	—
	1	+	+	+	+	+	+	—
	5	+	+	+	+	+	+	—[f]
Poly(vinyl butyral)	0	+	+	+	+	—	—	—
	5	+	+	+	+	+	+	—
Poly(methyl methacrylate)	0	+	—	—	—	—	—	—
	10	+	+	+	+	—	—	—
	20					+	+	+
Polystyrene	0	—	—	—	—	—	—	—
	10	+	+	+	+	+	+	—
Bisphenol A polycarbonate	0	—	—	—	—	—	—	—
	10	+	+	+	+	—	+	—
K-1 polycarbonate	0	+	—	—	—	—	—	—
	1	+	+	+	+	+	—	—
Aromatic polysulfone	0	—	—	—	—	—	—	—
	10	+	+	+	—	+	+	—
Poly(2,6-dimethyl-p-phenylene oxide)	0	+	—	—	—	—	—	—
	10	+	+	+	+	+	—	—
T50I(NPG) carboxylated[b]	100	+	+	+	+	+	+	+
T50I(NPG) uncarboxylated	100	+	+	—	+	—	—	+

[a] Determined by cross-hatched cellophane tape adhesion test (+ is pass; — is fail).

[b] Prepared by extending the polyester of neopentyl glycol and equimolar amounts of terephthalic and isophthalic acids (segment molecular weight 3000) with pyromellitic dianhydride; acid number was 39.

[c] Chrome-coated steel.

[d] Poly(ethylene terephthalate) film (Mylar 300A).

[e] EAB-381-0.5 has an appreciably lower molecular weight than EAB-381-20 (falling ball viscosities of 0.5 vs. 20 sec) (3).

[f] Passed adhesion test when solvent was chloroform.

The carboxyl content of the polymer is controlled by the length of the hydroxyl-terminated polyester segment (I). To extend the hydroxyl-terminated polyester segment, an equimolar amount of dianhydride was added. No crosslinking occurred when the reaction temperature was 175°C, but some crosslinking took place at higher temperatures (200°C), presumably because of esterification and perhaps acidolysis. Esterification could occur between the terminal hydroxyl groups and the carboxyl groups produced from the dianhydride, and acidolysis could occur if the carboxyl groups attacked the polymer chain.

In addition to pyromellitic dianhydride (PMDA), three other dianhydrides were used to extend the polyester segments. These anhydrides were prepared by heating trimellitic anhydride (IV) with glycol diacetates (V) (6):

2 (IV) + $CH_3COO-R-OOCCH_3$ (V) ⟶

(VI) + 2 CH_3COOH

a. $R' = $ —⟨C₆H₄⟩—C(CH₃)₂—⟨C₆H₄⟩—

b. $R = -CH_2-C(CH_3)_2-CH_2-$

c. $R = $ (cyclohexane with $-CH_2-$ and $-CH_2-$ substituents, trans)

For convenience, the letters T (terephthalic acid), I (isophthalic acid), H [50/50 cis/trans-hexahydroterephthalic acid (1,4-cyclohexanedicarboxylic acid)], and NPG [neopentyl glycol (2,2-dimethyl-1,3- propanediol)] are used to refer to the polyesters prepared from these intermediates; thus, T50I(NPG) is the copolyester from neopentyl glycol and equimolar amounts (50/50 molar ratio) of terephthalic and isophthalic acids.

The carboxylated polyesters based on examples 9-14 in Table I were soluble in butyl acetate and methyl ethyl ketone, but it was necessary to add up to 10% of an alcohol (methyl or butyl) to toluene to give solubility. Some of these carboxylated polyesters were also soluble in plasticizers. Carboxylated T50I(NPG) and T50H(NPG) (3000-molecular weight segments extended with PMDA) were soluble in dibutyl phthalate but not in dioctyl phthalate. The similar H(NPG) carboxylated polyester was soluble in both dibutyl and dioctyl phthalates, but it was inferior to the above two polymers in promoting adhesion.

Adhesion of Polymer Blends Containing Carboxylated Polyesters. The adhesion of blends containing a carboxylated polyester was affected by the structure of the polyester segments, the structure of the molecule providing the carboxyl groups, and the carboxyl content. Of the soluble polyesters listed in Table I, a carboxylated polyester which was particularly effective in imparting adhesion to coatings was obtained by extending hydroxyl-terminated T50I(NPG) (example 13) with PMDA:

$$\left[-OCH_2\underset{\underset{CH_3}{|}}{\overset{\overset{CH_3}{|}}{C}}CH_2O - \left(-\overset{O}{\underset{\|}{C}} - \underset{}{\bigcirc} - CCOCH_2\underset{\underset{CH_3}{|}}{\overset{\overset{CH_3}{|}}{C}}CH_2O - \right)_m - \overset{O}{\underset{\underset{O}{\|}}{\underset{\|}{C}}}\underset{HOC}{} \bigcirc \underset{COH}{\overset{O}{\underset{\|}{C}}}\underset{O}{\overset{\|}{}} - \right]_n$$

VII

The carboxyl content was determined by the molecular weight (length) of the polyester segment which was extended by the dianhydride—the greater the molecular weight (e.g., the value of m in formula VII), the lower the carboxyl content. The value of n was about 2 to 4 when the polyester segment molecular weight was about 3000 ($m = 13$).

Table II shows the effect of the carboxyl content on the adhesive characteristics of PMDA-extended polyesters in blends with K-1 polycarbonate [4,4'-(2-norbornylidene)diphenol polycarbonate] (VIII), cellulose acetate butyrate, and poly(vinyl chloride). K-1 polycarbonate is an experimental polymer which, like cellulose acetate butyrate and poly(vinyl chloride), is very sensitive to adhesive changes because only 1–2% of a carboxylated

VIII

polyester is required to give good adhesion. As shown in Table II, incorporation of carboxyl groups improved the adhesion significantly. About 1% of the carboxylated polyester was as effective in improving the adhesion of the three polymers on steel as 5–10% of the uncarboxylated polyester. On aluminum, 1% of the carboxylated polyester was more effective than 10% of the uncarboxylated polymer.

All the coated samples listed in the tables were heated in an oven for 2 hours at 115°C to ensure the removal of all solvent. Drying at room temperature, however, was sufficient for many of the coatings to pass the cellophane tape test. The cellulose acetate butyrate blends with 1% of each of the four carboxylated polyesters in Table II, for instance, passed the adhesion test on steel after the coatings had dried at 23°C for only 0.5 hour, and the blends with 1% of the polyesters having acid numbers of 39–126 passed the adhesion test on aluminum.

Polyesters extended with dianhydride VIa (from bisphenol A diacetate and trimellitic anhydride) instead of PMDA gave results similar to those shown in Table II. Polyesters extended with dianhydrides VIb or VIc gave inferior results.

Metal specimens coated with the various blends were immersed in water in an accelerated test to determine the effect of high humidity. Cellulose acetate butyrate blends containing carboxylated polyesters prepared with hexahydroterephthalic acid [H(NPG), T50H(NPG)] were particularly susceptible to moisture and failed the adhesion test after immersion for only 0.5 hour; similar blends containing T50I(NPG) extended with dianhydride VIa or with PMDA passed the adhesion test after immersion for 16 hours. When coatings on cold-rolled steel of cellulose acetate butyrate (EAB-381-0.5) blends containing 1% of each of the four T50I(NPG)/PMDA polyesters of Table II were scored (crosshatched) with a razor and immersed in water, all four passed the adhesion test after immersion for 8 hours. When these samples were returned to water for 3 days, the least rusting of the steel occurred when the polyester adhesion promoter having an acid number of 39 was used.

T50I(NPG)/PMDA was of particular interest because of its adhesive characteristics, oxidative stability, and polymer cost. Table III shows the improvement in adhesion obtained when various substrates were coated with blends containing this polyester (acid number 39). As indicated in the table, the ease of obtaining adhesion on the different substrates decreased approximately in the following order: brass > steel > copper > chrome-coated steel > aluminum > nylon 66 > poly(ethylene terephthalate). In spite of the wide differences in structure and polarity of the various polymers, the carboxylated polyester significantly improved the adhesion of the coatings.

The presence of a plasticizer in addition to the carboxylated polyester adversely affected the adhesion of some of the polymers—e.g., cellulose acetate butyrate and poly(vinyl chloride) plastisols.

Compatibility of Carboxylated Polyesters in Blends. Compatibility of the carboxylated polyester in the polymer blends did not appear to affect the adhesion significantly. T50I(NPG) (segment molecular weight 1400) extended with PMDA was compatible (clear coating) at a concentration of 20% in cellulose acetate butyrate, whereas the similar polyester with a segment molecular weight of 3000 was incompatible (hazy coating). Coatings of the latter blends had the best adhesion, however. The various carboxylated polyesters were incompatible in poly(methyl methacrylate) and polystyrene, but they were compatible in the vinyl chloride/vinyl acetate copolymer.

Carboxylated T50H(NPG) was superior in compatibility to carboxylated T50I(NPG) in some of the blends, but the adhesion was poorer after immersion in water.

Oxidative Stability of Carboxylated Polyesters. The polyesters which were extended with dianhydrides are those in Table I which were soluble in the lacquer solvents. Of these, theoretical considerations indicate that T50I(NPG) should be the most oxidatively and thermally stable because it is the only one with a completely aromatic acid component (terephthalic and isophthalic), and the glycol component has the stable neopentyl structure. When K-1 polycarbonate films containing 5% of this polyester extended with PMDA were heated in a forced-air oven at 200°C, the film life (time to brittleness when creased) was not lowered appreciably (compared with a control containing no carboxylated polyester). The incorporation of 5% of the similar carboxylated T50H(NPG) in the film, however, reduced the film life almost to one-half its normal life of 140 hours for a 2-mil film.

Surprisingly, coatings of 10% T50I(NPG)/PMDA in cellulose acetate butyrate prevented oxidation (discoloration) of the copper when heated in an air oven for 2 hours at 115°C (accelerated oxidation test). When the polyester was not present in the coating, the copper darkened under these conditions. Since the carboxylated polyester itself has a very high oxygen permeability (too high to measure), perhaps the copper was protected from oxidation because of the unusually strong adhesion of coatings containing the polyester.

Conclusions

In blends of 1–10% of the carboxylated polyesters with various coating materials, the incorporation of carboxyl groups was the most important structural feature which affected adhesion. Polyesters with acid numbers of about 25–125 appreciably improved the adhesion of coatings, and an acid number of about 40 was satisfactory for most of the coating materials. A carboxylated polyester which has good oxidative stability and is particularly effective as an adhesion promoter is obtained by extending with pyromellitic dianhydride the polyester of neopentyl glycol and equimolar amounts of terephthalic and isophthalic acids. This polymer shows promise as an adhesion promoter for air-dry automotive primers and is being evaluated for this application on an experimental basis by a number of companies.

Acknowledgment

We are pleased to acknowledge the excellent technical assistance of W. C. Cooper, H. F. Kuhfuss, H. G. Moore, and H. R. D. Spears.

Literature Cited

(1) Brown, H. P., Anderson, J. F., "Handbook of Adhesives," I. Skeist, Ed., p. 255, Reinhold, New York, 1962.
(2) Caldwell, J. R., U. S. Patents **3,459,584** and **3,484,339** (1969).
(3) Eastman Chemical Products, Inc., "Cellulose Acetate Butyrate for Protective Coatings," 2nd ed., pp. 9–10, 1968.
(4) Jackson, W. J., Jr., Caldwell, J. R., *Ind. Eng. Chem., Prod. Res. Develop.* **2**, 246 (1963).
(5) Kibler, C. J., Bell, A., Smith, J. G., *J. Polymer Sci., Pt. A* **2**, 2115 (1964).
(6) Loncrini, D. F., U. S. Patents **3,182,073** and **3,183,248** (1965).
(7) Smarook, W. H., Bonotto, S., *SPE Annual Tech. Conf. XIII, May 1967,* p. 119.
(8) Smith, J. G., Kibler, C. J., Sublett, B. J., *J. Polymer Sci. A-1,* **4**, 1851 (1966).

RECEIVED December 22, 1969.

36

Segregated Metallic Particles in Polymers

J. E. SCHEER and D. T. TURNER

Department of Metallurgical Engineering, Drexel Institute of Technology, Philadelphia, Pa. 19104

> *Mixtures of powders of poly(vinyl chloride) (PVC) and various metals were compacted at a pressure of 10,000 psig at 120–130°C. The compacts appear to be strong, and density measurements show the porosity to be <1.5%. Electrical resistivity is reduced, from a value for unloaded PVC of about 10^{15} Ωcm, to $<10^{-1}$ Ωcm by a fractional volume loading of nickel or copper as low as 0.06. Microscopic examination of polished sections of the compacts show the metallic particles to be segregated around zones of unpenetrated polymer which correspond in size to the initial particles of PVC. The pattern of segregation favors the formation of continuous chains of metallic particles at unusually low volume loadings.*

Metals have been introduced into polymers to increase electrical conductivity (2). A large volume loading has been found necessary unless special compounding techniques are used. In this latter category may be included the report that a marked increase in conductivity is achieved at a low volume loading if particles of polystyrene are coated with a thin layer of metal and then compacted (4). Apparently the metallic coating is broken partially, allowing a satisfactory compromise between attainment of a continuous polymer phase, which is necessary for mechanical strength, and an interpenetrating continuous metallic phase necessary for high electrical conductivity.

The onset of high electrical conductivity with increasing volume fraction of metallic particles has also been of interest in relation to theoretical treatments which consider the factors which control formation of a continuous disperse phase of randomly distributed particles. In pursuance of such work, the distribution of metallic particles was studied experimentally by quantitative microscopy of polished plane sections. A marked increase in conductivity was observed when the fractional volume loading of silver particles in Bakelite reached 0.36–0.38 (3).

The present report concerns a simple technique which provides electrically conducting samples of poly(vinyl chloride) (PVC) with fractional volume loadings of metallic particles as low as 0.06.

Experimental

The origin and particle size description of the powders used is given in Table I. The sample of PVC was unplasticized but contained initiator

Table I. Characteristics of Powders

Description	PVC	Dendritic Copper	Nickel	Aluminium
Manufacturer	Firestone Plastics Co. (EXON 9269)	United States Bronze Powders Inc. (D-100)	The International Nickel Co. (123 carbonyl nickel)	Alcan Metal Powders (MD 105)
Particle Size Distribution	0% 40–80 mesh 20% 80–100 mesh (180–150μ) 70% 100–200 mesh (150–75μ) 10% 200 mesh (75μ)	1% 60–100 mesh 10–20% 100–200 mesh 15–25% 200–325 mesh 60–75% 325 mesh	"uniform spiky equiaxed grains" Average particle 4–7μ	"uniform" particles of average size, 5μ

fragments and suspending agents left over from the polymerization reaction. The compacted polymer has an electrical resistivity of $>10^{14}$ Ωcm. Other physical properties which are pertinent in controlling compacting conditions, or eventual mechanical properties, are as follows: weight average molecular weight = 100,000; number average molecular weight = 58,000; bulk density = 0.48 gram/cc; specific gravity = 1.40; glass transition temperature, T_g = 76°C. A precise estimate of the melting temperature of crystalline regions in the polymer is not available, but the literature indicates a value near 180°C (1). Further information about the powders mentioned in Table I is available from the manufacturers.

Weighed amounts of PVC and metallic powders were mixed thoroughly for one-half hour in a rotary blender at room temperature. The mixture was compacted in a cavity mold for 10 minutes at 120°–130°C under a pressure of 10,000 psig. The mold was then allowed to cool under pressure to room temperature, and the cylindrical compact was removed. The dimensions are: length 3.5 cm; diameter 2 cm. Density measurements indicated that this procedure provided samples of porosity <1.5%.

Measurements of electrical resistivity were made at 25°C using the Van der Pauw technique (7) on discs 0.05 cm thick, which were cut from the cylindrical compacts. Four contacts were made to the perimeter

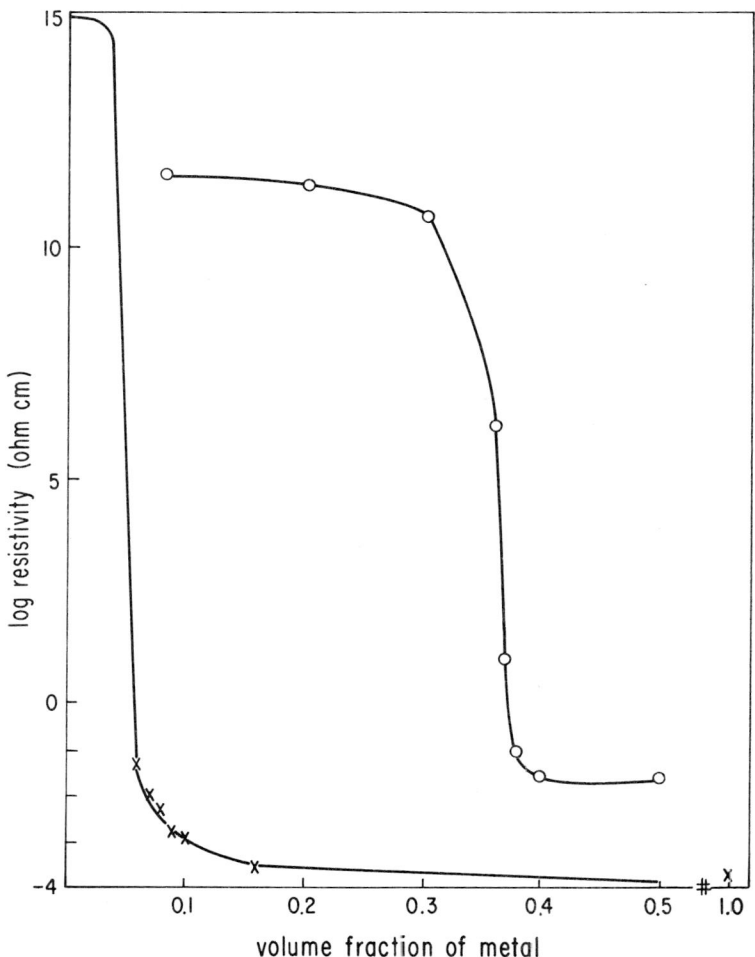

Figure 1. Resistivity of polymer–metal compacts as function of composition. X, PVC–copper; 0, Gurland's data on Bakelite–silver.

of the disc across two perpendicular lines. The resistivity, ρ, was calculated from the equation:

$$\rho = \frac{\pi d}{\ln 2} \frac{v}{i}$$

where v is the potential drop between two adjacent contacts when the current flowing between the remaining two contacts is i; d is the thickness of the disc. This is a preferred method for measuring resistivity and Hall coefficient on semiconductor lamellae of arbitrary shape, but its adoption in the present work was prompted by the inaccessibility of a more suitable technique.

Sections for microscopic examination were mounted in poly(methyl methacrylate) and polished to a 0.1μ finish. Examination of reflected light revealed the metallic particles as white features against a black continuous background of PVC. Evidence that the samples were homogeneous from one plane to another was provided by estimating the volume fraction of metal by quantitative metallography (6); in every case estimates were within 2% of the made-up values.

In some cases thin wafers, *ca.* 40μ, were examined with transmitted light under a stereoscopic microscope. This permitted observation of metallic particles below the surface and revealed the presence of continuous three-dimensional networks of metallic particles, for volume loadings in excess of *ca.* 6%, but it was not possible to document such observations with photomicrographs.

No attempt was made to prevent surface oxidation of the metals. The copper powder was initially free of oxide but oxidized noticeably during the experiments, which generally took about two weeks from compaction to measurement of resistivity. Optical examination was made immediately after polishing.

Discussion of Results

Preliminary experiments showed that a volume fraction of copper or nickel as low as 0.06 markedly reduces electrical resistivity. The decrease in resistivity with increasing loading of copper is shown in Figure 1. The technique used does not allow measurements of resistivities $>10^{13}$ Ωcm. Somewhat arbitrarily, the value of 10^{15} has been taken, and it is assumed that the resistivity would drop sharply at a volume loading approaching 0.06 since on-scale readings were not obtained for loadings <0.05. At a fractional volume loading of 0.06 the resistivity has a value of only 5×10^{-2} Ωcm and drops several orders of magnitude with increasing metal toward a value of 1.3×10^{-4} Ωcm for a sample of copper compacted alone. While the resistivity of sintered compacts of metallic and ceramic particles has been investigated extensively (5), the only previous well-documented data known for metallic particles in a polymeric matrix is for silver in Bakelite. The results of this previous investigation are included, for comparison, in Figure 1. The most striking difference is that in the previous work a much higher loading of metal was required to reduce markedly the resistivity; according to Gurland, the Bakelite aggregate becomes conducting at a volume loading of silver between 0.36 and 0.38 (3).

A reason for the difference in behavior noted above becomes apparent by comparing photomicrographs of sections. In Gurland's system the particles of silver are distributed more or less randomly throughout the polymer as may be seen from his results quoted in Figure 2. By contrast, in the present system the metallic particles are excluded from certain elements of volume and segregated into others. This was apparent in all

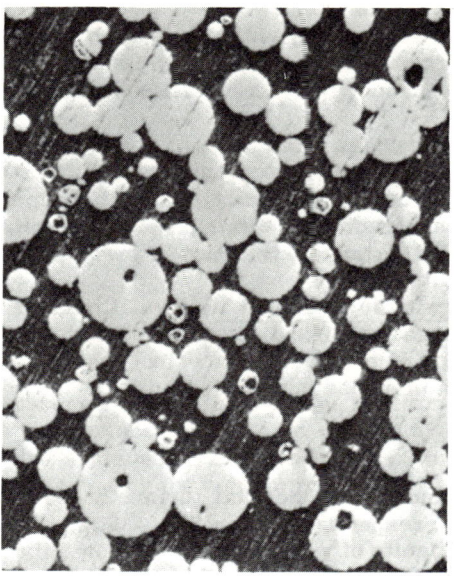

Transactions of the Metallurgical Society of AIME

Figure 2. Microstructure of compacts of Bakelite and silver (0.4) (3). Volume fraction of silver = 0.4 × 35 and enlarged 12%.

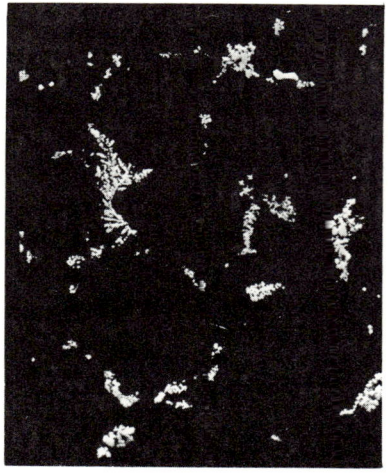

Figure 3. Microstructure of compact of PVC and copper (0.05) × 125

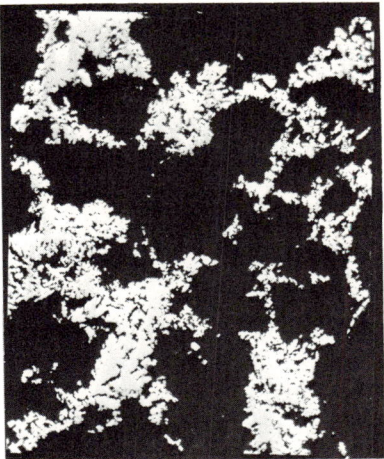

Figure 4. Microstructure of compact of PVC and copper (0.16) × 125

Figure 5. Microstructure of PVC compact and nickel (0.258) × 125

the samples prepared and is illustrated by a selection of photomicrographs in Figures 3 to 6. What is not immediately apparent is that the arrangements of metallic particles in Figures 4 to 6 do, indeed, correspond to continuous metallic paths throughout the samples. However, theoretical analyses of various packing arrangements of particles show that such paths are expected to be present when the average number of contacts per particle ranges from 1.42 to 1.55, the precise value depending on the model under consideration (3). Moreover, continuous paths can be observed experimentally in the electrically conducting samples by stereoscopic microscopy. Of course, electrical conductivity requires not merely the presence of chains of particles but also effective electrical contacts. This is not achieved in the case of aluminum, even with a volume loading as high as 0.6, presumably because of the presence of a surface layer of aluminum oxide on the particles which acts as an insulator.

The dark regions of the photomicrographs correspond in size approximately to the initial particles of PVC in the powder mixture. For example, the largest particle initially present is roughly spherical with a diameter of 180μ. This value corresponds to the larger dark regions in the photomicrographs. For example, the one marked X in Figure 5 is about 200μ long and about 80μ wide. Presumably, this anisotropy arises from compression during compaction and, consistently, the long axes of the particles are generally parallel to each other in Figures 5 and 6.

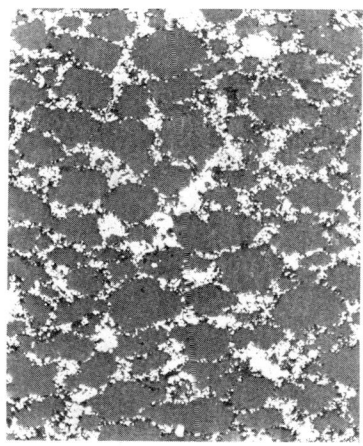

Figure 6. Microstructure of PVC compact and aluminum (0.34) × 50

The observed pattern of segregation might well be a consequence of the accommodation of metallic particles in the free volume available in an assembly of large approximately spherical particles of PVC. For example, close packing of spheres in a face centered cubic system would provide a continuous network of open channels which in the (100) planes would provide spaces of the kind shown in Figure 7. Some features corresponding to the filling of such spaces by small metallic particles

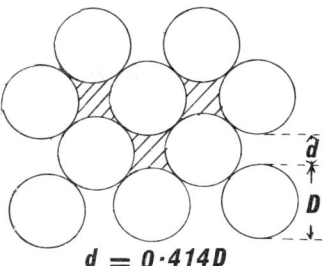

Figure 7. Spaces available in (100) planes of an FCC system

may be discerned in Figure 6 along with other irregular features which reflect the considerable departure of the real system from any simple model. In cases involving large dendrites, however, the resulting pattern of metallic particles is controlled more by their own geometry as can be seen in Figures 3 and 4.

During compaction the polymer flows sufficiently to exclude any free volume within the polymer particles and also to exclude free volume around the metallic particles. This flow is sufficiently facile to reduce the porosity of the compacted system to $<1.5\%$. On the other hand, the flow of the polymer particles is not so extensive as to cause the initial individuals to merge and lose their identity. This factor was taken into account in planning the present experiments by choosing a polymer with convenient values of T_g and T_m with the reasoning that choice of a compaction temperature, T, where $T_m > T > T_g$, would allow sufficient flow to exclude porosity and yet to preserve domains of polymer which would resist penetration, and hence wastage, of metallic particles. It remains to be seen from experiments with simple one-phase polymers, such as poly(methyl methacrylate) and polystyrene, whether this reasoning is pertinent. It also remains to be checked whether the supposed dependence of the pattern of segregation on packing of the powders can predict the optimum choice of particle size distributions necessary to achieve high electrical conductivity for a minimal volume fraction of metal.

Conclusions

Mixtures of PVC and nickel or copper powders have been compacted under heat and pressure to yield samples which conduct electricity, $\rho < 10^{-1}$ Ωcm, at volume loadings of metal as low as 6%. The compacts have a porosity of $<1.5\%$ and appear to be strong mechanically.

Acknowledgments

Richard Zaehring is thanked for help in taking the photomicrographs, and R. W. Heckel for advice in their interpretation. Manufacturers are thanked for donations of samples. One of us (DTT) was supported under a Departmental Development Grant from the National Science Foundation.

Literature Cited

(1) Bueche, F., "Physical Properties of Polymers," p. 110, Interscience, New York, 1962.
(2) Delmonte, J., "Metal-Filled Plastics," Reinhold, New York, 1961.
(3) Gurland, J., *Trans. Met. Soc. AIME* **236**, 642 (1966).
(4) Hochberg, F., U. S. Patent **2,721,357** (Oct. 25, 1955).
(5) Kingery, W. D., "Introduction to Ceramics," Wiley, New York, 1960.
(6) Underwood, E. E., *Metals Eng. Quart.* **1**, 70 (3), 62 (4) (1961).
(7) Van der Pauw, L. J., *Phillips Tech. Rev.* **20**, 220 (1958).

RECEIVED July 31, 1969.

37

Luminescence of Thin Plastic Scintillators

A Report on Energy Transfer in Plastic Scintillators

R. KOSFELD and K. MASCH

Institute of Physical Chemistry of the RWTH, Aachen, Germany

> Since the discovery of luminescence in plastics, the investigation of the action of plastic scintillators has developed rapidly during the last few years. Scintillators based on plastics can be conceived as solid solutions of organic, luminescent substances which are embedded with low concentrations in transparent plastics capable of luminescence. The scintillation mechanism of the system polystyrene–tetraphenylbutadiene was investigated.

The phenomenon of luminescence in plastics has received much attention in recent years (3). In particular the investigation of the scintillation mechanism of plastic scintillators of the system polystyrene (PST)–tetraphenylbutadiene (TPB) has been studied extensively. Hitherto known measurements (2, 4, 5. 10, 11, 12) on this system dealt only with samples about 10 mm thick. Carrying out these measurements leads to difficulties because of the self-absorption and the associated change of the observable characteristics of fluorescence. Until now temperature-dependent measurements have not been performed on the system mentioned above.

Experimental

Scintillation of PST/TPB scintillator samples 0.5 mm thick was measured. The preparation of very thin samples is not recommended because the scintillation characteristics depend strongly on the orientation of the polymer chains during the preparation of the film (*14*). The following samples were made with several concentrations of TPB for testing the dependence of the energy transfer of conditions of film preparation and of the individual plastic:

(1) PST and TPB were dissolved in a solvent, and the scintillators were obtained by evaporating the solvent.

(2) TPB was dissolved in the monomer before polymerization of PST.

(3) TPB was dissolved in the monomer mixture before polymerization of a poly(methyl methacrylate)–polystyrene copolymerizate.

The films contained between 0.1 and 1% residual monomer according to gas chromatographic analysis. The transmittance and fluorescence spectra were measured with a Spectronic 505 spectrophotometer by Bausch and Lomb. To obtain correct fluorescence spectra with respect to energy, the above mentioned apparatus was calibrated using a wolfram-band lamp and an aperture driver by a wavelength-dependent correction

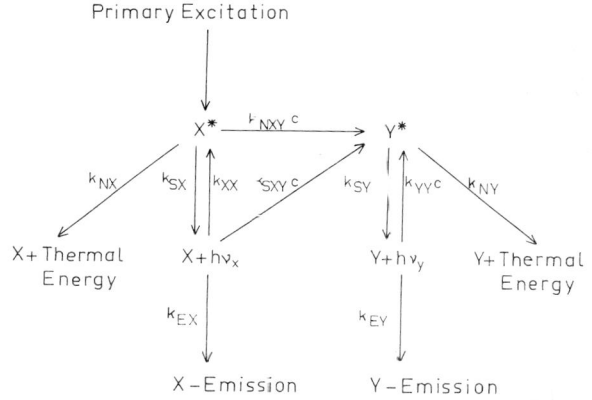

Process		Name of the Process	Probability
X^*	$\to X + h\nu_x$	Molecular X-Fluorescence	k_{SX}
X^*	$\to X + Th.E.$	Internal Quenching of X	k_{NX}
$X^* + Y$	$\to X + Y^*$	Non-Radiative Energy Transfer	$k_{NXY} c$
$h\nu_x + X$	$\to X^*$	Self-Absorption of X	k_{XX}
$h\nu_x + Y$	$\to Y^*$	Absorption of X-Fluorescence by Y	$k_{SXY} c$
$h\nu_x$	$\to X$-Emission	Technical X-Fluorescence	k_{EX}
Y^*	$\to Y + h\nu_y$	Molecular Y-Fluorescence	k_{SY}
Y^*	$\to Th.E.$	Internal Quenching of Y	k_{NY}
$h\nu_y + Y$	$\to Y^*$	Self-Absorption of Y	$k_{YY} c$
$h\nu_y$	$\to Y$-Emission	Technical Y-Fluorescence	k_{EY}

Figure 1. Scintillation processes and their probabilities in a binary system

curve. The temperature-dependent measurements of the scintillation intensity were performed with a Valvo-photomultiplier 153 AVP and a mirror galvanometer. For the excitation of the scintillators a ^{90}Sr source was used.

Theory

The processes occurring in a binary system capable of fluorescence which consists of the host molecules X and the guest molecules Y being embedded with a low concentration, c, are shown in Figure 1. The probabilities of the individual processes are marked by a parameter k. An energy quantum entering a binary system dissipates part of its energy on the way. Part of this energy loss produces excited states X* which may lead to fluorescence emission.

An excited molecule X* can pass into the basic level again. This can be performed by radiation—*i.e.*, by emission of the energy difference as a photon $h\nu_X$—or by nonradiation by internal quenching where the energy difference is dissipated as thermal energy. Furthermore there is the possibility that the excitation energy is not dissipated by radiation but by interaction with a guest molecule Y. This interaction can be described by diffusion of excitons and/or dipole–dipole resonance.

The photon $h\nu_X$ emitted in a radiating transition is reabsorbed with a certain probability by another molecule X or Y. By absorbing the photon the molecule makes a transition into the excited states X* and Y*, respectively. Thus the Y molecules are excited only by the radiating and nonradiating transfer of the excitation energy. The photons $h\nu_Y$ emitted in a radiating transition which do not undergo self-absorption leave the scintillator.

Therefore, the primary excitation causes the emission of X fluorescence or Y fluorescence depending on the kind of excitation. Under constant irradiation of the scintillator the fluorescence intensity L is proportional to the complete technical quantum efficiency Q:

$$L = K \times Q(c)$$

which consists of the technical quantum efficiency of the X fluorescence q_X and the technical quantum efficiency of the Y fluorescence q_Y multiplied by the efficiency of the radiating and nonradiating energy transfer f_{NXY} and f_{SXY}.

$$Q = q_X + (f_{NXY} + f_{SXY}) q_Y$$

The values in this equation which depend on the concentration c of the Y molecules are obtained by a simple calculation from the process scheme. On the condition that the radiating transitions are not temperature dependent and the nonradiating transitions satisfy a Boltzmann distribution concerning their temperature dependence, it follows for the nonradiating transition:

$$k_{Ni}(T) = k_{Ni_\infty} e^{-\frac{\Delta E_i}{RT}} \qquad i = X, Y, XY$$

ΔE_i are the activation energies for the respective process. The molecular quantum efficiencies q_{0X} and q_{0Y} then decrease with increasing temperature according to their definitions.

$$q_{0i}(T) = \frac{k_{Si}}{k_{Si} + k_{Ni}(T)} = \frac{1}{1 + \frac{k_{Ni_\infty}}{k_{Si}} e^{-\frac{\Delta E_i}{RT}}} \qquad i = X, Y$$

Experiments and Discussion

An important result of our experimental investigation is the fact that within experimental error the same results were obtained for the three series of sample mentioned above as well for the fluorescence quantum spectra and transmittance distributions as for the measurements of fluorescence intensity under β-irradiation. From this it follows that in the samples prepared by a solvent the remaining solvent in the scintillator does not influence the scintillation processes.

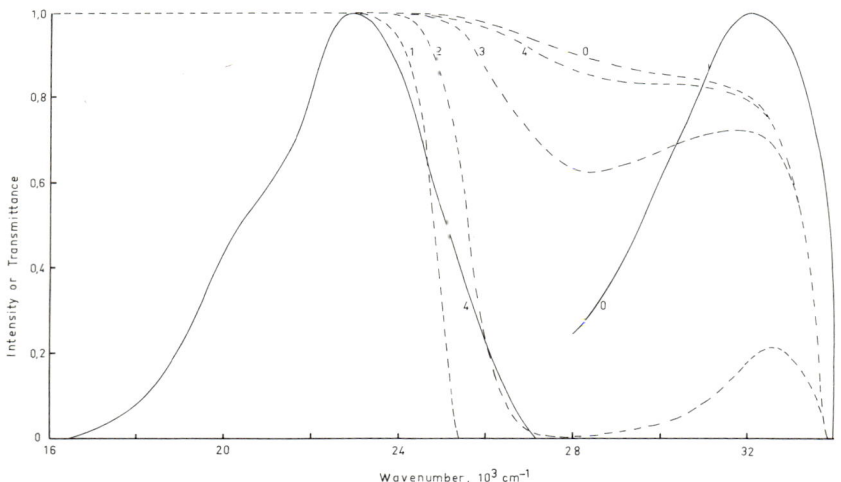

Figure 2. Solid curves: fluorescence spectra of 0.5 mm polystyrene (O) (11) and of TPB in 0.5 mm polystyrene (4)
Dashed curves: transmittance spectra

Parameter at the curves is the negative, common logarithm of the TPB concentration in mole %

Figure 2 shows in the full curves the relative fluorescence spectrum of PST with the maximum at 32,000 cm^{-1} and of 10^{-4} mole % TPB in PST with the maximum at 23,000 cm^{-1}. The spectra were drawn with maxima of the same height. The dashed curves show the spectral transmittance of the same samples. As seen from the overlapping of transmittance and the fluorescence quantum distribution, part of self-absorption is large, whereas it is much smaller in the case of TPB fluorescence.

By enlarging the guest concentration the transmittance decreases at about 28,000 cm^{-1}, and this causes an increase of reabsorption of the polystyrene fluorescence with growing TPB concentration by the TPB molecules. The matrix material, polystyrene, is handled as a uniform substance with respect to the scintillation processes. This remains the same during the polymerization. It has been shown (1) that the residual monomer in the polystyrene acts as a secondary solute. Yet because the emission spectra of polystyrene and styrene occur in the same wavelength range, the polystyrene–styrene matrix can be assumed to be a uniform material. The existence of excimer states in polystyrene has been discussed (8, 13, 14). In dilute solutions their formation has been checked but not in the solid state. The fact that the fluorescence spectrum of polystyrene occurs at a lower energy than the fluorescence spectrum of the monomer and that the absorption spectrum of polystyrene appears at higher energies than the absorption spectrum of the monomer could be a hint that in a solid material excimer formation occurs too.

From the measurements of scintillation intensity under constant irradiation of the scintillator the efficiency of the nonradiating energy transfer f_{NXY} and the efficiency of the radiating energy transfer f_{SXY} were calculated with reference to the reabsorption (Figure 3). The nonradiating energy transfer shows a strong increase in a certain concentration range and reaches a saturation value. This means that at high concentrations of guest molecules the excited host molecules are deactivated only by the nonradiating transfer of the excitation energy. Hence, the radiating energy transfer must decrease with increasing concentration because the plastic fluorescence emission decreases with increasing concentration.

Basile (2) has shown that the nonradiating energy transfer occurring in the system PST–TPB can be described well by the theory of dipole–dipole interaction developed by Forster (7). His calculations show the possibility of a nonradiating energy transfer over distances of 20–25 A.

The diffusion of excitons is possible as a nonradiating energy transfer over short distances. This means that the excited state can migrate through the matrix material till it is trapped by a guest molecule. This mechanism requires that excitable states of the same nature exist along an exciton migration way when it occurs—e.g., in the phenyl side groups of PST. If the nonradiating energy transfer is caused by exciton diffusion,

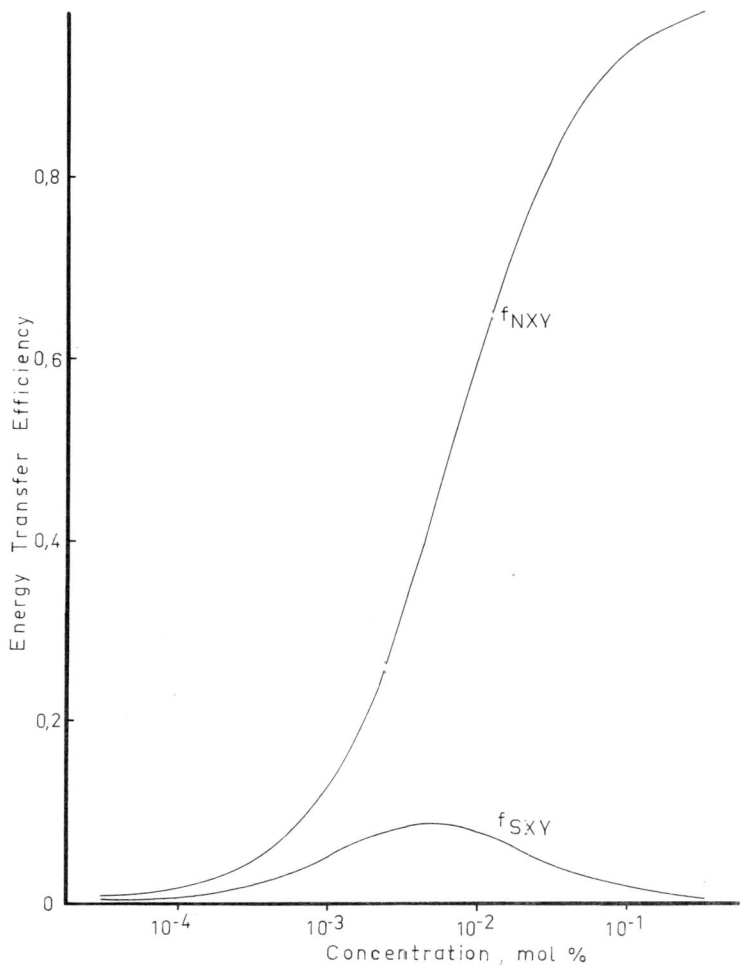

Figure 3. Nonradiative (f_{NXY}) and radiative (f_{SXY}) energy transfer in the system polystyrene–TPB

a lower effectiveness of the nonradiating energy transfer should result in the samples with poly(methyl methacrylate)–polystyrene as matrix material. As a result, the scintillation intensity with the same TPB concentration in a poly(methyl methacrylate)–polystyrene sample should be lower than in a sample with pure polystyrene as matrix material. The experiments do not show this phenomenon. Therefore, it is not possible to decide about the kind of nonradiative energy transfer.

Figure 4 shows the temperature dependence of the scintillation intensity for the samples PST/TPB in the glass temperature range, which lies at about 90°C for polystyrene. In pure polystyrene the intensity decrease is reduced when the molecular quantum efficiency q_{OX} increases. The mobility of the phenyl groups relative to the chain increases strongly above 50°C by which ability of the chromophorous groups to show fluorescence is influenced. Above the glass temperature the micro-Brownian mobility of the main chain begins next to the mobility of the phenyl groups, by which constant scintillation intensity is observed. The relative decrease of the scintillation intensity decreasing at high TPB concentration shows that the molecular quantum efficiency q_{OY} of TPB does not depend strongly on temperature in the investigated temperature

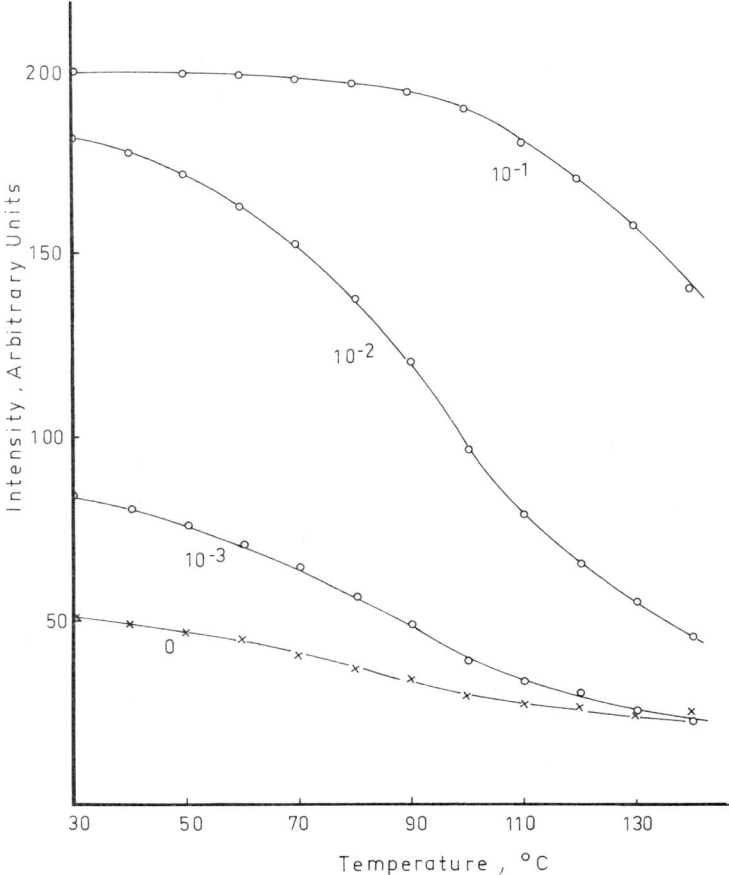

Figure 4. Fluorescence intensity vs. temperature for the system polystyrene–TPB

range. At high concentrations the scintillation intensity is proportional to the molecular quantum efficiency of q_{OY}.

At medium concentrations (10^{-2} mole %) the scintillation intensity is determined mainly by the product of the efficiency of the nonradiating energy transfer f_{NXY} and the molecular quantum efficiency of Y q_{OY} (3). Because the decrease of the scintillation intensity is maximum at this concentration and (12) because the molecular quantum efficiency q_{OY} does not depend strongly on temperature, the quantum efficiency of the nonradiating energy transfer must decrease. Therefore, the temperature-dependent behavior is determined mainly by the properties of the matrix material.

The process of the nonradiating deactivation in the matrix competes with the nonradiating energy transfer process for the activation energy of an excited molecule in the matrix. With increasing temperature the decision is made for the nonradiating deactivation of the polystyrene molecules (9).

Literature Cited

(1) Basile, L. J., *J. Chem. Phys.* **36/8,** 2204 (1962).
(2) Basile, L. J., *Trans. Faraday Soc.* **60,** 1702 (1964).
(3) Birks, J. B., "Theory and Practice of Scintillation Counting," Pergamon Press, New York, 1964.
(4) Birks, J. B., Kuchela, K. N., *Discussions Faraday Soc.* **27,** 57 (1959).
(5) Bothe, H. K., *Ann. Phys.* **7/5,** 339 (1960).
(6) *Ibid.,* **7/6,** 156 (1960).
(7) Forster, T., *Discussions Faraday Soc.* **27,** 7 (1959).
(8) Longworth, J. W., *Biopolymers* **4,** 1131 (1966).
(9) Masch, K., Ph.D. Thesis, Aachen, 1969.
(10) Pichat, L., Pesteil, P., Clement, J., *J. Chim. Phys.* **50,** 26 (1955).
(11) Rozman, I. M., Kilin, S. F., *Soviet Phys.* **2/6,** 856 (1960).
(12) Swank, R. K., Buck, W. L., *Phys. Rev.* **91,** 927 (1953).
(13) Vala, H. T., Haebig, J., Rice, S. A., *J. Chem. Phys.* **43/3,** 886 (1965).
(14) Yanari, S. S., Bovey, F. A., Lumry, R., *Nature* **200,** 242 (1963).

RECEIVED December 5, 1969.

INDEX

INDEX

A

ABA block polymers	508
ABS	
block copolymers	500
graft polymers	340
polymers	229
properties of	233
resins	86
Absorption, rubber energy	89
Acrylic polymers	251
properties of	232
rubber backbone	253
Acrylonitrile	
butadiene rubber	82, 229
copolymerization	229, 322
discoloration of	230
methylstyrene–	168
terpolymerization of cyclopentene, sulfur dioxide, and	211
Addition of mercaptans to carbon–carbon double bonds	381
Adhesion	529, 563
Adsorption	382
effect of temperature on	389
Aggregate	
definition of	352
shapes	363
size and shape as a function of molecular structure	365
Aggregation	
in block copolymers kinetics of	351
causes of	357
proof and characterization of	352
Aging of polymers	257
AIBN as initiator	218, 267
Allyl alcohol moieties	318
Ambifunctional silanes	464
Amorphous homopolymers	399
Anisotropy	527, 578
Area per grafted chain	346
Arrhenius	
behavior	408
equation	155, 398
plot	35, 212

B

Benzene	
and dichloroethane as swelling agents	285, 293
Benzoyl peroxide	267
BD–AN chains	294
Birefringence	410
Blend	
dry	194
mechanical	193
solution	193
Blending, polymer	191
Blends	
binary	190
chlorinated and normal PVC	119
compatibility of carboxylated polyesters in	570
dynamic properties of	201
impact-resistant polymer	237, 251
of resins with elastomers	237, 251
transparent poly–	248
Block copolymers	379, 425
Block and random copolymers	431
BN rubbers	260
Boltzmann distribution	583
Bonding	84
Boron fiber	485
Brittleness	
of PMMA	86
of polystyrene	86
of styrene/acrylonitrile	86
Bulk polymerization	221, 230
Butadiene	
–acrylonitrile copolymer	290, 534
grafting by *in situ* polymerization of	305
–methyl methacrylate	245
polymerization of	303
–styrene copolymers	189
Butyl acrylate	176

C

Capillary viscometer	55, 61
Carboxylated polyester	
additives	562
in blends, compatibility of	570
preparation of	563
Catalyst	
dialkylaluminum halide as	308
Et$_3$Al–cobalt chelate–benzyl chloride as	309
Et$_2$AlCl–cobalt stearate–*tert*-butyl chloride as	309
Cellulose	
density of polyacrylonitrile copolymers of	324
graft and block copolymers of fibrous	321
solubility of	322
Cellulosic textiles, properties of	333

Chain transfer agents	214
Charge transfer complex	211
Chemical grafting	300
Chemically homogeneous copolymers	178
Chlorinated	
butyl rubber	69
and normal PVC blends	119
polyethylenes	260
PVC	122
Chromophorous groups	587
Cloud-point curve	49
Coatings, adhesion of	563
Cohesive energy density	241
Compaction	579
Compatible high polymers	15
Compatibility	
of carboxylated polyesters in blends	570
of copolymers	175
limits of	190
in polyblends	2
Composites	
inorganic reinforcements for high performance structural	482
short fiber-elastomer	510
stress–strain properties of	519
Compression molding	305
Concrete	
properties of	548
monomer addition to premix	560
monomer impregnation of	549
–polymer applications	560
–polymer composites, preparation and characteristics of	547
polymerization of monomer-impregnated	551
–polymer vs. other construction materials	559
Cone-plate viscometer	55
Continuous phase	541
Copolymerization	302, 331, 494, 532
anionic	189
free-radical initiated	321
initiators	254
of maleic anhydride and methyl methacrylate	423
of α-methylstyrene	152
of olefins	211
in the presence of depolymerization	140
of styrene and acrylonitrile	229
of styrene and maleic anhydride, heterogeneous	421
Copolymer(s)	
ABS block	500
block and random	431
brittleness of	86
butadiene–styrene	189
chemically homogeneous	178
compatibility of	175
composition	448
damping function for	183
Copolymer(s) (Continued)	
demixing in	175
from DMP and DPP	431
of 2,6-dimethylphenol	432
emulsifiers	380
ethyl acrylate	494
ethylene–propylene	261
of fibrous cellulose, graft and block	321
graft	238, 294, 308
heterogeneity in	190
homogeneous	180, 237
of maleic anhydride and stearyl methacrylate	423
of maleic anhydride and vinyl monomers	418
morphology of cast block	415
natural	180
phase separation in natural	187
produced by oxidative coupling	432
random	191
resinous	240
simulated	181
structure, determination of	439
system, optical haze of	242
thermal properties of	447
time–temperature superposition in block	397
viscosity of	215
Copper	575
Cotton cellulose–polyvinyl copolymer yarns	332
Couette flow	59
Covalent bonds	323
CPVC	119
creep properties of	133
dynamic mechanical measurements of	127
thermal stability	123
Crack/craze dynamics	95
Crack propagation	509
Craze nucleation theory	92
Creep properties of CPVC	133
Critical	
concentration	45
mixtures, viscosity of	50
opalescence of polystyrene cyclohexane	43
phenomena in multicomponent polymer solutions	42
point	42
Crosslinked	
cellulose–poly(hydroxyethyl methacrylate) copolymer fabrics	334
cycloaliphatic epoxy resins	531
polymers	533
Crosslinking	300, 319, 507, 567
agents	68, 191, 245, 293
Crystallinity, reversal of	448
Curing	472, 532
Cyanoethylated cellulose–polyacrylonitrile	326
Cyclic peroxides, formation of	318

INDEX

Cycloaliphatic type epoxies 532
Cyclopentene, sulfur dioxide, and acrylonitrile, terpolymerization of 211

D

Degradation 300
Dehydrochlorination of PVC 303
Demixing 175, 187
Demulsification 223
Depolymerization, copolymerization in the presence of 140
Dialkylaluminum halide as catalyst 308
Dianhydrides 562
Diblock polymer 506
Dichloroethane
 as chain transfer agent 296
 as swelling agent 296
Diene elastomers 253
Differential
 solvent 85
 thermal analysis 3, 5, 305
Dilatation theory of yielding 90
Diphenoquinone formation 435
Dipole–dipole interaction 585
Discoloration 311
Disperse phase 541
Dispersion
 of solid particles in organic media 379
 stability 383, 389
 of titanium dioxide 379, 393
Distortion of the polystyrene domains 501
Divinylbenzene 288
DMP and DPP copolymers 431
tert-Dodecylmercaptan 293
Domains 399
Dynamic mechanical measurement of CPVC 127
Dynamic properties of blends 189, 195, 201

E

Effect of structure on the adhesion of coatings 562
E-glass 483
Elastic modulus 200
Elastomer(s) 327
 blends of resins
  with acrylic copolymer 251
  with diene copolymer 237
 composites 510
 crosslinked 71
 diene type 237
 modifier, structure and composition of 545
 phases, interfacial bonding between 68
 polarity of 532
 by polymeric fillers, mechanisms of reinforcements of 490

Elastomer(s) (*Continued*)
 reinforcement of epoxy systems with 531
 swollen 69
 vulcanizate, tensile strength of an 507
Electrical conductivity of polymers 572
Electrical resistivity 575
Elongation 534
Emulsifier 222
Emulsifying properties 379
Emulsion
 polymerization 230
 polymers 231, 494
Energy transfer in plastic scintillators 581
Epoxies
 cycloaliphatic type 532
 glycidyl ether type 532
Epoxy
 composite properties 485
 –elastomer incompatibility 537
 resins, crosslinked cycloaliphatic 531
  halogenated 471
  unhalogenated 477
 systems, toughness of 532
EPR techniques 425
Esterification 567
Et$_3$Al–cobalt chelate–benzyl chloride as catalyst 309
Et$_2$AlCl–cobalt stearate–*tert*-butyl chloride as catalyst 309
Ethyl acrylate 336
 copolymers 494
Ethylene–propylene copolymer ... 261
EVA 107
 copolymers 260
External plasticizer 302
Extrusion 275

F

Falling-ball viscometer 55–58
Fatigue of glass-fiber reinforced plastics 474
Fiber glass 513
 reinforced plastics 452
Fiber
 –matrix adhesion 510
 -reinforced plastics 471
 reinforcement 513
Fibrous
 cellulose–polyvinyl copolymers.. 327
 cyanoethylated cotton cellulose–polyacrylonitrile copolymer 332
Fibrous glass
 composites, physical and chemical concepts of 463
 composites, uses of 453
 properties of 463
 thermosets 454
 thermoplastics 369, 453
Filament-wound composites 471
Filled rubber 407

F

Fillers
  modulus of 495
  polystyrene and poly-2,6-
    dichlorostyrene 496
Film casting 10
Films 135
Flexibilizing agents 531
Fluorescence 581
Fractionation
  of raw polymerizates 262
  in solution 280
Free-radical initiation 330
Free-radical initiator 553
Functional end groups 532

G

Gardner impact test 535
Gel
  content 221
  permeation chromatograms 422
Glasses, polymeric 252
Glass
  fiber 483
  –fiber reinforced plastics, fatigue
    of 474
  shear effects on 466
  -to-matrix adhesion 465
  transition temperature ...10, 15, 130,
                               302, 398, 447
    of PMMA–PVDF blends 18
Glassy domains 400, 500
Glycidyl ether type epoxies 532
Graft and block copolymers of
  fibrous cellulose 321
Graft copolymerization 237
  of cis-1,4-polybutadiene 302
Graft copolymers 238, 294, 308
Grafting 115, 185, 253, 262
  in an aqueous suspension of vinyl
    chloride 260
  on ethylene–propylene
    copolymers 260
  by in situ polymerization of
    butadiene 305
  with PVC 261
  rubber 267
  temperature 283, 298
  with thermoplastic chains 531
Graphite fiber 484

H

Haze 544
Heat distortion temperature 535
Heterogeneity in copolymers 190
Heterogeneous
  copolymerization of styrene and
    maleic anhydride 421
  solution polymerization 420
Heterophase
  and phase inversion phenomena 542
  systems 508

High impact
  polystyrene 221
  strength 261
High polymers, compatible 15
Hildebrand's solubility parameters 420
Homogeneity 185, 189, 216, 261
Homogeneous copolymers 180, 237
  films 180
  liquid mixtures 8
  matrix 117
  polymerization 229
  polymers 239
Homopolymerization 149
Homopolymers 505
HRH system 510, 517
Hydrophile–lipophile balance 224

I

Impact-resistant polymer
  blends 237, 251
Impact strength 269, 340, 348
  vs. temperature 136
Improving the adhesion of coatings 562
Incompatible polymer blends 117
Incompatible polymers, viscosity of
  a solution of 53
Incompatibility 261
  epoxy–elastomer 537
  thermodynamic 399
Initiator, effect on 187
Injection-molded polyblends ...247, 254
Injection molding conditions 255
Inorganic fibrous reinforcements .. 482
Interfacial
  bonding between elastomer
    phases 68
  bonding, evidence of 72
  phase 415
  shear stress 510
  tension of emulsions 53
Intermolecular rearrangement of
  the ketal 434
Interphase adhesion 102
Isotropic model for polyblends ... 205
Izod impact strength of ABS 341

K

Kinetics
  of aggregation in block
    copolymers 351, 359
  of aggregation thermodynamics
    of 371
  of mixing 30
Kraton 401

L

Light scattering 44
Linear polymers from conjugated
  dienes 316
Loss maxima 196
Luminescence of thin plastic
  scintillators 581

M

Macromolecular network	291
Macroradicals	418
Magnetic environment of methyl protons	440
Maleic anhydride	
and methyl methacrylate, copolymerization of	423
and stearyl methacrylate, copolymers of	423
–vinyl monomer copolymers	418
Matrix resin	532
Mayo equation	154
MBS polymers	231
Mechanisms	
of reinforcements of elastomers by polymer fillers	490
of stabilization	316
Melt	
flow rheology	247, 257
viscosity	303
of PVC/EPR	274
Mercaptan	288
-terminated random copolymer	533
Metal composites	523
Methyl	
methacrylate	154, 176
–butadiene–styrene, graft copolymers	229
protons, magnetic environment of	440
resonance in polymers	440
Methylene chloride, solubility behavior in	442
Methylstyrene	
–acrylonitrile	168
block polymers	504
copolymerization of α-	152
Mixing	53
Model filler studies	492
Modulus of fillers	495
Moisture vs. adhesion	569
Molecular quantum efficiency	587
Monomer	
addition to premix concrete	560
impregnation of concrete	549
–polymer grafting	310
Morphology	
of cast block copolymers	415
of polymers	535
Multicomponent polymer solutions, critical phenomena in	42
Multifunctional silanes	510

N

Natural copolymers, multiphase nature of	182
NBR	82, 108
Network	
density of filled SBR vulcanizates	495
structure of block polymers	507
Nickel	575
Noncrystalline block copolymers	351
Nonpolymeric liquids, mixing of	4
Nonradiating energy transfer	585

O

Oil-based paints and printing inks	379
Opacity	537
Organosilanes, use in glass composites	465
Orientation of composites	526
Osmium tetroxide technique	539
Oxidation of DMP and DPP, sequential	437
Oxidative coupling of phenols	431

P

PEMA–PVdF polyblends	17
Permeability of PVC/EVA systems	107
Peroxide catalyst	551
Phase separation	399
in natural copolymers	187
in polyblends	502
in polymer mixtures	357
temperature	46
Phase structure of PVC/EVA systems	107
Phenolic mixed dimers, chemical shifts for	441
Phenols	
interaction of during oxidation	436
oxidative coupling of	431
rate of disappearance of during oxidation	436
simultaneous oxidation of a mixture of	435
Pigment dispersibility	383
Plastic	
composites	523
scintillators, luminescence of thin	581
Plastics	
fiber glass reinforced	452
fiber-reinforced	471
rigid	237
PMMA, brittleness of	86
PVdF blends, properties of	23
Polarity of elastomer	531, 532
Polyacrylonitrile (PAN)	54, 322
copolymers of cellulose, density of	324
Poly(arylene oxides)	431
Polyblends	3, 9, 190, 229, 244
Arrhenius plots of	35
compatibility in	2
dielectric relaxation measurements	33
dynamic-mechanical measurements of	11
dynamic and stress-optical properties of	189
glass transition temperature of	29
homogeneity of	29
interlocking network in	203

Polyblends *(Continued)*
  mechanical integrity of 29
  microscopy of 11
  a model for the behavior of ... 204
  multiphase nature of 199
  optical transparency of 29
  parallel coupling in 203
  phase separation in 502
  PMMA–PVdF and PEMA–PVdF 17
  preparation of 4
  transparency in 238
Polybutadiene 223, 229, 400, 533
cis-1,4-Polybutadiene
  graft copolymerization of 302
  unsaturation 310
Polydiene elastomers 69
Poly(2,6-dimethylphenylene
  oxide)–polystyrene 29
Polydispersity 49
Polyester, preparation and solubility
  of 565
Polyesters, carboxylated 562
Polymer *(see also* Polymers)
  blending 191
  carboxyl content of 567
  –fibrous glass composites 452
    potential of 468
  mixtures, phase separation in .. 357
  morphology 288
  structure, effect of polymerization
    method on 444
Polymeric
  additive 302
  carbonium ion 310
  fillers 508
    mechanisms of reinforcements
      of elastomers by 490
    properties of 494
  glasses 252
  oil-in-oil emulsions 379
  phenols, redistribution of 433
  systems as poly(vinyl chloride)
    modifiers 278
Polymerization
  in aqueous vinyl chloride 276
  bulk 230
  of butadiene 303
  of butadiene and acrylonitrile on
    a PVC latex 279
  emulsion 230
  heterogeneous solution 420
  kinetics 288
  method, effect on polymer
    structure 444
  of monomer-impregnated
    concrete 551
  temperature 283
  of vinyl chloride in the presence
    of EPR 263, 266
Polymers *(see also* Polymer) 53
  compatibility in 30
  crosslinked 71, 533
  determination of chlorine in ... 264
  homogeneous 239

Polymers *(Continued)*
  incompatibility of 1, 189
  methyl resonance in 440
  α-methylstyrene block 504
  mixed by grafting and milling .. 114
  morphology of 535
  network structure of block 507
  rubber toughening of brittle 86
  SIS and SBS block 504
  styrene–isoprene–styrene block.. 502
  tensile strengths of 501
Polymethacrylates 410
Poly(methyl methacrylate), light
  stability of 230
Polystyrene 193, 245, 398, 491
  brittleness of 86
  –co-maleic anhydride 426
  –cyclohexane 42
  domains 508
  /1,4-polybutadiene/polystyrene
    block copolymer 400
  and poly-2,6-dichlorostyrene
    fillers 496
  prepolymerization of 227
  –tetraphenylbutadiene 581
Polyurethane (PU) 54
Poly(vinyl chloride) 107, 119
  brittleness of 260
  –butadiene acrylonitrile 280
  dehydrochlorination of 303
  EPR, melt viscosity of 274
  EVA systems 109
  grafting with 261
  modifiers, polymeric systems as 278
  thermal stabilization of 302
PPO-PS, glass transition 37
Preparation and characteristics of
  concrete–polymer composites 547
Preformed blocks 446
Properties
  of cellulosic textiles 333
  of polymeric fillers 494, 495
Prepolymerization 221
Pressing temperature and
  permeability 117
PVC *(see* poly(vinyl chloride))
PVdF, ultraviolet degradation of .. 27
Pyrolysis gas chromatographic
  techniques 425

Q

Quinone ketal, formation and
  dissociation of 433

R

Radiating energy transfer 585
Radiation 583
  polymerization 551
  vs. thermal polymerization 556
Radiochemical polymerization ... 279
Random copolymers 399
Rayon–natural rubber composite .. 517
Reactivity ratios 241

INDEX

Redistribution
  in dimethylphenol polymerization 435
  of polymeric phenols 433
Reinforced
  concrete 556
  thermoplastics, dimensional stability of 461
  thermoplastics, heat distortion of 460
  thermoplastics, impact strength of 458
Reinforcement of epoxy systems with elastomers 531
Reinforcements of elastomers by polymeric fillers, mechanisms of 490
Relaxation modulus 406
Resin
  with acrylic copolymer elastomers, blends of 237
  matrix 532
  properties of cast 534
  and rubbers 237, 251
  water absorption of 473
Rubber 499, 543
  content, effect on toughness ... 103
  energy absorption 89
  fraction as a graft copolymer .. 261
  free 266
  grafting 267
  high strength 253
  -like behavior 327
  -like character 253
  particle size 222
  in styrene, solutions of 221
  toughening of brittle polymers.. 86

S

Sapphire fibers 486
SBR 82
  vulcanizates 490
    network density of filled 495
Schulz-Flory distribution function 148
Second-order transition temperatures 326
Segregated metallic particles in polymers 572
S-glass 483
Shear field 54
Shift factor 408
Short fiber–elastomer composites .. 510
Short fibers as fillers for elastomer compounds 510
Silicon carbide fibers 486
Silver 575
SIS and SBS block polymers 504
Solubility
  behavior in methylene chloride 442
  parameters, Hildebrand's 420
Solution behavior, thermodynamics of 7
Solvent
  pairs 69
  swelling 69
Solvents, effect on 185

Spherical domains 491
Spontaneous copolymerization ... 211
Stabilization
  mechanism of 316
  of solid particles in organic media 379
Stabilizers 314
Strain 415, 503
  sterically induced 173
Stress
  birefringence of polyblends ... 192
  -optical coefficient 199
  -optical measurements of polyblends 199
  –strain properties 504
  of composites 519
Structure and composition of elastomer modifier 545
Styrene 223
  /acrylonitrile, brittleness of 86
  and acrylonitrile, copolymerization of 229
  –butadiene 245
    block copolymers 381
    copolymers 189
  -isoprene–styrene block polymers 502
Substrate particle size 340
Sulfur dioxide, terpolymerization of cyclopentene, acrylonitrile, and 211
Supramolecular structure in block copolymers 351
Suspension polymerization 221, 224
Swelling 507
  agent 119, 285
  in benzene 496

T

Takayanagi model for polyblends 205
Teflon-filler vulcanizates 497
Temperature and initiator type vs. amount of grafted rubber .. 267
Tensile
  adhesion 94
  creep 271
  relaxation modulus 403
  strength 341, 349, 534
    of an elastomer vulcanizate .. 507
    of polymers 501
Terpolymerization of cyclopentene, sulfur dioxide, and acrylonitrile 211
Tetrahydrofuran 191
Textiles containing cellulose graft copolymers 333
Thermal
  degradation of PVC 303
  polymerization, radiation vs. ... 556
  stability 271
    of CPVC 123
    of grafted PVC 311
    of polysulfones 212
    tests for 304
  stabilization of poly(vinyl chloride) 302

Thermodynamic incompatibility .. 399
Thermoplastic elastomers 500
Thermoplastics, tensile and flexural
  properties of 456
Thermorheologically simple
  materials 398
Thermosets 468
Thermosetting
  cycloaliphatic epoxy systems .. 531
Thioglycolic acid 379
Time–temperature superposition in
  block copolymers 397
Toluene 212
Toughening
  craze/crack branching theory of 95
  theories of 89
Toughness of epoxy systems 532
Transparency in polyblends 238
Transparent
  acrylic 257
  polyblends 248
Triblock copolymer rubber 399

U

Ultraviolet light stability of
  polymers 236

V

Velocity profile 66
Vicat softening point 212, 269
Vinyl
  chloride in the presence of EPR,
    polymerization of 263, 266
  monomer 331
Viscosity of a solution in incom-
    patible polymers 53
Voids 517
Volume loading 572
Vulcanization 80, 510
  effect on adhesion 78

W

Water-in-oil emulsions 221
Weathering 261
Weather and rot resistance 335
Weather-o-meter 271
Whisker materials, inorganic 487

Y

Yield point 501
Yoffe mechanism 86
Young's modulus 199, 499